Student Solutions Manual

for
Stewart's

Multivariable Calculus:
Concepts and Contexts

Third Edition

Dan Clegg
Palomar College

THOMSON
™
BROOKS/COLE

Australia • Canada • Mexico • Singapore • Spain • United Kingdom • United States

Printed in the United States
1 2 3 4 5 6 7   08 07 06 05 04

Printer: Thomson/West

0-534-41005-7

For more information about our products,
contact us at:
Thomson Learning Academic Resource Center
1-800-423-0563

For permission to use material from this text or
product, submit a request online at
http://www.thomsonrights.com.
Any additional questions about permissions can be
submitted by email to **thomsonrights@thomson.com.**

Thomson Higher Education
10 Davis Drive
Belmont, CA 94002-3098
USA

Asia (including India)
Thomson Learning
5 Shenton Way
#01-01 UIC Building
Singapore 068808

Australia/New Zealand
Thomson Learning Australia
102 Dodds Street
Southbank, Victoria 3006
Australia

Canada
Thomson Nelson
1120 Birchmount Road
Toronto, Ontario M1K 5G4
Canada

UK/Europe/Middle East/Africa
Thomson Learning
High Holborn House
50–51 Bedford Road
London WC1R 4LR
United Kingdom

Latin America
Thomson Learning
Seneca, 53
Colonia Polanco
11560 Mexico
D.F. Mexico

Spain (including Portugal)
Thomson Paraninfo
Calle Magallanes, 25
28015 Madrid, Spain

☐ PREFACE

This *Student Solutions Manual* contains detailed solutions to selected exercises in the text *Multivariable Calculus: Concepts and Contexts,* Third Edition (Chapters 8—13 of *Calculus: Concepts and Contexts,* Third Edition) by James Stewart. Specifically, it contains solutions to the odd-numbered exercises in each chapter section, review section, True-False Quiz, and Focus on Problem Solving section as well as all solutions to the Concept Check questions.

Each solution is presented in the context of the corresponding section of the text. In general, solutions to the initial exercises involving a new concept illustrate that concept in more detail; this knowledge is then utilized in subsequent solutions. So, while the intermediate steps of a solution are given, you may need to refer back to earlier exercises in the section or prior sections for additional explanation of the concepts involved. Note that, in many cases, different routes to an answer may exist which are equally valid; also, answers can be expressed in different but equivalent forms. Thus, the goal of this manual is not to give the definitive solution to each exercise, but rather to assist you as a student in understanding the concepts of the text and learning how to apply them to the challenge of solving a problem.

I would like to thank James Stewart for entrusting me with the writing of this manual and offering suggestions, Kathi Townes and Jenny Turney of TECH-arts for typesetting and producing this manual, and Brian Betsill for creating the illustrations. Craig Chamberlin prepared solutions for comparison of accuracy and style in addition to proofreading manuscript; his assistance and suggestions were very helpful and much appreciated. Finally, I would like to thank Bob Pirtle and Stacy Green of Thomson Brooks/Cole for their trust, assistance, and patience.

dan clegg

Palomar College

□ CONTENTS

8 ☐ INFINITE SEQUENCES AND SERIES

8.1 Sequences

1. (a) A sequence is an ordered list of numbers. It can also be defined as a function whose domain is the set of positive integers.

   (b) The terms a_n approach 8 as n becomes large. In fact, we can make a_n as close to 8 as we like by taking n sufficiently large.

   (c) The terms a_n become large as n becomes large. In fact, we can make a_n as large as we like by taking n sufficiently large.

3. The first six terms of $a_n = \dfrac{n}{2n+1}$ are $\dfrac{1}{3}, \dfrac{2}{5}, \dfrac{3}{7}, \dfrac{4}{9}, \dfrac{5}{11}, \dfrac{6}{13}$. It appears that the sequence is approaching $\dfrac{1}{2}$.

$$\lim_{n\to\infty} \frac{n}{2n+1} = \lim_{n\to\infty} \frac{1}{2+1/n} = \frac{1}{2}$$

5. $\left\{1, -\dfrac{2}{3}, \dfrac{4}{9}, -\dfrac{8}{27}, \ldots\right\}$. Each term is $-\dfrac{2}{3}$ times the preceding one, so $a_n = \left(-\dfrac{2}{3}\right)^{n-1}$.

7. $\{2, 7, 12, 17, \ldots\}$. Each term is larger than the preceding one by 5, so $a_n = a_1 + d(n-1) = 2 + 5(n-1) = 5n - 3$.

9. $a_n = \dfrac{3+5n^2}{n+n^2} = \dfrac{(3+5n^2)/n^2}{(n+n^2)/n^2} = \dfrac{5+3/n^2}{1+1/n}$, so $a_n \to \dfrac{5+0}{1+0} = 5$ as $n \to \infty$. Converges

11. $a_n = \dfrac{2^n}{3^{n+1}} = \dfrac{1}{3}\left(\dfrac{2}{3}\right)^n$, so $\lim_{n\to\infty} a_n = \dfrac{1}{3}\lim_{n\to\infty}\left(\dfrac{2}{3}\right)^n = \dfrac{1}{3}\cdot 0 = 0$ by (8) with $r = \dfrac{2}{3}$. Converges

13. $a_n = \dfrac{(n+2)!}{n!} = \dfrac{(n+2)(n+1)n!}{n!} = (n+2)(n+1)$, so $a_n \to \infty$ as $n \to \infty$ and the sequence diverges.

15. $a_n = \dfrac{(-1)^{n-1}n}{n^2+1} = \dfrac{(-1)^{n-1}}{n+1/n}$, so $0 \le |a_n| = \dfrac{1}{n+1/n} \le \dfrac{1}{n} \to 0$ as $n \to \infty$, so $a_n \to 0$ by the Squeeze Theorem and Theorem 4. Converges

17. $a_n = \dfrac{e^n + e^{-n}}{e^{2n} - 1} \cdot \dfrac{e^{-n}}{e^{-n}} = \dfrac{1 + e^{-2n}}{e^n - e^{-n}} \to \dfrac{1+0}{e^n - 0} \to 0$ as $n \to \infty$. Converges

19. $a_n = n^2 e^{-n} = \dfrac{n^2}{e^n}$. Since $\lim_{x\to\infty} \dfrac{x^2}{e^x} \overset{\text{H}}{=} \lim_{x\to\infty} \dfrac{2x}{e^x} \overset{\text{H}}{=} \lim_{x\to\infty} \dfrac{2}{e^x} = 0$, it follows from Theorem 3 that $\lim_{n\to\infty} a_n = 0$. Converges

21. $0 \le \dfrac{\cos^2 n}{2^n} \le \dfrac{1}{2^n}$ [since $0 \le \cos^2 n \le 1$],   so since $\lim_{n\to\infty} \dfrac{1}{2^n} = 0$, $\left\{\dfrac{\cos^2 n}{2^n}\right\}$ converges to 0 by the Squeeze Theorem.

23. $y = \left(1 + \dfrac{2}{x}\right)^x \Rightarrow \ln y = x\ln\left(1 + \dfrac{2}{x}\right)$, so

$$\lim_{x\to\infty} \ln y = \lim_{x\to\infty} \frac{\ln(1+2/x)}{1/x} \overset{\text{H}}{=} \lim_{x\to\infty} \frac{\left(\dfrac{1}{1+2/x}\right)\left(-\dfrac{2}{x^2}\right)}{-1/x^2} = \lim_{x\to\infty} \frac{2}{1+2/x} = 2 \Rightarrow$$

$$\lim_{x\to\infty}\left(1 + \dfrac{2}{x}\right)^x = \lim_{x\to\infty} e^{\ln y} = e^2, \text{ so by Theorem 2, } \lim_{n\to\infty}\left(1 + \dfrac{2}{n}\right)^n = e^2. \text{ Convergent}$$

25. $\{0, 1, 0, 0, 1, 0, 0, 0, 1, \ldots\}$ diverges since the sequence takes on only two values, 0 and 1, and never stays arbitrarily close to either one (or any other value) for n sufficiently large.

27. $a_n = \ln(2n^2+1) - \ln(n^2+1) = \ln\left(\dfrac{2n^2+1}{n^2+1}\right) = \ln\left(\dfrac{2+1/n^2}{1+1/n^2}\right) \to \ln 2$ as $n \to \infty$. Convergent

29.

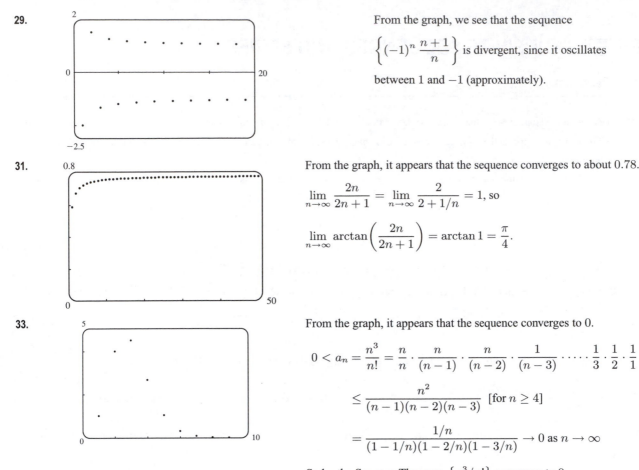

From the graph, we see that the sequence

$\left\{(-1)^n \dfrac{n+1}{n}\right\}$ is divergent, since it oscillates

between 1 and -1 (approximately).

31.

From the graph, it appears that the sequence converges to about 0.78.

$$\lim_{n\to\infty} \frac{2n}{2n+1} = \lim_{n\to\infty} \frac{2}{2+1/n} = 1, \text{ so}$$

$$\lim_{n\to\infty} \arctan\left(\frac{2n}{2n+1}\right) = \arctan 1 = \frac{\pi}{4}.$$

33.

From the graph, it appears that the sequence converges to 0.

$$0 < a_n = \frac{n^3}{n!} = \frac{n}{n} \cdot \frac{n}{(n-1)} \cdot \frac{n}{(n-2)} \cdot \frac{1}{(n-3)} \cdots \cdot \frac{1}{3} \cdot \frac{1}{2} \cdot \frac{1}{1}$$

$$\leq \frac{n^2}{(n-1)(n-2)(n-3)} \quad [\text{for } n \geq 4]$$

$$= \frac{1/n}{(1-1/n)(1-2/n)(1-3/n)} \to 0 \text{ as } n \to \infty$$

So by the Squeeze Theorem, $\{n^3/n!\}$ converges to 0.

35. (a) $a_n = 1000(1.06)^n \Rightarrow a_1 = 1060, a_2 = 1123.60, a_3 = 1191.02, a_4 = 1262.48$, and $a_5 = 1338.23$.

(b) $\lim_{n\to\infty} a_n = 1000 \lim_{n\to\infty} (1.06)^n$, so the sequence diverges by (6) with $r = 1.06 > 1$.

37. (a) $a_1 = 1, a_2 = 4 - a_1 = 4 - 1 = 3, a_3 = 4 - a_2 = 4 - 3 = 1, a_4 = 4 - a_3 = 4 - 1 = 3, a_5 = 4 - a_4 = 4 - 3 = 1$.
Since the terms of the sequence alternate between 1 and 3, the sequence is divergent.

(b) $a_1 = 2, a_2 = 4 - a_1 = 4 - 2 = 2, a_3 = 4 - a_2 = 4 - 2 = 2$. Since all of the terms are 2, $\lim_{n\to\infty} a_n = 2$ and hence, the
sequence is convergent.

39. (a) Let a_n be the number of rabbit pairs in the nth month. Clearly $a_1 = 1 = a_2$. In the nth month, each pair that is 2 or more
months old (that is, a_{n-2} pairs) will produce a new pair to add to the a_{n-1} pairs already present. Thus,
$a_n = a_{n-1} + a_{n-2}$, so that $\{a_n\} = \{f_n\}$, the Fibonacci sequence.

(b) $a_n = \dfrac{f_{n+1}}{f_n} \Rightarrow a_{n-1} = \dfrac{f_n}{f_{n-1}} = \dfrac{f_{n-1} + f_{n-2}}{f_{n-1}} = 1 + \dfrac{f_{n-2}}{f_{n-1}} = 1 + \dfrac{1}{f_{n-1}/f_{n-2}} = 1 + \dfrac{1}{a_{n-2}}$. If $L = \lim_{n\to\infty} a_n$,

then $L = \lim_{n\to\infty} a_{n-1}$ and $L = \lim_{n\to\infty} a_{n-2}$, so L must satisfy $L = 1 + \dfrac{1}{L} \Rightarrow L^2 - L - 1 = 0 \Rightarrow L = \dfrac{1+\sqrt{5}}{2}$
(since L must be positive).

41. $a_n = \dfrac{1}{2n+3}$ is decreasing since $a_{n+1} = \dfrac{1}{2(n+1)+3} = \dfrac{1}{2n+5} < \dfrac{1}{2n+3} = a_n$ for each $n \geq 1$. The sequence is
bounded since $0 < a_n \leq \frac{1}{5}$ for all $n \geq 1$. Note that $a_1 = \frac{1}{5}$.

43. $a_n = \cos(n\pi/2)$ is not monotonic. The first few terms are $0, -1, 0, 1, 0, -1, 0, 1, \ldots$. In fact, the sequence consists of the terms $0, -1, 0, 1$ repeated over and over again in that order. The sequence is bounded since $|a_n| \le 1$ for all $n \ge 1$.

45. Since $\{a_n\}$ is a decreasing sequence, $a_n > a_{n+1}$ for all $n \ge 1$. Because all of its terms lie between 5 and 8, $\{a_n\}$ is a bounded sequence. By the Monotonic Sequence Theorem, $\{a_n\}$ is convergent; that is, $\{a_n\}$ has a limit L. L must be less than 8 since $\{a_n\}$ is decreasing, so $5 \le L < 8$.

47. We show by induction that $\{a_n\}$ is increasing and bounded above by 3. Let P_n be the proposition that $a_{n+1} > a_n$ and $0 < a_n < 3$. Clearly P_1 is true. Assume that P_n is true.

Then $a_{n+1} > a_n \quad \Rightarrow \quad \dfrac{1}{a_{n+1}} < \dfrac{1}{a_n} \quad \Rightarrow \quad -\dfrac{1}{a_{n+1}} > -\dfrac{1}{a_n}$. Now $a_{n+2} = 3 - \dfrac{1}{a_{n+1}} > 3 - \dfrac{1}{a_n} = a_{n+1} \quad \Leftrightarrow \quad P_{n+1}$.

This proves that $\{a_n\}$ is increasing and bounded above by 3, so $1 = a_1 < a_n < 3$, that is, $\{a_n\}$ is bounded, and hence convergent by the Monotonic Sequence Theorem. If $L = \lim\limits_{n \to \infty} a_n$, then $\lim\limits_{n \to \infty} a_{n+1} = L$ also, so L must satisfy

$L = 3 - 1/L \quad \Rightarrow \quad L^2 - 3L + 1 = 0 \quad \Rightarrow \quad L = \frac{3 \pm \sqrt{5}}{2}$. But $L > 1$, so $L = \frac{3 + \sqrt{5}}{2}$.

49. $(0.8)^n < 0.000001 \quad \Rightarrow \quad \ln(0.8)^n < \ln(0.000001) \quad \Rightarrow \quad n \ln(0.8) < \ln(0.000001) \quad \Rightarrow$

$n > \dfrac{\ln(0.000001)}{\ln(0.8)} \quad \Rightarrow \quad n > 61.9$, so n must be at least 62 to satisfy the given inequality.

51. (a) Suppose $\{p_n\}$ converges to p. Then $p_{n+1} = \dfrac{bp_n}{a + p_n} \quad \Rightarrow \quad \lim\limits_{n \to \infty} p_{n+1} = \dfrac{b \lim\limits_{n \to \infty} p_n}{a + \lim\limits_{n \to \infty} p_n} \quad \Rightarrow \quad p = \dfrac{bp}{a + p} \quad \Rightarrow$

$p^2 + ap = bp \quad \Rightarrow \quad p(p + a - b) = 0 \quad \Rightarrow \quad p = 0 \text{ or } p = b - a.$

(b) $p_{n+1} = \dfrac{bp_n}{a + p_n} = \dfrac{\left(\dfrac{b}{a}\right)p_n}{1 + \dfrac{p_n}{a}} < \left(\dfrac{b}{a}\right)p_n$ since $1 + \dfrac{p_n}{a} > 1$.

(c) By part (b), $p_1 < \left(\dfrac{b}{a}\right)p_0, \; p_2 < \left(\dfrac{b}{a}\right)p_1 < \left(\dfrac{b}{a}\right)^2 p_0, \; p_3 < \left(\dfrac{b}{a}\right)p_2 < \left(\dfrac{b}{a}\right)^3 p_0$, etc. In general, $p_n < \left(\dfrac{b}{a}\right)^n p_0$,

so $\lim\limits_{n \to \infty} p_n \le \lim\limits_{n \to \infty} \left(\dfrac{b}{a}\right)^n \cdot p_0 = 0$ since $b < a$. $\left[\text{By result 8, } \lim\limits_{n \to \infty} r^n = 0 \text{ if } -1 < r < 1. \text{ Here } r = \dfrac{b}{a} \in (0, 1).\right]$

(d) Let $a < b$. We first show, by induction, that if $p_0 < b - a$, then $p_n < b - a$ and $p_{n+1} > p_n$.

For $n = 0$, we have $p_1 - p_0 = \dfrac{bp_0}{a + p_0} - p_0 = \dfrac{p_0(b - a - p_0)}{a + p_0} > 0$ since $p_0 < b - a$. So $p_1 > p_0$.

Now we suppose the assertion is true for $n = k$, that is, $p_k < b - a$ and $p_{k+1} > p_k$. Then

$b - a - p_{k+1} = b - a - \dfrac{bp_k}{a + p_k} = \dfrac{a(b - a) + bp_k - ap_k - bp_k}{a + p_k} = \dfrac{a(b - a - p_k)}{a + p_k} > 0$ because $p_k < b - a$. So

$p_{k+1} < b - a$. And $p_{k+2} - p_{k+1} = \dfrac{bp_{k+1}}{a + p_{k+1}} - p_{k+1} = \dfrac{p_{k+1}(b - a - p_{k+1})}{a + p_{k+1}} > 0$ since $p_{k+1} < b - a$. Therefore,

$p_{k+2} > p_{k+1}$. Thus, the assertion is true for $n = k + 1$. It is therefore true for all n by mathematical induction. A similar proof by induction shows that if $p_0 > b - a$, then $p_n > b - a$ and $\{p_n\}$ is decreasing.

In either case the sequence $\{p_n\}$ is bounded and monotonic, so it is convergent by the Monotonic Sequence Theorem. It then follows from part (a) that $\lim\limits_{n \to \infty} p_n = b - a$.

8.2 Series

1. (a) A sequence is an ordered list of numbers whereas a series is the *sum* of a list of numbers.

   (b) A series is convergent if the sequence of partial sums is a convergent sequence. A series is divergent if it is not convergent.

3.

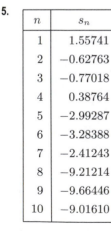

n	s_n
1	−2.40000
2	−1.92000
3	−2.01600
4	−1.99680
5	−2.00064
6	−1.99987
7	−2.00003
8	−1.99999
9	−2.00000
10	−2.00000

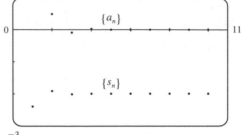

From the graph and the table, it seems that the series converges to −2. In fact, it is a geometric

series with $a = -2.4$ and $r = -\frac{1}{5}$, so its sum is $\displaystyle\sum_{n=1}^{\infty} \frac{12}{(-5)^n} = \frac{-2.4}{1 - \left(-\frac{1}{5}\right)} = \frac{-2.4}{1.2} = -2.$

Note that the dot corresponding to $n = 1$ is part of both $\{a_n\}$ and $\{s_n\}$.

TI-86 Note: To graph $\{a_n\}$ and $\{s_n\}$, set your calculator to Param mode and DrawDot mode. (DrawDot is under GRAPH,

MORE, FORMT (F3).) Now under E(t) = make the assignments: xt1=t, yt1=12/(-5)^t, xt2=t, yt2=sum

seq(yt1,t,1,t,1). (sum and seq are under LIST, OPS (F5), MORE.) Under WIND use 1,10,1,0,10,1,-3,1,1

to obtain a graph similar to the one above. Then use TRACE (F4) to see the values.

5.

n	s_n
1	1.55741
2	−0.62763
3	−0.77018
4	0.38764
5	−2.99287
6	−3.28388
7	−2.41243
8	−9.21214
9	−9.66446
10	−9.01610

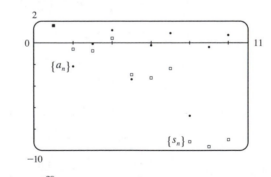

The series $\displaystyle\sum_{n=1}^{\infty} \tan n$ diverges, since its terms do not approach 0.

7.

n	s_n
1	0.64645
2	0.80755
3	0.87500
4	0.91056
5	0.93196
6	0.94601
7	0.95581
8	0.96296
9	0.96838
10	0.97259

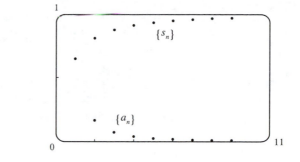

From the graph, it seems that the series converges to 1. To find the sum, we write

$$s_n = \sum_{i=1}^{n} \left(\frac{1}{i^{1.5}} - \frac{1}{(i+1)^{1.5}} \right)$$

$$= \left(1 - \frac{1}{2^{1.5}}\right) + \left(\frac{1}{2^{1.5}} - \frac{1}{3^{1.5}}\right) + \left(\frac{1}{3^{1.5}} - \frac{1}{4^{1.5}}\right) + \cdots + \left(\frac{1}{n^{1.5}} - \frac{1}{(n+1)^{1.5}}\right)$$

$$= 1 - \frac{1}{(n+1)^{1.5}}$$

So the sum is $\lim\limits_{n \to \infty} s_n = 1 - 0 = 1$.

9. (a) $\lim\limits_{n \to \infty} a_n = \lim\limits_{n \to \infty} \frac{2n}{3n+1} = \frac{2}{3}$, so the *sequence* $\{a_n\}$ is convergent by (8.1.1).

(b) Since $\lim\limits_{n \to \infty} a_n = \frac{2}{3} \neq 0$, the *series* $\sum\limits_{n=1}^{\infty} a_n$ is divergent by the Test for Divergence (7).

11. $5 - \frac{10}{3} + \frac{20}{9} - \frac{40}{27} + \cdots$ is a geometric series with $a = 5$ and $r = -\frac{2}{3}$. Since $|r| = \frac{2}{3} < 1$, the series converges to

$$\frac{a}{1-r} = \frac{5}{1-(-2/3)} = \frac{5}{5/3} = 3.$$

13. $\sum\limits_{n=1}^{\infty} 5\left(\frac{2}{3}\right)^{n-1}$ is a geometric series with $a = 5$ and $r = \frac{2}{3}$. Since $|r| = \frac{2}{3} < 1$, the series converges to

$$\frac{a}{1-r} = \frac{5}{1-2/3} = \frac{5}{1/3} = 15.$$

15. $\sum\limits_{n=0}^{\infty} \frac{\pi^n}{3^{n+1}} = \frac{1}{3} \sum\limits_{n=0}^{\infty} \left(\frac{\pi}{3}\right)^n$ is a geometric series with ratio $r = \frac{\pi}{3}$. Since $|r| > 1$, the series diverges.

17. $\sum\limits_{n=1}^{\infty} \frac{1}{2n} = \frac{1}{2} \sum\limits_{n=1}^{\infty} \frac{1}{n}$ diverges since each of its partial sums is $\frac{1}{2}$ times the corresponding partial sum of the harmonic series

$\sum\limits_{n=1}^{\infty} \frac{1}{n}$, which diverges. $\left[\text{If } \sum\limits_{n=1}^{\infty} \frac{1}{2n} \text{ were to converge, then } \sum\limits_{n=1}^{\infty} \frac{1}{n} \text{ would also have to converge by Theorem 8(i).}\right]$ In general,

constant multiples of divergent series are divergent.

19. $\sum\limits_{k=2}^{\infty} \frac{k^2}{k^2-1}$ diverges by the Test for Divergence since $\lim\limits_{k \to \infty} a_k = \lim\limits_{k \to \infty} \frac{k^2}{k^2-1} = 1 \neq 0$.

21. Converges.

$$\sum_{n=1}^{\infty} \frac{1+2^n}{3^n} = \sum_{n=1}^{\infty} \left(\frac{1}{3^n} + \frac{2^n}{3^n} \right) = \sum_{n=1}^{\infty} \left[\left(\frac{1}{3} \right)^n + \left(\frac{2}{3} \right)^n \right] \qquad \text{[sum of two convergent geometric series]}$$

$$= \frac{1/3}{1-1/3} + \frac{2/3}{1-2/3} = \frac{1}{2} + 2 = \frac{5}{2}$$

23. $\displaystyle\sum_{n=1}^{\infty} \sqrt[n]{2} = 2 + \sqrt{2} + \sqrt[3]{2} + \sqrt[4]{2} + \cdots$ diverges by the Test for Divergence since

$$\lim_{n \to \infty} a_n = \lim_{n \to \infty} \sqrt[n]{2} = \lim_{n \to \infty} 2^{1/n} = 2^0 = 1 \neq 0.$$

25. $\displaystyle\lim_{n \to \infty} a_n = \lim_{n \to \infty} \arctan n = \frac{\pi}{2} \neq 0$, so the series diverges by the Test for Divergence.

27. Using partial fractions, the partial sums of the series $\displaystyle\sum_{n=2}^{\infty} \frac{2}{n^2 - 1}$ are

$$s_n = \sum_{i=2}^{n} \frac{2}{(i-1)(i+1)} = \sum_{i=2}^{n} \left(\frac{1}{i-1} - \frac{1}{i+1} \right)$$

$$= \left(1 - \frac{1}{3} \right) + \left(\frac{1}{2} - \frac{1}{4} \right) + \left(\frac{1}{3} - \frac{1}{5} \right) + \cdots + \left(\frac{1}{n-3} - \frac{1}{n-1} \right) + \left(\frac{1}{n-2} - \frac{1}{n} \right)$$

This sum is a telescoping series and $s_n = 1 + \dfrac{1}{2} - \dfrac{1}{n-1} - \dfrac{1}{n}$.

Thus, $\displaystyle\sum_{n=2}^{\infty} \frac{2}{n^2 - 1} = \lim_{n \to \infty} s_n = \lim_{n \to \infty} \left(1 + \frac{1}{2} - \frac{1}{n-1} - \frac{1}{n} \right) = \frac{3}{2}$.

29. For the series $\displaystyle\sum_{n=1}^{\infty} \frac{3}{n(n+3)}$, $s_n = \displaystyle\sum_{i=1}^{n} \frac{3}{i(i+3)} = \sum_{i=1}^{n} \left(\frac{1}{i} - \frac{1}{i+3} \right)$ [using partial fractions]. The latter sum is

$$\left(1 - \tfrac{1}{4} \right) + \left(\tfrac{1}{2} - \tfrac{1}{5} \right) + \left(\tfrac{1}{3} - \tfrac{1}{6} \right) + \left(\tfrac{1}{4} - \tfrac{1}{7} \right) + \cdots + \left(\tfrac{1}{n-3} - \tfrac{1}{n} \right) + \left(\tfrac{1}{n-2} - \tfrac{1}{n+1} \right) + \left(\tfrac{1}{n-1} - \tfrac{1}{n+2} \right) + \left(\tfrac{1}{n} - \tfrac{1}{n+3} \right)$$

$$= 1 + \tfrac{1}{2} + \tfrac{1}{3} - \tfrac{1}{n+1} - \tfrac{1}{n+2} - \tfrac{1}{n+3} \qquad \text{[telescoping series]}$$

Thus, $\displaystyle\sum_{n=1}^{\infty} \frac{3}{n(n+3)} = \lim_{n \to \infty} s_n = \lim_{n \to \infty} \left(1 + \tfrac{1}{2} + \tfrac{1}{3} - \tfrac{1}{n+1} - \tfrac{1}{n+2} - \tfrac{1}{n+3} \right) = 1 + \tfrac{1}{2} + \tfrac{1}{3} = \tfrac{11}{6}$. Converges

31. $0.\overline{2} = \dfrac{2}{10} + \dfrac{2}{10^2} + \cdots$ is a geometric series with $a = \dfrac{2}{10}$ and $r = \dfrac{1}{10}$. It converges to $\dfrac{a}{1-r} = \dfrac{2/10}{1-1/10} = \dfrac{2}{9}$.

33. $3.\overline{417} = 3 + \dfrac{417}{10^3} + \dfrac{417}{10^6} + \cdots$. Now $\dfrac{417}{10^3} + \dfrac{417}{10^6} + \cdots$ is a geometric series with $a = \dfrac{417}{10^3}$ and $r = \dfrac{1}{10^3}$. It converges to

$$\frac{a}{1-r} = \frac{417/10^3}{1-1/10^3} = \frac{417/10^3}{999/10^3} = \frac{417}{999}. \text{ Thus, } 3.\overline{417} = 3 + \frac{417}{999} = \frac{3414}{999} = \frac{1138}{333}.$$

35. $\displaystyle\sum_{n=1}^{\infty} \frac{x^n}{3^n} = \sum_{n=1}^{\infty} \left(\frac{x}{3} \right)^n$ is a geometric series with $r = \dfrac{x}{3}$, so the series converges $\Leftrightarrow |r| < 1 \Leftrightarrow \dfrac{|x|}{3} < 1 \Leftrightarrow |x| < 3$;

that is, $-3 < x < 3$. In that case, the sum of the series is $\dfrac{a}{1-r} = \dfrac{x/3}{1-x/3} = \dfrac{x/3}{1-x/3} \cdot \dfrac{3}{3} = \dfrac{x}{3-x}$.

37. $\displaystyle\sum_{n=0}^{\infty} \frac{\cos^n x}{2^n}$ is a geometric series with first term 1 and ratio $r = \dfrac{\cos x}{2}$, so it converges $\Leftrightarrow$ $|r| < 1$. But $|r| = \dfrac{|\cos x|}{2} \le \dfrac{1}{2}$

for all x. Thus, the series converges for all real values of x and the sum of the series is $\dfrac{1}{1 - (\cos x)/2} = \dfrac{2}{2 - \cos x}$.

39. After defining f, We use `convert(f,parfrac);` in Maple, `Apart` in Mathematica, or `Expand Rational` and

`Simplify` in Derive to find that the general term is $\dfrac{1}{(4n+1)(4n-3)} = -\dfrac{1/4}{4n+1} + \dfrac{1/4}{4n-3}$. So the nth partial sum is

$$s_n = \sum_{k=1}^{n}\left(-\frac{1/4}{4k+1} + \frac{1/4}{4k-3}\right) = \frac{1}{4}\sum_{k=1}^{n}\left(\frac{1}{4k-3} - \frac{1}{4k+1}\right)$$

$$= \frac{1}{4}\left[\left(1 - \frac{1}{5}\right) + \left(\frac{1}{5} - \frac{1}{9}\right) + \left(\frac{1}{9} - \frac{1}{13}\right) + \cdots + \left(\frac{1}{4n-3} - \frac{1}{4n+1}\right)\right] = \frac{1}{4}\left(1 - \frac{1}{4n+1}\right)$$

The series converges to $\displaystyle\lim_{n\to\infty} s_n = \frac{1}{4}$. This can be confirmed by directly computing the sum using

`sum(f,1..infinity);` (in Maple), `Sum[f,{n,1,Infinity}]` (in Mathematica), or `Calculus Sum` (from 1 to ∞)

and `Simplify` (in Derive).

41. For $n = 1$, $a_1 = 0$ since $s_1 = 0$. For $n > 1$,

$$a_n = s_n - s_{n-1} = \frac{n-1}{n+1} - \frac{(n-1)-1}{(n-1)+1} = \frac{(n-1)n - (n+1)(n-2)}{(n+1)n} = \frac{2}{n(n+1)}$$

Also, $\displaystyle\sum_{n=1}^{\infty} a_n = \lim_{n\to\infty} s_n = \lim_{n\to\infty} \frac{1 - 1/n}{1 + 1/n} = 1$.

43. (a) The first step in the chain occurs when the local government spends D dollars. The people who receive it spend a fraction c

of those D dollars, that is, Dc dollars. Those who receive the Dc dollars spend a fraction c of it, that

is, Dc^2 dollars. Continuing in this way, we see that the total spending after n transactions is

$$S_n = D + Dc + Dc^2 + \cdots + Dc^{n-1} = \frac{D(1 - c^n)}{1 - c} \text{ by (3)}.$$

(b) $\displaystyle\lim_{n\to\infty} S_n = \lim_{n\to\infty} \frac{D(1 - c^n)}{1 - c} = \frac{D}{1 - c}\lim_{n\to\infty}(1 - c^n) = \frac{D}{1 - c}$ [since $0 < c < 1$ $\Rightarrow$ $\displaystyle\lim_{n\to\infty} c^n = 0$]

$= \dfrac{D}{s}$ [since $c + s = 1$] $= kD$ [since $k = 1/s$]

If $c = 0.8$, then $s = 1 - c = 0.2$ and the multiplier is $k = 1/s = 5$.

45. $\displaystyle\sum_{n=2}^{\infty}(1+c)^{-n}$ is a geometric series with $a = (1+c)^{-2}$ and $r = (1+c)^{-1}$, so the series converges when $\left|(1+c)^{-1}\right| < 1$

$\Leftrightarrow$ $|1 + c| > 1$ $\Leftrightarrow$ $1 + c > 1$ or $1 + c < -1$ $\Leftrightarrow$ $c > 0$ or $c < -2$. We calculate the sum of the series and set it equal

to 2: $\dfrac{(1+c)^{-2}}{1 - (1+c)^{-1}} = 2$ $\Leftrightarrow$ $\left(\dfrac{1}{1+c}\right)^2 = 2 - 2\left(\dfrac{1}{1+c}\right)$ $\Leftrightarrow$ $1 = 2(1+c)^2 - 2(1+c)$ $\Leftrightarrow$ $2c^2 + 2c - 1 = 0$

$\Leftrightarrow$ $c = \dfrac{-2 \pm \sqrt{12}}{4} = \dfrac{\pm\sqrt{3} - 1}{2}$. However, the negative root is inadmissible because $-2 < \dfrac{-\sqrt{3} - 1}{2} < 0$. So $c = \dfrac{\sqrt{3} - 1}{2}$.

47. Let d_n be the diameter of C_n. We draw lines from the centers of the C_i to the center of D (or C), and using the Pythagorean Theorem, we can write

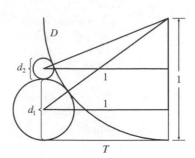

$1^2 + \left(1 - \tfrac{1}{2}d_1\right)^2 = \left(1 + \tfrac{1}{2}d_1\right)^2 \quad \Leftrightarrow$

$1 = \left(1 + \tfrac{1}{2}d_1\right)^2 - \left(1 - \tfrac{1}{2}d_1\right)^2 = 2d_1$ [difference of squares] $\Rightarrow d_1 = \tfrac{1}{2}$.

Similarly,

$1 = \left(1 + \tfrac{1}{2}d_2\right)^2 - \left(1 - d_1 - \tfrac{1}{2}d_2\right)^2 = 2d_2 + 2d_1 - d_1^2 - d_1 d_2$

$\quad = (2 - d_1)(d_1 + d_2) \quad \Leftrightarrow$

$d_2 = \dfrac{1}{2 - d_1} - d_1 = \dfrac{(1 - d_1)^2}{2 - d_1}, 1 = \left(1 + \tfrac{1}{2}d_3\right)^2 - \left(1 - d_1 - d_2 - \tfrac{1}{2}d_3\right)^2 \quad \Leftrightarrow \quad d_3 = \dfrac{[1 - (d_1 + d_2)]^2}{2 - (d_1 + d_2)}$, and in general,

$d_{n+1} = \dfrac{\left(1 - \sum_{i=1}^{n} d_i\right)^2}{2 - \sum_{i=1}^{n} d_i}$. If we actually calculate d_2 and d_3 from the formulas above, we find that they are $\dfrac{1}{6} = \dfrac{1}{2 \cdot 3}$ and

$\dfrac{1}{12} = \dfrac{1}{3 \cdot 4}$ respectively, so we suspect that in general, $d_n = \dfrac{1}{n(n + 1)}$. To prove this, we use induction: Assume that for all

$k \leq n, d_k = \dfrac{1}{k(k + 1)} = \dfrac{1}{k} - \dfrac{1}{k + 1}$. Then $\sum_{i=1}^{n} d_i = 1 - \dfrac{1}{n + 1} = \dfrac{n}{n + 1}$ [telescoping sum]. Substituting this into our

formula for d_{n+1}, we get $d_{n+1} = \dfrac{\left[1 - \dfrac{n}{n + 1}\right]^2}{2 - \left(\dfrac{n}{n + 1}\right)} = \dfrac{\dfrac{1}{(n + 1)^2}}{\dfrac{n + 2}{n + 1}} = \dfrac{1}{(n + 1)(n + 2)}$, and the induction is complete.

Now, we observe that the partial sums $\sum_{i=1}^{n} d_i$ of the diameters of the circles approach 1 as $n \to \infty$; that is,

$\sum_{n=1}^{\infty} a_n = \sum_{n=1}^{\infty} \dfrac{1}{n(n + 1)} = 1$, which is what we wanted to prove.

49. The series $1 - 1 + 1 - 1 + 1 - 1 + \cdots$ diverges (geometric series with $r = -1$) so we cannot say that
$0 = 1 - 1 + 1 - 1 + 1 - 1 + \cdots$.

51. Suppose on the contrary that $\sum(a_n + b_n)$ converges. Then $\sum(a_n + b_n)$ and $\sum a_n$ are convergent series. So by Theorem 8,
$\sum [(a_n + b_n) - a_n]$ would also be convergent. But $\sum [(a_n + b_n) - a_n] = \sum b_n$, a contradiction, since $\sum b_n$ is given to be
divergent.

53. The partial sums $\{s_n\}$ form an increasing sequence, since $s_n - s_{n-1} = a_n > 0$ for all n. Also, the sequence $\{s_n\}$ is bounded
since $s_n \leq 1000$ for all n. So by Theorem 8.1.7, the sequence of partial sums converges, that is, the series $\sum a_n$ is convergent.

55. (a) At the first step, only the interval $\left(\tfrac{1}{3}, \tfrac{2}{3}\right)$ (length $\tfrac{1}{3}$) is removed. At the second step, we remove the intervals $\left(\tfrac{1}{9}, \tfrac{2}{9}\right)$ and

$\left(\tfrac{7}{9}, \tfrac{8}{9}\right)$, which have a total length of $2 \cdot \left(\tfrac{1}{3}\right)^2$. At the third step, we remove 2^2 intervals, each of length $\left(\tfrac{1}{3}\right)^3$. In general, at

the nth step we remove 2^{n-1} intervals, each of length $\left(\tfrac{1}{3}\right)^n$, for a length of $2^{n-1} \cdot \left(\tfrac{1}{3}\right)^n = \tfrac{1}{3}\left(\tfrac{2}{3}\right)^{n-1}$. Thus, the total

length of all removed intervals is $\sum_{n=1}^{\infty} \tfrac{1}{3}\left(\tfrac{2}{3}\right)^{n-1} = \dfrac{1/3}{1 - 2/3} = 1$ [geometric series with $a = \tfrac{1}{3}$ and $r = \tfrac{2}{3}$]. Notice that at

the nth step, the leftmost interval that is removed is $\left(\left(\tfrac{1}{3}\right)^n, \left(\tfrac{2}{3}\right)^n\right)$, so we never remove 0, and 0 is in the Cantor set. Also,

the rightmost interval removed is $\left(1 - \left(\tfrac{2}{3}\right)^n, 1 - \left(\tfrac{1}{3}\right)^n\right)$, so 1 is never removed. Some other numbers in the Cantor set are

$\tfrac{1}{3}, \tfrac{2}{3}, \tfrac{1}{9}, \tfrac{2}{9}, \tfrac{7}{9}$, and $\tfrac{8}{9}$.

(b) The area removed at the first step is $\frac{1}{9}$; at the second step, $8 \cdot \left(\frac{1}{9}\right)^2$; at the third step, $(8)^2 \cdot \left(\frac{1}{9}\right)^3$. In general, the area

removed at the nth step is $(8)^{n-1}\left(\frac{1}{9}\right)^n = \frac{1}{9}\left(\frac{8}{9}\right)^{n-1}$, so the total area of all removed squares is

$$\sum_{n=1}^{\infty} \frac{1}{9}\left(\frac{8}{9}\right)^{n-1} = \frac{1/9}{1-8/9} = 1.$$

57. (a) For $\sum\limits_{n=1}^{\infty} \dfrac{n}{(n+1)!}$, $s_1 = \dfrac{1}{1\cdot 2} = \dfrac{1}{2}$, $s_2 = \dfrac{1}{2} + \dfrac{2}{1\cdot 2\cdot 3} = \dfrac{5}{6}$, $s_3 = \dfrac{5}{6} + \dfrac{3}{1\cdot 2\cdot 3\cdot 4} = \dfrac{23}{24}$,

$s_4 = \dfrac{23}{24} + \dfrac{4}{1\cdot 2\cdot 3\cdot 4\cdot 5} = \dfrac{119}{120}$. The denominators are $(n+1)!$, so a guess would be $s_n = \dfrac{(n+1)!-1}{(n+1)!}$.

(b) For $n = 1$, $s_1 = \dfrac{1}{2} = \dfrac{2!-1}{2!}$, so the formula holds for $n = 1$. Assume $s_k = \dfrac{(k+1)!-1}{(k+1)!}$. Then

$$s_{k+1} = \frac{(k+1)!-1}{(k+1)!} + \frac{k+1}{(k+2)!} = \frac{(k+1)!-1}{(k+1)!} + \frac{k+1}{(k+1)!(k+2)}$$

$$= \frac{(k+2)! - (k+2) + k + 1}{(k+2)!} = \frac{(k+2)!-1}{(k+2)!}$$

Thus, the formula is true for $n = k+1$. So by induction, the guess is correct.

(c) $\lim\limits_{n\to\infty} s_n = \lim\limits_{n\to\infty} \dfrac{(n+1)!-1}{(n+1)!} = \lim\limits_{n\to\infty}\left[1 - \dfrac{1}{(n+1)!}\right] = 1$ and so $\sum\limits_{n=1}^{\infty} \dfrac{n}{(n+1)!} = 1$.

8.3   The Integral and Comparison Tests; Estimating Sums

1. The picture shows that $a_2 = \dfrac{1}{2^{1.3}} < \displaystyle\int_1^2 \dfrac{1}{x^{1.3}}\,dx$,

$a_3 = \dfrac{1}{3^{1.3}} < \displaystyle\int_2^3 \dfrac{1}{x^{1.3}}\,dx$, and so on, so $\displaystyle\sum_{n=2}^{\infty} \dfrac{1}{n^{1.3}} < \int_1^{\infty} \dfrac{1}{x^{1.3}}\,dx$. The

integral converges by (5.10.2) with $p = 1.3 > 1$, so the series converges.

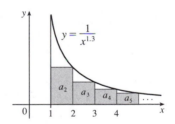

3. (a) We cannot say anything about $\sum a_n$. If $a_n > b_n$ for all n and $\sum b_n$ is convergent, then $\sum a_n$ could be convergent or divergent. (See the note after Example 4.)

(b) If $a_n < b_n$ for all n, then $\sum a_n$ is convergent. [This is part (i) of the Comparison Test.]

5. $\displaystyle\sum_{n=1}^{\infty} n^b$ is a p-series with $p = -b$. $\displaystyle\sum_{n=1}^{\infty} b^n$ is a geometric series. By (1), the p-series is convergent if $p > 1$. In this case,

$\displaystyle\sum_{n=1}^{\infty} n^b = \sum_{n=1}^{\infty}\left(1/n^{-b}\right)$, so $-b > 1 \iff b < -1$ are the values for which the series converge. A geometric series

$\displaystyle\sum_{n=1}^{\infty} ar^{n-1}$ converges if $|r| < 1$, so $\displaystyle\sum_{n=1}^{\infty} b^n$ converges if $|b| < 1 \iff -1 < b < 1$.

7. The function $f(x) = 1/x^4$ is continuous, positive, and decreasing on $[1, \infty)$, so the Integral Test applies.

$\displaystyle\int_1^{\infty} \frac{1}{x^4}\,dx = \lim_{t\to\infty}\int_1^t x^{-4}\,dx = \lim_{t\to\infty}\left[\frac{x^{-3}}{-3}\right]_1^t = \lim_{t\to\infty}\left(-\frac{1}{3t^3} + \frac{1}{3}\right) = \frac{1}{3}$. Since this improper integral is convergent, the

series $\displaystyle\sum_{n=1}^{\infty} \frac{1}{n^4}$ is also convergent by the Integral Test.

9. $\dfrac{1}{n^2+n+1} < \dfrac{1}{n^2}$ for all $n \geq 1$, so $\displaystyle\sum_{n=1}^{\infty} \frac{1}{n^2+n+1}$ converges by comparison with $\displaystyle\sum_{n=1}^{\infty} \frac{1}{n^2}$, which converges because it is a p-series with $p = 2 > 1$.

11. $1 + \dfrac{1}{8} + \dfrac{1}{27} + \dfrac{1}{64} + \dfrac{1}{125} + \cdots = \sum\limits_{n=1}^{\infty} \dfrac{1}{n^3}$. This is a p-series with $p = 3 > 1$, so it converges by (1).

13. $f(x) = xe^{-x}$ is continuous and positive on $[1, \infty)$. $f'(x) = -xe^{-x} + e^{-x} = e^{-x}(1 - x) < 0$ for $x > 1$, so f is decreasing on $[1, \infty)$. Thus, the Integral Test applies.

$$\int_1^{\infty} xe^{-x}\, dx = \lim_{b \to \infty} \int_1^b xe^{-x}\, dx = \lim_{b \to \infty} \left[-xe^{-x} - e^{-x}\right]_1^b \quad \text{[by parts]}$$

$$= \lim_{b \to \infty} \left[-be^{-b} - e^{-b} + e^{-1} + e^{-1}\right] = 2/e$$

since $\lim\limits_{b \to \infty} be^{-b} = \lim\limits_{b \to \infty} \left(b/e^b\right) \overset{\text{H}}{=} \lim\limits_{b \to \infty} \left(1/e^b\right) = 0$ and $\lim\limits_{b \to \infty} e^{-b} = 0$. Thus, $\sum\limits_{n=1}^{\infty} ne^{-n}$ converges.

15. $f(x) = \dfrac{1}{x \ln x}$ is continuous and positive on $[2, \infty)$, and also decreasing since $f'(x) = -\dfrac{1 + \ln x}{x^2 (\ln x)^2} < 0$ for $x > 2$, so we can

use the Integral Test. $\displaystyle\int_2^{\infty} \dfrac{1}{x \ln x}\, dx = \lim_{t \to \infty} \left[\ln(\ln x)\right]_2^t = \lim_{t \to \infty} \left[\ln(\ln t) - \ln(\ln 2)\right] = \infty$, so the series $\sum\limits_{n=2}^{\infty} \dfrac{1}{n \ln n}$ diverges.

17. $\dfrac{\cos^2 n}{n^2 + 1} \le \dfrac{1}{n^2 + 1} < \dfrac{1}{n^2}$, so the series $\sum\limits_{n=1}^{\infty} \dfrac{\cos^2 n}{n^2 + 1}$ converges by comparison with the p-series $\sum\limits_{n=1}^{\infty} \dfrac{1}{n^2}$ $(p = 2 > 1)$.

19. $\dfrac{n - 1}{n\, 4^n}$ is positive for $n > 1$ and $\dfrac{n - 1}{n\, 4^n} < \dfrac{n}{n\, 4^n} = \dfrac{1}{4^n} = \left(\dfrac{1}{4}\right)^n$, so $\sum\limits_{n=1}^{\infty} \dfrac{n - 1}{n\, 4^n}$ converges by comparison with the convergent

geometric series $\sum\limits_{n=1}^{\infty} \left(\dfrac{1}{4}\right)^n$.

21. Use the Limit Comparison Test with $a_n = \dfrac{1}{\sqrt{n^2 + 1}}$ and $b_n = \dfrac{1}{n}$:

$$\lim_{n \to \infty} \dfrac{a_n}{b_n} = \lim_{n \to \infty} \dfrac{n}{\sqrt{n^2 + 1}} = \lim_{n \to \infty} \dfrac{1}{\sqrt{1 + (1/n^2)}} = 1 > 0. \text{ Since the harmonic series } \sum_{n=1}^{\infty} \dfrac{1}{n} \text{ diverges, so does}$$

$$\sum_{n=1}^{\infty} \dfrac{1}{\sqrt{n^2 + 1}}.$$

23. $\dfrac{2 + (-1)^n}{n \sqrt{n}} \le \dfrac{3}{n \sqrt{n}}$, and $\sum\limits_{n=1}^{\infty} \dfrac{3}{n \sqrt{n}}$ converges because it is a constant multiple of the convergent p-series $\sum\limits_{n=1}^{\infty} \dfrac{1}{n \sqrt{n}}$

$\left(p = \dfrac{3}{2} > 1\right)$, so the given series converges by the Comparison Test.

25. Use the Limit Comparison Test with $a_n = \sin\left(\dfrac{1}{n}\right)$ and $b_n = \dfrac{1}{n}$. Then $\sum a_n$ and $\sum b_n$ are series with positive terms and

$$\lim_{n \to \infty} \dfrac{a_n}{b_n} = \lim_{n \to \infty} \dfrac{\sin(1/n)}{1/n} = \lim_{\theta \to 0} \dfrac{\sin \theta}{\theta} = 1 > 0. \text{ Since } \sum_{n=1}^{\infty} b_n \text{ is the divergent harmonic series, } \sum_{n=1}^{\infty} \sin(1/n) \text{ also}$$

diverges.

[Note that we could also use l'Hospital's Rule to evaluate the limit:

$$\lim_{x \to \infty} \dfrac{\sin(1/x)}{1/x} \overset{\text{H}}{=} \lim_{x \to \infty} \dfrac{\cos(1/x) \cdot (-1/x^2)}{-1/x^2} = \lim_{x \to \infty} \cos \dfrac{1}{x} = \cos 0 = 1.]$$

27. We have already shown (in Exercise 15) that when $p = 1$ the series $\sum\limits_{n=2}^{\infty} \dfrac{1}{n(\ln n)^p}$ diverges, so assume that $p \ne 1$.

$f(x) = \dfrac{1}{x(\ln x)^p}$ is continuous and positive on $[2, \infty)$, and $f'(x) = -\dfrac{p + \ln x}{x^2 (\ln x)^{p+1}} < 0$ if $x > e^{-p}$, so that f is eventually

decreasing and we can use the Integral Test.

$$\int_2^{\infty} \dfrac{1}{x(\ln x)^p}\, dx = \lim_{t \to \infty} \left[\dfrac{(\ln x)^{1-p}}{1 - p}\right]_2^t \quad \text{[for } p \ne 1\text{]} = \lim_{t \to \infty} \left[\dfrac{(\ln t)^{1-p}}{1 - p}\right] - \dfrac{(\ln 2)^{1-p}}{1 - p}$$

This limit exists whenever $1 - p < 0 \;\Leftrightarrow\; p > 1$, so the series converges for $p > 1$.

29. (a) $f(x) = \dfrac{1}{x^2}$ is positive and continuous and $f'(x) = -\dfrac{2}{x^3}$ is negative for $x > 0$, and so the Integral Test applies.

$$\sum_{n=1}^{\infty} \frac{1}{n^2} \approx s_{10} = \frac{1}{1^2} + \frac{1}{2^2} + \frac{1}{3^2} + \cdots + \frac{1}{10^2} \approx 1.549768.$$

$$R_{10} \leq \int_{10}^{\infty} \frac{1}{x^2}\,dx = \lim_{t \to \infty}\left[\frac{-1}{x}\right]_{10}^{t} = \lim_{t \to \infty}\left(-\frac{1}{t} + \frac{1}{10}\right) = \frac{1}{10}, \text{ so the error is at most } 0.1.$$

(b) $s_{10} + \displaystyle\int_{11}^{\infty} \frac{1}{x^2}\,dx \leq s \leq s_{10} + \int_{10}^{\infty} \frac{1}{x^2}\,dx \Rightarrow s_{10} + \frac{1}{11} \leq s \leq s_{10} + \frac{1}{10} \Rightarrow$

$1.549768 + 0.090909 = 1.640677 \leq s \leq 1.549768 + 0.1 = 1.649768$, so we get $s \approx 1.64522$ (the average of 1.640677 and 1.649768) with error ≤ 0.005 (the maximum of $1.649768 - 1.64522$ and $1.64522 - 1.640677$, rounded up).

(c) $R_n \leq \displaystyle\int_{n}^{\infty} \frac{1}{x^2}\,dx = \frac{1}{n}.$ So $R_n < 0.001$ if $\dfrac{1}{n} < \dfrac{1}{1000} \Leftrightarrow n > 1000.$

31. $f(x) = x^{-3/2}$ is positive and continuous and $f'(x) = -\frac{3}{2}x^{-5/2}$ is negative for $x > 0$, so the Integral Test applies. From the end of Example 7, we see that the error is at most half the length of the interval. From (4), the interval is

$\left(s_n + \int_{n+1}^{\infty} f(x)\,dx,\ s_n + \int_{n}^{\infty} f(x)\,dx\right)$, so its length is $\int_{n}^{\infty} f(x)\,dx - \int_{n+1}^{\infty} f(x)\,dx = \int_{n}^{n+1} f(x)\,dx$. Thus, we need n

such that $0.01 > \dfrac{1}{2}\displaystyle\int_{n}^{n+1} x^{-3/2}\,dx = \frac{1}{2}\left[\frac{-2}{\sqrt{x}}\right]_{n}^{n+1} = \frac{1}{\sqrt{n}} - \frac{1}{\sqrt{n+1}} \Leftrightarrow n > 13.08$ (use a graphing calculator to solve

$1/\sqrt{x} - 1/\sqrt{x+1} < 0.01$). Again from the end of Example 7, we approximate s by the midpoint of this interval. In general,

the midpoint is $\frac{1}{2}\left[\left(s_n + \int_{n+1}^{\infty} f(x)\,dx\right) + \left(s_n + \int_{n}^{\infty} f(x)\,dx\right)\right] = s_n + \frac{1}{2}\left(\int_{n+1}^{\infty} f(x)\,dx + \int_{n}^{\infty} f(x)\,dx\right)$. So using

$n = 14$, we have $s \approx s_{14} + \frac{1}{2}\left(\int_{14}^{\infty} x^{-3/2}\,dx + \int_{15}^{\infty} x^{-3/2}\,dx\right) \approx 2.0872 + \frac{1}{\sqrt{14}} + \frac{1}{\sqrt{15}} \approx 2.6127 \approx 2.61.$ Any larger value

of n will also work. For instance, $s \approx s_{30} + \frac{1}{\sqrt{30}} + \frac{1}{\sqrt{31}} \approx 2.6124.$

33. $\displaystyle\sum_{n=1}^{10} \frac{1}{n^4 + n^2} = \frac{1}{2} + \frac{1}{20} + \frac{1}{90} + \cdots + \frac{1}{10,100} \approx 0.567975.$ Now $\dfrac{1}{n^4 + n^2} < \dfrac{1}{n^4}$, so using the reasoning and notation of

Example 8, the error is $R_{10} \leq T_{10} = \displaystyle\sum_{n=11}^{\infty} \frac{1}{n^4} \leq \int_{10}^{\infty} \frac{dx}{x^4} = \lim_{t \to \infty}\left[-\frac{x^{-3}}{3}\right]_{10}^{t} = \frac{1}{3000} = 0.000\overline{3}.$

35. (a) From the figure, $a_2 + a_3 + \cdots + a_n \leq \int_{1}^{n} f(x)\,dx$, so with

$f(x) = \dfrac{1}{x}, \dfrac{1}{2} + \dfrac{1}{3} + \dfrac{1}{4} + \cdots + \dfrac{1}{n} \leq \displaystyle\int_{1}^{n} \frac{1}{x}\,dx = \ln n.$ Thus,

$s_n = 1 + \dfrac{1}{2} + \dfrac{1}{3} + \dfrac{1}{4} + \cdots + \dfrac{1}{n} \leq 1 + \ln n.$

(b) By part (a), $s_{10^6} \leq 1 + \ln 10^6 \approx 14.82 < 15$ and $s_{10^9} \leq 1 + \ln 10^9 \approx 21.72 < 22.$

37. Since $\dfrac{d_n}{10^n} \leq \dfrac{9}{10^n}$ for each n, and since $\displaystyle\sum_{n=1}^{\infty} \frac{9}{10^n}$ is a convergent geometric series $\left(|r| = \frac{1}{10} < 1\right)$, $0.d_1d_2d_3\ldots = \displaystyle\sum_{n=1}^{\infty} \frac{d_n}{10^n}$

will always converge by the Comparison Test.

39. Yes. Since $\sum a_n$ is a convergent series with positive terms, $\lim\limits_{n \to \infty} a_n = 0$ by Theorem 8.2.6, and $\sum b_n = \sum \sin(a_n)$ is a

series with positive terms (for large enough n). We have $\lim\limits_{n \to \infty} \dfrac{b_n}{a_n} = \lim\limits_{n \to \infty} \dfrac{\sin(a_n)}{a_n} = 1 > 0$ by Theorem 3.4.2. Thus, $\sum b_n$

is also convergent by the Limit Comparison Test.

8.4 Other Convergence Tests

1. (a) An alternating series is a series whose terms are alternately positive and negative.

   (b) An alternating series $\sum_{n=1}^{\infty}(-1)^{n-1}b_n$ converges if $0 < b_{n+1} \le b_n$ for all n and $\lim_{n \to \infty} b_n = 0$. (This is the Alternating Series Test.)

   (c) The error involved in using the partial sum s_n as an approximation to the total sum s is the remainder $R_n = s - s_n$ and the size of the error is smaller than b_{n+1}; that is, $|R_n| \le b_{n+1}$. (This is the Alternating Series Estimation Theorem.)

3. $\dfrac{4}{7} - \dfrac{4}{8} + \dfrac{4}{9} - \dfrac{4}{10} + \dfrac{4}{11} - \cdots = \sum_{n=1}^{\infty}(-1)^{n-1}\dfrac{4}{n+6}$. Now $b_n = \dfrac{4}{n+6} > 0$, $\{b_n\}$ is decreasing, and $\lim_{n \to \infty} b_n = 0$, so the

   series converges by the Alternating Series Test.

5. $b_n = \dfrac{1}{\sqrt{n}} > 0$, $\{b_n\}$ is decreasing, and $\lim_{n \to \infty} b_n = 0$, so the series $\sum_{n=1}^{\infty} \dfrac{(-1)^{n-1}}{\sqrt{n}}$ converges by the Alternating Series Test.

7. $\sum_{n=1}^{\infty} a_n = \sum_{n=1}^{\infty}(-1)^n \dfrac{3n-1}{2n+1} = \sum_{n=1}^{\infty}(-1)^n b_n$. Now $\lim_{n \to \infty} b_n = \lim_{n \to \infty}\dfrac{3-1/n}{2+1/n} = \dfrac{3}{2} \ne 0$. Since $\lim_{n \to \infty} a_n \ne 0$

   (in fact the limit does not exist), the series diverges by the Test for Divergence.

9. $\sum_{n=1}^{\infty} \dfrac{(-1)^{n-1}}{n} = 1 - \dfrac{1}{2} + \dfrac{1}{3} - \dfrac{1}{4} + \cdots + \dfrac{1}{49} - \dfrac{1}{50} + \dfrac{1}{51} - \dfrac{1}{52} + \cdots$. The 50th partial sum of this series is an

   underestimate, since $\sum_{n=1}^{\infty} \dfrac{(-1)^{n-1}}{n} = s_{50} + \left(\dfrac{1}{51} - \dfrac{1}{52}\right) + \left(\dfrac{1}{53} - \dfrac{1}{54}\right) + \cdots$, and the terms in parentheses are all positive.

   The result can be seen geometrically in Figure 1.

11. If $p > 0$, $\dfrac{1}{(n+1)^p} \le \dfrac{1}{n^p}$ ($\{1/n^p\}$ is decreasing) and $\lim_{n \to \infty} \dfrac{1}{n^p} = 0$, so the series converges by the Alternating Series Test.

    If $p \le 0$, $\lim_{n \to \infty} \dfrac{(-1)^{n-1}}{n^p}$ does not exist, so the series diverges by the Test for Divergence. Thus, $\sum_{n=1}^{\infty} \dfrac{(-1)^{n-1}}{n^p}$ converges

    $\Leftrightarrow \quad p > 0$.

13. The series $\sum_{n=1}^{\infty} \dfrac{(-1)^{n+1}}{n^6}$ satisfies (i) of the Alternating Series Test because $\dfrac{1}{(n+1)^6} < \dfrac{1}{n^6}$ and (ii) $\lim_{n \to \infty} \dfrac{1}{n^6} = 0$, so the

    series is convergent. Now $b_5 = \dfrac{1}{5^6} = 0.000064 > 0.00005$ and $b_6 = \dfrac{1}{6^6} \approx 0.00002 < 0.00005$, so by the Alternating Series

    Estimation Theorem, $n = 5$. (That is, since the 6th term is less than the desired error, we need to add the first 5 terms to get the

    sum to the desired accuracy.)

15. The graph gives us an estimate for the sum of the series

    $\sum_{n=1}^{\infty} \dfrac{(-1)^{n-1}}{(2n-1)!}$ of 0.84.

    $b_5 = \dfrac{1}{(2 \cdot 5 - 1)!} = \dfrac{1}{362,880} \approx 0.000\,003$, so

    $\sum_{n=1}^{\infty} \dfrac{(-1)^{n-1}}{(2n-1)!} \approx s_4 = \sum_{n=1}^{4} \dfrac{(-1)^{n-1}}{(2n-1)!} = 1 - \dfrac{1}{6} + \dfrac{1}{120} - \dfrac{1}{5040}$

    $\approx 0.841468.$

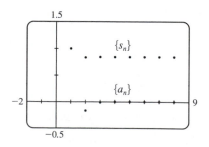

    Adding b_5 to s_4 does not change the fourth decimal place of s_4, so
    the sum of the series, correct to four decimal places, is 0.8415.

17. $b_7 = \dfrac{7^2}{10^7} = 0.000\,004\,9$, so

$$\sum_{n=1}^{\infty} \frac{(-1)^{n-1}n^2}{10^n} \approx s_6 = \sum_{n=1}^{6} \frac{(-1)^{n-1}n^2}{10^n} = \frac{1}{10} - \frac{4}{100} + \frac{9}{1000} - \frac{16}{10,000} + \frac{25}{100,000} - \frac{36}{1,000,000} = 0.067\,614.$$ Adding b_7 to s_6

does not change the fourth decimal place of s_6, so the sum of the series, correct to four decimal places, is 0.0676.

19. Using the Ratio Test, $\displaystyle\lim_{n\to\infty}\left|\frac{a_{n+1}}{a_n}\right| = \lim_{n\to\infty}\left|\frac{(-3)^{n+1}/(n+1)^3}{(-3)^n/n^3}\right| = \lim_{n\to\infty}\left|\frac{(-3)n^3}{(n+1)^3}\right| = 3\lim_{n\to\infty}\left(\frac{n}{n+1}\right)^3 = 3 > 1$,

so the series diverges.

21. $\displaystyle\sum_{n=0}^{\infty} \frac{(-10)^n}{n!}$. Using the Ratio Test, $\displaystyle\lim_{n\to\infty}\left|\frac{a_{n+1}}{a_n}\right| = \lim_{n\to\infty}\left|\frac{(-10)^{n+1}}{(n+1)!}\cdot\frac{n!}{(-10)^n}\right| = \lim_{n\to\infty}\left|\frac{-10}{n+1}\right| = 0 < 1$, so the series is

absolutely convergent.

23. Consider the series whose terms are the absolute values of the terms of the given series. $\displaystyle\sum_{n=1}^{\infty}\left|\frac{(-1)^{n-1}}{\sqrt{n}}\right| = \sum_{n=1}^{\infty}\frac{1}{n^{1/2}}$, which is

a divergent p-series $\left(p = \frac{1}{2} \le 1\right)$. Thus, $\displaystyle\sum_{n=1}^{\infty}\frac{(-1)^{n-1}}{\sqrt{n}}$ is *not* absolutely convergent.

25. $\displaystyle\lim_{n\to\infty}\left|\frac{a_{n+1}}{a_n}\right| = \lim_{n\to\infty}\left[\frac{10^{n+1}}{(n+2)4^{2(n+1)+1}}\cdot\frac{(n+1)4^{2n+1}}{10^n}\right] = \lim_{n\to\infty}\left[\frac{10^{n+1}}{(n+2)4^{2n+3}}\cdot\frac{(n+1)4^{2n+1}}{10^n}\right]$

$$= \lim_{n\to\infty}\left(\frac{10}{4^2}\cdot\frac{n+1}{n+2}\right) = \frac{5}{8} < 1,$$

so the series is absolutely convergent by the Ratio Test. Since the terms of this series are positive, absolute convergence is the

same as convergence.

27. $\displaystyle\left|\frac{(-1)^n\arctan n}{n^2}\right| < \frac{\pi/2}{n^2}$, so since $\displaystyle\sum_{n=1}^{\infty}\frac{\pi/2}{n^2} = \frac{\pi}{2}\sum_{n=1}^{\infty}\frac{1}{n^2}$ converges $(p = 2 > 1)$, the given series $\displaystyle\sum_{n=1}^{\infty}\frac{(-1)^n\arctan n}{n^2}$

converges absolutely by the Comparison Test.

29. Use the Ratio Test with the series

$$1 - \frac{1\cdot 3}{3!} + \frac{1\cdot 3\cdot 5}{5!} - \frac{1\cdot 3\cdot 5\cdot 7}{7!} + \cdots + (-1)^{n-1}\frac{1\cdot 3\cdot 5\cdots\cdots(2n-1)}{(2n-1)!} + \cdots = \sum_{n=1}^{\infty}(-1)^{n-1}\frac{1\cdot 3\cdot 5\cdots\cdots(2n-1)}{(2n-1)!}.$$

$$\lim_{n\to\infty}\left|\frac{a_{n+1}}{a_n}\right| = \lim_{n\to\infty}\left|\frac{(-1)^n\cdot 1\cdot 3\cdot 5\cdots\cdots(2n-1)[2(n+1)-1]}{[2(n+1)-1]!}\cdot\frac{(2n-1)!}{(-1)^{n-1}\cdot 1\cdot 3\cdot 5\cdots\cdots(2n-1)}\right|$$

$$= \lim_{n\to\infty}\left|\frac{(-1)(2n+1)(2n-1)!}{(2n+1)(2n)(2n-1)!}\right| = \lim_{n\to\infty}\frac{1}{2n} = 0 < 1,$$

so the given series is absolutely convergent and therefore convergent.

31. By the recursive definition, $\displaystyle\lim_{n\to\infty}\left|\frac{a_{n+1}}{a_n}\right| = \lim_{n\to\infty}\left|\frac{5n+1}{4n+3}\right| = \frac{5}{4} > 1$, so the series diverges by the Ratio Test.

33. (a) $\displaystyle\lim_{n\to\infty}\left|\frac{1/(n+1)^3}{1/n^3}\right| = \lim_{n\to\infty}\frac{n^3}{(n+1)^3} = \lim_{n\to\infty}\frac{1}{(1+1/n)^3} = 1$. Inconclusive

(b) $\displaystyle\lim_{n\to\infty}\left|\frac{(n+1)}{2^{n+1}}\cdot\frac{2^n}{n}\right| = \lim_{n\to\infty}\frac{n+1}{2n} = \lim_{n\to\infty}\left(\frac{1}{2}+\frac{1}{2n}\right) = \frac{1}{2}$. Conclusive (convergent)

(c) $\displaystyle\lim_{n\to\infty}\left|\frac{(-3)^n}{\sqrt{n+1}}\cdot\frac{\sqrt{n}}{(-3)^{n-1}}\right| = 3\lim_{n\to\infty}\sqrt{\frac{n}{n+1}} = 3\lim_{n\to\infty}\sqrt{\frac{1}{1+1/n}} = 3$. Conclusive (divergent)

(d) $\displaystyle\lim_{n\to\infty}\left|\frac{\sqrt{n+1}}{1+(n+1)^2}\cdot\frac{1+n^2}{\sqrt{n}}\right| = \lim_{n\to\infty}\left[\sqrt{1+\frac{1}{n}}\cdot\frac{1/n^2+1}{1/n^2+(1+1/n)^2}\right] = 1$. Inconclusive

35. (a) $\lim\limits_{n\to\infty}\left|\dfrac{a_{n+1}}{a_n}\right|=\lim\limits_{n\to\infty}\left|\dfrac{x^{n+1}}{(n+1)!}\cdot\dfrac{n!}{x^n}\right|=\lim\limits_{n\to\infty}\left|\dfrac{x}{n+1}\right|=|x|\lim\limits_{n\to\infty}\dfrac{1}{n+1}=|x|\cdot 0=0<1$, so by the Ratio Test the

series $\sum\limits_{n=0}^{\infty}\dfrac{x^n}{n!}$ converges for all x.

(b) Since the series of part (a) always converges, we must have $\lim\limits_{n\to\infty}\dfrac{x^n}{n!}=0$ by Theorem 8.2.6.

8.5   Power Series

1. A power series is a series of the form $\sum\limits_{n=0}^{\infty}c_n x^n=c_0+c_1 x+c_2 x^2+c_3 x^3+\cdots$, where x is a variable and the c_n's are

constants called the coefficients of the series.

More generally, a series of the form $\sum\limits_{n=0}^{\infty}c_n(x-a)^n=c_0+c_1(x-a)+c_2(x-a)^2+\cdots$ is called a power series in

$(x-a)$ or a power series centered at a or a power series about a, where a is a constant.

3. If $a_n=\dfrac{x^n}{\sqrt{n}}$, then $\lim\limits_{n\to\infty}\left|\dfrac{a_{n+1}}{a_n}\right|=\lim\limits_{n\to\infty}\left|\dfrac{x^{n+1}}{\sqrt{n+1}}\cdot\dfrac{\sqrt{n}}{x^n}\right|=\lim\limits_{n\to\infty}\left|\dfrac{x}{\sqrt{n+1}/\sqrt{n}}\right|=\lim\limits_{n\to\infty}\dfrac{|x|}{\sqrt{1+1/n}}=|x|.$

By the Ratio Test, the series $\sum\limits_{n=1}^{\infty}\dfrac{x^n}{\sqrt{n}}$ converges when $|x|<1$, so the radius of convergence $R=1$. Now we'll check the

endpoints, that is, $x=\pm 1$. When $x=1$, the series $\sum\limits_{n=1}^{\infty}\dfrac{1}{\sqrt{n}}$ diverges because it is a p-series with $p=\frac{1}{2}\le 1$. When $x=-1$,

the series $\sum\limits_{n=1}^{\infty}\dfrac{(-1)^n}{\sqrt{n}}$ converges by the Alternating Series Test. Thus, the interval of convergence is $I=[-1,1)$.

5. If $a_n=\dfrac{(-1)^{n-1}x^n}{n^3}$, then

$$\lim\limits_{n\to\infty}\left|\dfrac{a_{n+1}}{a_n}\right|=\lim\limits_{n\to\infty}\left|\dfrac{(-1)^n x^{n+1}}{(n+1)^3}\cdot\dfrac{n^3}{(-1)^{n-1}x^n}\right|=\lim\limits_{n\to\infty}\left|\dfrac{(-1)xn^3}{(n+1)^3}\right|$$

$$=\lim\limits_{n\to\infty}\left[\left(\dfrac{n}{n+1}\right)^3|x|\right]=1^3\cdot|x|=|x|$$

By the Ratio Test, the series $\sum\limits_{n=1}^{\infty}\dfrac{(-1)^{n-1}x^n}{n^3}$ converges when $|x|<1$, so the radius of convergence $R=1$. Now we'll check

the endpoints, that is, $x=\pm 1$. When $x=1$, the series $\sum\limits_{n=1}^{\infty}\dfrac{(-1)^{n-1}}{n^3}$ converges by the Alternating Series Test. When

$x=-1$, the series $\sum\limits_{n=1}^{\infty}\dfrac{(-1)^{n-1}(-1)^n}{n^3}=-\sum\limits_{n=1}^{\infty}\dfrac{1}{n^3}$ converges because it is a constant multiple of a convergent p-series

$(p=3>1)$. Thus, the interval of convergence is $I=[-1,1]$.

7. If $a_n=\dfrac{x^n}{n!}$, then $\lim\limits_{n\to\infty}\left|\dfrac{a_{n+1}}{a_n}\right|=\lim\limits_{n\to\infty}\left|\dfrac{x^{n+1}}{(n+1)!}\cdot\dfrac{n!}{x^n}\right|=\lim\limits_{n\to\infty}\left|\dfrac{x}{n+1}\right|=|x|\lim\limits_{n\to\infty}\dfrac{1}{n+1}=|x|\cdot 0=0<1$ for *all* real x.

So, by the Ratio Test, $R=\infty$, and $I=(-\infty,\infty)$.

9. $a_n=\dfrac{(-2)^n x^n}{\sqrt[4]{n}}$, so $\lim\limits_{n\to\infty}\left|\dfrac{a_{n+1}}{a_n}\right|=\lim\limits_{n\to\infty}\dfrac{2^{n+1}|x|^{n+1}}{\sqrt[4]{n+1}}\cdot\dfrac{\sqrt[4]{n}}{2^n|x|^n}=\lim\limits_{n\to\infty}2|x|\sqrt[4]{\dfrac{n}{n+1}}=2|x|$, so by the Ratio Test, the

series converges when $2|x|<1\quad\Leftrightarrow\quad|x|<\frac{1}{2}$, so $R=\frac{1}{2}$. When $x=-\frac{1}{2}$, we get the divergent p-series $\sum\limits_{n=1}^{\infty}\dfrac{1}{\sqrt[4]{n}}$

$(p=\frac{1}{4}\le 1)$. When $x=\frac{1}{2}$, we get the series $\sum\limits_{n=1}^{\infty}\dfrac{(-1)^n}{\sqrt[4]{n}}$, which converges by the Alternating Series Test. Thus,

$I=\left(-\frac{1}{2},\frac{1}{2}\right].$

11. If $a_n = (-1)^n \dfrac{x^n}{4^n \ln n}$, then $\lim\limits_{n \to \infty} \left| \dfrac{a_{n+1}}{a_n} \right| = \lim\limits_{n \to \infty} \left| \dfrac{x^{n+1}}{4^{n+1} \ln(n+1)} \cdot \dfrac{4^n \ln n}{x^n} \right| = \dfrac{|x|}{4} \lim\limits_{n \to \infty} \dfrac{\ln n}{\ln(n+1)} = \dfrac{|x|}{4} \cdot 1$

[by l'Hospital's Rule] $= \dfrac{|x|}{4}$. By the Ratio Test, the series converges when $\dfrac{|x|}{4} < 1 \iff |x| < 4$, so $R = 4$. When

$x = -4$, $\displaystyle\sum_{n=2}^{\infty} (-1)^n \dfrac{x^n}{4^n \ln n} = \sum_{n=2}^{\infty} \dfrac{[(-1)(-4)]^n}{4^n \ln n} = \sum_{n=2}^{\infty} \dfrac{1}{\ln n}$. Since $\ln n < n$ for $n \geq 2$, $\dfrac{1}{\ln n} > \dfrac{1}{n}$ and $\displaystyle\sum_{n=2}^{\infty} \dfrac{1}{n}$ is the

divergent harmonic series (without the $n = 1$ term), $\displaystyle\sum_{n=2}^{\infty} \dfrac{1}{\ln n}$ is divergent by the Comparison Test. When $x = 4$,

$\displaystyle\sum_{n=2}^{\infty} (-1)^n \dfrac{x^n}{4^n \ln n} = \sum_{n=2}^{\infty} (-1)^n \dfrac{1}{\ln n}$, which converges by the Alternating Series Test. Thus, $I = (-4, 4]$.

13. If $a_n = (-1)^n \dfrac{(x+2)^n}{n\, 2^n}$, then $\lim\limits_{n \to \infty} \left| \dfrac{a_{n+1}}{a_n} \right| = \lim\limits_{n \to \infty} \left[\dfrac{|x+2|^{n+1}}{(n+1)2^{n+1}} \cdot \dfrac{n 2^n}{|x+2|^n} \right] = \lim\limits_{n \to \infty} \dfrac{n}{n+1} \cdot \dfrac{|x+2|}{2} = \dfrac{|x+2|}{2}$. By the

Ratio Test, the series converges when $\dfrac{|x+2|}{2} < 1 \iff |x+2| < 2$ [so $R = 2$] $\iff -2 < x + 2 < 2 \iff$

$-4 < x < 0$. When $x = -4$, the series becomes $\displaystyle\sum_{n=1}^{\infty} (-1)^n \dfrac{(-2)^n}{n2^n} = \sum_{n=1}^{\infty} \dfrac{2^n}{n\, 2^n} = \sum_{n=1}^{\infty} \dfrac{1}{n}$, which is the divergent harmonic

series. When $x = 0$, the series is $\displaystyle\sum_{n=1}^{\infty} \dfrac{(-1)^n}{n}$, the alternating harmonic series, which converges by the Alternating Series Test.

Thus, $I = (-4, 0]$.

15. $a_n = \dfrac{n}{b^n} (x-a)^n$, where $b > 0$.

$\lim\limits_{n \to \infty} \left| \dfrac{a_{n+1}}{a_n} \right| = \lim\limits_{n \to \infty} \dfrac{(n+1)|x-a|^{n+1}}{b^{n+1}} \cdot \dfrac{b^n}{n|x-a|^n} = \lim\limits_{n \to \infty} \left(1 + \dfrac{1}{n}\right) \dfrac{|x-a|}{b} = \dfrac{|x-a|}{b}$.

By the Ratio Test, the series converges when $\dfrac{|x-a|}{b} < 1 \iff |x-a| < b$ [so $R = b$] $\iff -b < x - a < b \iff$

$a - b < x < a + b$. When $|x-a| = b$, $\lim\limits_{n \to \infty} |a_n| = \lim\limits_{n \to \infty} n = \infty$, so the series diverges. Thus, $I = (a-b, a+b)$.

17. If $a_n = n!(2x-1)^n$, then $\lim\limits_{n \to \infty} \left| \dfrac{a_{n+1}}{a_n} \right| = \lim\limits_{n \to \infty} \left| \dfrac{(n+1)!(2x-1)^{n+1}}{n!(2x-1)^n} \right| = \lim\limits_{n \to \infty} (n+1)\,|2x-1| \to \infty$ as $n \to \infty$ for all

$x \neq \frac{1}{2}$. Since the series diverges for all $x \neq \frac{1}{2}$, $R = 0$ and $I = \left\{ \frac{1}{2} \right\}$.

19. (a) We are given that the power series $\sum_{n=0}^{\infty} c_n x^n$ is convergent for $x = 4$. So by Theorem 3, it must converge for at least

$-4 < x \leq 4$. In particular, it converges when $x = -2$; that is, $\sum_{n=0}^{\infty} c_n(-2)^n$ is convergent.

(b) It does not follow that $\sum_{n=0}^{\infty} c_n(-4)^n$ is necessarily convergent. [See the comments after Theorem 3 about convergence at

the endpoint of an interval. An example is $c_n = (-1)^n/(n4^n)$.]

21. If $a_n = \dfrac{(n!)^k}{(kn)!} x^n$, then

$$\lim\limits_{n \to \infty} \left| \dfrac{a_{n+1}}{a_n} \right| = \lim\limits_{n \to \infty} \dfrac{[(n+1)!]^k (kn)!}{(n!)^k [k(n+1)]!} |x| = \lim\limits_{n \to \infty} \dfrac{(n+1)^k}{(kn+k)(kn+k-1)\cdots(kn+2)(kn+1)} |x|$$

$$= \lim\limits_{n \to \infty} \left[\dfrac{(n+1)}{(kn+1)} \dfrac{(n+1)}{(kn+2)} \cdots \dfrac{(n+1)}{(kn+k)} \right] |x|$$

$$= \lim\limits_{n \to \infty} \left[\dfrac{n+1}{kn+1} \right] \lim\limits_{n \to \infty} \left[\dfrac{n+1}{kn+2} \right] \cdots \lim\limits_{n \to \infty} \left[\dfrac{n+1}{kn+k} \right] |x|$$

$$= \left(\dfrac{1}{k} \right)^k |x| < 1 \iff |x| < k^k \text{ for convergence, and the radius of convergence is } R = k^k.$$

23. (a) If $a_n = \dfrac{(-1)^n \, x^{2n+1}}{n!(n+1)! \, 2^{2n+1}}$, then

$$\lim_{n \to \infty} \left| \frac{a_{n+1}}{a_n} \right| = \lim_{n \to \infty} \left| \frac{x^{2n+3}}{(n+1)!(n+2)! \, 2^{2n+3}} \cdot \frac{n!(n+1)! \, 2^{2n+1}}{x^{2n+1}} \right| = \left(\frac{x}{2} \right)^2 \lim_{n \to \infty} \frac{1}{(n+1)(n+2)} = 0 \text{ for all } x.$$

So $J_1(x)$ converges for all x and its domain is $(-\infty, \infty)$.

(b), (c) The initial terms of $J_1(x)$ up to $n = 5$ are $a_0 = \dfrac{x}{2}$,

$a_1 = -\dfrac{x^3}{16}$, $a_2 = \dfrac{x^5}{384}$, $a_3 = -\dfrac{x^7}{18{,}432}$, $a_4 = \dfrac{x^9}{1{,}474{,}560}$,

and $a_5 = -\dfrac{x^{11}}{176{,}947{,}200}$. The partial sums seem to

approximate $J_1(x)$ well near the origin, but as $|x|$ increases,

we need to take a large number of terms to get a good

approximation.

25. $s_{2n-1} = 1 + 2x + x^2 + 2x^3 + x^4 + 2x^5 + \cdots + x^{2n-2} + 2x^{2n-1}$

$= 1(1 + 2x) + x^2(1 + 2x) + x^4(1 + 2x) + \cdots + x^{2n-2}(1 + 2x) = (1 + 2x)(1 + x^2 + x^4 + \cdots + x^{2n-2})$

$= (1 + 2x)\dfrac{1 - x^{2n}}{1 - x^2}$ [by (8.2.3)] with $r = x^2$] $\to \dfrac{1 + 2x}{1 - x^2}$ as $n \to \infty$ [by (8.2.4)], when $|x| < 1$.

Also $s_{2n} = s_{2n-1} + x^{2n} \to \dfrac{1 + 2x}{1 - x^2}$ since $x^{2n} \to 0$ for $|x| < 1$. Therefore, $s_n \to \dfrac{1 + 2x}{1 - x^2}$ since s_{2n} and s_{2n-1} both

approach $\dfrac{1 + 2x}{1 - x^2}$ as $n \to \infty$. Thus, the interval of convergence is $(-1, 1)$ and $f(x) = \dfrac{1 + 2x}{1 - x^2}$.

27. For $2 < x < 3$, $\sum c_n x^n$ diverges and $\sum d_n x^n$ converges. By Exercise 8.2.51, $\sum(c_n + d_n) x^n$ diverges. Since both series converge for $|x| < 2$, the radius of convergence of $\sum(c_n + d_n) x^n$ is 2.

8.6 Representations of Functions as Power Series

1. If $f(x) = \displaystyle\sum_{n=0}^{\infty} c_n x^n$ has radius of convergence 10, then $f'(x) = \displaystyle\sum_{n=1}^{\infty} n c_n x^{n-1}$ also has radius of convergence 10 by

Theorem 2.

3. Our goal is to write the function in the form $\dfrac{1}{1 - r}$, and then use Equation (1) to represent the function as a sum of a power

series. $f(x) = \dfrac{1}{1 + x} = \dfrac{1}{1 - (-x)} = \displaystyle\sum_{n=0}^{\infty} (-x)^n = \sum_{n=0}^{\infty} (-1)^n x^n$ with $|-x| < 1 \iff |x| < 1$, so $R = 1$ and $I = (-1, 1)$.

5. Replacing x with x^3 in (1) gives $f(x) = \dfrac{1}{1 - x^3} = \displaystyle\sum_{n=0}^{\infty} (x^3)^n = \sum_{n=0}^{\infty} x^{3n}$. The series converges when $|x^3| < 1 \iff$

$|x|^3 < 1 \iff |x| < \sqrt[3]{1} \iff |x| < 1$. Thus, $R = 1$ and $I = (-1, 1)$.

7. $f(x) = \dfrac{1}{x-5} = -\dfrac{1}{5}\left(\dfrac{1}{1-x/5}\right) = -\dfrac{1}{5}\sum_{n=0}^{\infty}\left(\dfrac{x}{5}\right)^n$ or equivalently, $-\sum_{n=0}^{\infty}\dfrac{1}{5^{n+1}}x^n$. The series converges when $\left|\dfrac{x}{5}\right| < 1$;

that is, when $|x| < 5$, so $I = (-5, 5)$.

9. $f(x) = \dfrac{x}{9+x^2} = \dfrac{x}{9}\left[\dfrac{1}{1+(x/3)^2}\right] = \dfrac{x}{9}\left[\dfrac{1}{1-\{-(x/3)^2\}}\right] = \dfrac{x}{9}\sum_{n=0}^{\infty}\left[-\left(\dfrac{x}{3}\right)^2\right]^n = \dfrac{x}{9}\sum_{n=0}^{\infty}(-1)^n\dfrac{x^{2n}}{9^n}$

$= \sum_{n=0}^{\infty}(-1)^n\dfrac{x^{2n+1}}{9^{n+1}}$

The geometric series $\sum_{n=0}^{\infty}\left[-\left(\dfrac{x}{3}\right)^2\right]^n$ converges when $\left|-\left(\dfrac{x}{3}\right)^2\right| < 1 \iff \dfrac{|x^2|}{9} < 1 \iff$

$|x|^2 < 9 \iff |x| < 3$, so $R = 3$ and $I = (-3, 3)$.

11. (a) $f(x) = \dfrac{1}{(1+x)^2} = \dfrac{d}{dx}\left(\dfrac{-1}{1+x}\right) = -\dfrac{d}{dx}\left[\sum_{n=0}^{\infty}(-1)^n x^n\right]$ [from Exercise 3]

$= \sum_{n=1}^{\infty}(-1)^{n+1}nx^{n-1}$ [from Theorem 2(i)] $= \sum_{n=0}^{\infty}(-1)^n(n+1)x^n$ with $R = 1$.

In the last step, note that we *decreased* the initial value of the summation variable n by 1, and then *increased* each

occurrence of n in the term by 1 [also note that $(-1)^{n+2} = (-1)^n$].

(b) $f(x) = \dfrac{1}{(1+x)^3} = -\dfrac{1}{2}\dfrac{d}{dx}\left[\dfrac{1}{(1+x)^2}\right] = -\dfrac{1}{2}\dfrac{d}{dx}\left[\sum_{n=0}^{\infty}(-1)^n(n+1)x^n\right]$ [from part (a)]

$= -\dfrac{1}{2}\sum_{n=1}^{\infty}(-1)^n(n+1)nx^{n-1} = \dfrac{1}{2}\sum_{n=0}^{\infty}(-1)^n(n+2)(n+1)x^n$ with $R = 1$.

(c) $f(x) = \dfrac{x^2}{(1+x)^3} = x^2 \cdot \dfrac{1}{(1+x)^3} = x^2 \cdot \dfrac{1}{2}\sum_{n=0}^{\infty}(-1)^n(n+2)(n+1)x^n$ [from part (b)]

$= \dfrac{1}{2}\sum_{n=0}^{\infty}(-1)^n(n+2)(n+1)x^{n+2}$

To write the power series with x^n rather than x^{n+2}, we will *decrease* each occurrence of n in the term by 2 and *increase*

the initial value of the summation variable by 2. This gives us $\dfrac{1}{2}\sum_{n=2}^{\infty}(-1)^n(n)(n-1)x^n$.

13. $f(x) = \ln(5-x) = -\displaystyle\int\dfrac{dx}{5-x} = -\dfrac{1}{5}\int\dfrac{dx}{1-x/5} = -\dfrac{1}{5}\int\left[\sum_{n=0}^{\infty}\left(\dfrac{x}{5}\right)^n\right]dx = C - \dfrac{1}{5}\sum_{n=0}^{\infty}\dfrac{x^{n+1}}{5^n(n+1)} = C - \sum_{n=1}^{\infty}\dfrac{x^n}{n\,5^n}$

Putting $x = 0$, we get $C = \ln 5$. The series converges for $|x/5| < 1 \iff |x| < 5$, so $R = 5$.

15. $\dfrac{1}{2-x} = \dfrac{1}{2(1-x/2)} = \dfrac{1}{2}\sum_{n=0}^{\infty}\left(\dfrac{x}{2}\right)^n = \sum_{n=0}^{\infty}\dfrac{1}{2^{n+1}}x^n$ for $\left|\dfrac{x}{2}\right| < 1 \iff |x| < 2$. Now

$\dfrac{1}{(x-2)^2} = \dfrac{d}{dx}\left(\dfrac{1}{2-x}\right) = \dfrac{d}{dx}\left(\sum_{n=0}^{\infty}\dfrac{1}{2^{n+1}}x^n\right) = \sum_{n=1}^{\infty}\dfrac{n}{2^{n+1}}x^{n-1} = \sum_{n=0}^{\infty}\dfrac{n+1}{2^{n+2}}x^n$. So

$f(x) = \dfrac{x^3}{(x-2)^2} = x^3\sum_{n=0}^{\infty}\dfrac{n+1}{2^{n+2}}x^n = \sum_{n=0}^{\infty}\dfrac{n+1}{2^{n+2}}x^{n+3}$ or $\sum_{n=3}^{\infty}\dfrac{n-2}{2^{n-1}}x^n$ for $|x| < 2$. Thus, $R = 2$ and $I = (-2, 2)$.

17. $f(x) = \ln(3 + x) = \int \dfrac{dx}{3 + x} = \dfrac{1}{3} \int \dfrac{dx}{1 + x/3} = \dfrac{1}{3} \int \dfrac{dx}{1 - (-x/3)} = \dfrac{1}{3} \int \sum\limits_{n=0}^{\infty} \left(-\dfrac{x}{3}\right)^n dx$

$= C + \dfrac{1}{3} \sum\limits_{n=0}^{\infty} \dfrac{(-1)^n}{(n+1)3^n} x^{n+1} = \ln 3 + \dfrac{1}{3} \sum\limits_{n=1}^{\infty} \dfrac{(-1)^{n-1}}{n3^{n-1}} x^n \quad [C = f(0) = \ln 3]$

$= \ln 3 + \sum\limits_{n=1}^{\infty} \dfrac{(-1)^{n-1}}{n\, 3^n} x^n$. The series converges when $|-x/3| < 1 \quad \Leftrightarrow \quad |x| < 3$, so $R = 3$.

The terms of the series are $a_0 = \ln 3, a_1 = \dfrac{x}{3}, a_2 = -\dfrac{x^2}{18}, a_3 = \dfrac{x^3}{81}, a_4 = -\dfrac{x^4}{324}, a_5 = \dfrac{x^5}{1215}, \ldots$

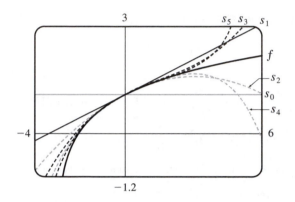

As n increases, $s_n(x)$ approximates f better on the interval of convergence, which is $(-3, 3)$.

19. $f(x) = \ln\left(\dfrac{1 + x}{1 - x}\right) = \ln(1 + x) - \ln(1 - x) = \int \dfrac{dx}{1 + x} + \int \dfrac{dx}{1 - x} = \int \dfrac{dx}{1 - (-x)} + \int \dfrac{dx}{1 - x}$

$= \int \left[\sum\limits_{n=0}^{\infty} (-1)^n x^n + \sum\limits_{n=0}^{\infty} x^n\right] dx = \int \left[\left(1 - x + x^2 - x^3 + x^4 - \cdots\right) + \left(1 + x + x^2 + x^3 + x^4 + \cdots\right)\right] dx$

$= \int \left(2 + 2x^2 + 2x^4 + \cdots\right) dx = \int \sum\limits_{n=0}^{\infty} 2x^{2n} \, dx = C + \sum\limits_{n=0}^{\infty} \dfrac{2x^{2n+1}}{2n + 1}$

But $f(0) = \ln \dfrac{1}{1} = 0$, so $C = 0$ and we have $f(x) = \sum\limits_{n=0}^{\infty} \dfrac{2x^{2n+1}}{2n + 1}$ with $R = 1$. If $x = \pm 1$, then $f(x) = \pm 2 \sum\limits_{n=0}^{\infty} \dfrac{1}{2n + 1}$,

which both diverge by the Limit Comparison Test with $b_n = \dfrac{1}{n}$.

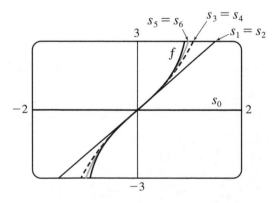

As n increases, $s_n(x)$ approximates f better on the interval of convergence, which is $(-1, 1)$.

21. $\dfrac{t}{1-t^8} = t \cdot \dfrac{1}{1-t^8} = t \sum\limits_{n=0}^{\infty} (t^8)^n = \sum\limits_{n=0}^{\infty} t^{8n+1} \quad \Rightarrow \quad \displaystyle\int \dfrac{t}{1-t^8}\,dt = C + \sum\limits_{n=0}^{\infty} \dfrac{t^{8n+2}}{8n+2}.$ The series for $\dfrac{1}{1-t^8}$ converges

when $\left| t^8 \right| < 1 \quad \Leftrightarrow \quad |t| < 1$, so $R = 1$ for that series and also the series for $t/(1-t^8)$. By Theorem 2, the series for

$\displaystyle\int \dfrac{t}{1-t^8}\,dt$ also has $R = 1$.

23. By Example 7, $\tan^{-1} x = \sum\limits_{n=0}^{\infty} (-1)^n \dfrac{x^{2n+1}}{2n+1}$ with $R = 1$, so

$$x - \tan^{-1} x = x - \left(x - \dfrac{x^3}{3} + \dfrac{x^5}{5} - \dfrac{x^7}{7} + \cdots \right) = \dfrac{x^3}{3} - \dfrac{x^5}{5} + \dfrac{x^7}{7} - \cdots = \sum\limits_{n=1}^{\infty} (-1)^{n+1} \dfrac{x^{2n+1}}{2n+1} \text{ and}$$

$$\dfrac{x - \tan^{-1} x}{x^3} = \sum\limits_{n=1}^{\infty} (-1)^{n+1} \dfrac{x^{2n-2}}{2n+1}, \text{ so}$$

$$\int \dfrac{x - \tan^{-1} x}{x^3}\,dx = C + \sum\limits_{n=1}^{\infty} (-1)^{n+1} \dfrac{x^{2n-1}}{(2n+1)(2n-1)} = C + \sum\limits_{n=1}^{\infty} (-1)^{n+1} \dfrac{x^{2n-1}}{4n^2-1}. \text{ By Theorem 2, } R = 1.$$

25. $\dfrac{1}{1+x^5} = \dfrac{1}{1-(-x^5)} = \sum\limits_{n=0}^{\infty} \left(-x^5 \right)^n = \sum\limits_{n=0}^{\infty} (-1)^n x^{5n} \quad \Rightarrow$

$$\int \dfrac{1}{1+x^5}\,dx = \int \sum\limits_{n=0}^{\infty} (-1)^n x^{5n}\,dx = C + \sum\limits_{n=0}^{\infty} (-1)^n \dfrac{x^{5n+1}}{5n+1}. \text{ Thus,}$$

$$I = \int_0^{0.2} \dfrac{1}{1+x^5}\,dx = \left[x - \dfrac{x^6}{6} + \dfrac{x^{11}}{11} - \cdots \right]_0^{0.2} = 0.2 - \dfrac{(0.2)^6}{6} + \dfrac{(0.2)^{11}}{11} - \cdots. \text{ The series is alternating, so if we use}$$

the first two terms, the error is at most $(0.2)^{11}/11 \approx 1.9 \times 10^{-9}$. So $I \approx 0.2 - (0.2)^6/6 \approx 0.199989$ to six decimal places.

27. We substitute $3x$ for x in Example 7, and find that

$$\int x \arctan(3x)\,dx = \int x \sum\limits_{n=0}^{\infty} (-1)^n \dfrac{(3x)^{2n+1}}{2n+1}\,dx = \int \sum\limits_{n=0}^{\infty} (-1)^n \dfrac{3^{2n+1}\,x^{2n+2}}{2n+1}\,dx$$

$$= C + \sum\limits_{n=0}^{\infty} (-1)^n \dfrac{3^{2n+1}\,x^{2n+3}}{(2n+1)(2n+3)}$$

So $\displaystyle\int_0^{0.1} x \arctan(3x)\,dx = \left[\dfrac{3x^3}{1 \cdot 3} - \dfrac{3^3 x^5}{3 \cdot 5} + \dfrac{3^5 x^7}{5 \cdot 7} - \dfrac{3^7 x^9}{7 \cdot 9} + \cdots \right]_0^{0.1}$

$$= \dfrac{1}{10^3} - \dfrac{9}{5 \times 10^5} + \dfrac{243}{35 \times 10^7} - \dfrac{2187}{63 \times 10^9} + \cdots.$$

The series is alternating, so if we use three terms, the error is at most $\dfrac{2187}{63 \times 10^9} \approx 3.5 \times 10^{-8}$. So

$$\int_0^{0.1} x \arctan(3x)\,dx \approx \dfrac{1}{10^3} - \dfrac{9}{5 \times 10^5} + \dfrac{243}{35 \times 10^7} \approx 0.000\,983 \text{ to six decimal places.}$$

29. Using the result of Example 6, $\ln(1-x) = -\sum\limits_{n=1}^{\infty} \dfrac{x^n}{n}$, with $x = -0.1$, we have

$$\ln 1.1 = \ln[1 - (-0.1)] = 0.1 - \dfrac{0.01}{2} + \dfrac{0.001}{3} - \dfrac{0.0001}{4} + \dfrac{0.00001}{5} - \cdots. \text{ The series is alternating, so if we use only the}$$

first four terms, the error is at most $\dfrac{0.00001}{5} = 0.000002$. So $\ln 1.1 \approx 0.1 - \dfrac{0.01}{2} + \dfrac{0.001}{3} - \dfrac{0.0001}{4} \approx 0.09531$.

31. (a) $J_0(x) = \sum\limits_{n=0}^{\infty} \dfrac{(-1)^n x^{2n}}{2^{2n}(n!)^2}$, $J_0'(x) = \sum\limits_{n=1}^{\infty} \dfrac{(-1)^n 2nx^{2n-1}}{2^{2n}(n!)^2}$, and $J_0''(x) = \sum\limits_{n=1}^{\infty} \dfrac{(-1)^n 2n(2n-1)x^{2n-2}}{2^{2n}(n!)^2}$, so

$$x^2 J_0''(x) + x J_0'(x) + x^2 J_0(x) = \sum_{n=1}^{\infty} \dfrac{(-1)^n 2n(2n-1)x^{2n}}{2^{2n}(n!)^2} + \sum_{n=1}^{\infty} \dfrac{(-1)^n 2nx^{2n}}{2^{2n}(n!)^2} + \sum_{n=0}^{\infty} \dfrac{(-1)^n x^{2n+2}}{2^{2n}(n!)^2}$$

$$= \sum_{n=1}^{\infty} \dfrac{(-1)^n 2n(2n-1)x^{2n}}{2^{2n}(n!)^2} + \sum_{n=1}^{\infty} \dfrac{(-1)^n 2nx^{2n}}{2^{2n}(n!)^2} + \sum_{n=1}^{\infty} \dfrac{(-1)^{n-1} x^{2n}}{2^{2n-2}[(n-1)!]^2}$$

$$= \sum_{n=1}^{\infty} \dfrac{(-1)^n 2n(2n-1)x^{2n}}{2^{2n}(n!)^2} + \sum_{n=1}^{\infty} \dfrac{(-1)^n 2nx^{2n}}{2^{2n}(n!)^2} + \sum_{n=1}^{\infty} \dfrac{(-1)^n(-1)^{-1}2^2 n^2 x^{2n}}{2^{2n}(n!)^2}$$

$$= \sum_{n=1}^{\infty} (-1)^n \left[\dfrac{2n(2n-1) + 2n - 2^2 n^2}{2^{2n}(n!)^2} \right] x^{2n}$$

$$= \sum_{n=1}^{\infty} (-1)^n \left[\dfrac{4n^2 - 2n + 2n - 4n^2}{2^{2n}(n!)^2} \right] x^{2n} = 0$$

(b) $\displaystyle\int_0^1 J_0(x)\,dx = \int_0^1 \left[\sum_{n=0}^{\infty} \dfrac{(-1)^n x^{2n}}{2^{2n}(n!)^2} \right] dx = \int_0^1 \left(1 - \dfrac{x^2}{4} + \dfrac{x^4}{64} - \dfrac{x^6}{2304} + \cdots \right) dx$

$$= \left[x - \dfrac{x^3}{3 \cdot 4} + \dfrac{x^5}{5 \cdot 64} - \dfrac{x^7}{7 \cdot 2304} + \cdots \right]_0^1 = 1 - \dfrac{1}{12} + \dfrac{1}{320} - \dfrac{1}{16{,}128} + \cdots$$

Since $\frac{1}{16{,}128} \approx 0.000062$, it follows from The Alternating Series Estimation Theorem that, correct to three decimal places, $\int_0^1 J_0(x)\,dx \approx 1 - \frac{1}{12} + \frac{1}{320} \approx 0.920$.

33. (a) $f(x) = \sum\limits_{n=0}^{\infty} \dfrac{x^n}{n!} \;\Rightarrow\; f'(x) = \sum\limits_{n=1}^{\infty} \dfrac{nx^{n-1}}{n!} = \sum\limits_{n=1}^{\infty} \dfrac{x^{n-1}}{(n-1)!} = \sum\limits_{n=0}^{\infty} \dfrac{x^n}{n!} = f(x)$

(b) By Theorem 7.4.2, the only solution to the differential equation $df(x)/dx = f(x)$ is $f(x) = Ke^x$, but $f(0) = 1$, so $K = 1$ and $f(x) = e^x$.

Or: We could solve the equation $df(x)/dx = f(x)$ as a separable differential equation.

35. If $a_n = \dfrac{x^n}{n^2}$, then by the Ratio Test, $\lim\limits_{n\to\infty} \left| \dfrac{a_{n+1}}{a_n} \right| = \lim\limits_{n\to\infty} \left| \dfrac{x^{n+1}}{(n+1)^2} \cdot \dfrac{n^2}{x^n} \right| = |x| \lim\limits_{n\to\infty} \left(\dfrac{n}{n+1} \right)^2 = |x| < 1$ for

convergence, so $R = 1$. When $x = \pm 1$, $\sum\limits_{n=1}^{\infty} \left| \dfrac{x^n}{n^2} \right| = \sum\limits_{n=1}^{\infty} \dfrac{1}{n^2}$ which is a convergent p-series ($p = 2 > 1$), so the interval of

convergence for f is $[-1, 1]$. By Theorem 2, the radii of convergence of f' and f'' are both 1, so we need only check the

endpoints. $f(x) = \sum\limits_{n=1}^{\infty} \dfrac{x^n}{n^2} \;\Rightarrow\; f'(x) = \sum\limits_{n=1}^{\infty} \dfrac{nx^{n-1}}{n^2} = \sum\limits_{n=0}^{\infty} \dfrac{x^n}{n+1}$, and this series diverges for $x = 1$ (harmonic series)

and converges for $x = -1$ (Alternating Series Test), so the interval of convergence is $[-1, 1)$. $f''(x) = \sum\limits_{n=1}^{\infty} \dfrac{nx^{n-1}}{n+1}$ diverges

at both 1 and -1 (Test for Divergence) since $\lim\limits_{n\to\infty} \dfrac{n}{n+1} = 1 \neq 0$, so its interval of convergence is $(-1, 1)$.

37. By Example 7, $\tan^{-1} x = \sum\limits_{n=0}^{\infty} (-1)^n \dfrac{x^{2n+1}}{2n+1}$ for $|x| < 1$. In particular, for $x = \dfrac{1}{\sqrt{3}}$, we

have $\dfrac{\pi}{6} = \tan^{-1}\left(\dfrac{1}{\sqrt{3}} \right) = \sum\limits_{n=0}^{\infty} (-1)^n \dfrac{(1/\sqrt{3})^{2n+1}}{2n+1} = \sum\limits_{n=0}^{\infty} (-1)^n \left(\dfrac{1}{3} \right)^n \dfrac{1}{\sqrt{3}} \dfrac{1}{2n+1}$, so

$$\pi = \dfrac{6}{\sqrt{3}} \sum_{n=0}^{\infty} \dfrac{(-1)^n}{(2n+1)3^n} = 2\sqrt{3} \sum_{n=0}^{\infty} \dfrac{(-1)^n}{(2n+1)3^n}.$$

8.7   Taylor and Maclaurin Series

1. Using Theorem 5 with $\sum\limits_{n=0}^{\infty} b_n(x-5)^n$, $b_n = \dfrac{f^{(n)}(a)}{n!}$, so $b_8 = \dfrac{f^{(8)}(5)}{8!}$.

3. Since $f^{(n)}(0) = (n+1)!$, Equation 7 gives the Maclaurin series

$$\sum_{n=0}^{\infty} \frac{f^{(n)}(0)}{n!}x^n = \sum_{n=0}^{\infty} \frac{(n+1)!}{n!}x^n = \sum_{n=0}^{\infty} (n+1)x^n.$$ Applying the Ratio Test with $a_n = (n+1)x^n$ gives us

$\lim\limits_{n\to\infty}\left|\dfrac{a_{n+1}}{a_n}\right| = \lim\limits_{n\to\infty}\left|\dfrac{(n+2)x^{n+1}}{(n+1)x^n}\right| = |x|\lim\limits_{n\to\infty}\dfrac{n+2}{n+1} = |x|\cdot 1 = |x|$. For convergence, we must have $|x| < 1$, so the

radius of convergence $R = 1$.

5.

n	$f^{(n)}(x)$	$f^{(n)}(0)$
0	$\cos x$	1
1	$-\sin x$	0
2	$-\cos x$	-1
3	$\sin x$	0
4	$\cos x$	1
$\vdots$	$\vdots$	$\vdots$

We use Equation 7 with $f(x) = \cos x$.

$$\cos x = f(0) + f'(0)x + \frac{f''(0)}{2!}x^2 + \frac{f^{(3)}(0)}{3!}x^3 + \frac{f^{(4)}(0)}{4!}x^4 + \cdots$$

$$= 1 - \frac{x^2}{2!} + \frac{x^4}{4!} - \cdots = \sum_{n=0}^{\infty} \frac{(-1)^n x^{2n}}{(2n)!}$$

If $a_n = \dfrac{(-1)^n x^{2n}}{(2n)!}$, then

$\lim\limits_{n\to\infty}\left|\dfrac{a_{n+1}}{a_n}\right| = \lim\limits_{n\to\infty}\left|\dfrac{x^{2n+2}}{(2n+2)!}\cdot\dfrac{(2n)!}{x^{2n}}\right| = x^2\lim\limits_{n\to\infty}\dfrac{1}{(2n+2)(2n+1)} = 0 < 1$ for all x. So $R = \infty$ (Ratio Test).

7.

n	$f^{(n)}(x)$	$f^{(n)}(0)$
0	e^{5x}	1
1	$5e^{5x}$	5
2	$5^2 e^{5x}$	25
3	$5^3 e^{5x}$	125
4	$5^4 e^{5x}$	625
$\vdots$	$\vdots$	$\vdots$

$$e^{5x} = \sum_{n=0}^{\infty} \frac{f^{(n)}(0)}{n!}x^n = \sum_{n=0}^{\infty} \frac{5^n}{n!}x^n.$$

$$\lim_{n\to\infty}\left|\frac{a_{n+1}}{a_n}\right| = \lim_{n\to\infty}\left[\frac{5^{n+1}|x|^{n+1}}{(n+1)!}\cdot\frac{n!}{5^n|x|^n}\right]$$

$$= \lim_{n\to\infty}\frac{5|x|}{n+1} = 0 < 1 \text{ for all } x, \text{ so } R = \infty.$$

9.

n	$f^{(n)}(x)$	$f^{(n)}(2)$
0	$1 + x + x^2$	7
1	$1 + 2x$	5
2	2	2
3	0	0
4	0	0
$\vdots$	$\vdots$	$\vdots$

$$f(x) = 7 + 5(x-2) + \frac{2}{2!}(x-2)^2 + \sum_{n=3}^{\infty}\frac{0}{n!}(x-2)^n$$

$$= 7 + 5(x-2) + (x-2)^2$$

Since $a_n = 0$ for large n, $R = \infty$.

11. Clearly, $f^{(n)}(x) = e^x$, so $f^{(n)}(3) = e^3$ and $e^x = \sum\limits_{n=0}^{\infty} \dfrac{e^3}{n!}(x-3)^n$. If $a_n = \dfrac{e^3}{n!}(x-3)^n$, then

$$\lim_{n\to\infty}\left|\frac{a_{n+1}}{a_n}\right| = \lim_{n\to\infty}\left|\frac{e^3(x-3)^{n+1}}{(n+1)!}\cdot\frac{n!}{e^3(x-3)^n}\right| = \lim_{n\to\infty}\frac{|x-3|}{n+1} = 0 < 1 \text{ for all } x, \text{ so } R = \infty.$$

13.

n	$f^{(n)}(x)$	$f^{(n)}(\pi)$
0	$\cos x$	-1
1	$-\sin x$	0
2	$-\cos x$	1
3	$\sin x$	0
4	$\cos x$	-1
⋮	⋮	⋮

$$\cos x = \sum_{k=0}^{\infty}\frac{f^{(k)}(\pi)}{k!}(x-\pi)^k$$

$$= -1 + \frac{(x-\pi)^2}{2!} - \frac{(x-\pi)^4}{4!} + \frac{(x-\pi)^6}{6!} - \cdots$$

$$= \sum_{n=0}^{\infty}(-1)^{n+1}\frac{(x-\pi)^{2n}}{(2n)!}$$

$$\lim_{n\to\infty}\left|\frac{a_{n+1}}{a_n}\right| = \lim_{n\to\infty}\left[\frac{|x-\pi|^{2n+2}}{(2n+2)!}\cdot\frac{(2n)!}{|x-\pi|^{2n}}\right] = \lim_{n\to\infty}\frac{|x-\pi|^2}{(2n+2)(2n+1)} = 0 < 1 \text{ for all } x, \text{ so } R = \infty.$$

15.

n	$f^{(n)}(x)$	$f^{(n)}(9)$
0	$x^{-1/2}$	$\frac{1}{3}$
1	$-\frac{1}{2}x^{-3/2}$	$-\frac{1}{2}\cdot\frac{1}{3^3}$
2	$\frac{3}{4}x^{-5/2}$	$-\frac{1}{2}\cdot\left(-\frac{3}{2}\right)\cdot\frac{1}{3^5}$
3	$-\frac{15}{8}x^{-7/2}$	$-\frac{1}{2}\cdot\left(-\frac{3}{2}\right)\cdot\left(-\frac{5}{2}\right)\cdot\frac{1}{3^7}$
⋮	⋮	⋮

$$\frac{1}{\sqrt{x}} = \frac{1}{3} - \frac{1}{2\cdot3^3}(x-9) + \frac{3}{2^2\cdot3^5}\frac{(x-9)^2}{2!} - \frac{3\cdot5}{2^3\cdot3^7}\frac{(x-9)^3}{3!} + \cdots$$

$$= \sum_{n=0}^{\infty}(-1)^n\frac{1\cdot3\cdot5\cdot\cdots\cdot(2n-1)}{2^n\cdot3^{2n+1}\cdot n!}(x-9)^n.$$

$$\lim_{n\to\infty}\left|\frac{a_{n+1}}{a_n}\right| = \lim_{n\to\infty}\left[\frac{1\cdot3\cdot5\cdot\cdots\cdot(2n-1)[2(n+1)-1]\,|x-9|^{n+1}}{2^{n+1}\cdot3^{[2(n+1)+1]}\cdot(n+1)!}\cdot\frac{2^n\cdot3^{2n+1}\cdot n!}{1\cdot3\cdot5\cdot\cdots\cdot(2n-1)\,|x-9|^n}\right]$$

$$= \lim_{n\to\infty}\left[\frac{(2n+1)\,|x-9|}{2\cdot3^2(n+1)}\right] = \frac{1}{9}\,|x-9| < 1$$

for convergence, so $|x-9| < 9$ and $R = 9$.

17. If $f(x) = \cos x$, then $f^{(n+1)}(x) = \pm\sin x$ or $\pm\cos x$. In each case, $\left|f^{(n+1)}(x)\right| \le 1$, so by Formula 9 with $a = 0$ and

$M = 1$, $|R_n(x)| \le \dfrac{1}{(n+1)!}\,|x|^{n+1}$. Thus, $|R_n(x)| \to 0$ as $n \to \infty$ by Equation 10. So $\lim\limits_{n\to\infty}R_n(x) = 0$ and, by

Theorem 8, the series in Exercise 5 represents $\cos x$ for all x.

19. $\cos x = \sum\limits_{n=0}^{\infty}(-1)^n\dfrac{x^{2n}}{(2n)!} \quad\Rightarrow\quad f(x) = \cos(\pi x) = \sum\limits_{n=0}^{\infty}\dfrac{(-1)^n(\pi x)^{2n}}{(2n)!} = \sum\limits_{n=0}^{\infty}\dfrac{(-1)^n\pi^{2n}x^{2n}}{(2n)!}, \quad R = \infty$

21. $\tan^{-1} x = \sum\limits_{n=0}^{\infty} (-1)^n \dfrac{x^{2n+1}}{2n+1}$ $\Rightarrow$ $f(x) = x \tan^{-1} x = x \sum\limits_{n=0}^{\infty} (-1)^n \dfrac{x^{2n+1}}{2n+1} = \sum\limits_{n=0}^{\infty} (-1)^n \dfrac{x^{2n+2}}{2n+1}$, $R = 1$

23. $e^x = \sum\limits_{n=0}^{\infty} \dfrac{x^n}{n!}$ $\Rightarrow$ $f(x) = x^2 e^{-x} = x^2 \sum\limits_{n=0}^{\infty} \dfrac{(-x)^n}{n!} = \sum\limits_{n=0}^{\infty} \dfrac{(-1)^n x^{n+2}}{n!}$, $R = \infty$

25. $\sin^2 x = \dfrac{1}{2}(1 - \cos 2x) = \dfrac{1}{2}\left[1 - \sum\limits_{n=0}^{\infty} \dfrac{(-1)^n (2x)^{2n}}{(2n)!}\right] = \dfrac{1}{2}\left[1 - 1 - \sum\limits_{n=1}^{\infty} \dfrac{(-1)^n (2x)^{2n}}{(2n)!}\right]$

$= \sum\limits_{n=1}^{\infty} \dfrac{(-1)^{n+1} 2^{2n-1} x^{2n}}{(2n)!}$, $R = \infty$

27.

n	$f^{(n)}(x)$	$f^{(n)}(0)$
0	$(1+x)^{1/2}$	1
1	$\frac{1}{2}(1+x)^{-1/2}$	$\frac{1}{2}$
2	$-\frac{1}{4}(1+x)^{-3/2}$	$-\frac{1}{4}$
3	$\frac{3}{8}(1+x)^{-5/2}$	$\frac{3}{8}$
4	$-\frac{15}{16}(1+x)^{-7/2}$	$-\frac{15}{16}$
⋮	⋮	⋮

So $f^{(n)}(0) = \dfrac{(-1)^{n-1} 1 \cdot 3 \cdot 5 \cdot \cdots \cdot (2n-3)}{2^n}$ for $n \geq 2$, and $\sqrt{1+x} = 1 + \dfrac{x}{2} + \sum\limits_{n=2}^{\infty} \dfrac{(-1)^{n-1} 1 \cdot 3 \cdot 5 \cdot \cdots \cdot (2n-3)}{2^n n!} x^n$.

If $a_n = \dfrac{(-1)^{n-1} 1 \cdot 3 \cdot 5 \cdot \cdots \cdot (2n-3)}{2^n n!} x^n$, then

$\lim\limits_{n \to \infty} \left| \dfrac{a_{n+1}}{a_n} \right| = \lim\limits_{n \to \infty} \left| \dfrac{1 \cdot 3 \cdot 5 \cdot \cdots \cdot (2n-3)(2n-1) x^{n+1}}{2^{n+1}(n+1)!} \cdot \dfrac{2^n n!}{1 \cdot 3 \cdot 5 \cdot \cdots \cdot (2n-3) x^n} \right|$

$= \dfrac{|x|}{2} \lim\limits_{n \to \infty} \dfrac{2n-1}{n+1} = \dfrac{|x|}{2} \cdot 2 = |x| < 1$ for convergence, so $R = 1$.

Notice that, as n increases, $T_n(x)$ becomes a better approximation to $f(x)$ for $-1 < x < 1$.

29. $\cos x = \sum_{n=0}^{\infty} (-1)^n \dfrac{x^{2n}}{(2n)!}$ $\Rightarrow$ $f(x) = \cos(x^2) = \sum_{n=0}^{\infty} \dfrac{(-1)^n \left(x^2\right)^{2n}}{(2n)!} = \sum_{n=0}^{\infty} \dfrac{(-1)^n \, x^{4n}}{(2n)!}$, $R = \infty$

Notice that, as n increases, $T_n(x)$

becomes a better approximation to $f(x)$.

$T_0 = T_1 = T_2 = T_3$

$T_8 = T_9 = T_{10} = T_{11}$

f

$T_4 = T_5 = T_6 = T_7$

31. $e^x = \sum_{n=0}^{\infty} \dfrac{x^n}{n!}$, so $e^{-0.2} = \sum_{n=0}^{\infty} \dfrac{(-0.2)^n}{n!} = 1 - 0.2 + \dfrac{1}{2!}(0.2)^2 - \dfrac{1}{3!}(0.2)^3 + \dfrac{1}{4!}(0.2)^4 - \dfrac{1}{5!}(0.2)^5 + \dfrac{1}{6!}(0.2)^6 - \cdots$.

But $\dfrac{1}{6!}(0.2)^6 = 8.\overline{8} \times 10^{-8}$, so by the Alternating Series Estimation Theorem, $e^{-0.2} \approx \sum_{n=0}^{5} \dfrac{(-0.2)^n}{n!} \approx 0.81873$, correct to

five decimal places.

33. $\cos x \overset{(16)}{=} \sum_{n=0}^{\infty} (-1)^n \dfrac{x^{2n}}{(2n)!}$ $\Rightarrow$ $\cos(x^3) = \sum_{n=0}^{\infty} (-1)^n \dfrac{(x^3)^{2n}}{(2n)!} = \sum_{n=0}^{\infty} (-1)^n \dfrac{x^{6n}}{(2n)!}$ $\Rightarrow$

$x \cos(x^3) = \sum_{n=0}^{\infty} (-1)^n \dfrac{x^{6n+1}}{(2n)!}$ $\Rightarrow$ $\displaystyle\int x \cos(x^3)\, dx = C + \sum_{n=0}^{\infty} (-1)^n \dfrac{x^{6n+2}}{(6n+2)(2n)!}$, with $R = \infty$.

35. Using the series from Exercise 27 and substituting x^3 for x, we get

$$\int \sqrt{x^3 + 1}\, dx = \int \left[1 + \dfrac{x^3}{2} + \sum_{n=2}^{\infty} \dfrac{(-1)^{n-1} \, 1 \cdot 3 \cdot 5 \cdots (2n-3)}{2^n n!} x^{3n} \right] dx$$

$$= C + x + \dfrac{x^4}{8} + \sum_{n=2}^{\infty} \dfrac{(-1)^{n-1} \, 1 \cdot 3 \cdot 5 \cdots (2n-3)}{2^n n! (3n+1)} x^{3n+1}$$

37. By Exercise 33, $\displaystyle\int x \cos(x^3)\, dx = C + \sum_{n=0}^{\infty} (-1)^n \dfrac{x^{6n+2}}{(6n+2)(2n)!}$, so

$$\int_0^1 x \cos(x^3)\, dx = \left[\sum_{n=0}^{\infty} (-1)^n \dfrac{x^{6n+2}}{(6n+2)(2n)!} \right]_0^1 = \sum_{n=0}^{\infty} \dfrac{(-1)^n}{(6n+2)(2n)!} = \dfrac{1}{2} - \dfrac{1}{8 \cdot 2!} + \dfrac{1}{14 \cdot 4!} - \dfrac{1}{20 \cdot 6!} + \cdots, \text{ but}$$

$\dfrac{1}{20 \cdot 6!} = \dfrac{1}{14{,}400} \approx 0.000\,069$, so $\displaystyle\int_0^1 x \cos(x^3)\, dx \approx \dfrac{1}{2} - \dfrac{1}{16} + \dfrac{1}{336} \approx 0.440$ (correct to three decimal places) by the

Alternating Series Estimation Theorem.

39. We first find a series representation for $f(x) = (1+x)^{-1/2}$, and then substitute.

n	$f^{(n)}(x)$	$f^{(n)}(0)$
0	$(1+x)^{-1/2}$	1
1	$-\frac{1}{2}(1+x)^{-3/2}$	$-\frac{1}{2}$
2	$\frac{3}{4}(1+x)^{-5/2}$	$\frac{3}{4}$
3	$-\frac{15}{8}(1+x)^{-7/2}$	$-\frac{15}{8}$
⋮	⋮	⋮

$$\frac{1}{\sqrt{1+x}} = 1 - \frac{x}{2} + \frac{3}{4}\left(\frac{x^2}{2!}\right) - \frac{15}{8}\left(\frac{x^3}{3!}\right) + \cdots \quad \Rightarrow \quad \frac{1}{\sqrt{1+x^3}} = 1 - \frac{1}{2}x^3 + \frac{3}{8}x^6 - \frac{5}{16}x^9 + \cdots \quad \Rightarrow$$

$$\int_0^{0.1} \frac{dx}{\sqrt{1+x^3}} = \left[x - \frac{1}{8}x^4 + \frac{3}{56}x^7 - \frac{1}{32}x^{10} + \cdots\right]_0^{0.1} \approx (0.1) - \frac{1}{8}(0.1)^4, \text{ by the Alternating Series Estimation}$$

Theorem, since $\frac{3}{56}(0.1)^7 \approx 0.000\,000\,005\,4 < 10^{-8}$, which is the maximum desired error. Therefore,

$$\int_0^{0.1} \frac{dx}{\sqrt{1+x^3}} \approx 0.099\,987\,50.$$

41. $\lim\limits_{x\to 0} \dfrac{x - \tan^{-1} x}{x^3} = \lim\limits_{x\to 0} \dfrac{x - \left(x - \frac{1}{3}x^3 + \frac{1}{5}x^5 - \frac{1}{7}x^7 + \cdots\right)}{x^3} = \lim\limits_{x\to 0} \dfrac{\frac{1}{3}x^3 - \frac{1}{5}x^5 + \frac{1}{7}x^7 - \cdots}{x^3}$

$$= \lim_{x\to 0}\left(\frac{1}{3} - \frac{1}{5}x^2 + \frac{1}{7}x^4 - \cdots\right) = \frac{1}{3}$$

since power series are continuous functions.

43. $\lim\limits_{x\to 0} \dfrac{\sin x - x + \frac{1}{6}x^3}{x^5} = \lim\limits_{x\to 0} \dfrac{\left(x - \frac{1}{3!}x^3 + \frac{1}{5!}x^5 - \frac{1}{7!}x^7 + \cdots\right) - x + \frac{1}{6}x^3}{x^5}$

$$= \lim_{x\to 0} \frac{\frac{1}{5!}x^5 - \frac{1}{7!}x^7 + \cdots}{x^5} = \lim_{x\to 0}\left(\frac{1}{5!} - \frac{x^2}{7!} + \frac{x^4}{9!} - \cdots\right) = \frac{1}{5!} = \frac{1}{120}$$

since power series are continuous functions.

45. As in Example 8(a), we have $e^{-x^2} = 1 - \dfrac{x^2}{1!} + \dfrac{x^4}{2!} - \dfrac{x^6}{3!} + \cdots$ and we know that $\cos x = 1 - \dfrac{x^2}{2!} + \dfrac{x^4}{4!} - \cdots$ from

Equation 16. Therefore, $e^{-x^2}\cos x = \left(1 - x^2 + \frac{1}{2}x^4 - \cdots\right)\left(1 - \frac{1}{2}x^2 + \frac{1}{24}x^4 - \cdots\right)$. Writing only the terms with

degree ≤ 4, we get $e^{-x^2}\cos x = 1 - \frac{1}{2}x^2 + \frac{1}{24}x^4 - x^2 + \frac{1}{2}x^4 + \frac{1}{2}x^4 + \cdots = 1 - \frac{3}{2}x^2 + \frac{25}{24}x^4 + \cdots$.

47. $\dfrac{x}{\sin x} \overset{(15)}{=} \dfrac{x}{x - \frac{1}{6}x^3 + \frac{1}{120}x^5 - \cdots}$.

$$
\begin{array}{r}
1 + \frac{1}{6}x^2 + \frac{7}{360}x^4 + \cdots \\
x - \frac{1}{6}x^3 + \frac{1}{120}x^5 - \cdots \overline{\big)\; x } \\
\underline{x - \frac{1}{6}x^3 + \frac{1}{120}x^5 - \cdots} \\
\frac{1}{6}x^3 - \frac{1}{120}x^5 + \cdots \\
\underline{\frac{1}{6}x^3 - \frac{1}{36}x^5 + \cdots} \\
\frac{7}{360}x^5 + \cdots \\
\underline{\frac{7}{360}x^5 + \cdots} \\
\cdots
\end{array}
$$

From the long division above, $\dfrac{x}{\sin x} = 1 + \frac{1}{6}x^2 + \frac{7}{360}x^4 + \cdots$.

49. $\displaystyle\sum_{n=0}^{\infty} (-1)^n \frac{x^{4n}}{n!} = \sum_{n=0}^{\infty} \frac{\left(-x^4\right)^n}{n!} = e^{-x^4}$, by (11).

51. $\displaystyle\sum_{n=0}^{\infty} \frac{(-1)^n\, \pi^{2n+1}}{4^{2n+1}(2n+1)!} = \sum_{n=0}^{\infty} \frac{(-1)^n \left(\frac{\pi}{4}\right)^{2n+1}}{(2n+1)!} = \sin\frac{\pi}{4} = \frac{1}{\sqrt{2}}$, by (15).

53. $3 + \dfrac{9}{2!} + \dfrac{27}{3!} + \dfrac{81}{4!} + \cdots = \dfrac{3^1}{1!} + \dfrac{3^2}{2!} + \dfrac{3^3}{3!} + \dfrac{3^4}{4!} + \cdots = \displaystyle\sum_{n=1}^{\infty} \frac{3^n}{n!} = \sum_{n=0}^{\infty} \frac{3^n}{n!} - 1 = e^3 - 1$, by (11).

55. Assume that $|f'''(x)| \leq M$, so $f'''(x) \leq M$ for $a \leq x \leq a + d$. Now $\int_a^x f'''(t)\,dt \leq \int_a^x M\,dt$ $\Rightarrow$

$f''(x) - f''(a) \leq M(x - a)$ $\Rightarrow$ $f''(x) \leq f''(a) + M(x - a)$. Thus, $\int_a^x f''(t)\,dt \leq \int_a^x [f''(a) + M(t - a)]\,dt$ $\Rightarrow$

$f'(x) - f'(a) \leq f''(a)(x - a) + \frac{1}{2}M(x - a)^2$ $\Rightarrow$ $f'(x) \leq f'(a) + f''(a)(x - a) + \frac{1}{2}M(x - a)^2$ $\Rightarrow$

$\int_a^x f'(t)\,dt \leq \int_a^x \left[f'(a) + f''(a)(t - a) + \frac{1}{2}M(t - a)^2 \right] dt$ $\Rightarrow$

$f(x) - f(a) \leq f'(a)(x - a) + \frac{1}{2}f''(a)(x - a)^2 + \frac{1}{6}M(x - a)^3$. So

$f(x) - f(a) - f'(a)(x - a) - \frac{1}{2}f''(a)(x - a)^2 \leq \frac{1}{6}M(x - a)^3$. But

$R_2(x) = f(x) - T_2(x) = f(x) - f(a) - f'(a)(x - a) - \frac{1}{2}f''(a)(x - a)^2$, so $R_2(x) \leq \frac{1}{6}M(x - a)^3$.

A similar argument using $f'''(x) \geq -M$ shows that $R_2(x) \geq -\frac{1}{6}M(x - a)^3$. So $|R_2(x_2)| \leq \frac{1}{6}M\,|x - a|^3$.

Although we have assumed that $x > a$, a similar calculation shows that this inequality is also true if $x < a$.

8.8   The Binomial Series

1. The general binomial series in (2) is

$$(1 + x)^k = \sum_{n=0}^{\infty} \binom{k}{n} x^n = 1 + kx + \frac{k(k - 1)}{2!}x^2 + \frac{k(k - 1)(k - 2)}{3!}x^3 + \cdots.$$

$$(1 + x)^{1/2} = \sum_{n=0}^{\infty} \binom{\frac{1}{2}}{n} x^n = 1 + \left(\tfrac{1}{2}\right)x + \frac{\left(\frac{1}{2}\right)\left(-\frac{1}{2}\right)}{2!}x^2 + \frac{\left(\frac{1}{2}\right)\left(-\frac{1}{2}\right)\left(-\frac{3}{2}\right)}{3!}x^3 + \cdots$$

$$= 1 + \frac{x}{2} - \frac{x^2}{2^2 \cdot 2!} + \frac{1 \cdot 3 \cdot x^3}{2^3 \cdot 3!} - \frac{1 \cdot 3 \cdot 5 \cdot x^4}{2^4 \cdot 4!} + \cdots$$

$$= 1 + \frac{x}{2} + \sum_{n=2}^{\infty} \frac{(-1)^{n-1}\,1 \cdot 3 \cdot 5 \cdots (2n - 3)x^n}{2^n \cdot n!} \quad \text{for } |x| < 1, \quad \text{so } R = 1.$$

3. $\dfrac{1}{(2 + x)^3} = \dfrac{1}{[2(1 + x/2)]^3} = \dfrac{1}{8}\left(1 + \dfrac{x}{2}\right)^{-3} = \dfrac{1}{8}\displaystyle\sum_{n=0}^{\infty}\binom{-3}{n}\left(\dfrac{x}{2}\right)^n$. The binomial coefficient is

$$\binom{-3}{n} = \frac{(-3)(-4)(-5) \cdots (-3 - n + 1)}{n!} = \frac{(-3)(-4)(-5) \cdots [-(n + 2)]}{n!}$$

$$= \frac{(-1)^n \cdot 2 \cdot 3 \cdot 4 \cdot 5 \cdots (n + 1)(n + 2)}{2 \cdot n!} = \frac{(-1)^n (n + 1)(n + 2)}{2}$$

Thus, $\dfrac{1}{(2 + x)^3} = \dfrac{1}{8}\displaystyle\sum_{n=0}^{\infty} \frac{(-1)^n (n + 1)(n + 2)}{2} \frac{x^n}{2^n} = \displaystyle\sum_{n=0}^{\infty} \frac{(-1)^n (n + 1)(n + 2)x^n}{2^{n+4}}$ for $\left|\dfrac{x}{2}\right| < 1$ $\Leftrightarrow$

$|x| < 2$, so $R = 2$.

5. We must write the binomial in the form (1+ expression), so we'll factor out a 4.

$$\frac{x}{\sqrt{4+x^2}} = \frac{x}{\sqrt{4(1+x^2/4)}} = \frac{x}{2\sqrt{1+x^2/4}} = \frac{x}{2}\left(1+\frac{x^2}{4}\right)^{-1/2} = \frac{x}{2}\sum_{n=0}^{\infty}\binom{-\frac{1}{2}}{n}\left(\frac{x^2}{4}\right)^n$$

$$= \frac{x}{2}\left[1 + \left(-\tfrac{1}{2}\right)\frac{x^2}{4} + \frac{\left(-\frac{1}{2}\right)\left(-\frac{3}{2}\right)}{2!}\left(\frac{x^2}{4}\right)^2 + \frac{\left(-\frac{1}{2}\right)\left(-\frac{3}{2}\right)\left(-\frac{5}{2}\right)}{3!}\left(\frac{x^2}{4}\right)^3 + \cdots\right]$$

$$= \frac{x}{2} + \frac{x}{2}\sum_{n=1}^{\infty}(-1)^n\frac{1\cdot3\cdot5\cdot\cdots\cdot(2n-1)}{2^n\cdot4^n\cdot n!}x^{2n}$$

$$= \frac{x}{2} + \sum_{n=1}^{\infty}(-1)^n\frac{1\cdot3\cdot5\cdot\cdots\cdot(2n-1)}{n!\,2^{3n+1}}x^{2n+1} \text{ and } \frac{x^2}{4} < 1 \;\Leftrightarrow\; \frac{|x|}{2} < 1 \;\Leftrightarrow\; |x| < 2, \text{ so } R = 2.$$

7. $(1+2x)^{3/4} = 1 + \frac{3}{4}(2x) + \frac{\left(\frac{3}{4}\right)\left(-\frac{1}{4}\right)}{2!}(2x)^2 + \frac{\left(\frac{3}{4}\right)\left(-\frac{1}{4}\right)\left(-\frac{5}{4}\right)}{3!}(2x)^3 + \cdots$

$$= 1 + \frac{3}{2}x + 3\sum_{n=2}^{\infty}(-1)^{n+1}\frac{1\cdot5\cdot9\cdot\cdots\cdot(4n-7)}{4^n\cdot n!}\cdot2^n x^n$$

$$= 1 + \frac{3}{2}x + 3\sum_{n=2}^{\infty}(-1)^{n+1}\frac{1\cdot5\cdot9\cdot\cdots\cdot(4n-7)}{2^n\cdot n!}x^n \text{ and } |2x| < 1 \;\Leftrightarrow\; |x| < \frac{1}{2}, \text{ so } R = \frac{1}{2}.$$

The three Taylor polynomials are $T_1(x) = 1 + \frac{3}{2}x$, $T_2(x) = 1 + \frac{3}{2}x - \frac{3}{8}x^2$, and $T_3(x) = 1 + \frac{3}{2}x - \frac{3}{8}x^2 + \frac{5}{16}x^3$.

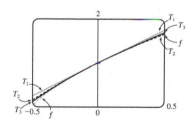

9. (a) $1/\sqrt{1-x^2} = \left[1+(-x^2)\right]^{-1/2} = 1 + \left(-\tfrac{1}{2}\right)(-x^2) + \frac{\left(-\frac{1}{2}\right)\left(-\frac{3}{2}\right)}{2!}(-x^2)^2 + \frac{\left(-\frac{1}{2}\right)\left(-\frac{3}{2}\right)\left(-\frac{5}{2}\right)}{3!}(-x^2)^3 + \cdots$

$$= 1 + \sum_{n=1}^{\infty}\frac{1\cdot3\cdot5\cdot\cdots\cdot(2n-1)}{2^n\cdot n!}x^{2n}$$

(b) $\sin^{-1}x = \displaystyle\int\frac{1}{\sqrt{1-x^2}}\,dx = C + x + \sum_{n=1}^{\infty}\frac{1\cdot3\cdot5\cdot\cdots\cdot(2n-1)}{(2n+1)2^n\cdot n!}x^{2n+1}$

$$= x + \sum_{n=1}^{\infty}\frac{1\cdot3\cdot5\cdot\cdots\cdot(2n-1)}{(2n+1)2^n\cdot n!}x^{2n+1} \quad \text{since } 0 = \sin^{-1}0 = C.$$

11. (a) $[1+(-x)]^{-2} = 1 + (-2)(-x) + \frac{(-2)(-3)}{2!}(-x)^2 + \frac{(-2)(-3)(-4)}{3!}(-x)^3 + \cdots$

$$= 1 + 2x + 3x^2 + 4x^3 + \cdots = \sum_{n=0}^{\infty}(n+1)x^n,$$

so $\dfrac{x}{(1-x)^2} = x\displaystyle\sum_{n=0}^{\infty}(n+1)x^n = \sum_{n=0}^{\infty}(n+1)x^{n+1} = \sum_{n=1}^{\infty}nx^n.$

(b) With $x = \frac{1}{2}$ in part (a), we have $\displaystyle\sum_{n=1}^{\infty}n\left(\tfrac{1}{2}\right)^n = \sum_{n=1}^{\infty}\frac{n}{2^n} = \frac{\frac{1}{2}}{\left(1-\frac{1}{2}\right)^2} = \frac{\frac{1}{2}}{\frac{1}{4}} = 2.$

13. (a) $\left(1+x^2\right)^{1/2} = 1 + \left(\frac{1}{2}\right)x^2 + \frac{\left(\frac{1}{2}\right)\left(-\frac{1}{2}\right)}{2!}\left(x^2\right)^2 + \frac{\left(\frac{1}{2}\right)\left(-\frac{1}{2}\right)\left(-\frac{3}{2}\right)}{3!}\left(x^2\right)^3 + \cdots$

$$= 1 + \frac{x^2}{2} + \sum_{n=2}^{\infty} \frac{(-1)^{n-1}\, 1 \cdot 3 \cdot 5 \cdot \cdots \cdot (2n-3)}{2^n \cdot n!} x^{2n}$$

(b) The coefficient of x^{10} (corresponding to $n=5$) in the above Maclaurin series is $\dfrac{f^{(10)}(0)}{10!}$, so

$$\frac{f^{(10)}(0)}{10!} = \frac{(-1)^4 \cdot 1 \cdot 3 \cdot 5 \cdot 7}{2^5 \cdot 5!} \quad \Rightarrow \quad f^{(10)}(0) = 10!\left(\frac{1 \cdot 3 \cdot 5 \cdot 7}{2^5 \cdot 5!}\right) = 99{,}225.$$

15. (a) $g(x) = \displaystyle\sum_{n=0}^{\infty} \binom{k}{n} x^n \quad \Rightarrow \quad g'(x) = \sum_{n=1}^{\infty} \binom{k}{n} n x^{n-1}$, so

$$(1+x)g'(x) = (1+x)\sum_{n=1}^{\infty}\binom{k}{n}nx^{n-1} = \sum_{n=1}^{\infty}\binom{k}{n}nx^{n-1} + \sum_{n=1}^{\infty}\binom{k}{n}nx^{n}$$

$$= \sum_{n=0}^{\infty}\binom{k}{n+1}(n+1)x^n + \sum_{n=0}^{\infty}\binom{k}{n}nx^n \qquad \begin{bmatrix}\text{Replace } n \text{ with } n+1 \\ \text{in the first series}\end{bmatrix}$$

$$= \sum_{n=0}^{\infty}(n+1)\frac{k(k-1)(k-2)\cdots(k-n+1)(k-n)}{(n+1)!}x^n + \sum_{n=0}^{\infty}\left[(n)\frac{k(k-1)(k-2)\cdots(k-n+1)}{n!}\right]x^n$$

$$= \sum_{n=0}^{\infty}\frac{(n+1)k(k-1)(k-2)\cdots(k-n+1)}{(n+1)!}\left[(k-n)+n\right]x^n$$

$$= k\sum_{n=0}^{\infty}\frac{k(k-1)(k-2)\cdots(k-n+1)}{n!}x^n = k\sum_{n=0}^{\infty}\binom{k}{n}x^n = kg(x)$$

Thus, $g'(x) = \dfrac{kg(x)}{1+x}$.

(b) $h(x) = (1+x)^{-k}g(x) \quad \Rightarrow$

$$h'(x) = -k(1+x)^{-k-1}g(x) + (1+x)^{-k}g'(x) \quad \text{[Product Rule]}$$

$$= -k(1+x)^{-k-1}g(x) + (1+x)^{-k}\frac{kg(x)}{1+x} \quad \text{[from part (a)]}$$

$$= -k(1+x)^{-k-1}g(x) + k(1+x)^{-k-1}g(x) = 0$$

(c) From part (b) we see that $h(x)$ must be constant for $x \in (-1, 1)$, so $h(x) = h(0) = 1$ for $x \in (-1, 1)$.

Thus, $h(x) = 1 = (1+x)^{-k}g(x) \quad \Leftrightarrow \quad g(x) = (1+x)^k$ for $x \in (-1, 1)$.

8.9   Applications of Taylor Polynomials

1. (a)

n	$f^{(n)}(x)$	$f^{(n)}(0)$	$T_n(x)$
0	$\cos x$	1	1
1	$-\sin x$	0	1
2	$-\cos x$	-1	$1 - \frac{1}{2}x^2$
3	$\sin x$	0	$1 - \frac{1}{2}x^2$
4	$\cos x$	1	$1 - \frac{1}{2}x^2 + \frac{1}{24}x^4$
5	$-\sin x$	0	$1 - \frac{1}{2}x^2 + \frac{1}{24}x^4$
6	$-\cos x$	-1	$1 - \frac{1}{2}x^2 + \frac{1}{24}x^4 - \frac{1}{720}x^6$

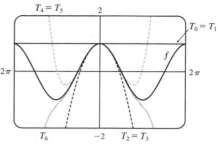

(b)

x	f	$T_0 = T_1$	$T_2 = T_3$	$T_4 = T_5$	T_6
$\frac{\pi}{4}$	0.7071	1	0.6916	0.7074	0.7071
$\frac{\pi}{2}$	0	1	-0.2337	0.0200	-0.0009
π	-1	1	-3.9348	0.1239	-1.2114

(c) As n increases, $T_n(x)$ is a good approximation to $f(x)$ on a larger and larger interval.

3.

n	$f^{(n)}(x)$	$f^{(n)}\left(\frac{\pi}{6}\right)$
0	$\sin x$	$\frac{1}{2}$
1	$\cos x$	$\frac{\sqrt{3}}{2}$
2	$-\sin x$	$-\frac{1}{2}$
3	$-\cos x$	$-\frac{\sqrt{3}}{2}$

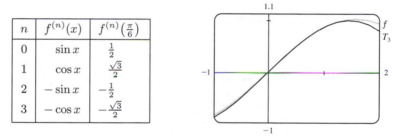

$$T_3(x) = \sum_{n=0}^{3} \frac{f^{(n)}\left(\frac{\pi}{6}\right)}{n!}\left(x - \frac{\pi}{6}\right)^n = \frac{1}{2} + \frac{\sqrt{3}}{2}\left(x - \frac{\pi}{6}\right) - \frac{1}{4}\left(x - \frac{\pi}{6}\right)^2 - \frac{\sqrt{3}}{12}\left(x - \frac{\pi}{6}\right)^3$$

5.

n	$f^{(n)}(x)$	$f^{(n)}(0)$
0	$\arcsin x$	0
1	$1/\sqrt{1 - x^2}$	1
2	$x/(1 - x^2)^{3/2}$	0
3	$(2x^2 + 1)/(1 - x^2)^{5/2}$	1

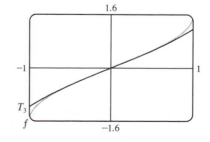

$$T_3(x) = \sum_{n=0}^{3} \frac{f^{(n)}(0)}{n!}x^n = x + \frac{x^3}{6}$$

7.

n	$f^{(n)}(x)$	$f^{(n)}(0)$
0	xe^{-2x}	0
1	$(1-2x)e^{-2x}$	1
2	$4(x-1)e^{-2x}$	-4
3	$4(3-2x)e^{-2x}$	12

$$T_3(x) = \sum_{n=0}^{3} \frac{f^{(n)}(0)}{n!} x^n = \tfrac{0}{1} \cdot 1 + \tfrac{1}{1}x^1 + \tfrac{-4}{2}x^2 + \tfrac{12}{6}x^3 = x - 2x^2 + 2x^3$$

9. In Maple, we can find the Taylor polynomials by the following method: first define f:=sec(x); and then set

T2:=convert(taylor(f,x=0,3),polynom);, T4:=convert(taylor(f,x=0,5),polynom);, etc.

(The third argument in the taylor function is one more than the degree of the desired polynomial). We must

convert to the type polynom because the output of the

taylor function contains an error term which we do not

want. In Mathematica, we use

Tn:=Normal[Series[f,{x,0,n}]], with n=2, 4, etc.

Note that in Mathematica, the "degree" argument is the same

as the degree of the desired polynomial. In Derive, author

sec x, then enter Calculus,Taylor,8,0; and then

simplify the expression. The eighth Taylor polynomial is

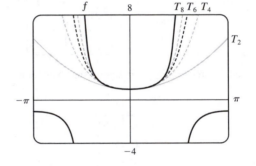

$$T_8(x) = 1 + \tfrac{1}{2}x^2 + \tfrac{5}{24}x^4 + \tfrac{61}{720}x^6 + \tfrac{277}{8064}x^8.$$

11.

n	$f^{(n)}(x)$	$f^{(n)}(4)$
0	$\sqrt{x}$	2
1	$\tfrac{1}{2}x^{-1/2}$	$\tfrac{1}{4}$
2	$-\tfrac{1}{4}x^{-3/2}$	$-\tfrac{1}{32}$
3	$\tfrac{3}{8}x^{-5/2}$	

(a) $f(x) = \sqrt{x} \approx T_2(x) = 2 + \tfrac{1}{4}(x-4) - \tfrac{1/32}{2!}(x-4)^2 = 2 + \tfrac{1}{4}(x-4) - \tfrac{1}{64}(x-4)^2$

(b) $|R_2(x)| \leq \dfrac{M}{3!}|x-4|^3$, where $|f'''(x)| \leq M$. Now $4 \leq x \leq 4.2 \;\Rightarrow\; |x-4| \leq 0.2 \;\Rightarrow\; |x-4|^3 \leq 0.008$.

Since $f'''(x)$ is decreasing on $[4, 4.2]$, we can take $M = |f'''(4)| = \tfrac{3}{8}4^{-5/2} = \tfrac{3}{256}$, so

$|R_2(x)| \leq \dfrac{3/256}{6}(0.008) = \dfrac{0.008}{512} = 0.000015625.$

(c)

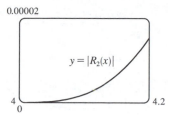

From the graph of $|R_2(x)| = |\sqrt{x} - T_2(x)|$, it seems that the error is less than 1.52×10^{-5} on $[4, 4.2]$.

13.

n	$f^{(n)}(x)$	$f^{(n)}(1)$
0	$x^{2/3}$	1
1	$\frac{2}{3}x^{-1/3}$	$\frac{2}{3}$
2	$-\frac{2}{9}x^{-4/3}$	$-\frac{2}{9}$
3	$\frac{8}{27}x^{-7/3}$	$\frac{8}{27}$
4	$-\frac{56}{81}x^{-10/3}$	

(a) $f(x) = x^{2/3} \approx T_3(x) = 1 + \frac{2}{3}(x-1) - \frac{2/9}{2!}(x-1)^2 + \frac{8/27}{3!}(x-1)^3 = 1 + \frac{2}{3}(x-1) - \frac{1}{9}(x-1)^2 + \frac{4}{81}(x-1)^3$

(b) $|R_3(x)| \le \frac{M}{4!}|x-1|^4$, where $\left|f^{(4)}(x)\right| \le M$. Now $0.8 \le x \le 1.2 \Rightarrow |x-1| \le 0.2 \Rightarrow |x-1|^4 \le 0.0016$.

Since $\left|f^{(4)}(x)\right|$ is decreasing on $[0.8, 1.2]$, we can take $M = \left|f^{(4)}(0.8)\right| = \frac{56}{81}(0.8)^{-10/3}$, so

$$|R_3(x)| \le \frac{\frac{56}{81}(0.8)^{-10/3}}{24}(0.0016) \approx 0.000\,096\,97.$$

(c)

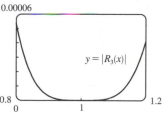

From the graph of $|R_3(x)| = \left|x^{2/3} - T_3(x)\right|$, it seems that the error is less than $0.000\,053\,3$ on $[0.8, 1.2]$.

15.

n	$f^{(n)}(x)$	$f^{(n)}(0)$
0	e^{x^2}	1
1	$e^{x^2}(2x)$	0
2	$e^{x^2}(2 + 4x^2)$	2
3	$e^{x^2}(12x + 8x^3)$	0
4	$e^{x^2}(12 + 48x^2 + 16x^4)$	

(a) $f(x) = e^{x^2} \approx T_3(x) = 1 + \frac{2}{2!}x^2 = 1 + x^2$

(b) $|R_3(x)| \le \frac{M}{4!}|x|^4$, where $\left|f^{(4)}(x)\right| \le M$. Now $0 \le x \le 0.1 \Rightarrow x^4 \le (0.1)^4$, and letting $x = 0.1$ gives

$$|R_3(x)| \le \frac{e^{0.01}(12 + 0.48 + 0.0016)}{24}(0.1)^4 \approx 0.00006.$$

(c)

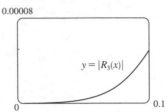

0.00008

$y = |R_3(x)|$

0

0.1

From the graph of $|R_3(x)| = \left| e^{x^2} - (1 + x^2) \right|$, it appears that the error is less than 0.000051 on $[0, 0.1]$.

17.

n	$f^{(n)}(x)$	$f^{(n)}(0)$
0	$x \sin x$	0
1	$\sin x + x \cos x$	0
2	$2 \cos x - x \sin x$	2
3	$-3 \sin x - x \cos x$	0
4	$-4 \cos x + x \sin x$	-4
5	$5 \sin x + x \cos x$	

(a) $f(x) = x \sin x \approx T_4(x) = \dfrac{2}{2!}(x - 0)^2 + \dfrac{-4}{4!}(x - 0)^4 = x^2 - \dfrac{1}{6}x^4$

(b) $|R_4(x)| \le \dfrac{M}{5!} |x|^5$, where $\left| f^{(5)}(x) \right| \le M$. Now $-1 \le x \le 1 \Rightarrow |x| \le 1$, and a graph of $f^{(5)}(x)$ shows that $\left| f^{(5)}(x) \right| \le 5$ for $-1 \le x \le 1$. Thus, we can take $M = 5$ and get $|R_4(x)| \le \dfrac{5}{5!} \cdot 1^5 = \dfrac{1}{24} = 0.041\overline{6}$.

(c)

0.009

$y = |R_4(x)|$

-1 1

0

From the graph of $|R_4(x)| = |x \sin x - T_4(x)|$, it seems that the error is less than 0.0082 on $[-1, 1]$.

19. From Exercise 3, $\sin x = \dfrac{1}{2} + \dfrac{\sqrt{3}}{2}\left(x - \dfrac{\pi}{6}\right) - \dfrac{1}{4}\left(x - \dfrac{\pi}{6}\right)^2 - \dfrac{\sqrt{3}}{12}\left(x - \dfrac{\pi}{6}\right)^3 + R_3(x)$, where $|R_3(x)| \le \dfrac{M}{4!} \left|x - \dfrac{\pi}{6}\right|^4$ with

$\left| f^{(4)}(x) \right| = |\sin x| \le M = 1$. Now $x = 35° = (30° + 5°) = \left(\dfrac{\pi}{6} + \dfrac{\pi}{36}\right)$ radians, so the error is

$\left| R_3\left(\dfrac{\pi}{36}\right) \right| \le \dfrac{\left(\dfrac{\pi}{36}\right)^4}{4!} < 0.000003$. Therefore, to five decimal places,

$\sin 35° \approx \dfrac{1}{2} + \dfrac{\sqrt{3}}{2}\left(\dfrac{\pi}{36}\right) - \dfrac{1}{4}\left(\dfrac{\pi}{36}\right)^2 - \dfrac{\sqrt{3}}{12}\left(\dfrac{\pi}{36}\right)^3 \approx 0.57358$.

21. All derivatives of e^x are e^x, so $|R_n(x)| \le \dfrac{e^x}{(n + 1)!} |x|^{n+1}$, where $0 < x < 0.1$. Letting $x = 0.1$,

$R_n(0.1) \le \dfrac{e^{0.1}}{(n + 1)!}(0.1)^{n+1} < 0.00001$, and by trial and error we find that $n = 3$ satisfies this inequality since

$R_3(0.1) < 0.0000046$. Thus, by adding the four terms of the Maclaurin series for e^x corresponding to $n = 0, 1, 2,$ and 3, we

can estimate $e^{0.1}$ to within 0.00001. (In fact, this sum is $1.10516\overline{6}$ and $e^{0.1} \approx 1.10517$.)

23. $\sin x = x - \dfrac{1}{3!}x^3 + \dfrac{1}{5!}x^5 - \cdots$. By the Alternating

Series Estimation Theorem, the error in the

approximation $\sin x = x - \dfrac{1}{3!}x^3$ is less than

$\left|\dfrac{1}{5!}x^5\right| < 0.01 \quad \Leftrightarrow \quad |x^5| < 120(0.01) \quad \Leftrightarrow$

$|x| < (1.2)^{1/5} \approx 1.037$. The curves $y = x - \dfrac{1}{6}x^3$ and

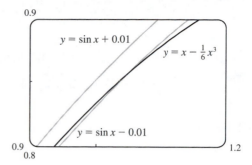

$y = \sin x - 0.01$ intersect at $x \approx 1.043$, so the graph confirms our estimate. Since both the sine function and the given

approximation are odd functions, we need to check the estimate only for $x > 0$. Thus, the desired range of values for x

is $-1.037 < x < 1.037$.

25. Let $s(t)$ be the position function of the car, and for convenience set $s(0) = 0$. The velocity of the car is $v(t) = s'(t)$ and the

acceleration is $a(t) = s''(t)$, so the second degree Taylor polynomial is $T_2(t) = s(0) + v(0)t + \dfrac{a(0)}{2}t^2 = 20t + t^2$. We

estimate the distance travelled during the next second to be $s(1) \approx T_2(1) = 20 + 1 = 21$ m. The function $T_2(t)$ would not be

accurate over a full minute, since the car could not possibly maintain an acceleration of 2 m/s^2 for that long (if it did, its final

speed would be 140 m/s ≈ 313 mi/h!)

27. $E = \dfrac{q}{D^2} - \dfrac{q}{(D+d)^2} = \dfrac{q}{D^2} - \dfrac{q}{D^2(1+d/D)^2} = \dfrac{q}{D^2}\left[1 - \left(1 + \dfrac{d}{D}\right)^{-2}\right]$.

We use the Binomial Series to expand $(1 + d/D)^{-2}$:

$$E = \dfrac{q}{D^2}\left[1 - \left(1 - 2\left(\dfrac{d}{D}\right) + \dfrac{2\cdot 3}{2!}\left(\dfrac{d}{D}\right)^2 - \dfrac{2\cdot 3\cdot 4}{3!}\left(\dfrac{d}{D}\right)^3 + \cdots\right)\right]$$

$$= \dfrac{q}{D^2}\left[2\left(\dfrac{d}{D}\right) - 3\left(\dfrac{d}{D}\right)^2 + 4\left(\dfrac{d}{D}\right)^3 - \cdots\right] \approx \dfrac{q}{D^2}\cdot 2\left(\dfrac{d}{D}\right) = 2qd\cdot\dfrac{1}{D^3}$$

when D is much larger than d; that is, when P is far away from the dipole.

29. (a) L is the length of the arc subtended by the angle θ, so $L = R\theta \quad \Rightarrow$

$\theta = L/R$. Now $\sec\theta = (R+C)/R \quad \Rightarrow \quad R\sec\theta = R+C \quad \Rightarrow$

$C = R\sec\theta - R = R\sec(L/R) - R$.

(b) From Exercise 9, $\sec x \approx T_4(x) = 1 + \frac{1}{2}x^2 + \frac{5}{24}x^4$. By part (a),

$$C \approx R\left[1 + \dfrac{1}{2}\left(\dfrac{L}{R}\right)^2 + \dfrac{5}{24}\left(\dfrac{L}{R}\right)^4\right] - R = R + \dfrac{1}{2}R\cdot\dfrac{L^2}{R^2} + \dfrac{5}{24}R\cdot\dfrac{L^4}{R^4} - R = \dfrac{L^2}{2R} + \dfrac{5L^4}{24R^3}.$$

(c) Taking $L = 100$ km and $R = 6370$ km, the formula in part (a) says that

$C = R\sec(L/R) - R = 6370\sec(100/6370) - 6370 \approx 0.785\,009\,965\,44$ km. The formula in part (b) says that

$$C \approx \dfrac{L^2}{2R} + \dfrac{5L^4}{24R^3} = \dfrac{100^2}{2\cdot 6370} + \dfrac{5\cdot 100^4}{24\cdot 6370^3} \approx 0.785\,009\,957\,36 \text{ km}.$$

The difference between these two results is only $0.000\,000\,008\,08$ km, or $0.000\,008\,08$ m!

31. Using $f(x) = T_n(x) + R_n(x)$ with $n = 1$ and $x = r$, we have $f(r) = T_1(r) + R_1(r)$, where T_1 is the first-degree Taylor polynomial of f at a. Because $a = x_n$, $f(r) = f(x_n) + f'(x_n)(r - x_n) + R_1(r)$. But r is a root of f, so $f(r) = 0$ and we have $0 = f(x_n) + f'(x_n)(r - x_n) + R_1(r)$. Taking the first two terms to the left side gives us

$f'(x_n)(x_n - r) - f(x_n) = R_1(r)$. Dividing by $f'(x_n)$, we get $x_n - r - \dfrac{f(x_n)}{f'(x_n)} = \dfrac{R_1(r)}{f'(x_n)}$. By the formula for Newton's

method, the left side of the preceding equation is $x_{n+1} - r$, so $|x_{n+1} - r| = \left| \dfrac{R_1(r)}{f'(x_n)} \right|$. Taylor's Inequality gives us

$|R_1(r)| \leq \dfrac{|f''(r)|}{2!} |r - x_n|^2$. Combining this inequality with the facts $|f''(x)| \leq M$ and $|f'(x)| \geq K$ gives us

$|x_{n+1} - r| \leq \dfrac{M}{2K} |x_n - r|^2$.

8 Review

CONCEPT CHECK

---

1. (a) See Definition 8.1.1.

   (b) See Definition 8.2.2.

   (c) The terms of the sequence $\{a_n\}$ approach 3 as n becomes large.

   (d) By adding sufficiently many terms of the series, we can make the partial sums as close to 3 as we like.

2. (a) See the definition on page 563.

   (b) A sequence is monotonic if it is either increasing or decreasing.

   (c) By Theorem 8.1.7, every bounded, monotonic sequence is convergent.

3. (a) See (4) in Section 8.2.

   (b) The p-series $\displaystyle\sum_{n=1}^{\infty} \dfrac{1}{n^p}$ is convergent if $p > 1$.

4. If $\sum a_n = 3$, then $\displaystyle\lim_{n \to \infty} a_n = 0$ and $\displaystyle\lim_{n \to \infty} s_n = 3$.

5. (a) See the Test for Divergence on page 572.

   (b) See the Integral Test on page 578.

   (c) See the Comparison Test on page 580.

   (d) See the Limit Comparison Test on page 582.

   (e) See the Alternating Series Test on page 587.

   (f) See the Ratio Test on page 591.

6. (a) A series $\sum a_n$ is called *absolutely convergent* if the series of absolute values $\sum |a_n|$ is convergent.

   (b) If a series $\sum a_n$ is absolutely convergent, then it is convergent.

7. (a) Use (4) in Section 8.3.

   (b) See Example 8 in Section 8.3.

   (c) By adding terms until you reach the desired accuracy given by the Alternating Series Estimation Theorem on page 588.

8. (a) $\sum_{n=0}^{\infty} c_n(x - a)^n$

   (b) Given the power series $\sum_{n=0}^{\infty} c_n(x - a)^n$, the radius of convergence is:

     (i) 0 if the series converges only when $x = a$

     (ii) ∞ if the series converges for all x, or

     (iii) a positive number R such that the series converges if $|x - a| < R$ and diverges if $|x - a| > R$.

(c) The interval of convergence of a power series is the interval that consists of all values of x for which the series converges. Corresponding to the cases in part (b), the interval of convergence is: (i) the single point $\{a\}$, (ii) all real numbers, that is, the real number line $(-\infty, \infty)$, or (iii) an interval with endpoints $a - R$ and $a + R$ which can contain neither, either, or both of the endpoints. In this case, we must test the series for convergence at each endpoint to determine the interval of convergence.

9. (a), (b) See Theorem 8.6.2.

10. (a) $T_n(x) = \sum\limits_{i=0}^{n} \dfrac{f^{(i)}(a)}{i!}(x - a)^i$

(b) $\sum\limits_{n=0}^{\infty} \dfrac{f^{(n)}(a)}{n!}(x - a)^n$

(c) $\sum\limits_{n=0}^{\infty} \dfrac{f^{(n)}(0)}{n!}x^n$ $[a = 0$ in part (b)]

(d) See Theorem 8.7.8.

(e) See Taylor's Inequality (8.7.9).

11. (a) – (e) See the table on page 612.

12. See the Binomial Series (8.8.2) for the expansion. The radius of convergence for the binomial series is 1.

TRUE-FALSE QUIZ

1. False.   See Note 2 after Theorem 8.2.6.

3. True.   If $\lim\limits_{n \to \infty} a_n = L$, then given any $\varepsilon > 0$, we can find a positive integer N such that $|a_n - L| < \varepsilon$ whenever $n > N$. If $n > N$, then $2n + 1 > N$ and $|a_{2n+1} - L| < \varepsilon$. Thus, $\lim\limits_{n \to \infty} a_{2n+1} = L$.

5. False.   For example, take $c_n = (-1)^n/(n6^n)$.

7. False, since $\lim\limits_{n \to \infty} \left| \dfrac{a_{n+1}}{a_n} \right| = \lim\limits_{n \to \infty} \left| \dfrac{1}{(n+1)^3} \cdot \dfrac{n^3}{1} \right| = \lim\limits_{n \to \infty} \left| \dfrac{n^3}{(n+1)^3} \cdot \dfrac{1/n^3}{1/n^3} \right| = \lim\limits_{n \to \infty} \dfrac{1}{(1 + 1/n)^3} = 1.$

9. False.   See the note after Example 4 in Section 8.3.

11. True.   See (6) in Section 8.1.

13. True.   By Theorem 8.7.5 the coefficient of x^3 is $\dfrac{f'''(0)}{3!} = \dfrac{1}{3}$ $\Rightarrow$ $f'''(0) = 2$.

   Or: Use Theorem 8.6.2 to differentiate f three times.

15. False.   For example, let $a_n = b_n = (-1)^n$. Then $\{a_n\}$ and $\{b_n\}$ are divergent, but $a_n b_n = 1$, so $\{a_n b_n\}$ is convergent.

17. True by Theorem 8.4.1. $\left[\sum (-1)^n\, a_n \text{ is absolutely convergent and hence convergent.} \right]$

EXERCISES

1. $\left\{ \dfrac{2 + n^3}{1 + 2n^3} \right\}$ converges since $\lim\limits_{n \to \infty} \dfrac{2 + n^3}{1 + 2n^3} = \lim\limits_{n \to \infty} \dfrac{2/n^3 + 1}{1/n^3 + 2} = \dfrac{1}{2}$.

3. $\lim\limits_{n \to \infty} a_n = \lim\limits_{n \to \infty} \dfrac{n^3}{1 + n^2} = \lim\limits_{n \to \infty} \dfrac{n}{1/n^2 + 1} = \infty$, so the sequence diverges.

5. $|a_n| = \left| \dfrac{n \sin n}{n^2 + 1} \right| \le \dfrac{n}{n^2 + 1} < \dfrac{1}{n}$, so $|a_n| \to 0$ as $n \to \infty$. Thus, $\lim\limits_{n \to \infty} a_n = 0$. The sequence $\{a_n\}$ is convergent.

7. $\left\{\left(1+\dfrac{3}{n}\right)^{4n}\right\}$ is convergent. Let $y = \left(1+\dfrac{3}{x}\right)^{4x}$. Then

$$\lim_{x\to\infty} \ln y = \lim_{x\to\infty} 4x \ln(1+3/x) = \lim_{x\to\infty} \frac{\ln(1+3/x)}{1/(4x)} \overset{\text{H}}{=} \lim_{x\to\infty} \frac{\dfrac{1}{1+3/x}\left(-\dfrac{3}{x^2}\right)}{-1/(4x^2)} = \lim_{x\to\infty} \frac{12}{1+3/x} = 12$$

so $\displaystyle\lim_{x\to\infty} y = \lim_{n\to\infty}\left(1+\frac{3}{n}\right)^{4n} = e^{12}$.

Or: Use Exercise 4.5.38.

9. $\dfrac{n}{n^3+1} < \dfrac{n}{n^3} = \dfrac{1}{n^2}$, so $\displaystyle\sum_{n=1}^{\infty} \frac{n}{n^3+1}$ converges by the Comparison Test with the convergent p-series $\displaystyle\sum_{n=1}^{\infty} \frac{1}{n^2}$ ($p = 2 > 1$).

11. $\displaystyle\lim_{n\to\infty}\left|\frac{a_{n+1}}{a_n}\right| = \lim_{n\to\infty}\left[\frac{(n+1)^3}{5^{n+1}}\cdot\frac{5^n}{n^3}\right] = \lim_{n\to\infty}\left(1+\frac{1}{n}\right)^3\cdot\frac{1}{5} = \frac{1}{5} < 1$, so $\displaystyle\sum_{n=1}^{\infty} \frac{n^3}{5^n}$ converges by the Ratio Test.

13. Let $f(x) = \dfrac{1}{x\sqrt{\ln x}}$. Then f is continuous, positive, and decreasing on $[2, \infty)$, so the Integral Test applies.

$$\int_2^\infty f(x)\,dx = \lim_{t\to\infty}\int_2^t \frac{1}{x\sqrt{\ln x}}\,dx \quad\begin{bmatrix}u = \ln x, \\ du = \dfrac{1}{x}\,dx\end{bmatrix} = \lim_{t\to\infty}\int_{\ln 2}^{\ln t} u^{-1/2}\,du$$

$$= \lim_{t\to\infty}\left[2\sqrt{u}\right]_{\ln 2}^{\ln t} = \lim_{t\to\infty}\left(2\sqrt{\ln t} - 2\sqrt{\ln 2}\right) = \infty, \text{ so the series } \sum_{n=2}^{\infty} \frac{1}{n\sqrt{\ln n}} \text{ diverges.}$$

15. $b_n = \dfrac{\sqrt{n}}{n+1} > 0$, $\{b_n\}$ is decreasing, and $\displaystyle\lim_{n\to\infty} b_n = 0$, so the series $\displaystyle\sum_{n=1}^{\infty}(-1)^{n-1}\frac{\sqrt{n}}{n+1}$ converges by the Alternating Series Test.

17. $\displaystyle\lim_{n\to\infty}\left|\frac{a_{n+1}}{a_n}\right| = \lim_{n\to\infty} \frac{1\cdot 3\cdot 5\cdots(2n-1)(2n+1)}{5^{n+1}(n+1)!}\cdot\frac{5^n n!}{1\cdot 3\cdot 5\cdots(2n-1)} = \lim_{n\to\infty}\frac{2n+1}{5(n+1)} = \frac{2}{5} < 1$, so the series converges by the Ratio Test.

19. $\dfrac{2^{2n+1}}{5^n} = \dfrac{2^{2n}\cdot 2^1}{5^n} = \dfrac{(2^2)^n\cdot 2}{5^n} = 2\left(\dfrac{4}{5}\right)^n$, so $\displaystyle\sum_{n=1}^{\infty}\frac{2^{2n+1}}{5^n} = 2\sum_{n=1}^{\infty}\left(\frac{4}{5}\right)^n$ is a geometric series with $a = \dfrac{8}{5}$ and $r = \dfrac{4}{5}$. Since $|r| = \dfrac{4}{5} < 1$, the series converges to $\dfrac{a}{1-r} = \dfrac{8/5}{1-4/5} = \dfrac{8/5}{1/5} = 8$.

21. $\displaystyle\sum_{n=1}^{\infty}\left[\tan^{-1}(n+1) - \tan^{-1} n\right] = \lim_{n\to\infty} s_n$

$$= \lim_{n\to\infty}\left[(\tan^{-1} 2 - \tan^{-1} 1) + (\tan^{-1} 3 - \tan^{-1} 2) + \cdots\right.$$
$$\left. + (\tan^{-1}(n+1) - \tan^{-1} n)\right]$$
$$= \lim_{n\to\infty}\left[\tan^{-1}(n+1) - \tan^{-1} 1\right] = \frac{\pi}{2} - \frac{\pi}{4} = \frac{\pi}{4}$$

23. $1.2345345345\ldots = 1.2 + 0.0\overline{345} = \dfrac{12}{10} + \dfrac{345/10{,}000}{1 - 1/1000} = \dfrac{12}{10} + \dfrac{345}{9990} = \dfrac{4111}{3330}$

25. $\displaystyle\sum_{n=1}^{\infty}\frac{(-1)^{n+1}}{n^5} = 1 - \frac{1}{32} + \frac{1}{243} - \frac{1}{1024} + \frac{1}{3125} - \frac{1}{7776} + \frac{1}{16{,}807} - \frac{1}{32{,}768} + \cdots$.

Since $b_8 = \dfrac{1}{8^5} = \dfrac{1}{32{,}768} < 0.000031$, $\displaystyle\sum_{n=1}^{\infty}\frac{(-1)^{n+1}}{n^5} \approx \sum_{n=1}^{7}\frac{(-1)^{n+1}}{n^5} \approx 0.9721$.

27. $\displaystyle\sum_{n=1}^{\infty}\frac{1}{2+5^n} \approx \sum_{n=1}^{8}\frac{1}{2+5^n} \approx 0.18976224$. To estimate the error, note that $\dfrac{1}{2+5^n} < \dfrac{1}{5^n}$, so the remainder term is

$$R_8 = \sum_{n=9}^{\infty}\frac{1}{2+5^n} < \sum_{n=9}^{\infty}\frac{1}{5^n} = \frac{1/5^9}{1-1/5} = 6.4\times 10^{-7} \quad [\text{geometric series with } a = \tfrac{1}{5^9} \text{ and } r = \tfrac{1}{5}].$$

29. Use the Limit Comparison Test. $\displaystyle\lim_{n\to\infty}\left|\frac{\left(\frac{n+1}{n}\right)a_n}{a_n}\right| = \lim_{n\to\infty}\frac{n+1}{n} = \lim_{n\to\infty}\left(1+\frac{1}{n}\right) = 1 > 0.$

Since $\sum |a_n|$ is convergent, so is $\displaystyle\sum\left|\left(\frac{n+1}{n}\right)a_n\right|$, by the Limit Comparison Test.

31. $\displaystyle\lim_{n\to\infty}\left|\frac{a_{n+1}}{a_n}\right| = \lim_{n\to\infty}\left[\frac{|x+2|^{n+1}}{(n+1)\,4^{n+1}}\cdot\frac{n\,4^n}{|x+2|^n}\right] = \lim_{n\to\infty}\left[\frac{n}{n+1}\cdot\frac{|x+2|}{4}\right] = \frac{|x+2|}{4} < 1 \quad\Leftrightarrow\quad |x+2| < 4,$ so $R = 4$.

$|x+2| < 4 \quad\Leftrightarrow\quad -4 < x+2 < 4 \quad\Leftrightarrow\quad -6 < x < 2.$ If $x = -6$, then the series $\displaystyle\sum_{n=1}^{\infty}\frac{(x+2)^n}{n4^n}$ becomes

$\displaystyle\sum_{n=1}^{\infty}\frac{(-4)^n}{n4^n} = \sum_{n=1}^{\infty}\frac{(-1)^n}{n}$, the alternating harmonic series, which converges by the Alternating Series Test. When $x = 2$, the

series becomes the harmonic series $\displaystyle\sum_{n=1}^{\infty}\frac{1}{n}$, which diverges. Thus, $I = [-6, 2)$.

33. $\displaystyle\lim_{n\to\infty}\left|\frac{a_{n+1}}{a_n}\right| = \lim_{n\to\infty}\left|\frac{2^{n+1}(x-3)^{n+1}}{\sqrt{n+4}}\cdot\frac{\sqrt{n+3}}{2^n(x-3)^n}\right| = 2\,|x-3|\lim_{n\to\infty}\sqrt{\frac{n+3}{n+4}} = 2\,|x-3| < 1 \quad\Leftrightarrow\quad |x-3| < \tfrac{1}{2},$ so

$R = \tfrac{1}{2}.\ |x-3| < \tfrac{1}{2} \quad\Leftrightarrow\quad -\tfrac{1}{2} < x-3 < \tfrac{1}{2} \quad\Leftrightarrow\quad \tfrac{5}{2} < x < \tfrac{7}{2}.$ For $x = \tfrac{7}{2}$, the series $\displaystyle\sum_{n=1}^{\infty}\frac{2^n(x-3)^n}{\sqrt{n+3}}$ becomes

$\displaystyle\sum_{n=0}^{\infty}\frac{1}{\sqrt{n+3}} = \sum_{n=3}^{\infty}\frac{1}{n^{1/2}}$, which diverges ($p = \tfrac{1}{2} \le 1$), but for $x = \tfrac{5}{2}$, we get $\displaystyle\sum_{n=0}^{\infty}\frac{(-1)^n}{\sqrt{n+3}}$, which is a convergent

alternating series, so $I = \left[\tfrac{5}{2}, \tfrac{7}{2}\right)$.

35.

n	$f^{(n)}(x)$	$f^{(n)}\left(\frac{\pi}{6}\right)$
0	$\sin x$	$\frac{1}{2}$
1	$\cos x$	$\frac{\sqrt{3}}{2}$
2	$-\sin x$	$-\frac{1}{2}$
3	$-\cos x$	$-\frac{\sqrt{3}}{2}$
4	$\sin x$	$\frac{1}{2}$
⋮	⋮	⋮

$$\sin x = f\left(\frac{\pi}{6}\right) + f'\left(\frac{\pi}{6}\right)\left(x-\frac{\pi}{6}\right) + \frac{f''\left(\frac{\pi}{6}\right)}{2!}\left(x-\frac{\pi}{6}\right)^2 + \frac{f^{(3)}\left(\frac{\pi}{6}\right)}{3!}\left(x-\frac{\pi}{6}\right)^3 + \frac{f^{(4)}\left(\frac{\pi}{6}\right)}{4!}\left(x-\frac{\pi}{6}\right)^4 + \cdots$$

$$= \frac{1}{2}\left[1 - \frac{1}{2!}\left(x-\frac{\pi}{6}\right)^2 + \frac{1}{4!}\left(x-\frac{\pi}{6}\right)^4 - \cdots\right] + \frac{\sqrt{3}}{2}\left[\left(x-\frac{\pi}{6}\right) - \frac{1}{3!}\left(x-\frac{\pi}{6}\right)^3 + \cdots\right]$$

$$= \frac{1}{2}\sum_{n=0}^{\infty}(-1)^n\frac{1}{(2n)!}\left(x-\frac{\pi}{6}\right)^{2n} + \frac{\sqrt{3}}{2}\sum_{n=0}^{\infty}(-1)^n\frac{1}{(2n+1)!}\left(x-\frac{\pi}{6}\right)^{2n+1}$$

37. $\displaystyle\frac{1}{1+x} = \frac{1}{1-(-x)} = \sum_{n=0}^{\infty}(-x)^n = \sum_{n=0}^{\infty}(-1)^n\,x^n$ for $|x| < 1 \quad\Rightarrow\quad \frac{x^2}{1+x} = \sum_{n=0}^{\infty}(-1)^n\,x^{n+2}$ with $R = 1$.

39. $\displaystyle\frac{1}{1-x} = \sum_{n=0}^{\infty}x^n$ for $|x| < 1 \quad\Rightarrow\quad \ln(1-x) = -\int\frac{dx}{1-x} = -\int\sum_{n=0}^{\infty}x^n\,dx = C - \sum_{n=0}^{\infty}\frac{x^{n+1}}{n+1}.$

$\ln(1-0) = C - 0 \quad\Rightarrow\quad C = 0 \quad\Rightarrow\quad \ln(1-x) = -\sum_{n=0}^{\infty}\frac{x^{n+1}}{n+1} = \sum_{n=1}^{\infty}\frac{-x^n}{n}$ with $R = 1$.

41. $\sin x = \sum\limits_{n=0}^{\infty} \dfrac{(-1)^n x^{2n+1}}{(2n+1)!}$ $\Rightarrow$ $\sin(x^4) = \sum\limits_{n=0}^{\infty} \dfrac{(-1)^n (x^4)^{2n+1}}{(2n+1)!} = \sum\limits_{n=0}^{\infty} \dfrac{(-1)^n x^{8n+4}}{(2n+1)!}$ for all x, so the radius of

convergence is ∞.

43. $f(x) = \dfrac{1}{\sqrt[4]{16-x}} = \dfrac{1}{\sqrt[4]{16(1-x/16)}} = \dfrac{1}{\sqrt[4]{16}\left(1-\frac{1}{16}x\right)^{1/4}} = \frac{1}{2}\left(1-\frac{1}{16}x\right)^{-1/4}$

$= \dfrac{1}{2}\left[1 + \left(-\frac{1}{4}\right)\left(-\frac{x}{16}\right) + \dfrac{\left(-\frac{1}{4}\right)\left(-\frac{5}{4}\right)}{2!}\left(-\frac{x}{16}\right)^2 + \dfrac{\left(-\frac{1}{4}\right)\left(-\frac{5}{4}\right)\left(-\frac{9}{4}\right)}{3!}\left(-\frac{x}{16}\right)^3 + \cdots\right]$

$= \dfrac{1}{2} + \sum\limits_{n=1}^{\infty} \dfrac{1\cdot5\cdot9\cdot\ \cdots\ \cdot(4n-3)}{2\cdot4^n\cdot n!\cdot16^n} x^n = \dfrac{1}{2} + \sum\limits_{n=1}^{\infty} \dfrac{1\cdot5\cdot9\cdot\ \cdots\ \cdot(4n-3)}{2^{6n+1}\,n!} x^n$

for $\left|-\dfrac{x}{16}\right| < 1$ $\Leftrightarrow$ $|x| < 16$, so $R = 16$.

45. $e^x = \sum\limits_{n=0}^{\infty} \dfrac{x^n}{n!}$, so $\dfrac{e^x}{x} = \dfrac{1}{x}\sum\limits_{n=0}^{\infty}\dfrac{x^n}{n!} = \sum\limits_{n=0}^{\infty}\dfrac{x^{n-1}}{n!} = x^{-1} + \sum\limits_{n=1}^{\infty}\dfrac{x^{n-1}}{n!} = \dfrac{1}{x} + \sum\limits_{n=1}^{\infty}\dfrac{x^{n-1}}{n!}$ and

$\displaystyle\int \dfrac{e^x}{x}\,dx = C + \ln|x| + \sum\limits_{n=1}^{\infty}\dfrac{x^n}{n\cdot n!}$.

47. (a)

n	$f^{(n)}(x)$	$f^{(n)}(1)$
0	$x^{1/2}$	1
1	$\frac{1}{2}x^{-1/2}$	$\frac{1}{2}$
2	$-\frac{1}{4}x^{-3/2}$	$-\frac{1}{4}$
3	$\frac{3}{8}x^{-5/2}$	$\frac{3}{8}$
4	$-\frac{15}{16}x^{-7/2}$	$-\frac{15}{16}$
⋮	⋮	⋮

$\sqrt{x} \approx T_3(x) = 1 + \dfrac{1/2}{1!}(x-1) - \dfrac{1/4}{2!}(x-1)^2 + \dfrac{3/8}{3!}(x-1)^3$

$= 1 + \frac{1}{2}(x-1) - \frac{1}{8}(x-1)^2 + \frac{1}{16}(x-1)^3$

(b)

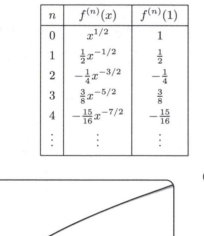

(c) $|R_3(x)| \le \dfrac{M}{4!}|x-1|^4$, where $\left|f^{(4)}(x)\right| \le M$ with

$f^{(4)}(x) = -\frac{15}{16}x^{-7/2}$. Now $0.9 \le x \le 1.1$ $\Rightarrow$

$-0.1 \le x - 1 \le 0.1$ $\Rightarrow$ $(x-1)^4 \le (0.1)^4$,

and letting $x = 0.9$ gives $M = \dfrac{15}{16(0.9)^{7/2}}$, so

$|R_3(x)| \le \dfrac{15}{16(0.9)^{7/2}\,4!}(0.1)^4 \approx 0.000\,005\,648$

$\approx 0.000\,006 = 6 \times 10^{-6}$

(d)

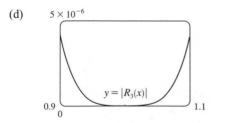

From the graph of $|R_3(x)| = |\sqrt{x} - T_3(x)|$, it appears

that the error is less than 5×10^{-6} on $[0.9, 1.1]$.

49. $\sin x = \sum_{n=0}^{\infty} (-1)^n \dfrac{x^{2n+1}}{(2n+1)!} = x - \dfrac{x^3}{3!} + \dfrac{x^5}{5!} - \dfrac{x^7}{7!} + \cdots$, so $\sin x - x = -\dfrac{x^3}{3!} + \dfrac{x^5}{5!} - \dfrac{x^7}{7!} + \cdots$ and

$\dfrac{\sin x - x}{x^3} = -\dfrac{1}{3!} + \dfrac{x^2}{5!} - \dfrac{x^4}{7!} + \cdots$. Thus, $\lim\limits_{x \to 0} \dfrac{\sin x - x}{x^3} = \lim\limits_{x \to 0} \left(-\dfrac{1}{6} + \dfrac{x^2}{120} - \dfrac{x^4}{5040} + \cdots \right) = -\dfrac{1}{6}$.

51. (a) From Formula 14a in Appendix C, with $x = y = \theta$, we get $\tan 2\theta = \dfrac{2\tan\theta}{1 - \tan^2\theta}$, so $\cot 2\theta = \dfrac{1 - \tan^2\theta}{2\tan\theta}$ $\Rightarrow$

$2\cot 2\theta = \dfrac{1 - \tan^2\theta}{\tan\theta} = \cot\theta - \tan\theta$. Replacing θ by $\tfrac{1}{2}x$, we get $2\cot x = \cot\tfrac{1}{2}x - \tan\tfrac{1}{2}x$, or

$\tan\tfrac{1}{2}x = \cot\tfrac{1}{2}x - 2\cot x$.

(b) From part (a) with $\dfrac{x}{2^{n-1}}$ in place of x, $\tan\dfrac{x}{2^n} = \cot\dfrac{x}{2^n} - 2\cot\dfrac{x}{2^{n-1}}$, so the nth partial sum of $\sum\limits_{n=1}^{\infty} \dfrac{1}{2^n}\tan\dfrac{x}{2^n}$ is

$$s_n = \dfrac{\tan(x/2)}{2} + \dfrac{\tan(x/4)}{4} + \dfrac{\tan(x/8)}{8} + \cdots + \dfrac{\tan(x/2^n)}{2^n}$$

$$= \left[\dfrac{\cot(x/2)}{2} - \cot x\right] + \left[\dfrac{\cot(x/4)}{4} - \dfrac{\cot(x/2)}{2}\right] + \left[\dfrac{\cot(x/8)}{8} - \dfrac{\cot(x/4)}{4}\right] + \cdots$$

$$+ \left[\dfrac{\cot(x/2^n)}{2^n} - \dfrac{\cot(x/2^{n-1})}{2^{n-1}}\right] = -\cot x + \dfrac{\cot(x/2^n)}{2^n} \quad \text{[telescoping sum]}$$

Now $\dfrac{\cot(x/2^n)}{2^n} = \dfrac{\cos(x/2^n)}{2^n \sin(x/2^n)} = \dfrac{\cos(x/2^n)}{x} \cdot \dfrac{x/2^n}{\sin(x/2^n)} \to \dfrac{1}{x} \cdot 1 = \dfrac{1}{x}$ as $n \to \infty$ since $x/2^n \to 0$

for $x \neq 0$. Therefore, if $x \neq 0$ and $x \neq k\pi$ where k is any integer, then

$$\sum_{n=1}^{\infty} \dfrac{1}{2^n}\tan\dfrac{x}{2^n} = \lim_{n\to\infty} s_n = \lim_{n\to\infty}\left(-\cot x + \dfrac{1}{2^n}\cot\dfrac{x}{2^n}\right) = -\cot x + \dfrac{1}{x}$$

If $x = 0$, then all terms in the series are 0, so the sum is 0.

FOCUS ON PROBLEM SOLVING

1. It would be far too much work to compute 15 derivatives of f. The key idea is to remember that $f^{(n)}(0)$ occurs in the coefficient of x^n in the Maclaurin series of f. We start with the Maclaurin series for sin: $\sin x = x - \dfrac{x^3}{3!} + \dfrac{x^5}{5!} - \cdots$

Then $\sin(x^3) = x^3 - \dfrac{x^9}{3!} + \dfrac{x^{15}}{5!} - \cdots$, and so the coefficient of x^{15} is $\dfrac{f^{(15)}(0)}{15!} = \dfrac{1}{5!}$. Therefore,

$f^{(15)}(0) = \dfrac{15!}{5!} = 6 \cdot 7 \cdot 8 \cdot 9 \cdot 10 \cdot 11 \cdot 12 \cdot 13 \cdot 14 \cdot 15 = 10{,}897{,}286{,}400.$

3. (a) At each stage, each side is replaced by four shorter sides, each of length $\frac{1}{3}$ of the side length at the preceding stage. Writing s_0 and ℓ_0 for the number of sides and the length of the side of the initial triangle, we generate the table at right. In general, we have

$s_0 = 3$	$\ell_0 = 1$
$s_1 = 3 \cdot 4$	$\ell_1 = 1/3$
$s_2 = 3 \cdot 4^2$	$\ell_2 = 1/3^2$
$s_3 = 3 \cdot 4^3$	$\ell_3 = 1/3^3$
$\vdots$	$\vdots$

$s_n = 3 \cdot 4^n$ and $\ell_n = \left(\frac{1}{3}\right)^n$, so the length of the perimeter at the nth stage of construction is $p_n = s_n \ell_n = 3 \cdot 4^n \cdot \left(\frac{1}{3}\right)^n = 3 \cdot \left(\frac{4}{3}\right)^n$.

(b) $p_n = \dfrac{4^n}{3^{n-1}} = 4\left(\dfrac{4}{3}\right)^{n-1}$. Since $\frac{4}{3} > 1$, $p_n \to \infty$ as $n \to \infty$.

(c) The area of each of the small triangles added at a given stage is one-ninth of the area of the triangle added at the preceding stage. Let a be the area of the original triangle. Then the area a_n of each of the small triangles added at stage n is

$a_n = a \cdot \dfrac{1}{9^n} = \dfrac{a}{9^n}$. Since a small triangle is added to each side at every stage, it follows that the total area A_n added to

the figure at the nth stage is $A_n = s_{n-1} \cdot a_n = 3 \cdot 4^{n-1} \cdot \dfrac{a}{9^n} = a \cdot \dfrac{4^{n-1}}{3^{2n-1}}$. Then the total area enclosed by the snowflake

curve is $A = a + A_1 + A_2 + A_3 + \cdots = a + a \cdot \dfrac{1}{3} + a \cdot \dfrac{4}{3^3} + a \cdot \dfrac{4^2}{3^5} + a \cdot \dfrac{4^3}{3^7} + \cdots$. After the first term, this is a

geometric series with common ratio $\dfrac{4}{9}$, so $A = a + \dfrac{a/3}{1 - \frac{4}{9}} = a + \dfrac{a}{3} \cdot \dfrac{9}{5} = \dfrac{8a}{5}$. But the area of the original equilateral

triangle with side 1 is $a = \dfrac{1}{2} \cdot 1 \cdot \sin \dfrac{\pi}{3} = \dfrac{\sqrt{3}}{4}$. So the area enclosed by the snowflake curve is $\dfrac{8}{5} \cdot \dfrac{\sqrt{3}}{4} = \dfrac{2\sqrt{3}}{5}$.

5. $\ln\left(1 - \dfrac{1}{n^2}\right) = \ln\left(\dfrac{n^2 - 1}{n^2}\right) = \ln \dfrac{(n+1)(n-1)}{n^2} = \ln[(n+1)(n-1)] - \ln n^2$

$\qquad = \ln(n+1) + \ln(n-1) - 2\ln n = \ln(n-1) - \ln n - \ln n + \ln(n+1)$

$\qquad = \ln \dfrac{n-1}{n} - [\ln n - \ln(n+1)] = \ln \dfrac{n-1}{n} - \ln \dfrac{n}{n+1}.$

Let $s_k = \displaystyle\sum_{n=2}^{k} \ln\left(1 - \dfrac{1}{n^2}\right) = \sum_{n=2}^{k}\left(\ln \dfrac{n-1}{n} - \ln \dfrac{n}{n+1}\right)$ for $k \geq 2$. Then

$s_k = \left(\ln \dfrac{1}{2} - \ln \dfrac{2}{3}\right) + \left(\ln \dfrac{2}{3} - \ln \dfrac{3}{4}\right) + \cdots + \left(\ln \dfrac{k-1}{k} - \ln \dfrac{k}{k+1}\right) = \ln \dfrac{1}{2} - \ln \dfrac{k}{k+1}$, so

$\displaystyle\sum_{n=2}^{\infty} \ln\left(1 - \dfrac{1}{n^2}\right) = \lim_{k \to \infty} s_k = \lim_{k \to \infty}\left(\ln \dfrac{1}{2} - \ln \dfrac{k}{k+1}\right) = \ln \dfrac{1}{2} - \ln 1 = \ln 1 - \ln 2 - \ln 1 = -\ln 2.$

7. $u = 1 + \dfrac{x^3}{3!} + \dfrac{x^6}{6!} + \dfrac{x^9}{9!} + \cdots,\ v = x + \dfrac{x^4}{4!} + \dfrac{x^7}{7!} + \dfrac{x^{10}}{10!} + \cdots,\ w = \dfrac{x^2}{2!} + \dfrac{x^5}{5!} + \dfrac{x^8}{8!} + \cdots.$

Use the Ratio Test to show that the series for u, v, and w have positive radii of convergence (∞ in each case), so Theorem 8.6.2 applies, and hence, we may differentiate each of these series:

$$\frac{du}{dx} = \frac{3x^2}{3!} + \frac{6x^5}{6!} + \frac{9x^8}{9!} + \cdots = \frac{x^2}{2!} + \frac{x^5}{5!} + \frac{x^8}{8!} + \cdots = w$$

Similarly, $\dfrac{dv}{dx} = 1 + \dfrac{x^3}{3!} + \dfrac{x^6}{6!} + \dfrac{x^9}{9!} + \cdots = u$, and $\dfrac{dw}{dx} = x + \dfrac{x^4}{4!} + \dfrac{x^7}{7!} + \dfrac{x^{10}}{10!} + \cdots = v.$

So $u' = w$, $v' = u$, and $w' = v$. Now differentiate the left hand side of the desired equation:

$$\frac{d}{dx}\left(u^3 + v^3 + w^3 - 3uvw\right) = 3u^2 u' + 3v^2 v' + 3w^2 w' - 3(u'vw + uv'w + uvw')$$

$$= 3u^2 w + 3v^2 u + 3w^2 v - 3(vw^2 + u^2 w + uv^2) = 0 \quad \Rightarrow$$

$u^3 + v^3 + w^3 - 3uvw = C$. To find the value of the constant C, we put $x = 0$ in the last equation and get

$1^3 + 0^3 + 0^3 - 3(1 \cdot 0 \cdot 0) = C \quad \Rightarrow \quad C = 1$, so $u^3 + v^3 + w^3 - 3uvw = 1$.

9. If L is the length of a side of the equilateral triangle, then the area is $A = \frac{1}{2}L \cdot \frac{\sqrt{3}}{2}L = \frac{\sqrt{3}}{4}L^2$ and so $L^2 = \frac{4}{\sqrt{3}}A$.

Let r be the radius of one of the circles. When there are n rows of circles, the figure shows that

$$L = \sqrt{3}r + r + (n-2)(2r) + r + \sqrt{3}r = r\left(2n - 2 + 2\sqrt{3}\right),\text{ so } r = \frac{L}{2\left(n + \sqrt{3} - 1\right)}.$$

The number of circles is $1 + 2 + \cdots + n = \dfrac{n(n+1)}{2}$, and so the total area of the circles is

$$A_n = \frac{n(n+1)}{2}\pi r^2 = \frac{n(n+1)}{2}\pi \frac{L^2}{4\left(n + \sqrt{3} - 1\right)^2} = \frac{n(n+1)}{2}\pi \frac{4A/\sqrt{3}}{4\left(n + \sqrt{3} - 1\right)^2}$$

$$= \frac{n(n+1)}{\left(n + \sqrt{3} - 1\right)^2}\frac{\pi A}{2\sqrt{3}} \quad \Rightarrow$$

$$\frac{A_n}{A} = \frac{n(n+1)}{\left(n + \sqrt{3} - 1\right)^2}\frac{\pi}{2\sqrt{3}}$$

$$= \frac{1 + 1/n}{\left[1 + (\sqrt{3} - 1)/n\right]^2}\frac{\pi}{2\sqrt{3}} \to \frac{\pi}{2\sqrt{3}} \text{ as } n \to \infty$$

11. As in Section 8.6 we have to integrate the function x^x by integrating series. Writing $x^x = (e^{\ln x})^x = e^{x \ln x}$ and using the Maclaurin series for e^x, we have $x^x = (e^{\ln x})^x = e^{x \ln x} = \displaystyle\sum_{n=0}^{\infty} \frac{(x \ln x)^n}{n!} = \sum_{n=0}^{\infty} \frac{x^n (\ln x)^n}{n!}$. As with power series, we can

integrate this series term-by-term: $\displaystyle\int_0^1 x^x\, dx = \sum_{n=0}^{\infty}\int_0^1 \frac{x^n (\ln x)^n}{n!}\, dx = \sum_{n=0}^{\infty}\frac{1}{n!}\int_0^1 x^n(\ln x)^n\, dx$. We integrate by parts with $u = (\ln x)^n$, $dv = x^n\,dx$, so $du = \dfrac{n(\ln x)^{n-1}}{x}\,dx$ and $v = \dfrac{x^{n+1}}{n+1}$:

$$\int_0^1 x^n(\ln x)^n\,dx = \lim_{t\to 0^+}\int_t^1 x^n(\ln x)^n\,dx = \lim_{t\to 0^+}\left[\frac{x^{n+1}}{n+1}(\ln x)^n\right]_t^1 - \lim_{t\to 0^+}\int_t^1 \frac{n}{n+1}x^n(\ln x)^{n-1}\,dx$$

$$= 0 - \frac{n}{n+1}\int_0^1 x^n(\ln x)^{n-1}\,dx$$

(where l'Hospital's Rule was used to help evaluate the first limit). Further integration by parts gives

$$\int_0^1 x^n(\ln x)^k\,dx = -\frac{k}{n+1}\int_0^1 x^n(\ln x)^{k-1}\,dx \text{ and, combining these steps, we get}$$

$$\int_0^1 x^n(\ln x)^n\,dx = \frac{(-1)^n\,n!}{(n+1)^n}\int_0^1 x^n\,dx = \frac{(-1)^n\,n!}{(n+1)^{n+1}} \quad \Rightarrow$$

$$\int_0^1 x^x\,dx = \sum_{n=0}^\infty \frac{1}{n!}\int_0^1 x^n(\ln x)^n\,dx = \sum_{n=0}^\infty \frac{1}{n!}\frac{(-1)^n\,n!}{(n+1)^{n+1}} = \sum_{n=0}^\infty \frac{(-1)^n}{(n+1)^{n+1}} = \sum_{n=1}^\infty \frac{(-1)^{n-1}}{n^n}.$$

13. Let $f(x) = \sum_{m=0}^\infty c_m x^m$ and $g(x) = e^{f(x)} = \sum_{n=0}^\infty d_n x^n$. Then $g'(x) = \sum_{n=0}^\infty n d_n x^{n-1}$, so $n d_n$ occurs as the coefficient of

x^{n-1}. But also

$$g'(x) = e^{f(x)}f'(x) = \left(\sum_{n=0}^\infty d_n x^n\right)\left(\sum_{m=1}^\infty m c_m x^{m-1}\right)$$

$$= (d_0 + d_1 x + d_2 x^2 + \cdots + d_{n-1}x^{n-1} + \cdots)(c_1 + 2c_2 x + 3c_3 x^2 + \cdots + n c_n x^{n-1} + \cdots)$$

so the coefficient of x^{n-1} is $c_1 d_{n-1} + 2c_2 d_{n-2} + 3c_3 d_{n-3} + \cdots + n c_n d_0 = \sum_{i=1}^n i c_i d_{n-i}$. Therefore, $n d_n = \sum_{i=1}^n i c_i d_{n-i}$.

15. Call the series S. We group the terms according to the number of digits in their denominators:

$$S = \underbrace{\left(\tfrac{1}{1} + \tfrac{1}{2} + \cdots + \tfrac{1}{8} + \tfrac{1}{9}\right)}_{g_1} + \underbrace{\left(\tfrac{1}{11} + \cdots + \tfrac{1}{99}\right)}_{g_2} + \underbrace{\left(\tfrac{1}{111} + \cdots + \tfrac{1}{999}\right)}_{g_3} + \cdots$$

Now in the group g_n, since we have 9 choices for each of the n digits in the denominator, there are 9^n terms. Furthermore,

each term in g_n is less than $\frac{1}{10^{n-1}}$ [except for the first term in g_1]. So $g_n < 9^n \cdot \frac{1}{10^{n-1}} = 9\left(\frac{9}{10}\right)^{n-1}$. Now

$\sum_{n=1}^\infty 9\left(\frac{9}{10}\right)^{n-1}$ is a geometric series with $a = 9$ and $r = \frac{9}{10} < 1$. Therefore, by the Comparison Test,

$S = \sum_{n=1}^\infty g_n < \sum_{n=1}^\infty 9\left(\frac{9}{10}\right)^{n-1} = \frac{9}{1 - 9/10} = 90.$

9 □   VECTORS AND THE GEOMETRY OF SPACE

9.1   Three-Dimensional Coordinate Systems

1. We start at the origin, which has coordinates $(0, 0, 0)$. First we move 4 units along the positive x-axis, affecting only the x-coordinate, bringing us to the point $(4, 0, 0)$. We then move 3 units straight downward, in the negative z-direction. Thus only the z-coordinate is affected, and we arrive at $(4, 0, -3)$.

3. The distance from a point to the xz-plane is the absolute value of the y-coordinate of the point. $Q(-5, -1, 4)$ has the y-coordinate with the smallest absolute value, so Q is the point closest to the xz-plane. $R(0, 3, 8)$ must lie in the yz-plane since the distance from R to the yz-plane, given by the x-coordinate of R, is 0.

5. The equation $x + y = 2$ represents the set of all points in $\mathbb{R}^3$ whose x- and y-coordinates have a sum of 2, or equivalently where $y = 2 - x$. This is the set $\{(x, 2 - x, z) \mid x \in \mathbb{R}, z \in \mathbb{R}\}$ which is a vertical plane that intersects the xy-plane in the line $y = 2 - x$, $z = 0$.

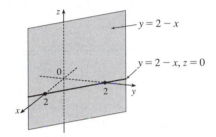

7. (a) We can find the lengths of the sides of the triangle by using the distance formula between pairs of vertices:

$$|PQ| = \sqrt{(7 - 3)^2 + [0 - (-2)]^2 + [1 - (-3)]^2} = \sqrt{16 + 4 + 16} = 6$$

$$|QR| = \sqrt{(1 - 7)^2 + (2 - 0)^2 + (1 - 1)^2} = \sqrt{36 + 4 + 0} = \sqrt{40} = 2\sqrt{10}$$

$$|RP| = \sqrt{(3 - 1)^2 + (-2 - 2)^2 + (-3 - 1)^2} = \sqrt{4 + 16 + 16} = 6$$

The longest side is QR, but the Pythagorean Theorem is not satisfied: $|PQ|^2 + |RP|^2 \neq |QR|^2$. Thus PQR is not a right triangle. PQR is isosceles, as two sides have the same length.

(b) Compute the lengths of the sides of the triangle by using the distance formula between pairs of vertices:

$$|PQ| = \sqrt{(4 - 2)^2 + [1 - (-1)]^2 + (1 - 0)^2} = \sqrt{4 + 4 + 1} = 3$$

$$|QR| = \sqrt{(4 - 4)^2 + (-5 - 1)^2 + (4 - 1)^2} = \sqrt{0 + 36 + 9} = \sqrt{45} = 3\sqrt{5}$$

$$|RP| = \sqrt{(2 - 4)^2 + [-1 - (-5)]^2 + (0 - 4)^2} = \sqrt{4 + 16 + 16} = 6$$

Since the Pythagorean Theorem is satisfied by $|PQ|^2 + |RP|^2 = |QR|^2$, PQR is a right triangle. PQR is not isosceles, as no two sides have the same length.

9. (a) First we find the distances between points:

$$|AB| = \sqrt{(3-2)^2 + (7-4)^2 + (-2-2)^2} = \sqrt{26}$$

$$|BC| = \sqrt{(1-3)^2 + (3-7)^2 + [3-(-2)]^2} = \sqrt{45} = 3\sqrt{5}$$

$$|AC| = \sqrt{(1-2)^2 + (3-4)^2 + (3-2)^2} = \sqrt{3}$$

In order for the points to lie on a straight line, the sum of the two shortest distances must be equal to the longest distance. Since $\sqrt{26} + \sqrt{3} \neq 3\sqrt{5}$, the three points do not lie on a straight line.

(b) First we find the distances between points:

$$|DE| = \sqrt{(1-0)^2 + [-2-(-5)]^2 + (4-5)^2} = \sqrt{11}$$

$$|EF| = \sqrt{(3-1)^2 + [4-(-2)]^2 + (2-4)^2} = \sqrt{44} = 2\sqrt{11}$$

$$|DF| = \sqrt{(3-0)^2 + [4-(-5)]^2 + (2-5)^2} = \sqrt{99} = 3\sqrt{11}$$

Since $|DE| + |EF| = |DF|$, the three points lie on a straight line.

11. The radius of the sphere is the distance between $(4, 3, -1)$ and $(3, 8, 1)$:

$$r = \sqrt{(3-4)^2 + (8-3)^2 + [1-(-1)]^2} = \sqrt{30}. \text{ Thus, an equation of the sphere is } (x-3)^2 + (y-8)^2 + (z-1)^2 = 30.$$

13. Completing squares in the equation $x^2 + y^2 + z^2 - 6x + 4y - 2z = 11$ gives

$$(x^2 - 6x + 9) + (y^2 + 4y + 4) + (z^2 - 2z + 1) = 11 + 9 + 4 + 1 \quad \Rightarrow$$

$(x-3)^2 + (y+2)^2 + (z-1)^2 = 25$ which we recognize as an equation of a sphere with center $(3, -2, 1)$ and radius 5.

15. (a) If the midpoint of the line segment from $P_1(x_1, y_1, z_1)$ to $P_2(x_2, y_2, z_2)$ is $Q = \left(\dfrac{x_1 + x_2}{2}, \dfrac{y_1 + y_2}{2}, \dfrac{z_1 + z_2}{2} \right)$, then the distances $|P_1Q|$ and $|QP_2|$ are equal, and each is half of $|P_1P_2|$. We verify that this is the case:

$$|P_1P_2| = \sqrt{(x_2 - x_1)^2 + (y_2 - y_1)^2 + (z_2 - z_1)^2}$$

$$|P_1Q| = \sqrt{\left[\tfrac{1}{2}(x_1 + x_2) - x_1 \right]^2 + \left[\tfrac{1}{2}(y_1 + y_2) - y_1 \right]^2 + \left[\tfrac{1}{2}(z_1 + z_2) - z_1 \right]^2}$$

$$= \sqrt{\left(\tfrac{1}{2}x_2 - \tfrac{1}{2}x_1 \right)^2 + \left(\tfrac{1}{2}y_2 - \tfrac{1}{2}y_1 \right)^2 + \left(\tfrac{1}{2}z_2 - \tfrac{1}{2}z_1 \right)^2}$$

$$= \sqrt{\left(\tfrac{1}{2} \right)^2 \left[(x_2 - x_1)^2 + (y_2 - y_1)^2 + (z_2 - z_1)^2 \right]}$$

$$= \tfrac{1}{2} \sqrt{(x_2 - x_1)^2 + (y_2 - y_1)^2 + (z_2 - z_1)^2}$$

$$= \tfrac{1}{2} |P_1P_2|$$

$$|QP_2| = \sqrt{\left[x_2 - \tfrac{1}{2}(x_1 + x_2)\right]^2 + \left[y_2 - \tfrac{1}{2}(y_1 + y_2)\right]^2 + \left[z_2 - \tfrac{1}{2}(z_1 + z_2)\right]^2}$$

$$= \sqrt{\left(\tfrac{1}{2}x_2 - \tfrac{1}{2}x_1\right)^2 + \left(\tfrac{1}{2}y_2 - \tfrac{1}{2}y_1\right)^2 + \left(\tfrac{1}{2}z_2 - \tfrac{1}{2}z_1\right)^2}$$

$$= \sqrt{\left(\tfrac{1}{2}\right)^2 \left[(x_2 - x_1)^2 + (y_2 - y_1)^2 + (z_2 - z_1)^2\right]}$$

$$= \tfrac{1}{2}\sqrt{(x_2 - x_1)^2 + (y_2 - y_1)^2 + (z_2 - z_1)^2}$$

$$= \tfrac{1}{2}|P_1 P_2|$$

So Q is indeed the midpoint of $P_1 P_2$.

(b) By part (a), the midpoints of sides AB, BC and CA are $P_1\left(-\tfrac{1}{2}, 1, 4\right)$, $P_2\left(1, \tfrac{1}{2}, 5\right)$ and $P_3\left(\tfrac{5}{2}, \tfrac{3}{2}, 4\right)$. (Recall that a median of a triangle is a line segment from a vertex to the midpoint of the opposite side.) Then the lengths of the medians are:

$$|AP_2| = \sqrt{0^2 + \left(\tfrac{1}{2} - 2\right)^2 + (5 - 3)^2} = \sqrt{\tfrac{9}{4} + 4} = \sqrt{\tfrac{25}{4}} = \tfrac{5}{2}$$

$$|BP_3| = \sqrt{\left(\tfrac{5}{2} + 2\right)^2 + \left(\tfrac{3}{2}\right)^2 + (4 - 5)^2} = \sqrt{\tfrac{81}{4} + \tfrac{9}{4} + 1} = \sqrt{\tfrac{94}{4}} = \tfrac{1}{2}\sqrt{94}$$

$$|CP_1| = \sqrt{\left(-\tfrac{1}{2} - 4\right)^2 + (1 - 1)^2 + (4 - 5)^2} = \sqrt{\tfrac{81}{4} + 1} = \tfrac{1}{2}\sqrt{85}$$

17. (a) Since the sphere touches the xy-plane, its radius is the distance from its center, $(2, -3, 6)$, to the xy-plane, namely 6. Therefore $r = 6$ and an equation of the sphere is $(x - 2)^2 + (y + 3)^2 + (z - 6)^2 = 6^2 = 36$.

(b) The radius of this sphere is the distance from its center $(2, -3, 6)$ to the yz-plane, which is 2. Therefore, an equation is $(x - 2)^2 + (y + 3)^2 + (z - 6)^2 = 4$.

(c) Here the radius is the distance from the center $(2, -3, 6)$ to the xz-plane, which is 3. Therefore, an equation is $(x - 2)^2 + (y + 3)^2 + (z - 6)^2 = 9$.

19. The equation $y = -4$ represents a plane parallel to the xz-plane and 4 units to the left of it.

21. The inequality $x > 3$ represents a half-space consisting of all points in front of the plane $x = 3$.

23. The inequality $0 \le z \le 6$ represents all points on or between the horizontal planes $z = 0$ (the xy-plane) and $z = 6$.

25. The inequality $x^2 + y^2 + z^2 \le 3$ is equivalent to $\sqrt{x^2 + y^2 + z^2} \le \sqrt{3}$, so the region consists of those points whose distance from the origin is at most $\sqrt{3}$. This is the set of all points on or inside the sphere with radius $\sqrt{3}$ and center $(0, 0, 0)$.

27. Here $x^2 + z^2 \le 9$ or equivalently $\sqrt{x^2 + z^2} \le 3$ which describes the set of all points in $\mathbb{R}^3$ whose distance from the y-axis is at most 3. Thus, the inequality represents the region consisting of all points on or inside a circular cylinder of radius 3 with axis the y-axis.

29. This describes all points with negative y-coordinates, that is, $y < 0$.

31. This describes a region all of whose points have a distance to the origin which is greater than r, but smaller than R. So inequalities describing the region are $r < \sqrt{x^2 + y^2 + z^2} < R$, or $r^2 < x^2 + y^2 + z^2 < R^2$.

33. (a) To find the x- and y-coordinates of the point P, we project it onto

L_2 and project the resulting point Q onto the x- and y-axes. To

find the z-coordinate, we project P onto either the xz-plane or the

yz-plane (using our knowledge of its x- or

y-coordinate) and then project the resulting point onto the z-axis.

(Or, we could draw a line parallel to QO from P to the z-axis.)

The coordinates of P are $(2, 1, 4)$.

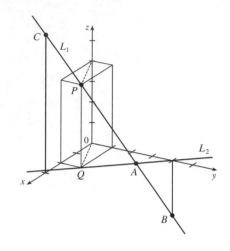

(b) A is the intersection of L_1 and L_2, B is directly below

the y-intercept of L_2, and C is directly above the x-intercept

of L_2.

35. We need to find a set of points $\{P(x, y, z) \mid |AP| = |BP|\}$.

$$\sqrt{(x+1)^2 + (y-5)^2 + (z-3)^2} = \sqrt{(x-6)^2 + (y-2)^2 + (z+2)^2} \Rightarrow$$

$$(x+1)^2 + (y-5) + (z-3)^2 = (x-6)^2 + (y-2)^2 + (z+2)^2 \Rightarrow$$

$$x^2 + 2x + 1 + y^2 - 10y + 25 + z^2 - 6z + 9 = x^2 - 12x + 36 + y^2 - 4y + 4 + z^2 + 4z + 4 \Rightarrow \quad 14x - 6y - 10z = 9.$$

Thus the set of points is a plane perpendicular to the line segment joining A and B (since this plane must contain the

perpendicular bisector of the line segment AB).

9.2 Vectors

1. (a) The cost of a theater ticket is a scalar, because it has only magnitude.

(b) The current in a river is a vector, because it has both magnitude (the speed of the current) and direction at any given

location.

(c) If we assume that the initial path is linear, the initial flight path from Houston to Dallas is a vector, because it has both

magnitude (distance) and direction.

(d) The population of the world is a scalar, because it has only magnitude.

3. Vectors are equal when they share the same length and direction (but not necessarily location). Using the symmetry of the

parallelogram as a guide, we see that $\overrightarrow{AB} = \overrightarrow{DC}$, $\overrightarrow{DA} = \overrightarrow{CB}$, $\overrightarrow{DE} = \overrightarrow{EB}$, and $\overrightarrow{EA} = \overrightarrow{CE}$.

5. (a)   (b)

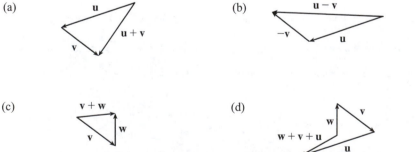

(c)   (d)

7. $\mathbf{a} = \langle -2 - 2, 1 - 3 \rangle = \langle -4, -2 \rangle$

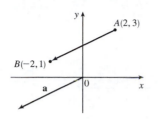

9. $\mathbf{a} = \langle 2 - 0, 3 - 3, -1 - 1 \rangle = \langle 2, 0, -2 \rangle$

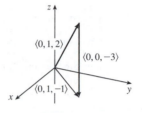

11. $\langle 3, -1 \rangle + \langle -2, 4 \rangle = \langle 3 + (-2), -1 + 4 \rangle$
$$= \langle 1, 3 \rangle$$

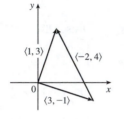

13. $\langle 0, 1, 2 \rangle + \langle 0, 0, -3 \rangle = \langle 0 + 0, 1 + 0, 2 + (-3) \rangle$
$$= \langle 0, 1, -1 \rangle$$

15. $\mathbf{a} + \mathbf{b} = \langle 5 + (-3), -12 + (-6) \rangle = \langle 2, -18 \rangle$

$2\mathbf{a} + 3\mathbf{b} = \langle 10, -24 \rangle + \langle -9, -18 \rangle = \langle 1, -42 \rangle$

$|\mathbf{a}| = \sqrt{5^2 + (-12)^2} = \sqrt{169} = 13$

$|\mathbf{a} - \mathbf{b}| = |\langle 5 - (-3), -12 - (-6) \rangle| = |\langle 8, -6 \rangle| = \sqrt{8^2 + (-6)^2} = \sqrt{100} = 10$

17. $\mathbf{a} + \mathbf{b} = (\mathbf{i} + 2\mathbf{j} - 3\mathbf{k}) + (-2\mathbf{i} - \mathbf{j} + 5\mathbf{k}) = -\mathbf{i} + \mathbf{j} + 2\mathbf{k}$

$2\mathbf{a} + 3\mathbf{b} = 2(\mathbf{i} + 2\mathbf{j} - 3\mathbf{k}) + 3(-2\mathbf{i} - \mathbf{j} + 5\mathbf{k}) = 2\mathbf{i} + 4\mathbf{j} - 6\mathbf{k} - 6\mathbf{i} - 3\mathbf{j} + 15\mathbf{k} = -4\mathbf{i} + \mathbf{j} + 9\mathbf{k}$

$|\mathbf{a}| = \sqrt{1^2 + 2^2 + (-3)^2} = \sqrt{14}$

$|\mathbf{a} - \mathbf{b}| = |(\mathbf{i} + 2\mathbf{j} - 3\mathbf{k}) - (-2\mathbf{i} - \mathbf{j} + 5\mathbf{k})| = |3\mathbf{i} + 3\mathbf{j} - 8\mathbf{k}| = \sqrt{3^2 + 3^2 + (-8)^2} = \sqrt{82}$

19. The vector $8\mathbf{i} - \mathbf{j} + 4\mathbf{k}$ has length $|8\mathbf{i} - \mathbf{j} + 4\mathbf{k}| = \sqrt{8^2 + (-1)^2 + 4^2} = \sqrt{81} = 9$, so by Equation 4 the unit vector with

the same direction is $\frac{1}{9}(8\mathbf{i} - \mathbf{j} + 4\mathbf{k}) = \frac{8}{9}\mathbf{i} - \frac{1}{9}\mathbf{j} + \frac{4}{9}\mathbf{k}$.

21. From the figure, we see that the x-component of $\mathbf{v}$ is

$v_1 = |\mathbf{v}| \cos(\pi/3) = 4 \cdot \frac{1}{2} = 2$ and the y-component is

$v_2 = |\mathbf{v}| \sin(\pi/3) = 4 \cdot \frac{\sqrt{3}}{2} = 2\sqrt{3}$. Thus

$\mathbf{v} = \langle v_1, v_2 \rangle = \langle 2, 2\sqrt{3} \rangle$.

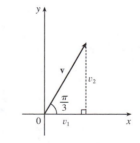

23. $|\mathbf{F}_1| = 10$ lb and $|\mathbf{F}_2| = 12$ lb.

$\mathbf{F}_1 = -|\mathbf{F}_1| \cos 45\,^{\circ}\mathbf{i} + |\mathbf{F}_1| \sin 45\,^{\circ}\mathbf{j} = -10 \cos 45\,^{\circ}\mathbf{i} + 10 \sin 45\,^{\circ}\mathbf{j} = -5\sqrt{2}\,\mathbf{i} + 5\sqrt{2}\,\mathbf{j}$

$\mathbf{F}_2 = |\mathbf{F}_2| \cos 30\,^{\circ}\mathbf{i} + |\mathbf{F}_2| \sin 30\,^{\circ}\mathbf{j} = 12 \cos 30\,^{\circ}\mathbf{i} + 12 \sin 30\,^{\circ}\mathbf{j} = 6\sqrt{3}\,\mathbf{i} + 6\,\mathbf{j}$

$\mathbf{F} = \mathbf{F}_1 + \mathbf{F}_2 = \left(6\sqrt{3} - 5\sqrt{2}\right)\mathbf{i} + \left(6 + 5\sqrt{2}\right)\mathbf{j} \approx 3.32\,\mathbf{i} + 13.07\,\mathbf{j}$

$|\mathbf{F}| \approx \sqrt{(3.32)^2 + (13.07)^2} \approx 13.5$ lb. $\tan\theta = \dfrac{6 + 5\sqrt{2}}{6\sqrt{3} - 5\sqrt{2}} \;\Rightarrow\; \theta = \tan^{-1}\dfrac{6 + 5\sqrt{2}}{6\sqrt{3} - 5\sqrt{2}} \approx 76\,^{\circ}.$

25. With respect to the water's surface, the woman's velocity is the vector sum of the velocity of the ship with respect to the water, and the woman's velocity with respect to the ship. If we let north be the positive y-direction, then $\mathbf{v} = \langle 0, 22 \rangle + \langle -3, 0 \rangle = \langle -3, 22 \rangle$. The woman's speed is $|\mathbf{v}| = \sqrt{9 + 484} \approx 22.2$ mi/h. The vector $\mathbf{v}$ makes an angle θ with the east, where $\theta = \tan^{-1}\left(\frac{22}{-3}\right) \approx 98\,^{\circ}$. Therefore, the woman's direction is about $N(98 - 90)\,^{\circ}W = N8\,^{\circ}W$.

27. Let $\mathbf{T}_1$ and $\mathbf{T}_2$ represent the tension vectors in each side of the clothesline as shown in the figure. $\mathbf{T}_1$ and $\mathbf{T}_2$ have equal vertical components and opposite horizontal components, so we can write

$\mathbf{T}_1 = -a\,\mathbf{i} + b\,\mathbf{j}$ and $\mathbf{T}_2 = a\,\mathbf{i} + b\,\mathbf{j}$ $(a, b > 0)$. By similar triangles, $\dfrac{b}{a} = \dfrac{0.08}{4} \;\Rightarrow\; a = 50b$. The force due to gravity acting on the shirt has magnitude $0.8g \approx (0.8)(9.8) = 7.84$ N, hence we have $\mathbf{w} = -7.84\,\mathbf{j}$. The resultant $\mathbf{T}_1 + \mathbf{T}_2$ of the tensile forces counterbalances $\mathbf{w}$, so $\mathbf{T}_1 + \mathbf{T}_2 = -\mathbf{w} \;\Rightarrow\; (-a\,\mathbf{i} + b\,\mathbf{j}) + (a\,\mathbf{i} + b\,\mathbf{j}) = 7.84\,\mathbf{j} \;\Rightarrow\;$

$(-50b\,\mathbf{i} + b\,\mathbf{j}) + (50b\,\mathbf{i} + b\,\mathbf{j}) = 2b\,\mathbf{j} = 7.84\,\mathbf{j} \;\Rightarrow\; b = \frac{7.84}{2} = 3.92$ and $a = 50b = 196$. Thus the tensions are $\mathbf{T}_1 = -a\,\mathbf{i} + b\,\mathbf{j} = -196\,\mathbf{i} + 3.92\,\mathbf{j}$ and $\mathbf{T}_2 = a\,\mathbf{i} + b\,\mathbf{j} = 196\,\mathbf{i} + 3.92\,\mathbf{j}$.

Alternatively, we can find the value of θ and proceed as in Example 7.

29. (a), (b)

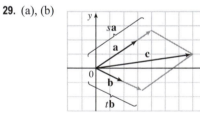

(c) From the sketch, we estimate that $s \approx 1.3$ and $t \approx 1.6$.

(d) $\mathbf{c} = s\,\mathbf{a} + t\,\mathbf{b} \;\Leftrightarrow\; 7 = 3s + 2t$ and $1 = 2s - t$.

Solving these equations gives $s = \frac{9}{7}$ and $t = \frac{11}{7}$.

31.

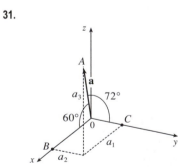

Let $\mathbf{a} = \langle a_1, a_2, a_3 \rangle$, as shown in the figure. Since $|\mathbf{a}| = 1$ and triangle ABO is a right triangle, we have $\cos 60\,^{\circ} = \dfrac{a_1}{1} \;\Rightarrow\; a_1 = \cos 60\,^{\circ}$. Similarly, triangle ACO is a right triangle, so $a_2 = \cos 72\,^{\circ}$. Finally, since $|\mathbf{a}| = 1$ we have

$\sqrt{(\cos 60\,^{\circ})^2 + (\cos 72\,^{\circ})^2 + a_3^2} = 1 \;\Rightarrow\;$

$a_3^2 = 1 - (\cos 60\,^{\circ})^2 - (\cos 72\,^{\circ})^2 \;\Rightarrow\;$

$a_3 = \sqrt{1 - (\cos 60\,^{\circ})^2 - (\cos 72\,^{\circ})^2}.$ Thus

$\mathbf{a} = \left\langle \cos 60\,^{\circ}, \cos 72\,^{\circ}, \sqrt{1 - (\cos 60\,^{\circ})^2 - (\cos 72\,^{\circ})^2} \right\rangle \approx \langle 0.50, 0.31, 0.81 \rangle.$

33. $|\mathbf{r} - \mathbf{r}_0|$ is the distance between the points (x, y, z) and (x_0, y_0, z_0), so the set of points is a sphere with radius 1 and

center (x_0, y_0, z_0).

Alternate method: $|\mathbf{r} - \mathbf{r}_0| = 1$ $\Leftrightarrow$ $\sqrt{(x - x_0)^2 + (y - y_0)^2 + (z - z_0)^2} = 1$ $\Leftrightarrow$

$(x - x_0)^2 + (y - y_0)^2 + (z - z_0)^2 = 1$, which is the equation of a sphere with radius 1 and center (x_0, y_0, z_0).

35. $\mathbf{a} + (\mathbf{b} + \mathbf{c}) = \langle a_1, a_2 \rangle + (\langle b_1, b_2 \rangle + \langle c_1, c_2 \rangle) = \langle a_1, a_2 \rangle + \langle b_1 + c_1, b_2 + c_2 \rangle$

$$= \langle a_1 + b_1 + c_1, a_2 + b_2 + c_2 \rangle = \langle (a_1 + b_1) + c_1, (a_2 + b_2) + c_2 \rangle$$

$$= \langle a_1 + b_1, a_2 + b_2 \rangle + \langle c_1, c_2 \rangle = (\langle a_1, a_2 \rangle + \langle b_1, b_2 \rangle) + \langle c_1, c_2 \rangle$$

$$= (\mathbf{a} + \mathbf{b}) + \mathbf{c}$$

37. Consider triangle ABC, where D and E are the midpoints of AB and BC. We know that $\overrightarrow{AB} + \overrightarrow{BC} = \overrightarrow{AC}$ (1) and

$\overrightarrow{DB} + \overrightarrow{BE} = \overrightarrow{DE}$ (2). However, $\overrightarrow{DB} = \frac{1}{2}\overrightarrow{AB}$, and $\overrightarrow{BE} = \frac{1}{2}\overrightarrow{BC}$. Substituting these expressions for $\overrightarrow{DB}$ and $\overrightarrow{BE}$ into (2)

gives $\frac{1}{2}\overrightarrow{AB} + \frac{1}{2}\overrightarrow{BC} = \overrightarrow{DE}$. Comparing this with (1) gives $\overrightarrow{DE} = \frac{1}{2}\overrightarrow{AC}$. Therefore $\overrightarrow{AC}$ and $\overrightarrow{DE}$ are parallel and

$\left|\overrightarrow{DE}\right| = \frac{1}{2}\left|\overrightarrow{AC}\right|$.

9.3   The Dot Product

1. (a) $\mathbf{a} \cdot \mathbf{b}$ is a scalar, and the dot product is defined only for vectors, so $(\mathbf{a} \cdot \mathbf{b}) \cdot \mathbf{c}$ has no meaning.

(b) $(\mathbf{a} \cdot \mathbf{b})\,\mathbf{c}$ is a scalar multiple of a vector, so it does have meaning.

(c) Both $|\mathbf{a}|$ and $\mathbf{b} \cdot \mathbf{c}$ are scalars, so $|\mathbf{a}|\,(\mathbf{b} \cdot \mathbf{c})$ is an ordinary product of real numbers, and has meaning.

(d) Both $\mathbf{a}$ and $\mathbf{b} + \mathbf{c}$ are vectors, so the dot product $\mathbf{a} \cdot (\mathbf{b} + \mathbf{c})$ has meaning.

(e) $\mathbf{a} \cdot \mathbf{b}$ is a scalar, but $\mathbf{c}$ is a vector, and so the two quantities cannot be added and this expression has no meaning.

(f) $|\mathbf{a}|$ is a scalar, and the dot product is defined only for vectors, so $|\mathbf{a}| \cdot (\mathbf{b} + \mathbf{c})$ has no meaning.

3. $\mathbf{a} \cdot \mathbf{b} = |\mathbf{a}|\,|\mathbf{b}|\cos\theta = (6)(5)\cos\frac{2\pi}{3} = 30\left(-\frac{1}{2}\right) = -15$

5. $\mathbf{a} \cdot \mathbf{b} = \left\langle 4, 1, \frac{1}{4} \right\rangle \cdot \langle 6, -3, -8 \rangle = (4)(6) + (1)(-3) + \left(\frac{1}{4}\right)(-8) = 19$

7. $\mathbf{a} \cdot \mathbf{b} = (\mathbf{i} - 2\mathbf{j} + 3\mathbf{k}) \cdot (5\mathbf{i} + 9\mathbf{k}) = (1)(5) + (-2)(0) + (3)(9) = 32$

9. $\mathbf{u}$, $\mathbf{v}$, and $\mathbf{w}$ are all unit vectors, so the triangle is an equilateral triangle. Thus the angle between $\mathbf{u}$ and $\mathbf{v}$ is $60°$ and

$\mathbf{u} \cdot \mathbf{v} = |\mathbf{u}|\,|\mathbf{v}|\cos 60° = (1)(1)\left(\frac{1}{2}\right) = \frac{1}{2}$. If $\mathbf{w}$ is moved so it has the same initial point as $\mathbf{u}$, we can see that the angle

between them is $120°$ and we have $\mathbf{u} \cdot \mathbf{w} = |\mathbf{u}|\,|\mathbf{w}|\cos 120° = (1)(1)\left(-\frac{1}{2}\right) = -\frac{1}{2}$.

11. (a) $\mathbf{i} \cdot \mathbf{j} = \langle 1, 0, 0 \rangle \cdot \langle 0, 1, 0 \rangle = (1)(0) + (0)(1) + (0)(0) = 0$. Similarly $\mathbf{j} \cdot \mathbf{k} = (0)(0) + (1)(0) + (0)(1) = 0$ and
$\mathbf{k} \cdot \mathbf{i} = (0)(1) + (0)(0) + (1)(0) = 0$.

Another method: Because $\mathbf{i}$, $\mathbf{j}$, and $\mathbf{k}$ are mutually perpendicular, the cosine factor in each dot product is $\cos\frac{\pi}{2} = 0$.

(b) By Property 1 of the dot product, $\mathbf{i} \cdot \mathbf{i} = |\mathbf{i}|^2 = 1^2 = 1$ since $\mathbf{i}$ is a unit vector. Similarly, $\mathbf{j} \cdot \mathbf{j} = |\mathbf{j}|^2 = 1$ and
$\mathbf{k} \cdot \mathbf{k} = |\mathbf{k}|^2 = 1$.

13. $|\mathbf{a}| = \sqrt{(-8)^2 + 6^2} = 10$, $|\mathbf{b}| = \sqrt{\left(\sqrt{7}\right)^2 + 3^2} = 4$, and $\mathbf{a} \cdot \mathbf{b} = (-8)(\sqrt{7}) + (6)(3) = 18 - 8\sqrt{7}$. From the definition of

the dot product, we have $\cos\theta = \dfrac{\mathbf{a} \cdot \mathbf{b}}{|\mathbf{a}|\,|\mathbf{b}|} = \dfrac{18 - 8\sqrt{7}}{10 \cdot 4} = \dfrac{9 - 4\sqrt{7}}{20}$. So the angle between $\mathbf{a}$ and $\mathbf{b}$ is

$$\theta = \cos^{-1}\left(\dfrac{9 - 4\sqrt{7}}{20}\right) \approx 95°.$$

15. $|\mathbf{a}| = \sqrt{0^2 + 1^2 + 1^2} = \sqrt{2}$, $|\mathbf{b}| = \sqrt{1^2 + 2^2 + (-3)^2} = \sqrt{14}$, and $\mathbf{a} \cdot \mathbf{b} = (0)(1) + (1)(2) + (1)(-3) = -1$. Then

$$\cos\theta = \dfrac{\mathbf{a} \cdot \mathbf{b}}{|\mathbf{a}|\,|\mathbf{b}|} = \dfrac{-1}{\sqrt{2} \cdot \sqrt{14}} = \dfrac{-1}{2\sqrt{7}} \text{ and } \theta = \cos^{-1}\left(-\dfrac{1}{2\sqrt{7}}\right) \approx 101°.$$

17. (a) $\mathbf{a} \cdot \mathbf{b} = (-5)(6) + (3)(-8) + (7)(2) = -40 \neq 0$, so $\mathbf{a}$ and $\mathbf{b}$ are not orthogonal. Also, since $\mathbf{a}$ is not a scalar multiple of
   $\mathbf{b}$, $\mathbf{a}$ and $\mathbf{b}$ are not parallel.

   (b) $\mathbf{a} \cdot \mathbf{b} = (4)(-3) + (6)(2) = 0$, so $\mathbf{a}$ and $\mathbf{b}$ are orthogonal (and not parallel).

   (c) $\mathbf{a} \cdot \mathbf{b} = (-1)(3) + (2)(4) + (5)(-1) = 0$, so $\mathbf{a}$ and $\mathbf{b}$ are orthogonal (and not parallel).

   (d) Because $\mathbf{a} = -\frac{2}{3}\,\mathbf{b}$, $\mathbf{a}$ and $\mathbf{b}$ are parallel.

19. $\overrightarrow{QP} = \langle -1, -3, 2 \rangle$, $\overrightarrow{QR} = \langle 4, -2, -1 \rangle$, and $\overrightarrow{QP} \cdot \overrightarrow{QR} = -4 + 6 - 2 = 0$. Thus $\overrightarrow{QP}$ and $\overrightarrow{QR}$ are orthogonal, so the angle of
   the triangle at vertex Q is a right angle.

21. Let $\mathbf{a} = a_1\,\mathbf{i} + a_2\,\mathbf{j} + a_3\,\mathbf{k}$ be a vector orthogonal to both $\mathbf{i} + \mathbf{j}$ and $\mathbf{i} + \mathbf{k}$. Then $\mathbf{a} \cdot (\mathbf{i} + \mathbf{j}) = 0 \iff a_1 + a_2 = 0$ and
   $\mathbf{a} \cdot (\mathbf{i} + \mathbf{k}) = 0 \iff a_1 + a_3 = 0$, so $a_1 = -a_2 = -a_3$. Furthermore $\mathbf{a}$ is to be a unit vector, so $1 = a_1^2 + a_2^2 + a_3^2 = 3a_1^2$
   implies $a_1 = \pm\frac{1}{\sqrt{3}}$. Thus $\mathbf{a} = \frac{1}{\sqrt{3}}\,\mathbf{i} - \frac{1}{\sqrt{3}}\,\mathbf{j} - \frac{1}{\sqrt{3}}\,\mathbf{k}$ and $\mathbf{a} = -\frac{1}{\sqrt{3}}\,\mathbf{i} + \frac{1}{\sqrt{3}}\,\mathbf{j} + \frac{1}{\sqrt{3}}\,\mathbf{k}$ are two such unit vectors.

23. $|\mathbf{a}| = \sqrt{3^2 + (-4)^2} = 5$. The scalar projection of $\mathbf{b}$ onto $\mathbf{a}$ is $\text{comp}_{\mathbf{a}}\,\mathbf{b} = \dfrac{\mathbf{a} \cdot \mathbf{b}}{|\mathbf{a}|} = \dfrac{3 \cdot 5 + (-4) \cdot 0}{5} = 3$ and the vector

   projection of $\mathbf{b}$ onto $\mathbf{a}$ is $\text{proj}_{\mathbf{a}}\,\mathbf{b} = \left(\dfrac{\mathbf{a} \cdot \mathbf{b}}{|\mathbf{a}|}\right)\dfrac{\mathbf{a}}{|\mathbf{a}|} = 3 \cdot \frac{1}{5}\langle 3, -4 \rangle = \langle \frac{9}{5}, -\frac{12}{5} \rangle$.

25. $|\mathbf{a}| = \sqrt{9 + 36 + 4} = 7$ so the scalar projection of $\mathbf{b}$ onto $\mathbf{a}$ is $\text{comp}_{\mathbf{a}}\mathbf{b} = \dfrac{\mathbf{a} \cdot \mathbf{b}}{|\mathbf{a}|} = \frac{1}{7}\,(3 + 12 - 6) = \frac{9}{7}$. The vector

   projection of $\mathbf{b}$ onto $\mathbf{a}$ is $\text{proj}_{\mathbf{a}}\mathbf{b} = \dfrac{9}{7}\dfrac{\mathbf{a}}{|\mathbf{a}|} = \frac{9}{7} \cdot \frac{1}{7}\langle 3, 6, -2 \rangle = \frac{9}{49}\langle 3, 6, -2 \rangle = \langle \frac{27}{49}, \frac{54}{49}, -\frac{18}{49} \rangle$.

27. $(\text{orth}_{\mathbf{a}}\,\mathbf{b}) \cdot \mathbf{a} = (\mathbf{b} - \text{proj}_{\mathbf{a}}\,\mathbf{b}) \cdot \mathbf{a} = \mathbf{b} \cdot \mathbf{a} - (\text{proj}_{\mathbf{a}}\,\mathbf{b}) \cdot \mathbf{a} = \mathbf{b} \cdot \mathbf{a} - \dfrac{\mathbf{a} \cdot \mathbf{b}}{|\mathbf{a}|^2}\,\mathbf{a} \cdot \mathbf{a}$

$$= \mathbf{b} \cdot \mathbf{a} - \dfrac{\mathbf{a} \cdot \mathbf{b}}{|\mathbf{a}|^2}\,|\mathbf{a}|^2 = \mathbf{b} \cdot \mathbf{a} - \mathbf{a} \cdot \mathbf{b} = 0$$

   So they are orthogonal by (2).

29. $\text{comp}_{\mathbf{a}}\,\mathbf{b} = \dfrac{\mathbf{a} \cdot \mathbf{b}}{|\mathbf{a}|} = 2 \iff \mathbf{a} \cdot \mathbf{b} = 2\,|\mathbf{a}| = 2\sqrt{10}$. If $\mathbf{b} = \langle b_1, b_2, b_3 \rangle$, then we need $3b_1 + 0b_2 - 1b_3 = 2\sqrt{10}$.

   One possible solution is obtained by taking $b_1 = 0$, $b_2 = 0$, $b_3 = -2\sqrt{10}$. In general, $\mathbf{b} = \langle s, t, 3s - 2\sqrt{10} \rangle$, $s, t \in \mathbb{R}$.

31. Here $\mathbf{D} = (4 - 2)\,\mathbf{i} + (9 - 3)\,\mathbf{j} + (15 - 0)\,\mathbf{k} = 2\,\mathbf{i} + 6\,\mathbf{j} + 15\,\mathbf{k}$ so $W = \mathbf{F} \cdot \mathbf{D} = 20 + 108 - 90 = 38$ joules.

33. $W = |\mathbf{F}|\,|\mathbf{D}|\cos\theta = (25)(10)\cos 20° \approx 235$ ft-lb

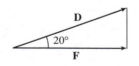

35. First note that $\mathbf{n} = \langle a, b \rangle$ is perpendicular to the line, because if $Q_1 = (a_1, b_1)$ and $Q_2 = (a_2, b_2)$ lie on the line, then

$\mathbf{n} \cdot \overrightarrow{Q_1 Q_2} = aa_2 - aa_1 + bb_2 - bb_1 = 0$, since $aa_2 + bb_2 = -c = aa_1 + bb_1$ from the equation of the line.

Let $P_2 = (x_2, y_2)$ lie on the line. Then the distance from P_1 to the line is the absolute value of the scalar projection

of $\overrightarrow{P_1 P_2}$ onto $\mathbf{n}$. $\text{comp}_{\mathbf{n}}\left(\overrightarrow{P_1 P_2}\right) = \dfrac{|\mathbf{n} \cdot \langle x_2 - x_1, y_2 - y_1 \rangle|}{|\mathbf{n}|} = \dfrac{|ax_2 - ax_1 + by_2 - by_1|}{\sqrt{a^2 + b^2}} = \dfrac{|ax_1 + by_1 + c|}{\sqrt{a^2 + b^2}}$ since

$ax_2 + by_2 = -c$. The required distance is $\dfrac{|3 \cdot -2 + -4 \cdot 3 + 5|}{\sqrt{3^2 + 4^2}} = \dfrac{13}{5}$.

37. For convenience, consider the unit cube positioned so that its back left corner is at the origin, and its edges lie along the coordinate axes. The diagonal of the cube that begins at the origin and ends at $(1, 1, 1)$ has vector representation $\langle 1, 1, 1 \rangle$. The angle θ between this vector and the vector of the edge which also begins at the origin and runs along the x-axis [that is,

$\langle 1, 0, 0 \rangle$] is given by $\cos \theta = \dfrac{\langle 1, 1, 1 \rangle \cdot \langle 1, 0, 0 \rangle}{|\langle 1, 1, 1 \rangle| |\langle 1, 0, 0 \rangle|} = \dfrac{1}{\sqrt{3}}$ $\Rightarrow$ $\theta = \cos^{-1}\left(\dfrac{1}{\sqrt{3}}\right) \approx 55°$.

39. Consider the H-C-H combination consisting of the sole carbon atom and the two hydrogen atoms that are at $(1, 0, 0)$ and $(0, 1, 0)$ (or any H-C-H combination, for that matter). Vector representations of the line segments emanating from the carbon atom and extending to these two hydrogen atoms are $\left\langle 1 - \frac{1}{2}, 0 - \frac{1}{2}, 0 - \frac{1}{2} \right\rangle = \left\langle \frac{1}{2}, -\frac{1}{2}, -\frac{1}{2} \right\rangle$ and

$\left\langle 0 - \frac{1}{2}, 1 - \frac{1}{2}, 0 - \frac{1}{2} \right\rangle = \left\langle -\frac{1}{2}, \frac{1}{2}, -\frac{1}{2} \right\rangle$. The bond angle, θ, is therefore given by

$\cos \theta = \dfrac{\left\langle \frac{1}{2}, -\frac{1}{2}, -\frac{1}{2} \right\rangle \cdot \left\langle -\frac{1}{2}, \frac{1}{2}, -\frac{1}{2} \right\rangle}{\left|\left\langle \frac{1}{2}, -\frac{1}{2}, -\frac{1}{2} \right\rangle\right| \left|\left\langle -\frac{1}{2}, \frac{1}{2}, -\frac{1}{2} \right\rangle\right|} = \dfrac{-\frac{1}{4} - \frac{1}{4} + \frac{1}{4}}{\sqrt{\frac{3}{4}}\sqrt{\frac{3}{4}}} = -\dfrac{1}{3}$ $\Rightarrow$ $\theta = \cos^{-1}\left(-\frac{1}{3}\right) \approx 109.5°$.

41. If $c = 0$ then $c\mathbf{a} = \mathbf{0}$, so $(c\mathbf{a}) \cdot \mathbf{b} = \mathbf{0} \cdot \mathbf{b} = 0$ by Property 5. Similarly, $\mathbf{a} \cdot (c\mathbf{b}) = \mathbf{a} \cdot \mathbf{0} = 0$, and $c(\mathbf{a} \cdot \mathbf{b}) = 0(|\mathbf{a}| |\mathbf{b}| \cos \theta) = 0$, thus $(c\mathbf{a}) \cdot \mathbf{b} = c(\mathbf{a} \cdot \mathbf{b}) = \mathbf{a} \cdot (c\mathbf{b})$. If $c > 0$, the angle θ between $\mathbf{a}$ and $\mathbf{b}$ coincides with the angle between $c\mathbf{a}$ and $\mathbf{b}$, so by definition of the dot product, $(c\mathbf{a}) \cdot \mathbf{b} = |c\mathbf{a}| |\mathbf{b}| \cos \theta = |c| |\mathbf{a}| |\mathbf{b}| \cos \theta = c |\mathbf{a}| |\mathbf{b}| \cos \theta$. Similarly, $\mathbf{a} \cdot (c\mathbf{b}) = |\mathbf{a}| |c\mathbf{b}| \cos \theta = |\mathbf{a}| |c| |\mathbf{b}| \cos \theta = c |\mathbf{a}| |\mathbf{b}| \cos \theta$, and $c(\mathbf{a} \cdot \mathbf{b}) = c |\mathbf{a}| |\mathbf{b}| \cos \theta$. Thus, $(c\mathbf{a}) \cdot \mathbf{b} = c(\mathbf{a} \cdot \mathbf{b}) = \mathbf{a} \cdot (c\mathbf{b})$. The case for $c < 0$ is similar. Using components, let $\mathbf{a} = \langle a_1, a_2, a_3 \rangle$ and $\mathbf{b} = \langle b_1, b_2, b_3 \rangle$. Then

$$(c\mathbf{a}) \cdot \mathbf{b} = \langle ca_1, ca_2, ca_3 \rangle \cdot \langle b_1, b_2, b_3 \rangle = (ca_1)b_1 + (ca_2)b_2 + (ca_3)b_3$$

$$= c(a_1 b_1 + a_2 b_2 + a_3 b_3) = c(\mathbf{a} \cdot \mathbf{b})$$

$$= a_1(cb_1) + a_2(cb_2) + a_3(cb_3) = \langle a_1, a_2, a_3 \rangle \cdot \langle cb_1, cb_2, cb_3 \rangle = \mathbf{a} \cdot (c\mathbf{b})$$

43. $|\mathbf{a} \cdot \mathbf{b}| = \big| |\mathbf{a}| |\mathbf{b}| \cos \theta \big| = |\mathbf{a}| |\mathbf{b}| |\cos \theta|$. Since $|\cos \theta| \le 1$, $|\mathbf{a} \cdot \mathbf{b}| = |\mathbf{a}| |\mathbf{b}| |\cos \theta| \le |\mathbf{a}| |\mathbf{b}|$.

Note: We have equality in the case of $\cos \theta = \pm 1$, so $\theta = 0$ or $\theta = \pi$, thus equality when $\mathbf{a}$ and $\mathbf{b}$ are parallel.

45. (a)

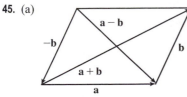

The Parallelogram Law states that the sum of the squares of the lengths of the diagonals of a parallelogram equals the sum of the squares of its (four) sides.

(b) $|\mathbf{a} + \mathbf{b}|^2 = (\mathbf{a} + \mathbf{b}) \cdot (\mathbf{a} + \mathbf{b}) = |\mathbf{a}|^2 + 2(\mathbf{a} \cdot \mathbf{b}) + |\mathbf{b}|^2$ and $|\mathbf{a} - \mathbf{b}|^2 = (\mathbf{a} - \mathbf{b}) \cdot (\mathbf{a} - \mathbf{b}) = |\mathbf{a}|^2 - 2(\mathbf{a} \cdot \mathbf{b}) + |\mathbf{b}|^2$.

Adding these two equations gives $|\mathbf{a} + \mathbf{b}|^2 + |\mathbf{a} - \mathbf{b}|^2 = 2 |\mathbf{a}|^2 + 2 |\mathbf{b}|^2$.

9.4 The Cross Product

1. (a) Since $\mathbf{b} \times \mathbf{c}$ is a vector, the dot product $\mathbf{a} \cdot (\mathbf{b} \times \mathbf{c})$ is meaningful and is a scalar.

(b) $\mathbf{b} \cdot \mathbf{c}$ is a scalar, so $\mathbf{a} \times (\mathbf{b} \cdot \mathbf{c})$ is meaningless, as the cross product is defined only for two *vectors*.

(c) Since $\mathbf{b} \times \mathbf{c}$ is a vector, the cross product $\mathbf{a} \times (\mathbf{b} \times \mathbf{c})$ is meaningful and results in another vector.

(d) $\mathbf{a} \cdot \mathbf{b}$ is a scalar, so the cross product $(\mathbf{a} \cdot \mathbf{b}) \times \mathbf{c}$ is meaningless.

(e) Since $(\mathbf{a} \cdot \mathbf{b})$ and $(\mathbf{c} \cdot \mathbf{d})$ are both scalars, the cross product $(\mathbf{a} \cdot \mathbf{b}) \times (\mathbf{c} \cdot \mathbf{d})$ is meaningless.

(f) $\mathbf{a} \times \mathbf{b}$ and $\mathbf{c} \times \mathbf{d}$ are both vectors, so the dot product $(\mathbf{a} \times \mathbf{b}) \cdot (\mathbf{c} \times \mathbf{d})$ is meaningful and is a scalar.

3. If we sketch $\mathbf{u}$ and $\mathbf{v}$ starting from the same initial point,

we see that the angle between them is $30\,^\circ$, so

$$|\mathbf{u} \times \mathbf{v}| = |\mathbf{u}|\,|\mathbf{v}| \sin 30\,^\circ = (6)(8)\left(\tfrac{1}{2}\right) = 24$$

By the right-hand rule, $\mathbf{u} \times \mathbf{v}$ is directed into the page.

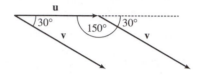

5. The magnitude of the torque is $|\boldsymbol{\tau}| = |\mathbf{r} \times \mathbf{F}| = |\mathbf{r}|\,|\mathbf{F}| \sin\theta = (0.18\text{ m})(60\text{ N}) \sin(70 + 10)\,^\circ = 10.8 \sin 80\,^\circ \approx 10.6\text{ J}.$

7. $\mathbf{a} \times \mathbf{b} = \begin{vmatrix} \mathbf{i} & \mathbf{j} & \mathbf{k} \\ 1 & 2 & 0 \\ 0 & 3 & 1 \end{vmatrix} = \begin{vmatrix} 2 & 0 \\ 3 & 1 \end{vmatrix} \mathbf{i} - \begin{vmatrix} 1 & 0 \\ 0 & 1 \end{vmatrix} \mathbf{j} + \begin{vmatrix} 1 & 2 \\ 0 & 3 \end{vmatrix} \mathbf{k} = (2 - 0)\,\mathbf{i} - (1 - 0)\,\mathbf{j} + (3 - 0)\,\mathbf{k} = 2\,\mathbf{i} - \mathbf{j} + 3\,\mathbf{k}$

Now $(\mathbf{a} \times \mathbf{b}) \cdot \mathbf{a} = \langle 2, -1, 3\rangle \cdot \langle 1, 2, 0\rangle = 2 - 2 + 0 = 0$ and $(\mathbf{a} \times \mathbf{b}) \cdot \mathbf{b} = \langle 2, -1, 3\rangle \cdot \langle 0, 3, 1\rangle = 0 - 3 + 3 = 0$, so $\mathbf{a} \times \mathbf{b}$

is orthogonal to both $\mathbf{a}$ and $\mathbf{b}$.

9. $\mathbf{a} \times \mathbf{b} = \begin{vmatrix} \mathbf{i} & \mathbf{j} & \mathbf{k} \\ t & t^2 & t^3 \\ 1 & 2t & 3t^2 \end{vmatrix} = \begin{vmatrix} t^2 & t^3 \\ 2t & 3t^2 \end{vmatrix} \mathbf{i} - \begin{vmatrix} t & t^3 \\ 1 & 3t^2 \end{vmatrix} \mathbf{j} + \begin{vmatrix} t & t^2 \\ 1 & 2t \end{vmatrix} \mathbf{k}$

$$= (3t^4 - 2t^4)\,\mathbf{i} - (3t^3 - t^3)\,\mathbf{j} + (2t^2 - t^2)\,\mathbf{k} = t^4\,\mathbf{i} - 2t^3\,\mathbf{j} + t^2\,\mathbf{k}$$

Since $(\mathbf{a} \times \mathbf{b}) \cdot \mathbf{a} = \langle t^4, -2t^3, t^2\rangle \cdot \langle t, t^2, t^3\rangle = t^5 - 2t^5 + t^5 = 0$, $\mathbf{a} \times \mathbf{b}$ is orthogonal to $\mathbf{a}$.

Since $(\mathbf{a} \times \mathbf{b}) \cdot \mathbf{b} = \langle t^4, -2t^3, t^2\rangle \cdot \langle 1, 2t, 3t^2\rangle = t^4 - 4t^4 + 3t^4 = 0$, $\mathbf{a} \times \mathbf{b}$ is orthogonal to $\mathbf{b}$.

11. $\mathbf{a} \times \mathbf{b} = \begin{vmatrix} \mathbf{i} & \mathbf{j} & \mathbf{k} \\ 3 & 2 & 4 \\ 1 & -2 & -3 \end{vmatrix} = \begin{vmatrix} 2 & 4 \\ -2 & -3 \end{vmatrix} \mathbf{i} - \begin{vmatrix} 3 & 4 \\ 1 & -3 \end{vmatrix} \mathbf{j} + \begin{vmatrix} 3 & 2 \\ 1 & -2 \end{vmatrix} \mathbf{k}$

$$= [-6 - (-8)]\,\mathbf{i} - (-9 - 4)\,\mathbf{j} + (-6 - 2)\,\mathbf{k} = 2\,\mathbf{i} + 13\,\mathbf{j} - 8\,\mathbf{k}$$

Since $(\mathbf{a} \times \mathbf{b}) \cdot \mathbf{a} = (2\,\mathbf{i} + 13\,\mathbf{j} - 8\,\mathbf{k}) \cdot (3\,\mathbf{i} + 2\,\mathbf{j} + 4\,\mathbf{k}) = 6 + 26 - 32 = 0$, $\mathbf{a} \times \mathbf{b}$ is orthogonal to $\mathbf{a}$.

Since $(\mathbf{a} \times \mathbf{b}) \cdot \mathbf{b} = (2\,\mathbf{i} + 13\,\mathbf{j} - 8\,\mathbf{k}) \cdot (\mathbf{i} - 2\,\mathbf{j} - 3\,\mathbf{k}) = 2 - 26 + 24 = 0$, $\mathbf{a} \times \mathbf{b}$ is orthogonal to $\mathbf{b}$.

13. We know that the cross product of two vectors is orthogonal to both. So we calculate

$$\langle 2, 0, -3\rangle \times \langle -1, 4, 2\rangle = \begin{vmatrix} \mathbf{i} & \mathbf{j} & \mathbf{k} \\ 2 & 0 & -3 \\ -1 & 4 & 2 \end{vmatrix} = \begin{vmatrix} 0 & -3 \\ 4 & 2 \end{vmatrix} \mathbf{i} - \begin{vmatrix} 2 & -3 \\ -1 & 2 \end{vmatrix} \mathbf{j} + \begin{vmatrix} 2 & 0 \\ -1 & 4 \end{vmatrix} \mathbf{k} = 12\,\mathbf{i} - \mathbf{j} + 8\,\mathbf{k}$$

So two unit vectors orthogonal to both are $\pm\dfrac{\langle 12,-1,8\rangle}{\sqrt{144+1+64}}=\pm\dfrac{\langle 12,-1,8\rangle}{\sqrt{209}}$, that is, $\left\langle \dfrac{12}{\sqrt{209}},-\dfrac{1}{\sqrt{209}},\dfrac{8}{\sqrt{209}}\right\rangle$

and $\left\langle -\dfrac{12}{\sqrt{209}},\dfrac{1}{\sqrt{209}},-\dfrac{8}{\sqrt{209}}\right\rangle$.

15. By plotting the vertices, we can see that the parallelogram is determined

by the vectors $\overrightarrow{AB}=\langle 2,3\rangle$ and $\overrightarrow{AD}=\langle 4,-2\rangle$. We know that the area

of the parallelogram determined by two vectors is equal to the length of

the cross product of these vectors. In order to compute the cross product,

we consider the vector $\overrightarrow{AB}$ as the three-dimensional vector $\langle 2,3,0\rangle$

(and similarly for $\overrightarrow{AD}$), and then the area of parallelogram $ABCD$ is

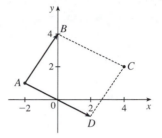

$$\left|\overrightarrow{AB}\times\overrightarrow{AD}\right|=\begin{vmatrix} \mathbf{i} & \mathbf{j} & \mathbf{k} \\ 2 & 3 & 0 \\ 4 & -2 & 0 \end{vmatrix}=|(0)\mathbf{i}-(0)\mathbf{j}+(-4-12)\mathbf{k}|=|-16\,\mathbf{k}|=16$$

17. (a) $\overrightarrow{PQ}=\langle 4,3,-2\rangle$ and $\overrightarrow{PR}=\langle 5,5,1\rangle$, so a vector orthogonal to the plane through P, Q, and R is

$\overrightarrow{PQ}\times\overrightarrow{PR}=\langle (3)(1)-(-2)(5),(-2)(5)-(4)(1),(4)(5)-(3)(5)\rangle=\langle 13,-14,5\rangle$

(or any scalar mutiple thereof).

(b) The area of the parallelogram determined by $\overrightarrow{PQ}$ and $\overrightarrow{PR}$ is

$\left|\overrightarrow{PQ}\times\overrightarrow{PR}\right|=|\langle 13,-14,5\rangle|=\sqrt{13^2+(-14)^2+5^2}=\sqrt{390}$, so the area of triangle PQR is $\frac{1}{2}\sqrt{390}$.

19. Using the notation of (1), $\mathbf{r}=\langle 0,0.3,0\rangle$ and $\mathbf{F}$ has direction $\langle 0,3,-4\rangle$. The angle θ between them can be determined by

$\cos\theta=\dfrac{\langle 0,0.3,0\rangle\cdot\langle 0,3,-4\rangle}{|\langle 0,0.3,0\rangle|\,|\langle 0,3,-4\rangle|}\quad\Rightarrow\quad \cos\theta=\dfrac{0.9}{(0.3)(5)}\quad\Rightarrow\quad \cos\theta=0.6\quad\Rightarrow\quad \theta\approx 53.1°$. Then $|\boldsymbol{\tau}|=|\mathbf{r}|\,|\mathbf{F}|\sin\theta\quad\Rightarrow$

$100=0.3\,|\mathbf{F}|\sin 53.1°\quad\Rightarrow\quad |\mathbf{F}|\approx 417\text{ N}$.

21. We know that the volume of the parallelepiped determined by $\mathbf{a}$, $\mathbf{b}$, and $\mathbf{c}$ is the magnitude of their scalar triple product, which

is

$$\mathbf{a}\cdot(\mathbf{b}\times\mathbf{c})=\begin{vmatrix} 6 & 3 & -1 \\ 0 & 1 & 2 \\ 4 & -2 & 5 \end{vmatrix}=6\begin{vmatrix} 1 & 2 \\ -2 & 5 \end{vmatrix}-3\begin{vmatrix} 0 & 2 \\ 4 & 5 \end{vmatrix}+(-1)\begin{vmatrix} 0 & 1 \\ 4 & -2 \end{vmatrix}$$

$$=6(5+4)-3(0-8)-(0-4)=82$$

Thus the volume of the parallelepiped is 82 cubic units.

23. $\mathbf{a}=\overrightarrow{PQ}=\langle 2,1,1\rangle$, $\mathbf{b}=\overrightarrow{PR}=\langle 1,-1,2\rangle$, and $\mathbf{c}=\overrightarrow{PS}=\langle 0,-2,3\rangle$.

$$\mathbf{a}\cdot(\mathbf{b}\times\mathbf{c})=\begin{vmatrix} 2 & 1 & 1 \\ 1 & -1 & 2 \\ 0 & -2 & 3 \end{vmatrix}=2\begin{vmatrix} -1 & 2 \\ -2 & 3 \end{vmatrix}-1\begin{vmatrix} 1 & 2 \\ 0 & 3 \end{vmatrix}+1\begin{vmatrix} 1 & -1 \\ 0 & -2 \end{vmatrix}=2-3-2=-3,$$

so the volume of the parallelepiped is 3 cubic units.

25. $\mathbf{u} \cdot (\mathbf{v} \times \mathbf{w}) = \begin{vmatrix} 1 & 5 & -2 \\ 3 & -1 & 0 \\ 5 & 9 & -4 \end{vmatrix} = 1\begin{vmatrix} -1 & 0 \\ 9 & -4 \end{vmatrix} - 5\begin{vmatrix} 3 & 0 \\ 5 & -4 \end{vmatrix} + (-2)\begin{vmatrix} 3 & -1 \\ 5 & 9 \end{vmatrix} = 4 + 60 - 64 = 0$, which says that the volume

of the parallelepiped determined by $\mathbf{u}$, $\mathbf{v}$ and $\mathbf{w}$ is 0, and thus these three vectors are coplanar.

27. (a)

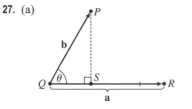

The distance between a point and a line is the length of the perpendicular from the

point to the line, here $\left|\overrightarrow{PS}\right| = d$. But referring to triangle PQS,

$d = \left|\overrightarrow{PS}\right| = \left|\overrightarrow{QP}\right| \sin\theta = |\mathbf{b}| \sin\theta$. But θ is the angle between $\overrightarrow{QP} = \mathbf{b}$ and

$\overrightarrow{QR} = \mathbf{a}$. Thus by the definition of the cross product, $\sin\theta = \dfrac{|\mathbf{a} \times \mathbf{b}|}{|\mathbf{a}|\,|\mathbf{b}|}$ and so

$$d = |\mathbf{b}| \sin\theta = \frac{|\mathbf{b}|\,|\mathbf{a} \times \mathbf{b}|}{|\mathbf{a}|\,|\mathbf{b}|} = \frac{|\mathbf{a} \times \mathbf{b}|}{|\mathbf{a}|}.$$

(b) $\mathbf{a} = \overrightarrow{QR} = \langle -1, -2, -1 \rangle$ and $\mathbf{b} = \overrightarrow{QP} = \langle 1, -5, -7 \rangle$. Then

$\mathbf{a} \times \mathbf{b} = \langle (-2)(-7) - (-1)(-5), (-1)(1) - (-1)(-7), (-1)(-5) - (-2)(1) \rangle = \langle 9, -8, 7 \rangle$.

Thus the distance is $d = \dfrac{|\mathbf{a} \times \mathbf{b}|}{|\mathbf{a}|} = \dfrac{1}{\sqrt{6}}\sqrt{81 + 64 + 49} = \sqrt{\dfrac{194}{6}} = \sqrt{\dfrac{97}{3}}$.

29. $(\mathbf{a} - \mathbf{b}) \times (\mathbf{a} + \mathbf{b}) = (\mathbf{a} - \mathbf{b}) \times \mathbf{a} + (\mathbf{a} - \mathbf{b}) \times \mathbf{b}$ by Property 3 of the cross product

$\qquad\qquad\qquad\qquad = \mathbf{a} \times \mathbf{a} + (-\mathbf{b}) \times \mathbf{a} + \mathbf{a} \times \mathbf{b} + (-\mathbf{b}) \times \mathbf{b}$ by Property 4

$\qquad\qquad\qquad\qquad = (\mathbf{a} \times \mathbf{a}) - (\mathbf{b} \times \mathbf{a}) + (\mathbf{a} \times \mathbf{b}) - (\mathbf{b} \times \mathbf{b})$ by Property 2 (with $c = -1$)

$\qquad\qquad\qquad\qquad = \mathbf{0} - (\mathbf{b} \times \mathbf{a}) + (\mathbf{a} \times \mathbf{b}) - \mathbf{0}$ by the margin note on page 658

$\qquad\qquad\qquad\qquad = (\mathbf{a} \times \mathbf{b}) + (\mathbf{a} \times \mathbf{b})$ by Property 1

$\qquad\qquad\qquad\qquad = 2(\mathbf{a} \times \mathbf{b})$

31. $\mathbf{a} \times (\mathbf{b} \times \mathbf{c}) + \mathbf{b} \times (\mathbf{c} \times \mathbf{a}) + \mathbf{c} \times (\mathbf{a} \times \mathbf{b})$

$\qquad = [(\mathbf{a} \cdot \mathbf{c})\mathbf{b} - (\mathbf{a} \cdot \mathbf{b})\mathbf{c}] + [(\mathbf{b} \cdot \mathbf{a})\mathbf{c} - (\mathbf{b} \cdot \mathbf{c})\mathbf{a}] + [(\mathbf{c} \cdot \mathbf{b})\mathbf{a} - (\mathbf{c} \cdot \mathbf{a})\mathbf{b}]$ by Exercise 30

$\qquad = (\mathbf{a} \cdot \mathbf{c})\mathbf{b} - (\mathbf{a} \cdot \mathbf{b})\mathbf{c} + (\mathbf{a} \cdot \mathbf{b})\mathbf{c} - (\mathbf{b} \cdot \mathbf{c})\mathbf{a} + (\mathbf{b} \cdot \mathbf{c})\mathbf{a} - (\mathbf{a} \cdot \mathbf{c})\mathbf{b} = \mathbf{0}$

33. (a) No. If $\mathbf{a} \cdot \mathbf{b} = \mathbf{a} \cdot \mathbf{c}$, then $\mathbf{a} \cdot (\mathbf{b} - \mathbf{c}) = 0$, so $\mathbf{a}$ is perpendicular to $\mathbf{b} - \mathbf{c}$, which can happen if $\mathbf{b} \neq \mathbf{c}$. For example,

     let $\mathbf{a} = \langle 1, 1, 1 \rangle$, $\mathbf{b} = \langle 1, 0, 0 \rangle$ and $\mathbf{c} = \langle 0, 1, 0 \rangle$.

(b) No. If $\mathbf{a} \times \mathbf{b} = \mathbf{a} \times \mathbf{c}$ then $\mathbf{a} \times (\mathbf{b} - \mathbf{c}) = \mathbf{0}$, which implies that $\mathbf{a}$ is parallel to $\mathbf{b} - \mathbf{c}$, which of course can happen if

     $\mathbf{b} \neq \mathbf{c}$.

(c) Yes. Since $\mathbf{a} \cdot \mathbf{c} = \mathbf{a} \cdot \mathbf{b}$, $\mathbf{a}$ is perpendicular to $\mathbf{b} - \mathbf{c}$, by part (a). From part (b), $\mathbf{a}$ is also parallel to $\mathbf{b} - \mathbf{c}$. Thus since

     $\mathbf{a} \neq \mathbf{0}$ but is both parallel and perpendicular to $\mathbf{b} - \mathbf{c}$, we have $\mathbf{b} - \mathbf{c} = \mathbf{0}$, so $\mathbf{b} = \mathbf{c}$.

9.5   Equations of Lines and Planes

1. (a) True; each of the first two lines has a direction vector parallel to the direction vector of the third line, so these vectors are

     each scalar multiples of the third direction vector. Then the first two direction vectors are also scalar multiples of each

     other, so these vectors, and hence the two lines, are parallel.

(b) False; for example, the x- and y-axes are both perpendicular to the z-axis, yet the x- and y-axes are not parallel.

(c) True; each of the first two planes has a normal vector parallel to the normal vector of the third plane, so these two normal

     vectors are parallel to each other and the planes are parallel.

(d) False; for example, the xy- and yz-planes are not parallel, yet they are both perpendicular to the xz-plane.

(e) False; the x- and y-axes are not parallel, yet they are both parallel to the plane $z = 1$.

(f) True; if each line is perpendicular to a plane, then the lines' direction vectors are both parallel to a normal vector for the plane. Thus, the direction vectors are parallel to each other and the lines are parallel.

(g) False; the planes $y = 1$ and $z = 1$ are not parallel, yet they are both parallel to the x-axis.

(h) True; if each plane is perpendicular to a line, then any normal vector for each plane is parallel to a direction vector for the line. Thus, the normal vectors are parallel to each other and the planes are parallel.

(i) True; see Figure 9 and the accompanying discussion.

(j) False; they can be skew, as in Example 3.

(k) True. Consider any normal vector for the plane and any direction vector for the line. If the normal vector is perpendicular to the direction vector, the line and plane are parallel. Otherwise, the vectors meet at an angle $\theta, 0° \le \theta < 90°$, and the line will intersect the plane at an angle $90° - \theta$.

3. For this line, we have $\mathbf{r}_0 = -2\,\mathbf{i} + 4\,\mathbf{j} + 10\,\mathbf{k}$ and $\mathbf{v} = 3\,\mathbf{i} + \mathbf{j} - 8\,\mathbf{k}$, so a vector equation is

$\mathbf{r} = \mathbf{r}_0 + t\,\mathbf{v} = (-2\,\mathbf{i} + 4\,\mathbf{j} + 10\,\mathbf{k}) + t(3\,\mathbf{i} + \mathbf{j} - 8\,\mathbf{k}) = (-2 + 3t)\,\mathbf{i} + (4 + t)\,\mathbf{j} + (10 - 8t)\,\mathbf{k}$ and parametric equations are $x = -2 + 3t,\, y = 4 + t,\, z = 10 - 8t$.

5. A line perpendicular to the given plane has the same direction as a normal vector to the plane, such as

$\mathbf{n} = \langle 1, 3, 1 \rangle$. So $\mathbf{r}_0 = \mathbf{i} + 6\,\mathbf{k}$, and we can take $\mathbf{v} = \mathbf{i} + 3\,\mathbf{j} + \mathbf{k}$. Then a vector equation is

$\mathbf{r} = (\mathbf{i} + 6\,\mathbf{k}) + t(\mathbf{i} + 3\,\mathbf{j} + \mathbf{k}) = (1 + t)\,\mathbf{i} + 3t\,\mathbf{j} + (6 + t)\,\mathbf{k}$, and parametric equations are $x = 1 + t,\, y = 3t,\, z = 6 + t$.

7. $\mathbf{v} = \langle 2 - 0, 1 - \frac{1}{2}, -3 - 1 \rangle = \langle 2, \frac{1}{2}, -4 \rangle$, and letting $P_0 = (2, 1, -3)$, parametric equations are $x = 2 + 2t,\, y = 1 + \frac{1}{2}t$,

$z = -3 - 4t$, while symmetric equations are $\dfrac{x - 2}{2} = \dfrac{y - 1}{1/2} = \dfrac{z + 3}{-4}$ or $\dfrac{x - 2}{2} = 2y - 2 = \dfrac{z + 3}{-4}$.

9. The line has direction $\mathbf{v} = \langle 1, 2, 1 \rangle$. Letting $P_0 = (1, -1, 1)$, parametric equations are $x = 1 + t,\, y = -1 + 2t,\, z = 1 + t$

and symmetric equations are $x - 1 = \dfrac{y + 1}{2} = z - 1$.

11. Direction vectors of the lines are $\mathbf{v}_1 = \langle -2 - (-4), 0 - (-6), -3 - 1 \rangle = \langle 2, 6, -4 \rangle$ and

$\mathbf{v}_2 = \langle 5 - 10, 3 - 18, 14 - 4 \rangle = \langle -5, -15, 10 \rangle$, and since $\mathbf{v}_2 = -\frac{5}{2}\mathbf{v}_1$, the direction vectors and thus the lines are parallel.

13. (a) A direction vector of the line with parametric equations $x = 1 + 2t,\, y = 3t,\, z = 5 - 7t$ is $\mathbf{v} = \langle 2, 3, -7 \rangle$ and the desired parallel line must also have $\mathbf{v}$ as a direction vector. Here $P_0 = (0, 2, -1)$, so symmetric equations for the line are

$\dfrac{x}{2} = \dfrac{y - 2}{3} = \dfrac{z + 1}{-7}$.

(b) The line intersects the xy-plane when $z = 0$, so we need $\dfrac{x}{2} = \dfrac{y - 2}{3} = \dfrac{1}{-7}$ or $x = -\frac{2}{7},\, y = \frac{11}{7}$. Thus the point of

intersection with the xy-plane is $\left(-\frac{2}{7}, \frac{11}{7}, 0\right)$. Similarly for the yz-plane, we need $x = 0 \iff 0 = \dfrac{y - 2}{3} = \dfrac{z + 1}{-7} \iff$

$y = 2,\, z = -1$. Thus the line intersects the yz-plane at $(0, 2, -1)$. For the xz-plane, we need $y = 0 \iff$

$\dfrac{x}{2} = -\dfrac{2}{3} = \dfrac{z + 1}{-7} \iff x = -\frac{4}{3},\, z = \frac{11}{3}$. So the line intersects the xz-plane at $\left(-\frac{4}{3}, 0, \frac{11}{3}\right)$.

15. From Equation 4, the line segment from $\mathbf{r}_0 = 2\,\mathbf{i} - \mathbf{j} + 4\,\mathbf{k}$ to $\mathbf{r}_1 = 4\,\mathbf{i} + 6\,\mathbf{j} + \mathbf{k}$ is

$\mathbf{r}(t) = (1 - t)\,\mathbf{r}_0 + t\,\mathbf{r}_1 = (1 - t)(2\,\mathbf{i} - \mathbf{j} + 4\,\mathbf{k}) + t(4\,\mathbf{i} + 6\,\mathbf{j} + \mathbf{k}) = (2\,\mathbf{i} - \mathbf{j} + 4\,\mathbf{k}) + t(2\,\mathbf{i} + 7\,\mathbf{j} - 3\,\mathbf{k}), 0 \le t \le 1$.

17. Since the direction vectors are $\mathbf{v}_1 = \langle -6, 9, -3 \rangle$ and $\mathbf{v}_2 = \langle 2, -3, 1 \rangle$, we have $\mathbf{v}_1 = -3\mathbf{v}_2$ so the lines are parallel.

19. Since the direction vectors $\langle 1, 2, 3 \rangle$ and $\langle -4, -3, 2 \rangle$ are not scalar multiples of each other, the lines are not parallel, so we check to see if the lines intersect. The parametric equations of the lines are L_1: $x = t$, $y = 1 + 2t$, $z = 2 + 3t$ and L_2: $x = 3 - 4s$, $y = 2 - 3s$, $z = 1 + 2s$. For the lines to intersect, we must be able to find one value of t and one value of s that produce the same point from the respective parametric equations. Thus we need to satisfy the following three equations: $t = 3 - 4s$, $1 + 2t = 2 - 3s$, $2 + 3t = 1 + 2s$. Solving the first two equations we get $t = -1$, $s = 1$ and checking, we see that these values don't satisfy the third equation. Thus the lines aren't parallel and don't intersect, so they must be skew lines.

21. Since the plane is perpendicular to the vector $\langle -2, 1, 5 \rangle$, we can take $\langle -2, 1, 5 \rangle$ as a normal vector to the plane. $(6, 3, 2)$ is a point on the plane, so setting $a = -2$, $b = 1$, $c = 5$ and $x_0 = 6$, $y_0 = 3$, $z_0 = 2$ in Equation 7 gives $-2(x - 6) + 1(y - 3) + 5(z - 2) = 0$ or $-2x + y + 5z = 1$ to be an equation of the plane.

23. Since the two planes are parallel, they will have the same normal vectors. So we can take $\mathbf{n} = \langle 2, -1, 3 \rangle$, and an equation of the plane is $2(x - 0) - 1(y - 0) + 3(z - 0) = 0$ or $2x - y + 3z = 0$.

25. Here the vectors $\mathbf{a} = \langle 1 - 0, 0 - 1, 1 - 1 \rangle = \langle 1, -1, 0 \rangle$ and $\mathbf{b} = \langle 1 - 0, 1 - 1, 0 - 1 \rangle = \langle 1, 0, -1 \rangle$ lie in the plane, so $\mathbf{a} \times \mathbf{b}$ is a normal vector to the plane. Thus, we can take $\mathbf{n} = \mathbf{a} \times \mathbf{b} = \langle 1 - 0, 0 + 1, 0 + 1 \rangle = \langle 1, 1, 1 \rangle$. If P_0 is the point $(0, 1, 1)$, an equation of the plane is $1(x - 0) + 1(y - 1) + 1(z - 1) = 0$ or $x + y + z = 2$.

27. If we first find two nonparallel vectors in the plane, their cross product will be a normal vector to the plane. Since the given line lies in the plane, its direction vector $\mathbf{a} = \langle -2, 5, 4 \rangle$ is one vector in the plane. We can verify that the given point $(6, 0, -2)$ does not lie on this line, so to find another nonparallel vector $\mathbf{b}$ which lies in the plane, we can pick any point on the line and find a vector connecting the points. If we put $t = 0$, we see that $(4, 3, 7)$ is on the line, so $\mathbf{b} = \langle 6 - 4, 0 - 3, -2 - 7 \rangle = \langle 2, -3, -9 \rangle$ and $\mathbf{n} = \mathbf{a} \times \mathbf{b} = \langle -45 + 12, 8 - 18, 6 - 10 \rangle = \langle -33, -10, -4 \rangle$. Thus, an equation of the plane is $-33(x - 6) - 10(y - 0) - 4[z - (-2)] = 0$ or $33x + 10y + 4z = 190$.

29. A direction vector for the line of intersection is $\mathbf{a} = \mathbf{n}_1 \times \mathbf{n}_2 = \langle 1, 1, -1 \rangle \times \langle 2, -1, 3 \rangle = \langle 2, -5, -3 \rangle$, and $\mathbf{a}$ is parallel to the desired plane. Another vector parallel to the plane is the vector connecting any point on the line of intersection to the given point $(-1, 2, 1)$ in the plane. Setting $x = 0$, the equations of the planes reduce to $y - z = 2$ and $-y + 3z = 1$ with simultaneous solution $y = \frac{7}{2}$ and $z = \frac{3}{2}$. So a point on the line is $\left(0, \frac{7}{2}, \frac{3}{2}\right)$ and another vector parallel to the plane is $\left\langle -1, -\frac{3}{2}, -\frac{1}{2} \right\rangle$. Then a normal vector to the plane is $\mathbf{n} = \langle 2, -5, -3 \rangle \times \left\langle -1, -\frac{3}{2}, -\frac{1}{2} \right\rangle = \langle -2, 4, -8 \rangle$ and an equation of the plane is $-2(x + 1) + 4(y - 2) - 8(z - 1) = 0$ or $x - 2y + 4z = -1$.

31. Substitute the parametric equations of the line into the equation of the plane: $(3 - t) - (2 + t) + 2(5t) = 9 \;\; \Rightarrow$ $8t = 8 \;\; \Rightarrow \;\; t = 1$. Therefore, the point of intersection of the line and the plane is given by $x = 3 - 1 = 2$, $y = 2 + 1 = 3$, and $z = 5(1) = 5$, that is, the point $(2, 3, 5)$.

33. Normal vectors for the planes are $\mathbf{n}_1 = \langle 1, 1, 1 \rangle$ and $\mathbf{n}_2 = \langle 1, -1, 1 \rangle$. The normals are not parallel, so neither are the planes. Furthermore, $\mathbf{n}_1 \cdot \mathbf{n}_2 = 1 - 1 + 1 = 1 \neq 0$, so the planes aren't perpendicular. The angle between them is given by

$$\cos\theta = \frac{\mathbf{n}_1 \cdot \mathbf{n}_2}{|\mathbf{n}_1|\,|\mathbf{n}_2|} = \frac{1}{\sqrt{3}\,\sqrt{3}} = \frac{1}{3} \;\; \Rightarrow \;\; \theta = \cos^{-1}\left(\tfrac{1}{3}\right) \approx 70.5°.$$

35. The normals are $\mathbf{n}_1 = \langle 1, -4, 2 \rangle$ and $\mathbf{n}_2 = \langle 2, -8, 4 \rangle$. Since $\mathbf{n}_2 = 2\mathbf{n}_1$, the normals (and thus the planes) are parallel.

37. (a) To find a point on the line of intersection, set one of the variables equal to a constant, say $z = 0$. (This will only work if the line of intersection crosses the xy-plane; otherwise, try setting x or y equal to 0.) Then the equations of the planes reduce to $x + y = 2$ and $3x - 4y = 6$. Solving these two equations gives $x = 2$, $y = 0$. So a point on the line of intersection is $(2, 0, 0)$. The direction of the line is $\mathbf{v} = \mathbf{n}_1 \times \mathbf{n}_2 = \langle 5 - 4, -3 - 5, -4 - 3 \rangle = \langle 1, -8, -7 \rangle$, and symmetric equations for the line are $x - 2 = \dfrac{y}{-8} = \dfrac{z}{-7}$.

(b) The angle between the planes satisfies $\cos \theta = \dfrac{\mathbf{n}_1 \cdot \mathbf{n}_2}{|\mathbf{n}_1| \, |\mathbf{n}_2|} = \dfrac{3 - 4 - 5}{\sqrt{3} \sqrt{50}} = -\dfrac{\sqrt{6}}{5}$. Therefore $\theta = \cos^{-1}\left(-\frac{\sqrt{6}}{5}\right) \approx 119\,°$ (or $61\,°$).

39. The plane contains the points $(a, 0, 0)$, $(0, b, 0)$ and $(0, 0, c)$. Thus the vectors $\mathbf{a} = \langle -a, b, 0 \rangle$ and $\mathbf{b} = \langle -a, 0, c \rangle$ lie in the plane, and $\mathbf{n} = \mathbf{a} \times \mathbf{b} = \langle bc - 0, 0 + ac, 0 + ab \rangle = \langle bc, ac, ab \rangle$ is a normal vector to the plane. The equation of the plane is therefore $bcx + acy + abz = abc + 0 + 0$ or $bcx + acy + abz = abc$. Notice that if $a \neq 0$, $b \neq 0$ and $c \neq 0$ then we can rewrite the equation as $\dfrac{x}{a} + \dfrac{y}{b} + \dfrac{z}{c} = 1$. This is a good equation to remember!

41. Two vectors which are perpendicular to the required line are the normal of the given plane, $\langle 1, 1, 1 \rangle$, and a direction vector for the given line, $\langle 1, -1, 2 \rangle$. So a direction vector for the required line is $\langle 1, 1, 1 \rangle \times \langle 1, -1, 2 \rangle = \langle 3, -1, -2 \rangle$. Thus L is given by $\langle x, y, z \rangle = \langle 0, 1, 2 \rangle + t\langle 3, -1, -2 \rangle$, or in parametric form, $x = 3t$, $y = 1 - t$, $z = 2 - 2t$.

43. Let P_i have normal vector $\mathbf{n}_i$. Then $\mathbf{n}_1 = \langle 4, -2, 6 \rangle$, $\mathbf{n}_2 = \langle 4, -2, -2 \rangle$, $\mathbf{n}_3 = \langle -6, 3, -9 \rangle$, $\mathbf{n}_4 = \langle 2, -1, -1 \rangle$. Now $\mathbf{n}_1 = -\frac{2}{3}\mathbf{n}_3$, so $\mathbf{n}_1$ and $\mathbf{n}_3$ are parallel, and hence P_1 and P_3 are parallel; similarly P_2 and P_4 are parallel because $\mathbf{n}_2 = 2\mathbf{n}_4$. However, $\mathbf{n}_1$ and $\mathbf{n}_2$ are not parallel. $\left(0, 0, \frac{1}{2}\right)$ lies on P_1, but not on P_3, so they are not the same plane, but both P_2 and P_4 contain the point $(0, 0, -3)$, so these two planes are identical.

45. Let $Q = (2, 2, 0)$ and $R = (3, -1, 5)$, points on the line corresponding to $t = 0$ and $t = 1$.

Let $P = (1, 2, 3)$. Then $\mathbf{a} = \overrightarrow{QR} = \langle 1, -3, 5 \rangle$, $\mathbf{b} = \overrightarrow{QP} = \langle -1, 0, 3 \rangle$. The distance is

$$d = \frac{|\mathbf{a} \times \mathbf{b}|}{|\mathbf{a}|} = \frac{|\langle 1, -3, 5 \rangle \times \langle -1, 0, 3 \rangle|}{|\langle 1, -3, 5 \rangle|} = \frac{|\langle -9, -8, -3 \rangle|}{|\langle 1, -3, 5 \rangle|} = \frac{\sqrt{9^2 + 8^2 + 3^2}}{\sqrt{1^2 + 3^2 + 5^2}} = \frac{\sqrt{154}}{\sqrt{35}} = \sqrt{\frac{22}{5}}.$$

47. By Equation 9, the distance is $D = \dfrac{1}{\sqrt{1 + 4 + 4}}\, [(1)(2) + (-2)(8) + (-2)(5) - 1] = \dfrac{25}{3}$.

49. Put $y = z = 0$ in the equation of the first plane to get the point $(-1, 0, 0)$ on the plane. Because the planes are parallel, the distance D between them is the distance from $(-1, 0, 0)$ to the second plane. By Equation 9,

$$D = \frac{|3(-1) + 6(0) - 3(0) - 4|}{\sqrt{3^2 + 6^2 + (-3)^2}} = \frac{7}{3\sqrt{6}} \text{ or } \frac{7\sqrt{6}}{18}.$$

51. The distance between two parallel planes is the same as the distance between a point on one of the planes and the other plane.

Let $P_0 = (x_0, y_0, z_0)$ be a point on the plane given by $ax + by + cz + d_1 = 0$. Then $ax_0 + by_0 + cz_0 + d_1 = 0$ and the distance between P_0 and the plane given by $ax + by + cz + d_2 = 0$ is, from Equation 9,

$$D = \frac{|ax_0 + by_0 + cz_0 + d_2|}{\sqrt{a^2 + b^2 + c^2}} = \frac{|-d_1 + d_2|}{\sqrt{a^2 + b^2 + c^2}} = \frac{|d_1 - d_2|}{\sqrt{a^2 + b^2 + c^2}}.$$

53. L_1: $x = y = z$ $\Rightarrow$ $x = y$ (1). L_2: $x + 1 = y/2 = z/3$ $\Rightarrow$ $x + 1 = y/2$ (2). The solution of (1) and (2) is

$x = y = -2$. However, when $x = -2$, $x = z$ $\Rightarrow$ $z = -2$, but $x + 1 = z/3$ $\Rightarrow$ $z = -3$, a contradiction. Hence the

lines do not intersect. For L_1, $\mathbf{v}_1 = \langle 1, 1, 1 \rangle$, and for L_2, $\mathbf{v}_2 = \langle 1, 2, 3 \rangle$, so the lines are not parallel. Thus the lines are skew

lines. If two lines are skew, they can be viewed as lying in two parallel planes and so the distance between the skew lines

would be the same as the distance between these parallel planes. The common normal vector to the planes must be

perpendicular to both $\langle 1, 1, 1 \rangle$ and $\langle 1, 2, 3 \rangle$, the direction vectors of the two lines. So set

$\mathbf{n} = \langle 1, 1, 1 \rangle \times \langle 1, 2, 3 \rangle = \langle 3 - 2, -3 + 1, 2 - 1 \rangle = \langle 1, -2, 1 \rangle$. From above, we know that $(-2, -2, -2)$ and $(-2, -2, -3)$

are points of L_1 and L_2 respectively. So in the notation of Equation 8, $1(-2) - 2(-2) + 1(-2) + d_1 = 0$ $\Rightarrow$ $d_1 = 0$ and

$1(-2) - 2(-2) + 1(-3) + d_2 = 0$ $\Rightarrow$ $d_2 = 1$.

By Exercise 51, the distance between these two skew lines is $D = \dfrac{|0 - 1|}{\sqrt{1 + 4 + 1}} = \dfrac{1}{\sqrt{6}}$.

Alternate solution (without reference to planes): A vector which is perpendicular to both of the lines is

$\mathbf{n} = \langle 1, 1, 1 \rangle \times \langle 1, 2, 3 \rangle = \langle 1, -2, 1 \rangle$. Pick any point on each of the lines, say $(-2, -2, -2)$ and $(-2, -2, -3)$, and form the

vector $\mathbf{b} = \langle 0, 0, 1 \rangle$ connecting the two points. The distance between the two skew lines is the absolute value of the scalar

projection of $\mathbf{b}$ along $\mathbf{n}$, that is, $D = \dfrac{|\mathbf{n} \cdot \mathbf{b}|}{|\mathbf{n}|} = \dfrac{|1 \cdot 0 - 2 \cdot 0 + 1 \cdot 1|}{\sqrt{1 + 4 + 1}} = \dfrac{1}{\sqrt{6}}$.

55. If $a \neq 0$, then $ax + by + cz + d = 0$ $\Rightarrow$ $a(x + d/a) + b(y - 0) + c(z - 0) = 0$ which by (7) is the scalar equation of the

plane through the point $(-d/a, 0, 0)$ with normal vector $\langle a, b, c \rangle$. Similarly, if $b \neq 0$ (or if $c \neq 0$) the equation of the plane can

be rewritten as $a(x - 0) + b(y + d/b) + c(z - 0) = 0$ [or as $a(x - 0) + b(y - 0) + c(z + d/c) = 0$] which by (7) is the

scalar equation of a plane through the point $(0, -d/b, 0)$ [or the point $(0, 0, -d/c)$] with normal vector $\langle a, b, c \rangle$.

9.6 Functions and Surfaces

1. (a) According to Table 1, $f(40, 15) = 25$, which means that if a 40-knot wind has been blowing in the open sea for 15 hours,
it will create waves with estimated heights of 25 feet.

(b) $h = f(30, t)$ means we fix v at 30 and allow t to vary, resulting in a function of one variable. Thus here, $h = f(30, t)$
gives the wave heights produced by 30-knot winds blowing for t hours. From the table (look at the row corresponding to
$v = 30$), the function increases but at a declining rate as t increases. In fact, the function values appear to be approaching a
limiting value of approximately 19, which suggests that 30-knot winds cannot produce waves higher than about 19 feet.

(c) $h = f(v, 30)$ means we fix t at 30, again giving a function of one variable. So, $h = f(v, 30)$ gives the wave heights
produced by winds of speed v blowing for 30 hours. From the table (look at the column corresponding to $t = 30$), the
function appears to increase at an increasing rate, with no apparent limiting value. This suggests that faster winds (lasting
30 hours) always create higher waves.

3. (a) $f(2, 0) = 2^2 e^{3(2)(0)} = 4(1) = 4$

(b) Since both x^2 and the exponential function are defined everywhere, $x^2 e^{3xy}$ is defined for all choices of values for x and y.
Thus, the domain of f is $\mathbb{R}^2$.

(c) Because the range of $g(x, y) = 3xy$ is $\mathbb{R}$, and the range of e^x is $(0, \infty)$, the range of $e^{g(x,y)} = e^{3xy}$ is $(0, \infty)$. The range
of x^2 is $[0, \infty)$, so the range of the product $x^2 e^{3xy}$ is $[0, \infty)$.

5. $\sqrt{y - x^2}$ is defined only when $y - x^2 \geq 0$, or $y \geq x^2$. In addition, f is not defined if $1 - x^2 = 0 \quad \Rightarrow \quad x = \pm 1$. Thus the domain of f is $\{(x, y) \mid y \geq x^2, x \neq \pm 1\}$.

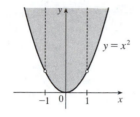

7. $\sqrt{1 - x^2}$ is defined only when $1 - x^2 \geq 0$, or $x^2 \leq 1 \quad \Leftrightarrow \quad -1 \leq x \leq 1$, and $\sqrt{1 - y^2}$ is defined only when $1 - y^2 \geq 0$, or $y^2 \leq 1 \quad \Leftrightarrow \quad -1 \leq y \leq 1$. Thus the domain of f is $\{(x, y) \mid -1 \leq x \leq 1, -1 \leq y \leq 1\}$.

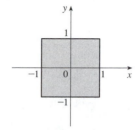

9. $z = 3$, a horizontal plane through the point $(0, 0, 3)$.

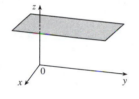

11. $z = 6 - 3x - 2y$ or $3x + 2y + z = 6$, a plane with intercepts 2, 3, and 6.

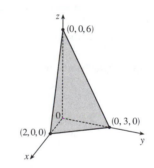

13. $z = y^2 + 1$, a parabolic cylinder.

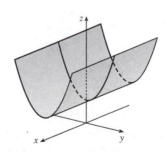

15. All six graphs have different traces in the planes $x = 0$ and $y = 0$, so we investigate these for each function.

(a) $f(x, y) = |x| + |y|$. The trace in $x = 0$ is $z = |y|$, and in $y = 0$ is $z = |x|$, so it must be graph VI.

(b) $f(x, y) = |xy|$. The trace in $x = 0$ is $z = 0$, and in $y = 0$ is $z = 0$, so it must be graph V.

(c) $f(x, y) = \dfrac{1}{1 + x^2 + y^2}$. The trace in $x = 0$ is $z = \dfrac{1}{1 + y^2}$, and in $y = 0$ is $z = \dfrac{1}{1 + x^2}$. In addition, we can see that f is close to 0 for large values of x and y, so this is graph I.

(d) $f(x, y) = \left(x^2 - y^2\right)^2$. The trace in $x = 0$ is $z = y^4$, and in $y = 0$ is $z = x^4$. Both graph II and graph IV seem plausible; notice the trace in $z = 0$ is $0 = \left(x^2 - y^2\right)^2 \quad \Rightarrow \quad y = \pm x$, so it must be graph IV.

(e) $f(x, y) = (x - y)^2$. The trace in $x = 0$ is $z = y^2$, and in $y = 0$ is $z = x^2$. Both graph II and graph IV seem plausible; notice the trace in $z = 0$ is $0 = (x - y)^2 \;\Rightarrow\; y = x$, so it must be graph II.

(f) $f(x, y) = \sin(|x| + |y|)$. The trace in $x = 0$ is $z = \sin|y|$, and in $y = 0$ is $z = \sin|x|$. In addition, notice that the oscillating nature of the graph is characteristic of trigonometric functions. So this is graph III.

17. The equation of the graph is $z = \sqrt{4x^2 + y^2}$ or equivalently

$4x^2 + y^2 = z^2$, $z \geq 0$. Traces in $x = k$ are $z^2 - y^2 = 4k^2$, $z \geq 0$, a

family of hyperbolas where we have only the upper branch. Traces in

$y = k$ are $z^2 - 4x^2 = k^2$, $z \geq 0$, again a family of half-hyperbolas.

Traces in $z = k$, $k \geq 0$, are $4x^2 + y^2 = k^2$ or $x^2 + \dfrac{y^2}{4} = \dfrac{k^2}{4}$, a family

of ellipses. Note that the original equation can be written as

$x^2 + \dfrac{y^2}{4} = \dfrac{z^2}{4}$, $z \geq 0$, which we recognize as the upper half of an

elliptical cone.

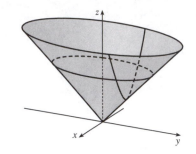

19. $y = z^2 - x^2$. The traces in $x = k$ are the parabolas $y = z^2 - k^2$;

the traces in $y = k$ are $k = z^2 - x^2$, which are hyperbolas (note the

hyperbolas are oriented differently for $k > 0$ than for $k < 0$); and the

traces in $z = k$ are the parabolas $y = k^2 - x^2$. Thus, $\dfrac{y}{1} = \dfrac{z^2}{1^2} - \dfrac{x^2}{1^2}$

is a hyperbolic paraboloid.

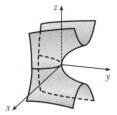

21. Completing squares in y and z gives $4x^2 + (y - 2)^2 + 4(z - 3)^2 = 4$ or $x^2 + \dfrac{(y - 2)^2}{4} + (z - 3)^2 = 1$, an ellipsoid with

center $(0, 2, 3)$.

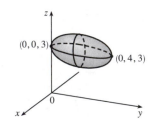

23. (a) In $\mathbb{R}^2$, $x^2 + y^2 = 1$ represents a circle of radius 1 centered at the origin.

(b) In $\mathbb{R}^3$, the equation doesn't involve z, which means that any horizontal plane $z = k$ intersects the surface in a circle $x^2 + y^2 = 1$, $z = k$. Thus the surface is a circular cylinder, made up of infinitely many shifted copies of the circle $x^2 + y^2 = 1$, with axis the z-axis.

(c) In $\mathbb{R}^3$, $x^2 + z^2 = 1$ also represents a circular cylinder of radius 1, this time with axis the y-axis.

25. (a) The traces of $x^2 + y^2 - z^2 = 1$ in $x = k$ are $y^2 - z^2 = 1 - k^2$, a family of hyperbolas. (Note that the hyperbolas are oriented differently for $-1 < k < 1$ than for $k < -1$ or $k > 1$.) The traces in $y = k$ are $x^2 - z^2 = 1 - k^2$, a similar family of hyperbolas. The traces in $z = k$ are $x^2 + y^2 = 1 + k^2$, a family of circles. For $k = 0$, the trace in the xy-plane, the circle is of radius 1. As $|k|$ increases, so does the radius of the circle. This behavior, combined with the hyperbolic vertical traces, gives the graph of the hyperboloid of one sheet in Table 2.

(b) The shape of the surface is unchanged, but the hyperboloid is rotated so that its axis is the y-axis. Traces in $y = k$ are circles, while traces in $x = k$ and $z = k$ are hyperbolas.

(c) Completing the square in y gives $x^2 + (y+1)^2 - z^2 = 1$. The surface is a hyperboloid identical to the one in part (a) but shifted one unit in the negative y-direction.

27. $f(x, y) = 3x - x^4 - 4y^2 - 10xy$

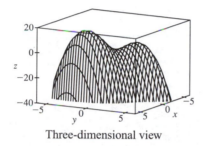

Three-dimensional view

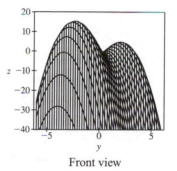

Front view

It does appear that the function has a maximum value, at the higher of the two "hilltops." From the front view graph, the maximum value appears to be approximately 15. Both hilltops could be considered local maximum points, as the values of f there are larger than at the neighboring points. There does not appear to be any local minimum point; although the valley shape between the two peaks looks like a minimum of some kind, some neighboring points have lower function values.

29.

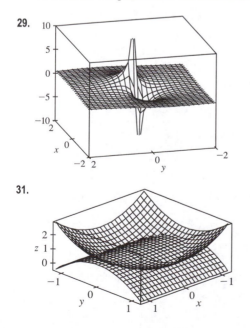

$f(x, y) = \dfrac{x + y}{x^2 + y^2}$. As both x and y become large, the function values appear to approach 0, regardless of which direction is considered. As (x, y) approaches the origin, the graph exhibits asymptotic behavior. From some directions, $f(x, y) \to \infty$, while in others $f(x, y) \to -\infty$. (These are the vertical spikes visible in the graph.) If the graph is examined carefully, however, one can see that $f(x, y)$ approaches 0 along the line $y = -x$.

31.

The curve of intersection looks like a bent ellipse. The projection of this curve onto the xy-plane is the set of points $(x, y, 0)$ which satisfy $x^2 + y^2 = 1 - y^2 \iff x^2 + 2y^2 = 1 \iff$

$x^2 + \dfrac{y^2}{\left(1/\sqrt{2}\right)^2} = 1$. This is an equation of an ellipse.

33. If (a, b, c) satisfies $z = y^2 - x^2$, then $c = b^2 - a^2$. L_1: $x = a + t$, $y = b + t$, $z = c + 2(b - a)t$,

L_2: $x = a + t$, $y = b - t$, $z = c - 2(b + a)t$. Substitute the parametric equations of L_1 into the equation

of the hyperbolic paraboloid in order to find the points of intersection: $z = y^2 - x^2 \Rightarrow$

$c + 2(b - a)t = (b + t)^2 - (a + t)^2 = b^2 - a^2 + 2(b - a)t \Rightarrow c = b^2 - a^2$. As this is true for all values of t,

L_1 lies on $z = y^2 - x^2$. Performing similar operations with L_2 gives: $z = y^2 - x^2 \Rightarrow$

$c - 2(b + a)t = (b - t)^2 - (a + t)^2 = b^2 - a^2 - 2(b + a)t \Rightarrow c = b^2 - a^2$. This tells us that all of L_2 also lies on

$z = y^2 - x^2$.

9.7 Cylindrical and Spherical Coordinates

1. See Figure 2 and the accompanying discussion on page 685; see the paragraph preceding Example 2 on page 686.

3. (a)

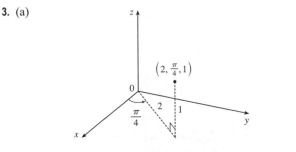

(b)

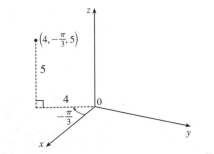

$x = 2 \cos \dfrac{\pi}{4} = \sqrt{2}$, $y = 2 \sin \dfrac{\pi}{4} = \sqrt{2}$, $z = 1$, so

the point is $\left(\sqrt{2}, \sqrt{2}, 1\right)$ in rectangular coordinates.

$x = 4 \cos\left(-\dfrac{\pi}{3}\right) = 2$, $y = 4 \sin\left(-\dfrac{\pi}{3}\right) = -2\sqrt{3}$, and

$z = 5$, so the point is $\left(2, -2\sqrt{3}, 5\right)$ in rectangular

coordinates.

5. (a) $r^2 = x^2 + y^2 = 1^2 + (-1)^2 = 2$ so $r = \sqrt{2}$; $\tan\theta = \dfrac{y}{x} = \dfrac{-1}{1} = -1$ and the point $(1, -1)$ is in the fourth quadrant of

the xy-plane, so $\theta = \dfrac{7\pi}{4} + 2n\pi$; $z = 4$. Thus, one set of cylindrical coordinates is $\left(\sqrt{2}, \dfrac{7\pi}{4}, 4\right)$.

(b) $r^2 = (-1)^2 + \left(-\sqrt{3}\right)^2 = 4$ so $r = 2$; $\tan\theta = \dfrac{-\sqrt{3}}{-1} = \sqrt{3}$ and the point $\left(-1, -\sqrt{3}\right)$ is in the third quadrant of the

xy-plane, so $\theta = \dfrac{4\pi}{3} + 2n\pi$; $z = 2$. Thus, one set of cylindrical coordinates is $\left(2, \dfrac{4\pi}{3}, 2\right)$.

7. (a)

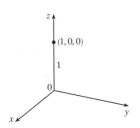

(b)

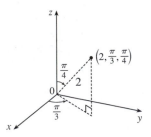

$x = \rho \sin\phi \cos\theta = (1) \sin 0 \cos 0 = 0$,

$y = \rho \sin\phi \sin\theta = (1) \sin 0 \sin 0 = 0$, and

$z = \rho \cos\phi = (1) \cos 0 = 1$ so the point is

$(0, 0, 1)$ in rectangular coordinates.

$x = 2 \sin\dfrac{\pi}{4} \cos\dfrac{\pi}{3} = \dfrac{\sqrt{2}}{2}$, $y = 2 \sin\dfrac{\pi}{4} \sin\dfrac{\pi}{3} = \dfrac{\sqrt{6}}{2}$,

$z = 2 \cos\dfrac{\pi}{4} = \sqrt{2}$ so the point is $\left(\dfrac{\sqrt{2}}{2}, \dfrac{\sqrt{6}}{2}, \sqrt{2}\right)$ in

rectangular coordinates.

9. (a) $\rho = \sqrt{x^2 + y^2 + z^2} = \sqrt{1 + 3 + 12} = 4$, $\cos\phi = \dfrac{z}{\rho} = \dfrac{2\sqrt{3}}{4} = \dfrac{\sqrt{3}}{2}$ $\Rightarrow$ $\phi = \dfrac{\pi}{6}$, and

$\cos\theta = \dfrac{x}{\rho\sin\phi} = \dfrac{1}{4\sin(\pi/6)} = \dfrac{1}{2}$ $\Rightarrow$ $\theta = \dfrac{\pi}{3}$ (since $y > 0$). Thus spherical coordinates are $\left(4, \dfrac{\pi}{3}, \dfrac{\pi}{6}\right)$.

(b) $\rho = \sqrt{0 + 1 + 1} = \sqrt{2}$, $\cos\phi = \dfrac{-1}{\sqrt{2}}$ $\Rightarrow$ $\phi = \dfrac{3\pi}{4}$, and $\cos\theta = \dfrac{0}{\sqrt{2}\sin(3\pi/4)} = 0$ $\Rightarrow$ $\theta = \dfrac{3\pi}{2}$

(since $y < 0$). Thus spherical coordinates are $\left(\sqrt{2}, \dfrac{3\pi}{2}, \dfrac{3\pi}{4}\right)$.

11. Since $r = 3$, $x^2 + y^2 = 9$ and the surface is a circular cylinder with radius 3 and axis the z-axis.

13. Since $\phi = \dfrac{\pi}{3}$, the surface is the top half of the right circular cone with vertex at the origin and axis the positive z-axis.

15. $z = r^2 = x^2 + y^2$, so the surface is a circular paraboloid with vertex at the origin and axis the positive z-axis.

17. $r = 2\cos\theta$ $\Rightarrow$ $r^2 = x^2 + y^2 = 2r\cos\theta = 2x$ $\Leftrightarrow$ $(x-1)^2 + y^2 = 1$, which is the equation of a circular cylinder with

radius 1, whose axis is the vertical line $x = 1$, $y = 0$, $z = z$.

19. Since $r^2 + z^2 = 25$ and $r^2 = x^2 + y^2$, we have $x^2 + y^2 + z^2 = 25$, a sphere with radius 5 and center at the origin.

21. (a) $x^2 + y^2 = r^2$, so the equation becomes $z = r^2$.

(b) $x = \rho\sin\phi\cos\theta$, $y = \rho\sin\phi\sin\theta$, and $z = \rho\cos\phi$, so the equation becomes

$\rho\cos\phi = (\rho\sin\phi\cos\theta)^2 + (\rho\sin\phi\sin\theta)^2$ or $\rho\cos\phi = \rho^2\sin^2\phi$ or $\rho\sin^2\phi = \cos\phi$.

23. (a) $r^2 = 2r\sin\theta$ or $r = 2\sin\theta$.

(b) $\rho^2\sin^2\phi(\cos^2\theta + \sin^2\theta) = 2\rho\sin\phi\sin\theta$ or $\rho\sin^2\phi = 2\sin\phi\sin\theta$ or $\rho\sin\phi = 2\sin\theta$.

25.

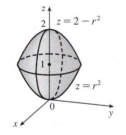

$z = r^2 = x^2 + y^2$ is a circular paraboloid with vertex

$(0, 0, 0)$, opening upward. $z = 2 - r^2$ $\Rightarrow$

$z - 2 = -(x^2 + y^2)$ is a circular paraboloid with

vertex $(0, 0, 2)$ opening downward. Thus

$r^2 \le z \le 2 - r^2$ is the solid region enclosed by these

two surfaces.

27.

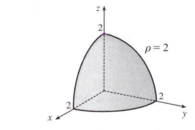

$\rho = 2$ represents a sphere of radius 2, centered at the

origin, so $\rho \le 2$ is this sphere and its interior.

$0 \le \phi \le \dfrac{\pi}{2}$ restricts the solid to that portion of the

region that lies on or above the xy-plane, and

$0 \le \theta \le \dfrac{\pi}{2}$ further restricts the solid to the first octant.

Thus the solid is the portion in the first octant of the

solid ball centered at the origin with radius 2.

29. $-\dfrac{\pi}{2} \le \theta \le \dfrac{\pi}{2}$ restricts the solid to the 4 octants in which x is positive. $\rho = \sec\phi$

$\Rightarrow$ $\rho\cos\phi = z = 1$, which is the equation of a horizontal plane. $0 \le \phi \le \dfrac{\pi}{6}$

describes a cone, opening upward. So the solid lies above the cone $\phi = \dfrac{\pi}{6}$ and

below the plane $z = 1$.

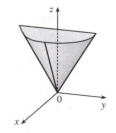

31. We can position the cylindrical shell vertically so that its axis coincides with the z-axis and its base lies in the xy-plane. If we use centimeters as the unit of measurement, then cylindrical coordinates conveniently describe the shell as $6 \leq r \leq 7$, $0 \leq \theta \leq 2\pi, 0 \leq z \leq 20$.

33. $z \geq \sqrt{x^2 + y^2}$ because the solid lies above the cone. Squaring both sides of this inequality gives $z^2 \geq x^2 + y^2$ $\Rightarrow$ $2z^2 \geq x^2 + y^2 + z^2 = \rho^2$ $\Rightarrow$ $z^2 = \rho^2 \cos^2 \phi \geq \frac{1}{2}\rho^2$ $\Rightarrow$ $\cos^2 \phi \geq \frac{1}{2}$. The cone opens upward so that the inequality is $\cos \phi \geq \frac{1}{\sqrt{2}}$, or equivalently $0 \leq \phi \leq \frac{\pi}{4}$. In spherical coordinates the sphere $z = x^2 + y^2 + z^2$ is $\rho \cos \phi = \rho^2$ $\Rightarrow$ $\rho = \cos \phi$. $0 \leq \rho \leq \cos \phi$ because the solid lies below the sphere. The solid can therefore be described as the region in spherical coordinates satisfying $0 \leq \rho \leq \cos \phi, 0 \leq \phi \leq \frac{\pi}{4}$.

35. In cylindrical coordinates, the equation of the cylinder is $r = 3, 0 \leq z \leq 10$. The hemisphere is the upper part of the sphere radius 3, center $(0, 0, 10)$, equation $r^2 + (z - 10)^2 = 3^2, z \geq 10$. In Maple, we can use either the `coords=cylindrical` option in a regular `plot` command, or the `plots[cylinderplot]` command. In Mathematica, we can use `ParametricPlot3d`.

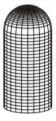

9 Review

CONCEPT CHECK

1. A scalar is a real number, while a vector is a quantity that has both a real-valued magnitude and a direction.

2. To add two vectors geometrically, we can use either the Triangle Law or the Parallelogram Law, as illustrated in Figures 3 and 4 in Section 9.2. (See also the definition of vector addition on page 643.) Algebraically, we add the corresponding components of the vectors.

3. For $c > 0$, $c\mathbf{a}$ is a vector with the same direction as $\mathbf{a}$ and length c times the length of $\mathbf{a}$. If $c < 0$, $c\mathbf{a}$ points in the opposite direction as $\mathbf{a}$ and has length $|c|$ times the length of $\mathbf{a}$. (See Figures 7 and 15 in Section 9.2.) Algebraically, to find $c\mathbf{a}$ we multiply each component of $\mathbf{a}$ by c.

4. See (1) in Section 9.2.

5. See the definition on page 651 and the boxed equation on page 653.

6. The dot product can be used to determine the work done moving an object given the force and displacement vectors. The dot product can also be used to find the angle between two vectors and the scalar projection of one vector onto another. In particular, the dot product can determine if two vectors are orthogonal.

7. See the boxed equations on page 655 as well as Figures 5 and 6 and the accompanying discussion on pages 654–55.

8. See the definition on page 658; use either (2) or (4) in Section 9.4.

9. The cross product can be used to determine torque if the force and position vectors are known. In addition, the cross product can be used to create a vector orthogonal to two given vectors as well as to determine if two vectors are parallel. The cross product can also be used to find the area of a parallelogram determined by two vectors.

10. (a) The area of the parallelogram determined by $\mathbf{a}$ and $\mathbf{b}$ is the length of the cross product: $|\mathbf{a} \times \mathbf{b}|$.

(b) The volume of the parallelepiped determined by $\mathbf{a}$, $\mathbf{b}$, and $\mathbf{c}$ is the magnitude of their scalar triple product: $|\mathbf{a} \cdot (\mathbf{b} \times \mathbf{c})|$.

11. If an equation of the plane is known, it can be written as $ax + by + cz + d = 0$. A normal vector, which is perpendicular to the plane, is $\langle a, b, c \rangle$ (or any scalar multiple of $\langle a, b, c \rangle$). If an equation is not known, we can use points on the plane to find two non-parallel vectors which lie in the plane. The cross product of these vectors is a vector perpendicular to the plane.

12. The angle between two intersecting planes is defined as the acute angle between their normal vectors. We can find this angle using the definition of the dot product on page 651.

13. See (1), (2), and (3) in Section 9.5.

14. See (5), (6), and (7) in Section 9.5.

15. (a) Two (nonzero) vectors are parallel if and only if one is a scalar multiple of the other. In addition, two nonzero vectors are parallel if and only if their cross product is $\mathbf{0}$.

    (b) Two vectors are perpendicular if and only if their dot product is 0.

    (c) Two planes are parallel if and only if their normal vectors are parallel.

16. (a) Determine the vectors $\overrightarrow{PQ} = \langle a_1, a_2, a_3 \rangle$ and $\overrightarrow{PR} = \langle b_1, b_2, b_3 \rangle$. If there is a scalar t such that $\langle a_1, a_2, a_3 \rangle = t \langle b_1, b_2, b_3 \rangle$, then the vectors are parallel and the points must all lie on the same line. Alternatively, if $\overrightarrow{PQ} \times \overrightarrow{PR} = \mathbf{0}$, then $\overrightarrow{PQ}$ and $\overrightarrow{PR}$ are parallel, so P, Q, and R are collinear.

    Thirdly, an algebraic method is to determine an equation of the line joining two of the points, and then check whether or not the third point satisfies this equation.

    (b) Find the vectors $\overrightarrow{PQ} = \mathbf{a}$, $\overrightarrow{PR} = \mathbf{b}$, $\overrightarrow{PS} = \mathbf{c}$. $\mathbf{a} \times \mathbf{b}$ is normal to the plane formed by P, Q and R, and so S lies on this plane if $\mathbf{a} \times \mathbf{b}$ and $\mathbf{c}$ are orthogonal, that is, if $(\mathbf{a} \times \mathbf{b}) \cdot \mathbf{c} = 0$. (Or use the reasoning in Example 6 in Section 9.4.) Alternatively, find an equation for the plane determined by three of the points and check whether or not the fourth point satisfies this equation.

17. (a) See Exercise 9.4.27.

    (b) See Example 8 in Section 9.5.

    (c) See Example 10 in Section 9.5.

18. One method of graphing a function of two variables is to first find traces (see Example 6 in Section 9.6 and the discussion preceding it).

19. See Table 2 in Section 9.6.

20. (a) See (1) and the discussion accompanying Figure 4 in Section 9.7.

    (b) See (3) and Figures 7–9, and the accompanying discussion, in Section 9.7.

TRUE-FALSE QUIZ

1. True, by Property 2 of the dot product. (See page 654.)

3. True. If θ is the angle between $\mathbf{u}$ and $\mathbf{v}$, then by the definition of the cross product,
$|\mathbf{u} \times \mathbf{v}| = |\mathbf{u}|\,|\mathbf{v}| \sin\theta = |\mathbf{v}|\,|\mathbf{u}| \sin\theta = |\mathbf{v} \times \mathbf{u}|$.
(Or, by Properties 1 and 2 of the cross product, $|\mathbf{u} \times \mathbf{v}| = |-\mathbf{v} \times \mathbf{u}| = |-1|\,|\mathbf{v} \times \mathbf{u}| = |\mathbf{v} \times \mathbf{u}|$.)

5. Property 2 of the cross product tells us that this is true.

7. This is true by (6) in Section 9.4.

9. This is true because $\mathbf{u} \times \mathbf{v}$ is orthogonal to $\mathbf{u}$ (see page 658), and the dot product of two orthogonal vectors is 0.

11. If $|\mathbf{u}| = 1$, $|\mathbf{v}| = 1$ and θ is the angle between these two vectors (so $0 \le \theta \le \pi$), then by the definition of the cross product, $|\mathbf{u} \times \mathbf{v}| = |\mathbf{u}| \, |\mathbf{v}| \sin \theta = \sin \theta$, which is equal to 1 if and only if $\theta = \frac{\pi}{2}$ (that is, if and only if the two vectors are orthogonal). Therefore, the assertion that the cross product of two unit vectors is a unit vector is false.

13. This is false. In $\mathbb{R}^2$, $x^2 + y^2 = 1$ represents a circle, but $\{(x, y, z) \mid x^2 + y^2 = 1\}$ represents a *three-dimensional surface*, namely, a circular cylinder with axis the z-axis.

15. False. For example, $\mathbf{i} \cdot \mathbf{j} = 0$ but $\mathbf{i} \ne \mathbf{0}$ and $\mathbf{j} \ne \mathbf{0}$.

EXERCISES

1. (a) The radius of the sphere is the distance between the points $(-1, 2, 1)$ and $(6, -2, 3)$, namely
$\sqrt{[6 - (-1)]^2 + (-2 - 2)^2 + (3 - 1)^2} = \sqrt{69}$. By the formula for an equation of a sphere (see page 640), an equation of the sphere with center $(-1, 2, 1)$ and radius $\sqrt{69}$ is $(x + 1)^2 + (y - 2)^2 + (z - 1)^2 = 69$.

    (b) The intersection of this sphere with the yz-plane is the set of points on the sphere whose x-coordinate is 0. Putting $x = 0$ into the equation, we have $(y - 2)^2 + (z - 1)^2 = 68$, $x = 0$ which represents a circle in the yz-plane with center $(0, 2, 1)$ and radius $\sqrt{68}$.

    (c) Completing squares gives $(x - 4)^2 + (y + 1)^2 + (z + 3)^2 = -1 + 16 + 1 + 9 = 25$. Thus, the sphere is centered at $(4, -1, -3)$ and has radius 5.

3. $\mathbf{u} \cdot \mathbf{v} = |\mathbf{u}| \, |\mathbf{v}| \cos 45\,° = (2)(3)\frac{\sqrt{2}}{2} = 3\sqrt{2}$. $|\mathbf{u} \times \mathbf{v}| = |\mathbf{u}| \, |\mathbf{v}| \sin 45\,° = (2)(3)\frac{\sqrt{2}}{2} = 3\sqrt{2}$. By the right-hand rule, $\mathbf{u} \times \mathbf{v}$ is directed out of the page.

5. For the two vectors to be orthogonal, we need $\langle 3, 2, x \rangle \cdot \langle 2x, 4, x \rangle = 0 \quad \Leftrightarrow \quad (3)(2x) + (2)(4) + (x)(x) = 0 \quad \Leftrightarrow$
$x^2 + 6x + 8 = 0 \quad \Leftrightarrow \quad (x + 2)(x + 4) = 0 \quad \Leftrightarrow \quad x = -2$ or $x = -4$.

7. (a) $(\mathbf{u} \times \mathbf{v}) \cdot \mathbf{w} = \mathbf{u} \cdot (\mathbf{v} \times \mathbf{w}) = 2$

    (b) $\mathbf{u} \cdot (\mathbf{w} \times \mathbf{v}) = \mathbf{u} \cdot [-(\mathbf{v} \times \mathbf{w})] = -\mathbf{u} \cdot (\mathbf{v} \times \mathbf{w}) = -2$

    (c) $\mathbf{v} \cdot (\mathbf{u} \times \mathbf{w}) = (\mathbf{v} \times \mathbf{u}) \cdot \mathbf{w} = -(\mathbf{u} \times \mathbf{v}) \cdot \mathbf{w} = -2$

    (d) $(\mathbf{u} \times \mathbf{v}) \cdot \mathbf{v} = \mathbf{u} \cdot (\mathbf{v} \times \mathbf{v}) = \mathbf{u} \cdot \mathbf{0} = 0$

9. For simplicity, consider a unit cube positioned with its back left corner at the origin. Vector representations of the diagonals joining the points $(0, 0, 0)$ to $(1, 1, 1)$ and $(1, 0, 0)$ to $(0, 1, 1)$ are $\langle 1, 1, 1 \rangle$ and $\langle -1, 1, 1 \rangle$. Let θ be the angle between these two vectors. $\langle 1, 1, 1 \rangle \cdot \langle -1, 1, 1 \rangle = -1 + 1 + 1 = 1 = |\langle 1, 1, 1 \rangle| \, |\langle -1, 1, 1 \rangle| \cos \theta = 3 \cos \theta \quad \Rightarrow \quad \cos \theta = \frac{1}{3} \quad \Rightarrow$
$\theta = \cos^{-1}\left(\frac{1}{3}\right) \approx 71\,°$.

11. $\overrightarrow{AB} = \langle 1, 0, -1 \rangle$, $\overrightarrow{AC} = \langle 0, 4, 3 \rangle$, so

    (a) a vector perpendicular to the plane is $\overrightarrow{AB} \times \overrightarrow{AC} = \langle 0 + 4, -(3 + 0), 4 - 0 \rangle = \langle 4, -3, 4 \rangle$.

    (b) $\frac{1}{2} \left| \overrightarrow{AB} \times \overrightarrow{AC} \right| = \frac{1}{2}\sqrt{16 + 9 + 16} = \frac{\sqrt{41}}{2}$.

13. Let F_1 be the magnitude of the force directed $20\,°$ away from the direction of shore, and let F_2 be the magnitude of the other force. Separating these forces into components parallel to the direction of the resultant force and perpendicular to it gives

$F_1 \cos 20\,° + F_2 \cos 30\,° = 255$ **(1)**, and $F_1 \sin 20\,° - F_2 \sin 30\,° = 0 \quad \Rightarrow \quad F_1 = F_2 \dfrac{\sin 30\,°}{\sin 20\,°}$ **(2)**. Substituting **(2)** into

(1) gives $F_2(\sin 30\,° \cot 20\,° + \cos 30\,°) = 255 \quad \Rightarrow \quad F_2 \approx 114$ N. Substituting this into **(2)** gives $F_1 \approx 166$ N.

15. The line has direction $\mathbf{v} = \langle -3, 2, 3 \rangle$. Letting $P_0 = (4, -1, 2)$, parametric equations are $x = 4 - 3t$, $y = -1 + 2t$, $z = 2 + 3t$.

17. A direction vector for the line is a normal vector for the plane, $\mathbf{n} = \langle 2, -1, 5 \rangle$, and parametric equations for the line are $x = -2 + 2t$, $y = 2 - t$, $z = 4 + 5t$.

19. Here the vectors $\mathbf{a} = \langle 4 - 3, 0 - (-1), 2 - 1 \rangle = \langle 1, 1, 1 \rangle$ and $\mathbf{b} = \langle 6 - 3, 3 - (-1), 1 - 1 \rangle = \langle 3, 4, 0 \rangle$ lie in the plane, so $\mathbf{n} = \mathbf{a} \times \mathbf{b} = \langle -4, 3, 1 \rangle$ is a normal vector to the plane and an equation of the plane is
$-4(x - 3) + 3(y - (-1)) + 1(z - 1) = 0$ or $-4x + 3y + z = -14$.

21. $\mathbf{n}_1 = \langle 1, 0, -1 \rangle$ and $\mathbf{n}_2 = \langle 0, 1, 2 \rangle$. Setting $z = 0$, it is easy to see that $(1, 3, 0)$ is a point on the line of intersection of $x - z = 1$ and $y + 2z = 3$. The direction of this line is $\mathbf{v}_1 = \mathbf{n}_1 \times \mathbf{n}_2 = \langle 1, -2, 1 \rangle$. A second vector parallel to the desired plane is $\mathbf{v}_2 = \langle 1, 1, -2 \rangle$, since it is perpendicular to $x + y - 2z = 1$. Therefore, the normal of the plane in question is $\mathbf{n} = \mathbf{v}_1 \times \mathbf{v}_2 = \langle 4 - 1, 1 + 2, 1 + 2 \rangle = 3 \langle 1, 1, 1 \rangle$. Taking $(x_0, y_0, z_0) = (1, 3, 0)$, the equation we are looking for is
$(x - 1) + (y - 3) + z = 0 \quad \Leftrightarrow \quad x + y + z = 4$.

23. Since the direction vectors $\langle 2, 3, 4 \rangle$ and $\langle 6, -1, 2 \rangle$ aren't parallel, neither are the lines. For the lines to intersect, the three equations $1 + 2t = -1 + 6s$, $2 + 3t = 3 - s$, $3 + 4t = -5 + 2s$ must be satisfied simultaneously. Solving the first two equations gives $t = \frac{1}{5}$, $s = \frac{2}{5}$ and checking we see these values don't satisfy the third equation. Thus the lines aren't parallel and they don't intersect, so they must be skew.

25. By Exercise 9.5.51, $D = \dfrac{|2 - 24|}{\sqrt{26}} = \dfrac{22}{\sqrt{26}}$.

27. $\ln(x - y^2)$ is defined only when $x - y^2 > 0$, or $x > y^2$, and x is defined for all real numbers, so the domain of the product $x \ln(x - y^2)$ is $\{(x, y) \mid x > y^2\}$.

29. The graph is the plane $z = 6 - 2x - 3y \quad \Rightarrow \quad 2x + 3y + z = 6$. The intercepts with the coordinate axes are $(3, 0, 0)$, $(0, 2, 0)$, and $(0, 0, 6)$ which enable us to sketch the portion of the plane that lies in the first octant.

31. The equation is $z = 4 - x^2 - 4y^2$. The traces in $x = k$ are $z = 4 - k^2 - 4y^2$, a family of parabolas opening downward, as are the traces in $y = k$, $z = 4 - 4k^2 - x^2$. The traces in $z = k$ are $x^2 + 4y^2 = 4 - k$, a family of ellipses, so the surface is an elliptic paraboloid.

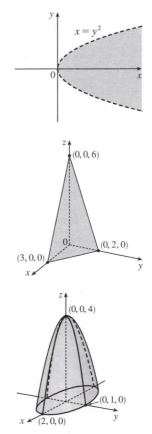

33. An equivalent equation is $\dfrac{x^2}{(1/2)^2} + y^2 + z^2 = 1$, an ellipsoid centered at

the origin with intercepts $\pm\frac{1}{2}$, ± 1,

and ± 1.

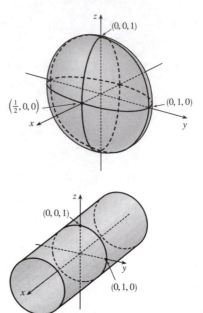

35. $y^2 + z^2 = 1$ is the equation of a circular cylinder with axis the x-axis.

37. $x = r\cos\theta = 2\sqrt{3}\cos\frac{\pi}{3} = 2\sqrt{3}\cdot\frac{1}{2} = \sqrt{3}$, $y = r\sin\theta = 2\sqrt{3}\sin\frac{\pi}{3} = 2\sqrt{3}\cdot\frac{\sqrt{3}}{2} = 3$, $z = 2$, so in rectangular

coordinates the point is $\left(\sqrt{3}, 3, 2\right)$. $\rho = \sqrt{r^2 + z^2} = \sqrt{12 + 4} = 4$, $\theta = \frac{\pi}{3}$, and $\cos\phi = \frac{z}{\rho} = \frac{1}{2}$, so $\phi = \frac{\pi}{3}$ and spherical

coordinates are $\left(4, \frac{\pi}{3}, \frac{\pi}{3}\right)$.

39. $x = \rho\sin\phi\cos\theta = 8\sin\frac{\pi}{6}\cos\frac{\pi}{4} = 8\cdot\frac{1}{2}\cdot\frac{\sqrt{2}}{2} = 2\sqrt{2}$, $y = \rho\sin\phi\sin\theta = 8\sin\frac{\pi}{6}\sin\frac{\pi}{4} = 2\sqrt{2}$, and

$z = \rho\cos\phi = 8\cos\frac{\pi}{6} = 8\cdot\frac{\sqrt{3}}{2} = 4\sqrt{3}$. Thus rectangular coordinates for the point are $\left(2\sqrt{2}, 2\sqrt{2}, 4\sqrt{3}\right)$.

$r^2 = x^2 + y^2 = 8 + 8 = 16 \quad\Rightarrow\quad r = 4$, $\theta = \frac{\pi}{4}$, and $z = 4\sqrt{3}$, so cylindrical coordinates are $\left(4, \frac{\pi}{4}, 4\sqrt{3}\right)$.

41. $x^2 + y^2 + z^2 = 4$. In cylindrical coordinates, this becomes $r^2 + z^2 = 4$. In spherical coordinates, it becomes $\rho^2 = 4$
or $\rho = 2$.

43. The resulting surface is a circular paraboloid with equation $z = 4x^2 + 4y^2$. Changing to cylindrical coordinates we have
$z = 4\left(x^2 + y^2\right) = 4r^2$.

FOCUS ON PROBLEM SOLVING

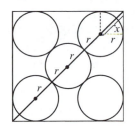

1. Since three-dimensional situations are often difficult to visualize and work with, let us first try to find an analogous problem in two dimensions. The analogue of a cube is a square and the analogue of a sphere is a circle. Thus a similar problem in two dimensions is the following: if five circles with the same radius r are contained in a square of side 1 m so that the circles touch each other and four of the circles touch two sides of the square, find r.

The diagonal of the square is $\sqrt{2}$. The diagonal is also $4r + 2x$. But x is the diagonal of a smaller square of side r. Therefore

$$x = \sqrt{2}\,r \quad \Rightarrow \quad \sqrt{2} = 4r + 2x = 4r + 2\sqrt{2}\,r = \left(4 + 2\sqrt{2}\right)r \quad \Rightarrow \quad r = \frac{\sqrt{2}}{4 + 2\sqrt{2}}.$$

Let's use these ideas to solve the original three-dimensional problem. The diagonal of the cube is $\sqrt{1^2 + 1^2 + 1^2} = \sqrt{3}$. The diagonal of the cube is also $4r + 2x$ where x is the diagonal of a smaller cube with edge r. Therefore

$$x = \sqrt{r^2 + r^2 + r^2} = \sqrt{3}\,r \quad \Rightarrow \quad \sqrt{3} = 4r + 2x = 4r + 2\sqrt{3}\,r = \left(4 + 2\sqrt{3}\right)r. \text{ Thus } r = \frac{\sqrt{3}}{4 + 2\sqrt{3}} = \frac{2\sqrt{3} - 3}{2}.$$

The radius of each ball is $\left(\sqrt{3} - \frac{3}{2}\right)$ m.

3. (a) We find the line of intersection L as in Example 9.5.7(b). Observe that the point $(-1, c, c)$ lies on both planes. Now since L lies in both planes, it is perpendicular to both of the normal vectors $\mathbf{n}_1$ and $\mathbf{n}_2$, and thus parallel to their cross product

$$\mathbf{n}_1 \times \mathbf{n}_2 = \begin{vmatrix} \mathbf{i} & \mathbf{j} & \mathbf{k} \\ c & 1 & 1 \\ 1 & -c & c \end{vmatrix} = \langle 2c, -c^2 + 1, -c^2 - 1 \rangle. \text{ So symmetric equations of } L \text{ can be written as}$$

$\dfrac{x + 1}{-2c} = \dfrac{y - c}{c^2 - 1} = \dfrac{z - c}{c^2 + 1}$, provided that $c \neq 0, \pm 1$.

If $c = 0$, then the two planes are given by $y + z = 0$ and $x = -1$, so symmetric equations of L are $x = -1$, $y = -z$. If $c = -1$, then the two planes are given by $-x + y + z = -1$ and $x + y + z = -1$, and they intersect in the line $x = 0$, $y = -z - 1$. If $c = 1$, then the two planes are given by $x + y + z = 1$ and $x - y + z = 1$, and they intersect in the line $y = 0$, $x = 1 - z$.

(b) If we set $z = t$ in the symmetric equations and solve for x and y separately, we get $x + 1 = \dfrac{(t - c)(-2c)}{c^2 + 1}$,

$y - c = \dfrac{(t - c)(c^2 - 1)}{c^2 + 1} \quad \Rightarrow \quad x = \dfrac{-2ct + (c^2 - 1)}{c^2 + 1}, \; y = \dfrac{(c^2 - 1)t + 2c}{c^2 + 1}$. Eliminating c from these equations, we have $x^2 + y^2 = t^2 + 1$. So the curve traced out by L in the plane $z = t$ is a circle with center at $(0, 0, t)$ and radius $\sqrt{t^2 + 1}$.

(c) The area of a horizontal cross-section of the solid is $A(z) = \pi(z^2 + 1)$, so $V = \int_0^1 A(z)\,dz = \pi\left[\frac{1}{3}z^3 + z\right]_0^1 = \frac{4\pi}{3}$.

5. (a) When $\theta = \theta_s$, the block is not moving, so the sum of the forces on the block

must be $\mathbf{0}$, thus $\mathbf{N} + \mathbf{F} + \mathbf{W} = \mathbf{0}$. This relationship is illustrated

geometrically in the figure. Since the vectors form a right triangle, we have

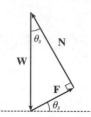

$$\tan(\theta_s) = \frac{|\mathbf{F}|}{|\mathbf{N}|} = \frac{\mu_s n}{n} = \mu_s.$$

(b) We place the block at the origin and sketch the force vectors acting on the block, including the additional horizontal force

$\mathbf{H}$, with initial points at the origin. We then rotate this system so that $\mathbf{F}$ lies along the positive x-axis and the inclined plane

is parallel to the x-axis.

$|\mathbf{F}|$ is maximal, so $|\mathbf{F}| = \mu_s n$ for $\theta > \theta_s$. Then the vectors, in terms of components parallel and perpendicular to the

inclined plane, are

$$\mathbf{N} = n\,\mathbf{j} \qquad \mathbf{F} = (\mu_s n)\,\mathbf{i}$$

$$\mathbf{W} = (-mg\sin\theta)\,\mathbf{i} + (-mg\cos\theta)\,\mathbf{j} \qquad \mathbf{H} = (h_{\min}\cos\theta)\,\mathbf{i} + (-h_{\min}\sin\theta)\,\mathbf{j}$$

Equating components, we have

$$\mu_s n - mg\sin\theta + h_{\min}\cos\theta = 0 \quad \Rightarrow \quad h_{\min}\cos\theta + \mu_s n = mg\sin\theta \tag{1}$$

$$n - mg\cos\theta - h_{\min}\sin\theta = 0 \quad \Rightarrow \quad h_{\min}\sin\theta + mg\cos\theta = n \tag{2}$$

(c) Since **(2)** is solved for n, we substitute into **(1)**:

$$h_{\min}\cos\theta + \mu_s(h_{\min}\sin\theta + mg\cos\theta) = mg\sin\theta \quad \Rightarrow$$

$$h_{\min}\cos\theta + h_{\min}\mu_s\sin\theta = mg\sin\theta - mg\mu_s\cos\theta \quad \Rightarrow$$

$$h_{\min} = mg\left(\frac{\sin\theta - \mu_s\cos\theta}{\cos\theta + \mu_s\sin\theta}\right) = mg\left(\frac{\tan\theta - \mu_s}{1 + \mu_s\tan\theta}\right)$$

From part (a) we know $\mu_s = \tan\theta_s$, so this becomes $h_{\min} = mg\left(\dfrac{\tan\theta - \tan\theta_s}{1 + \tan\theta_s\tan\theta}\right)$ and using a trigonometric identity,

this is $mg\tan(\theta - \theta_s)$ as desired.

Note for $\theta = \theta_s$, $h_{\min} = mg\tan 0 = 0$, which makes sense since the block is at rest for θ_s, thus no additional force $\mathbf{H}$

is necessary to prevent it from moving. As θ increases, the factor $\tan(\theta - \theta_s)$, and hence the value of $h_{\min}$, increases

slowly for small values of $\theta - \theta_s$ but much more rapidly as $\theta - \theta_s$ becomes significant. This seems reasonable, as the steeper the inclined plane, the less the horizontal components of the various forces affect the movement of the block, so we would need a much larger magnitude of horizontal force to keep the block motionless. If we allow $\theta \to 90°$, corresponding to the inclined plane being placed vertically, the value of $h_{\min}$ is quite large; this is to be expected, as it takes a great amount of horizontal force to keep an object from moving vertically. In fact, without friction (so $\theta_s = 0$), we would have $\theta \to 90°$ $\Rightarrow$ $h_{\min} \to \infty$, and it would be impossible to keep the block from slipping.

(d) Since $h_{\max}$ is the largest value of h that keeps the block from slipping, the force of friction is keeping the block from moving *up* the inclined plane; thus, $\mathbf{F}$ is directed *down* the plane. Our system of forces is similar to that in part (b), then, except that we have $\mathbf{F} = -(\mu_s n)\,\mathbf{i}$. (Note that $|\mathbf{F}|$ is again maximal.) Following our procedure in parts (b) and (c), we equate components:

$$-\mu_s n - mg\sin\theta + h_{\max}\cos\theta = 0 \quad \Rightarrow \quad h_{\max}\cos\theta - \mu_s n = mg\sin\theta$$

$$n - mg\cos\theta - h_{\max}\sin\theta = 0 \quad \Rightarrow \quad h_{\max}\sin\theta + mg\cos\theta = n$$

Then substituting,

$$h_{\max}\cos\theta - \mu_s(h_{\max}\sin\theta + mg\cos\theta) = mg\sin\theta \quad \Rightarrow$$

$$h_{\max}\cos\theta - h_{\max}\mu_s\sin\theta = mg\sin\theta + mg\mu_s\cos\theta \quad \Rightarrow$$

$$h_{\max} = mg\left(\frac{\sin\theta + \mu_s\cos\theta}{\cos\theta - \mu_s\sin\theta}\right) = mg\left(\frac{\tan\theta + \mu_s}{1 - \mu_s\tan\theta}\right)$$

$$= mg\left(\frac{\tan\theta + \tan\theta_s}{1 - \tan\theta_s\tan\theta}\right) = mg\tan(\theta + \theta_s)$$

We would expect $h_{\max}$ to increase as θ increases, with similar behavior as we established for $h_{\min}$, but with $h_{\max}$ values always larger than $h_{\min}$. We can see that this is the case if we graph $h_{\max}$ as a function of θ, as the curve is the graph of $h_{\min}$ translated $2\theta_s$ to the left, so the equation does seem reasonable. Notice that the equation predicts $h_{\max} \to \infty$ as $\theta \to (90° - \theta_s)$. In fact, as $h_{\max}$ increases, the normal force increases as well. When $(90° - \theta_s) \le \theta \le 90°$, the horizontal force is completely counteracted by the sum of the normal and frictional forces, so no part of the horizontal force contributes to moving the block up the plane no matter how large its magnitude.

10 □ VECTOR FUNCTIONS

10.1 Vector Functions and Space Curves

1. The component functions t^2, $\sqrt{t-1}$, and $\sqrt{5-t}$ are all defined when $t - 1 \geq 0 \Rightarrow t \geq 1$ and $5 - t \geq 0 \Rightarrow t \leq 5$, so the domain of $\mathbf{r}(t)$ is $[1, 5]$.

3. $\lim\limits_{t \to 0^+} \cos t = \cos 0 = 1$, $\lim\limits_{t \to 0^+} \sin t = \sin 0 = 0$, $\lim\limits_{t \to 0^+} t \ln t = \lim\limits_{t \to 0^+} \dfrac{\ln t}{1/t} = \lim\limits_{t \to 0^+} \dfrac{1/t}{-1/t^2} = \lim\limits_{t \to 0^+} -t = 0$ [by l'Hospital's

Rule]. Thus $\lim\limits_{t \to 0^+} \langle \cos t, \sin t, t \ln t \rangle = \left\langle \lim\limits_{t \to 0^+} \cos t, \lim\limits_{t \to 0^+} \sin t, \lim\limits_{t \to 0^+} t \ln t \right\rangle = \langle 1, 0, 0 \rangle$.

5. The corresponding parametric equations for this curve are

$x = \sin t$, $y = t$. We can make a table of values, or we can

eliminate the parameter: $t = y \Rightarrow x = \sin y$, with $y \in \mathbb{R}$.

By comparing different values of t, we find the direction in

which t increases as indicated in the graph.

7. The corresponding parametric equations are $x = t$, $y = \cos 2t$,

$z = \sin 2t$. Note that $y^2 + z^2 = \cos^2 2t + \sin^2 2t = 1$, so the

curve lies on the circular cylinder $y^2 + z^2 = 1$. Since $x = t$, the

curve is a helix.

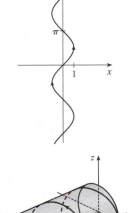

9. The corresponding parametric equations are $x = 1$, $y = \cos t$, $z = 2 \sin t$.

Eliminating the parameter in y and z gives $y^2 + (z/2)^2 = \cos^2 t + \sin^2 t = 1$ or

$y^2 + z^2/4 = 1$. Since $x = 1$, the curve is an ellipse centered at $(1, 0, 0)$ in the

plane $x = 1$.

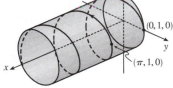

11. The parametric equations are $x = t^2$, $y = t^4$, $z = t^6$. These are positive for $t \neq 0$

and 0 when $t = 0$. So the curve lies entirely in the first quadrant. The projection of

the graph onto the xy-plane is $y = x^2$, $y > 0$, a half parabola. On the xz-plane

$z = x^3$, $z > 0$, a half cubic, and the yz-plane, $y^3 = z^2$.

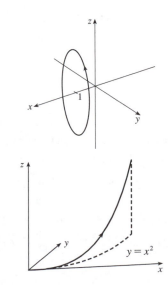

13. Taking $\mathbf{r}_0 = \langle 0, 0, 0 \rangle$ and $\mathbf{r}_1 = \langle 1, 2, 3 \rangle$, we have from Equation 9.5.4
$\mathbf{r}(t) = (1 - t)\,\mathbf{r}_0 + t\,\mathbf{r}_1 = (1 - t)\,\langle 0, 0, 0 \rangle + t\,\langle 1, 2, 3 \rangle, 0 \le t \le 1$ or $\mathbf{r}(t) = \langle t, 2t, 3t \rangle, 0 \le t \le 1$.
Parametric equations are $x = t, y = 2t, z = 3t, 0 \le t \le 1$.

15. Taking $\mathbf{r}_0 = \langle 1, -1, 2 \rangle$ and $\mathbf{r}_1 = \langle 4, 1, 7 \rangle$, we have $\mathbf{r}(t) = (1 - t)\,\mathbf{r}_0 + t\,\mathbf{r}_1 = (1 - t)\,\langle 1, -1, 2 \rangle + t\,\langle 4, 1, 7 \rangle, 0 \le t \le 1$ or
$\mathbf{r}(t) = \langle 1 + 3t, -1 + 2t, 2 + 5t \rangle, 0 \le t \le 1$. Parametric equations are $x = 1 + 3t, y = -1 + 2t, z = 2 + 5t, 0 \le t \le 1$.

17. $x = \cos 4t, y = t, z = \sin 4t$. At any point (x, y, z) on the curve, $x^2 + z^2 = \cos^2 4t + \sin^2 4t = 1$. So the curve lies on a circular cylinder with axis the y-axis. Since $y = t$, this is a helix. So the graph is VI.

19. $x = t, y = 1/(1 + t^2), z = t^2$. Note that y and z are positive for all t. The curve passes through $(0, 1, 0)$ when $t = 0$. As $t \to \infty, (x, y, z) \to (\infty, 0, \infty)$, and as $t \to -\infty, (x, y, z) \to (-\infty, 0, \infty)$. So the graph is IV.

21. $x = \cos t, y = \sin t, z = \sin 5t$. $x^2 + y^2 = \cos^2 t + \sin^2 t = 1$, so the curve lies on a circular cylinder with axis the z-axis. Each of x, y and z is periodic, and at $t = 0$ and $t = 2\pi$ the curve passes through the same point, so the curve repeats itself and the graph is V.

23. If $x = t \cos t, y = t \sin t$, and $z = t$, then

$x^2 + y^2 = t^2 \cos^2 t + t^2 \sin^2 t = t^2 = z^2$, so the curve lies on the

cone $z^2 = x^2 + y^2$. Since $z = t$, the curve is a spiral on this cone.

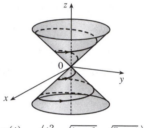

25. Parametric equations for the curve are $x = t, y = 0, z = 2t - t^2$.

Substituting into the equation of the paraboloid gives

$2t - t^2 = t^2 \quad \Rightarrow \quad 2t = 2t^2 \quad \Rightarrow \quad t = 0, 1$. Since $\mathbf{r}(0) = \mathbf{0}$

and $\mathbf{r}(1) = \mathbf{i} + \mathbf{k}$, the points of intersection are $(0, 0, 0)$

and $(1, 0, 1)$.

27. $\mathbf{r}(t) = \langle t^2, \sqrt{t - 1}, \sqrt{5 - t} \rangle$

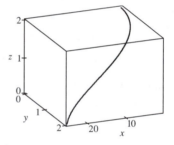

29.

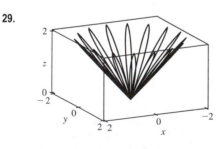

$x = (1 + \cos 16t) \cos t, y = (1 + \cos 16t) \sin t, z = 1 + \cos 16t$. At any

point on the graph,

$x^2 + y^2 = (1 + \cos 16t)^2 \cos^2 t + (1 + \cos 16t)^2 \sin^2 t$
$= (1 + \cos 16t)^2 = z^2$, so the graph lies on the cone $x^2 + y^2 = z^2$.

From the graph at left, we see that this curve looks like the projection of a

leaved two-dimensional curve onto a cone.

31. If $t = -1$, then $x = 1, y = 4, z = 0$, so the curve passes through the point $(1, 4, 0)$. If $t = 3$, then $x = 9, y = -8, z = 28$, so the curve passes through the point $(9, -8, 28)$. For the point $(4, 7, -6)$ to be on the curve, we require $y = 1 - 3t = 7 \quad \Rightarrow \quad t = -2$. But then $z = 1 + (-2)^3 = -7 \ne -6$, so $(4, 7, -6)$ is not on the curve.

33. Both equations are solved for z, so we can substitute to eliminate z: $\sqrt{x^2 + y^2} = 1 + y \quad \Rightarrow \quad x^2 + y^2 = 1 + 2y + y^2 \quad \Rightarrow$
$x^2 = 1 + 2y \quad \Rightarrow \quad y = \frac{1}{2}(x^2 - 1)$. We can form parametric equations for the curve C of intersection by choosing a
parameter $x = t$, then $y = \frac{1}{2}(t^2 - 1)$ and $z = 1 + y = 1 + \frac{1}{2}(t^2 - 1) = \frac{1}{2}(t^2 + 1)$. Thus a vector function representing C is
$\mathbf{r}(t) = t\,\mathbf{i} + \frac{1}{2}(t^2 - 1)\,\mathbf{j} + \frac{1}{2}(t^2 + 1)\,\mathbf{k}$.

35.

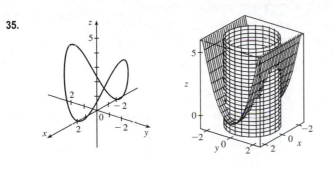

The projection of the curve C of intersection onto the xy-plane is the circle $x^2 + y^2 = 4, z = 0$. Then we can write $x = 2\cos t, y = 2\sin t, 0 \le t \le 2\pi$. Since C also lies on the surface $z = x^2$, we have $z = x^2 = (2\cos t)^2 = 4\cos^2 t$. Then parametric equations for C are $x = 2\cos t, y = 2\sin t$, $z = 4\cos^2 t, 0 \le t \le 2\pi$.

37. For the particles to collide, we require $\mathbf{r}_1(t) = \mathbf{r}_2(t) \;\Leftrightarrow\; \langle t^2, 7t - 12, t^2 \rangle = \langle 4t - 3, t^2, 5t - 6 \rangle$. Equating components gives $t^2 = 4t - 3$, $7t - 12 = t^2$, and $t^2 = 5t - 6$. From the first equation, $t^2 - 4t + 3 = 0 \;\Leftrightarrow\; (t - 3)(t - 1) = 0$ so $t = 1$ or $t = 3$. $t = 1$ does not satisfy the other two equations, but $t = 3$ does. The particles collide when $t = 3$, at the point $(9, 9, 9)$.

39. Let $\mathbf{u}(t) = \langle u_1(t), u_2(t), u_3(t) \rangle$ and $\mathbf{v}(t) = \langle v_1(t), v_2(t), v_3(t) \rangle$. In each part of this problem the basic procedure is to use Equation 1 and then analyze the individual component functions using the limit properties we have already developed for real-valued functions.

(a) $\displaystyle\lim_{t \to a} \mathbf{u}(t) + \lim_{t \to a} \mathbf{v}(t) = \left\langle \lim_{t \to a} u_1(t), \lim_{t \to a} u_2(t), \lim_{t \to a} u_3(t) \right\rangle + \left\langle \lim_{t \to a} v_1(t), \lim_{t \to a} v_2(t), \lim_{t \to a} v_3(t) \right\rangle$ and the limits of these

component functions must each exist since the vector functions both possess limits as $t \to a$. Then adding the two vectors and using the addition property of limits for real-valued functions, we have that

$$\lim_{t \to a} \mathbf{u}(t) + \lim_{t \to a} \mathbf{v}(t) = \left\langle \lim_{t \to a} u_1(t) + \lim_{t \to a} v_1(t), \lim_{t \to a} u_2(t) + \lim_{t \to a} v_2(t), \lim_{t \to a} u_3(t) + \lim_{t \to a} v_3(t) \right\rangle$$

$$= \left\langle \lim_{t \to a} [u_1(t) + v_1(t)], \lim_{t \to a} [u_2(t) + v_2(t)], \lim_{t \to a} [u_3(t) + v_3(t)] \right\rangle$$

$$= \lim_{t \to a} \langle u_1(t) + v_1(t), u_2(t) + v_2(t), u_3(t) + v_3(t) \rangle \qquad \text{[using (1) backward]}$$

$$= \lim_{t \to a} [\mathbf{u}(t) + \mathbf{v}(t)]$$

(b) $\displaystyle\lim_{t \to a} c\mathbf{u}(t) = \lim_{t \to a} \langle cu_1(t), cu_2(t), cu_3(t) \rangle = \left\langle \lim_{t \to a} cu_1(t), \lim_{t \to a} cu_2(t), \lim_{t \to a} cu_3(t) \right\rangle$

$$= \left\langle c\lim_{t \to a} u_1(t), c\lim_{t \to a} u_2(t), c\lim_{t \to a} u_3(t) \right\rangle = c\left\langle \lim_{t \to a} u_1(t), \lim_{t \to a} u_2(t), \lim_{t \to a} u_3(t) \right\rangle$$

$$= c\lim_{t \to a} \langle u_1(t), u_2(t), u_3(t) \rangle = c\lim_{t \to a} \mathbf{u}(t)$$

(c) $\displaystyle\lim_{t \to a} \mathbf{u}(t) \cdot \lim_{t \to a} \mathbf{v}(t) = \left\langle \lim_{t \to a} u_1(t), \lim_{t \to a} u_2(t), \lim_{t \to a} u_3(t) \right\rangle \cdot \left\langle \lim_{t \to a} v_1(t), \lim_{t \to a} v_2(t), \lim_{t \to a} v_3(t) \right\rangle$

$$= \left[\lim_{t \to a} u_1(t)\right]\left[\lim_{t \to a} v_1(t)\right] + \left[\lim_{t \to a} u_2(t)\right]\left[\lim_{t \to a} v_2(t)\right] + \left[\lim_{t \to a} u_3(t)\right]\left[\lim_{t \to a} v_3(t)\right]$$

$$= \lim_{t \to a} u_1(t)v_1(t) + \lim_{t \to a} u_2(t)v_2(t) + \lim_{t \to a} u_3(t)v_3(t)$$

$$= \lim_{t \to a} [u_1(t)v_1(t) + u_2(t)v_2(t) + u_3(t)v_3(t)] = \lim_{t \to a} [\mathbf{u}(t) \cdot \mathbf{v}(t)]$$

(d) $\lim_{t \to a} \mathbf{u}(t) \times \lim_{t \to a} \mathbf{v}(t) = \left\langle \lim_{t \to a} u_1(t), \lim_{t \to a} u_2(t), \lim_{t \to a} u_3(t) \right\rangle \times \left\langle \lim_{t \to a} v_1(t), \lim_{t \to a} v_2(t), \lim_{t \to a} v_3(t) \right\rangle$

$$= \left\langle \left[\lim_{t \to a} u_2(t)\right]\left[\lim_{t \to a} v_3(t)\right] - \left[\lim_{t \to a} u_3(t)\right]\left[\lim_{t \to a} v_2(t)\right],\right.$$

$$\left[\lim_{t \to a} u_3(t)\right]\left[\lim_{t \to a} v_1(t)\right] - \left[\lim_{t \to a} u_1(t)\right]\left[\lim_{t \to a} v_3(t)\right],$$

$$\left.\left[\lim_{t \to a} u_1(t)\right]\left[\lim_{t \to a} v_2(t)\right] - \left[\lim_{t \to a} u_2(t)\right]\left[\lim_{t \to a} v_1(t)\right]\right\rangle$$

$$= \left\langle \lim_{t \to a}\left[u_2(t)v_3(t) - u_3(t)v_2(t)\right], \lim_{t \to a}\left[u_3(t)v_1(t) - u_1(t)v_3(t)\right],\right.$$

$$\left.\lim_{t \to a}\left[u_1(t)v_2(t) - u_2(t)v_1(t)\right]\right\rangle$$

$$= \lim_{t \to a} \left\langle u_2(t)v_3(t) - u_3(t)v_2(t), u_3(t)\,v_1(t) - u_1(t)v_3(t), u_1(t)v_2(t) - u_2(t)v_1(t) \right\rangle$$

$$= \lim_{t \to a} \left[\mathbf{u}(t) \times \mathbf{v}(t)\right]$$

10.2 Derivatives and Integrals of Vector Functions

1. (a)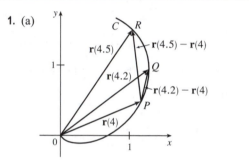

(b) $\dfrac{\mathbf{r}(4.5) - \mathbf{r}(4)}{0.5} = 2[\mathbf{r}(4.5) - \mathbf{r}(4)]$, so we draw a vector in the same direction but with twice the length of the vector $\mathbf{r}(4.5) - \mathbf{r}(4)$. $\dfrac{\mathbf{r}(4.2) - \mathbf{r}(4)}{0.2} = 5[\mathbf{r}(4.2) - \mathbf{r}(4)]$, so we draw a vector in the same direction but with 5 times the length of the vector $\mathbf{r}(4.2) - \mathbf{r}(4)$.

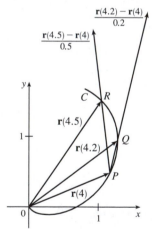

(c) By Definition 1, $\mathbf{r}'(4) = \lim_{h \to 0} \dfrac{\mathbf{r}(4+h) - \mathbf{r}(4)}{h}$.

$$\mathbf{T}(4) = \frac{\mathbf{r}'(4)}{|\mathbf{r}'(4)|}.$$

(d) $\mathbf{T}(4)$ is a unit vector in the same direction as $\mathbf{r}'(4)$, that is, parallel to the tangent line to the curve at $\mathbf{r}(4)$ with length 1.

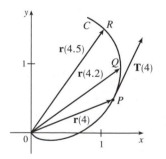

3. (a), (c)

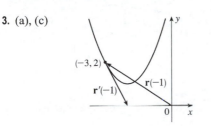

5. (a), (c)

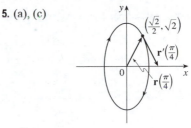

(b) $\mathbf{r}'(t) = \langle 1, 2t \rangle$

(b) $\mathbf{r}'(t) = \cos t\,\mathbf{i} - 2\sin t\,\mathbf{j}$

7. (a), (c)

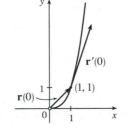

(b) $\mathbf{r}'(t) = e^t\,\mathbf{i} + 3e^{3t}\,\mathbf{j}$

9. $\mathbf{r}'(t) = \left\langle \dfrac{d}{dt}\,[t^2],\ \dfrac{d}{dt}\,[1-t],\ \dfrac{d}{dt}\,[\sqrt{t}] \right\rangle = \left\langle 2t, -1, \dfrac{1}{2\sqrt{t}} \right\rangle$

11. $\mathbf{r}(t) = e^{t^2}\,\mathbf{i} - \mathbf{j} + \ln(1+3t)\,\mathbf{k} \quad\Rightarrow\quad \mathbf{r}'(t) = 2te^{t^2}\,\mathbf{i} + \dfrac{3}{1+3t}\,\mathbf{k}$

13. $\mathbf{r}'(t) = \mathbf{0} + \mathbf{b} + 2t\,\mathbf{c} = \mathbf{b} + 2t\,\mathbf{c}$ by Formulas 1 and 3 of Theorem 3.

15. $\mathbf{r}'(t) = -\sin t\,\mathbf{i} + 3\,\mathbf{j} + 4\cos 2t\,\mathbf{k} \quad\Rightarrow\quad \mathbf{r}'(0) = 3\,\mathbf{j} + 4\,\mathbf{k}.$ Thus

$$\mathbf{T}(0) = \frac{\mathbf{r}'(0)}{|\mathbf{r}'(0)|} = \frac{1}{\sqrt{0^2 + 3^2 + 4^2}}\,(3\,\mathbf{j} + 4\,\mathbf{k}) = \tfrac{1}{5}(3\,\mathbf{j} + 4\,\mathbf{k}) = \tfrac{3}{5}\,\mathbf{j} + \tfrac{4}{5}\,\mathbf{k}.$$

17. $\mathbf{r}(t) = \langle t, t^2, t^3 \rangle \quad\Rightarrow\quad \mathbf{r}'(t) = \langle 1, 2t, 3t^2 \rangle.$ Then $\mathbf{r}'(1) = \langle 1, 2, 3 \rangle$ and $|\mathbf{r}'(1)| = \sqrt{1^2 + 2^2 + 3^2} = \sqrt{14}$, so

$$\mathbf{T}(1) = \frac{\mathbf{r}'(1)}{|\mathbf{r}'(1)|} = \tfrac{1}{\sqrt{14}}\,\langle 1, 2, 3 \rangle = \left\langle \tfrac{1}{\sqrt{14}}, \tfrac{2}{\sqrt{14}}, \tfrac{3}{\sqrt{14}} \right\rangle. \quad \mathbf{r}''(t) = \langle 0, 2, 6t \rangle,\ \text{so}$$

$$\mathbf{r}'(t) \times \mathbf{r}''(t) = \begin{vmatrix} \mathbf{i} & \mathbf{j} & \mathbf{k} \\ 1 & 2t & 3t^2 \\ 0 & 2 & 6t \end{vmatrix} = \begin{vmatrix} 2t & 3t^2 \\ 2 & 6t \end{vmatrix}\mathbf{i} - \begin{vmatrix} 1 & 3t^2 \\ 0 & 6t \end{vmatrix}\mathbf{j} + \begin{vmatrix} 1 & 2t \\ 0 & 2 \end{vmatrix}\mathbf{k}$$

$$= (12t^2 - 6t^2)\,\mathbf{i} - (6t - 0)\,\mathbf{j} + (2 - 0)\,\mathbf{k} = \langle 6t^2, -6t, 2 \rangle.$$

19. The vector equation for the curve is $\mathbf{r}(t) = \langle t^5, t^4, t^3 \rangle$, so $\mathbf{r}'(t) = \langle 5t^4, 4t^3, 3t^2 \rangle.$ The point $(1, 1, 1)$ corresponds to $t = 1$, so the tangent vector there is $\mathbf{r}'(1) = \langle 5, 4, 3 \rangle.$ Thus, the tangent line goes through the point $(1, 1, 1)$ and is parallel to the vector $\langle 5, 4, 3 \rangle.$ Parametric equations are $x = 1 + 5t,\ y = 1 + 4t,\ z = 1 + 3t.$

21. The vector equation for the curve is $\mathbf{r}(t) = \langle e^{-t}\cos t,\ e^{-t}\sin t,\ e^{-t} \rangle$, so

$$\mathbf{r}'(t) = \langle e^{-t}(-\sin t) + (\cos t)(-e^{-t}),\ e^{-t}\cos t + (\sin t)(-e^{-t}),\ (-e^{-t}) \rangle$$
$$= \langle -e^{-t}(\cos t + \sin t),\ e^{-t}(\cos t - \sin t),\ -e^{-t} \rangle.$$

The point $(1, 0, 1)$ corresponds to $t = 0$, so the tangent vector there is

$$\mathbf{r}'(0) = \langle -e^0(\cos 0 + \sin 0),\ e^0(\cos 0 - \sin 0),\ -e^0 \rangle = \langle -1, 1, -1 \rangle.$$ Thus, the tangent line is parallel to the vector

$\langle -1, 1, -1 \rangle$ and parametric equations are $x = 1 + (-1)t = 1 - t,\ y = 0 + 1 \cdot t = t,\ z = 1 + (-1)t = 1 - t.$

23. $\mathbf{r}(t) = \langle t, e^{-t}, 2t - t^2 \rangle \;\Rightarrow\; \mathbf{r}'(t) = \langle 1, -e^{-t}, 2 - 2t \rangle$. At

$(0, 1, 0)$, $t = 0$ and $\mathbf{r}'(0) = \langle 1, -1, 2 \rangle$. Thus, parametric equations
of the tangent line are $x = t$, $y = 1 - t$, $z = 2t$.

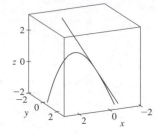

25. (a) $\mathbf{r}(t) = \langle t^3, t^4, t^5 \rangle \;\Rightarrow\; \mathbf{r}'(t) = \langle 3t^2, 4t^3, 5t^4 \rangle$, and since $\mathbf{r}'(0) = \langle 0, 0, 0 \rangle = \mathbf{0}$, the curve is not smooth.

(b) $\mathbf{r}(t) = \langle t^3 + t, t^4, t^5 \rangle \;\Rightarrow\; \mathbf{r}'(t) = \langle 3t^2 + 1, 4t^3, 5t^4 \rangle$. $\mathbf{r}'(t)$ is continuous since its component functions are
continuous. Also, $\mathbf{r}'(t) \neq \mathbf{0}$, as the y- and z-components are 0 only for $t = 0$, but $\mathbf{r}'(0) = \langle 1, 0, 0 \rangle \neq \mathbf{0}$. Thus, the curve is
smooth.

(c) $\mathbf{r}(t) = \langle \cos^3 t, \sin^3 t \rangle \;\Rightarrow\; \mathbf{r}'(t) = \langle -3\cos^2 t \sin t, 3 \sin^2 t \cos t \rangle$. Since
$\mathbf{r}'(0) = \langle -3 \cos^2 0 \sin 0, 3 \sin^2 0 \cos 0 \rangle = \langle 0, 0 \rangle = \mathbf{0}$, the curve is not smooth.

27. The angle of intersection of the two curves is the angle between the two tangent vectors to the curves at the point of

intersection. Since $\mathbf{r}_1'(t) = \langle 1, 2t, 3t^2 \rangle$ and $t = 0$ at $(0, 0, 0)$, $\mathbf{r}_1'(0) = \langle 1, 0, 0 \rangle$ is a tangent vector to $\mathbf{r}_1$ at $(0, 0, 0)$. Similarly,

$\mathbf{r}_2'(t) = \langle \cos t, 2 \cos 2t, 1 \rangle$ and since $\mathbf{r}_2(0) = \langle 0, 0, 0 \rangle$, $\mathbf{r}_2'(0) = \langle 1, 2, 1 \rangle$ is a tangent vector to $\mathbf{r}_2$ at $(0, 0, 0)$. If θ is the angle

between these two tangent vectors, then $\cos \theta = \frac{1}{\sqrt{1}\sqrt{6}} \langle 1, 0, 0 \rangle \cdot \langle 1, 2, 1 \rangle = \frac{1}{\sqrt{6}}$ and $\theta = \cos^{-1}\left(\frac{1}{\sqrt{6}}\right) \approx 66\,°$.

29. $\int_0^1 (16t^3\,\mathbf{i} - 9t^2\,\mathbf{j} + 25t^4\,\mathbf{k})\,dt = \left(\int_0^1 16t^3\,dt\right)\mathbf{i} - \left(\int_0^1 9t^2\,dt\right)\mathbf{j} + \left(\int_0^1 25t^4\,dt\right)\mathbf{k}$

$$= \left[4t^4\right]_0^1 \mathbf{i} - \left[3t^3\right]_0^1 \mathbf{j} + \left[5t^5\right]_0^1 \mathbf{k} = 4\,\mathbf{i} - 3\,\mathbf{j} + 5\,\mathbf{k}$$

31. $\int_0^{\pi/2} (3 \sin^2 t \cos t\,\mathbf{i} + 3 \sin t \cos^2 t\,\mathbf{j} + 2 \sin t \cos t\,\mathbf{k})\,dt$

$$= \left(\int_0^{\pi/2} 3 \sin^2 t \cos t\,dt\right)\mathbf{i} + \left(\int_0^{\pi/2} 3 \sin t \cos^2 t\,dt\right)\mathbf{j} + \left(\int_0^{\pi/2} 2 \sin t \cos t\,dt\right)\mathbf{k}$$

$$= \left[\sin^3 t\right]_0^{\pi/2}\mathbf{i} + \left[-\cos^3 t\right]_0^{\pi/2}\mathbf{j} + \left[\sin^2 t\right]_0^{\pi/2}\mathbf{k} = (1 - 0)\,\mathbf{i} + (0 + 1)\,\mathbf{j} + (1 - 0)\,\mathbf{k} = \mathbf{i} + \mathbf{j} + \mathbf{k}$$

33. $\int (e^t\,\mathbf{i} + 2t\,\mathbf{j} + \ln t\,\mathbf{k})\,dt = \left(\int e^t\,dt\right)\mathbf{i} + \left(\int 2t\,dt\right)\mathbf{j} + \left(\int \ln t\,dt\right)\mathbf{k}$

$$= e^t\,\mathbf{i} + t^2\,\mathbf{j} + (t \ln t - t)\,\mathbf{k} + \mathbf{C}, \text{ where } \mathbf{C} \text{ is a vector constant of integration.}$$

35. $\mathbf{r}'(t) = 2t\,\mathbf{i} + 3t^2\,\mathbf{j} + \sqrt{t}\,\mathbf{k} \;\Rightarrow\; \mathbf{r}(t) = t^2\,\mathbf{i} + t^3\,\mathbf{j} + \frac{2}{3}t^{3/2}\,\mathbf{k} + \mathbf{C}$, where $\mathbf{C}$ is a constant vector. But

$\mathbf{i} + \mathbf{j} = \mathbf{r}(1) = \mathbf{i} + \mathbf{j} + \frac{2}{3}\mathbf{k} + \mathbf{C}$. Thus $\mathbf{C} = -\frac{2}{3}\mathbf{k}$ and $\mathbf{r}(t) = t^2\,\mathbf{i} + t^3\,\mathbf{j} + \left(\frac{2}{3}t^{3/2} - \frac{2}{3}\right)\mathbf{k}$.

For Exercises 37–39, let $\mathbf{u}(t) = \langle u_1(t), u_2(t), u_3(t) \rangle$ and $\mathbf{v}(t) = \langle v_1(t), v_2(t), v_3(t) \rangle$. In each of these exercises, the procedure is to apply
Theorem 2 so that the corresponding properties of derivatives of real-valued functions can be used.

37. $\dfrac{d}{dt}\left[\mathbf{u}(t) + \mathbf{v}(t)\right] = \dfrac{d}{dt}\langle u_1(t) + v_1(t), u_2(t) + v_2(t), u_3(t) + v_3(t) \rangle$

$$= \left\langle \frac{d}{dt}\left[u_1(t) + v_1(t)\right], \frac{d}{dt}\left[u_2(t) + v_2(t)\right], \frac{d}{dt}\left[u_3(t) + v_3(t)\right] \right\rangle$$

$$= \langle u_1'(t) + v_1'(t), u_2'(t) + v_2'(t), u_3'(t) + v_3'(t) \rangle$$

$$= \langle u_1'(t), u_2'(t), u_3'(t) \rangle + \langle v_1'(t), v_2'(t), v_3'(t) \rangle = \mathbf{u}'(t) + \mathbf{v}'(t).$$

39. $\dfrac{d}{dt}[\mathbf{u}(t) \times \mathbf{v}(t)]$ $= \dfrac{d}{dt}\langle u_2(t)v_3(t) - u_3(t)v_2(t),\, u_3(t)v_1(t) - u_1(t)v_3(t),\, u_1(t)v_2(t) - u_2(t)v_1(t)\rangle$

$$= \langle u_2'v_3(t) + u_2(t)v_3'(t) - u_3'(t)v_2(t) - u_3(t)v_2'(t),$$

$$u_3'(t)v_1(t) + u_3(t)v_1'\,(t) - u_1'(t)v_3(t) - u_1(t)v_3'(t),$$

$$u_1'(t)v_2(t) + u_1(t)v_2'(t) - u_2'(t)v_1(t) - u_2(t)v_1'(t)\rangle$$

$$= \langle u_2'(t)v_3(t) - u_3'(t)v_2\,(t)\,,\, u_3'(t)v_1(t) - u_1'(t)v_3(t),\, u_1'(t)v_2(t) - u_2'(t)v_1(t)\rangle$$

$$+ \langle u_2(t)v_3'(t) - u_3(t)v_2'(t),\, u_3(t)v_1'\,(t) - u_1(t)v_3'(t),\, u_1(t)v_2'(t) - u_2(t)v_1'(t)\rangle$$

$$= \mathbf{u}'(t) \times \mathbf{v}(t) + \mathbf{u}(t) \times \mathbf{v}'(t)$$

Alternate solution: Let $\mathbf{r}(t) = \mathbf{u}(t) \times \mathbf{v}(t)$. Then

$$\mathbf{r}(t + h) - \mathbf{r}(t) = [\mathbf{u}(t + h) \times \mathbf{v}(t + h)] - [\mathbf{u}(t) \times \mathbf{v}(t)]$$

$$= [\mathbf{u}(t + h) \times \mathbf{v}(t + h)] - [\mathbf{u}(t) \times \mathbf{v}(t)] + [\mathbf{u}(t + h) \times \mathbf{v}(t)] - [\mathbf{u}(t + h) \times \mathbf{v}(t)]$$

$$= \mathbf{u}(t + h) \times [\mathbf{v}(t + h) - \mathbf{v}(t)] + [\mathbf{u}(t + h) - \mathbf{u}(t)] \times \mathbf{v}(t)$$

(Be careful of the order of the cross product.)

Dividing through by h and taking the limit as $h \to 0$ we have

$$\mathbf{r}'(t) = \lim_{h \to 0} \frac{\mathbf{u}(t + h) \times [\mathbf{v}(t + h) - \mathbf{v}(t)]}{h} + \lim_{h \to 0} \frac{[\mathbf{u}(t + h) - \mathbf{u}(t)] \times \mathbf{v}(t)}{h}$$

$$= \mathbf{u}(t) \times \mathbf{v}'(t) + \mathbf{u}'(t) \times \mathbf{v}(t)$$

by Exercise 10.1.39(a) and Definition 1.

41. $\dfrac{d}{dt}[\mathbf{u}(t) \cdot \mathbf{v}(t)] = \mathbf{u}'(t) \cdot \mathbf{v}(t) + \mathbf{u}(t) \cdot \mathbf{v}'(t)$ [by Formula 4 of Theorem 3]

$$= (-4t\,\mathbf{j} + 9t^2\,\mathbf{k}) \cdot (t\,\mathbf{i} + \cos t\,\mathbf{j} + \sin t\,\mathbf{k}) + (\mathbf{i} - 2t^2\,\mathbf{j} + 3t^3\,\mathbf{k}) \cdot (\mathbf{i} - \sin t\,\mathbf{j} + \cos t\,\mathbf{k})$$

$$= -4t\cos t + 9t^2 \sin t + 1 + 2t^2 \sin t + 3t^3 \cos t$$

$$= 1 - 4t\cos t + 11t^2 \sin t + 3t^3 \cos t$$

43. $\dfrac{d}{dt}[\mathbf{r}(t) \times \mathbf{r}'(t)] = \mathbf{r}'(t) \times \mathbf{r}'(t) + \mathbf{r}(t) \times \mathbf{r}''(t)$ by Formula 5 of Theorem 3. But $\mathbf{r}'(t) \times \mathbf{r}'(t) = \mathbf{0}$ [by the margin note on

page 658]. Thus, $\dfrac{d}{dt}[\mathbf{r}(t) \times \mathbf{r}'(t)] = \mathbf{r}(t) \times \mathbf{r}''(t)$.

45. $\dfrac{d}{dt}|\mathbf{r}(t)| = \dfrac{d}{dt}[\mathbf{r}(t) \cdot \mathbf{r}(t)]^{1/2} = \tfrac{1}{2}[\mathbf{r}(t) \cdot \mathbf{r}(t)]^{-1/2}[2\mathbf{r}(t) \cdot \mathbf{r}'(t)] = \dfrac{1}{|\mathbf{r}(t)|}\mathbf{r}(t) \cdot \mathbf{r}'(t)$

47. Since $\mathbf{u}(t) = \mathbf{r}(t) \cdot [\mathbf{r}'(t) \times \mathbf{r}''(t)]$,

$$\mathbf{u}'(t) = \mathbf{r}'(t) \cdot [\mathbf{r}'(t) \times \mathbf{r}''(t)] + \mathbf{r}(t) \cdot \dfrac{d}{dt}[\mathbf{r}'(t) \times \mathbf{r}''(t)]$$

$$= 0 + \mathbf{r}(t) \cdot [\mathbf{r}''(t) \times \mathbf{r}''(t) + \mathbf{r}'(t) \times \mathbf{r}'''(t)] \qquad [\text{since } \mathbf{r}'(t) \perp \mathbf{r}'(t) \times \mathbf{r}''(t)]$$

$$= \mathbf{r}(t) \cdot [\mathbf{r}'(t) \times \mathbf{r}'''(t)] \qquad [\text{since } \mathbf{r}''(t) \times \mathbf{r}''(t) = \mathbf{0}]$$

10.3 Arc Length and Curvature

1. $\mathbf{r}'(t) = \langle 2\cos t, 5, -2\sin t\rangle \;\Rightarrow\; |\mathbf{r}'(t)| = \sqrt{(2\cos t)^2 + 5^2 + (-2\sin t)^2} = \sqrt{29}$. Then using Formula 3, we have
$L = \int_{-10}^{10} |\mathbf{r}'(t)|\, dt = \int_{-10}^{10} \sqrt{29}\, dt = \sqrt{29}\, t\Big]_{-10}^{10} = 20\sqrt{29}.$

3. $\mathbf{r}'(t) = 2t\mathbf{j} + 3t^2\mathbf{k} \;\Rightarrow\; |\mathbf{r}'(t)| = \sqrt{4t^2 + 9t^4} = t\sqrt{4 + 9t^2}$ (since $t \geq 0$). Then
$L = \int_0^1 |\mathbf{r}'(t)|\, dt = \int_0^1 t\sqrt{4 + 9t^2}\, dt = \frac{1}{18} \cdot \frac{2}{3}(4 + 9t^2)^{3/2}\Big]_0^1 = \frac{1}{27}(13^{3/2} - 4^{3/2}) = \frac{1}{27}(13^{3/2} - 8).$

5. The point $(2, 4, 8)$ corresponds to $t = 2$, so by Equation 2, $L = \int_0^2 \sqrt{(1)^2 + (2t)^2 + (3t^2)^2}\, dt$. If $f(t) = \sqrt{1 + 4t^2 + 9t^4}$,
then Simpson's Rule gives $L \approx \dfrac{2 - 0}{10 \cdot 3}\, [f(0) + 4f(0.2) + 2f(0.4) + \cdots + 4f(1.8) + f(2)] \approx 9.5706.$

7. $\mathbf{r}'(t) = 2\mathbf{i} - 3\mathbf{j} + 4\mathbf{k}$ and $\dfrac{ds}{dt} = |\mathbf{r}'(t)| = \sqrt{4 + 9 + 16} = \sqrt{29}$. Then $s = s(t) = \int_0^t |\mathbf{r}'(u)|\, du = \int_0^t \sqrt{29}\, du = \sqrt{29}\, t$.
Therefore, $t = \frac{1}{\sqrt{29}}\, s$, and substituting for t in the original equation, we have
$\mathbf{r}(t(s)) = \frac{2}{\sqrt{29}}\, s\, \mathbf{i} + \left(1 - \frac{3}{\sqrt{29}}\, s\right)\mathbf{j} + \left(5 + \frac{4}{\sqrt{29}}\, s\right)\mathbf{k}.$

9. Here $\mathbf{r}(t) = \langle 3\sin t, 4t, 3\cos t\rangle$, so $\mathbf{r}'(t) = \langle 3\cos t, 4, -3\sin t\rangle$ and $|\mathbf{r}'(t)| = \sqrt{9\cos^2 t + 16 + 9\sin^2 t} = \sqrt{25} = 5$.
The point $(0, 0, 3)$ corresponds to $t = 0$, so the arc length function beginning at $(0, 0, 3)$ and measuring in the positive
direction is given by $s(t) = \int_0^t |\mathbf{r}'(u)|\, du = \int_0^t 5\, du = 5t$. $s(t) = 5 \;\Rightarrow\; 5t = 5 \;\Rightarrow\; t = 1$, thus your location after
moving 5 units along the curve is $(3\sin 1, 4, 3\cos 1)$.

11. (a) $\mathbf{r}'(t) = \langle 2\cos t, 5, -2\sin t\rangle \;\Rightarrow\; |\mathbf{r}'(t)| = \sqrt{4\cos^2 t + 25 + 4\sin^2 t} = \sqrt{29}$. Then
$\mathbf{T}(t) = \dfrac{\mathbf{r}'(t)}{|\mathbf{r}'(t)|} = \frac{1}{\sqrt{29}}\, \langle 2\cos t, 5, -2\sin t\rangle$ or $\left\langle \frac{2}{\sqrt{29}}\cos t, \frac{5}{\sqrt{29}}, -\frac{2}{\sqrt{29}}\sin t\right\rangle.$
$\mathbf{T}'(t) = \frac{1}{\sqrt{29}}\, \langle -2\sin t, 0, -2\cos t\rangle \;\Rightarrow\; |\mathbf{T}'(t)| = \frac{1}{\sqrt{29}}\sqrt{4\sin^2 t + 0 + 4\cos^2 t} = \frac{2}{\sqrt{29}}.$ Thus
$\mathbf{N}(t) = \dfrac{\mathbf{T}'(t)}{|\mathbf{T}'(t)|} = \dfrac{1/\sqrt{29}}{2/\sqrt{29}}\, \langle -2\sin t, 0, -2\cos t\rangle = \langle -\sin t, 0, -\cos t\rangle.$

(b) $\kappa(t) = \dfrac{|\mathbf{T}'(t)|}{|\mathbf{r}'(t)|} = \dfrac{2/\sqrt{29}}{\sqrt{29}} = \dfrac{2}{29}.$

13. (a) $\mathbf{r}'(t) = \langle \sqrt{2}, e^t, -e^{-t}\rangle \;\Rightarrow\; |\mathbf{r}'(t)| = \sqrt{2 + e^{2t} + e^{-2t}} = \sqrt{(e^t + e^{-t})^2} = e^t + e^{-t}$. Then
$\mathbf{T}(t) = \dfrac{\mathbf{r}'(t)}{|\mathbf{r}'(t)|} = \dfrac{1}{e^t + e^{-t}}\, \langle \sqrt{2}, e^t, -e^{-t}\rangle = \dfrac{1}{e^{2t} + 1}\, \langle \sqrt{2}e^t, e^{2t}, -1\rangle \quad \left(\text{after multiplying by } \dfrac{e^t}{e^t}\right)$ and

$$\mathbf{T}'(t) = \frac{1}{e^{2t} + 1}\, \left\langle \sqrt{2}e^t, 2e^{2t}, 0\right\rangle - \frac{2e^{2t}}{(e^{2t} + 1)^2}\, \left\langle \sqrt{2}e^t, e^{2t}, -1\right\rangle$$

$$= \frac{1}{(e^{2t} + 1)^2}\, \left[(e^{2t} + 1)\left\langle \sqrt{2}e^t, 2e^{2t}, 0\right\rangle - 2e^{2t}\left\langle \sqrt{2}e^t, e^{2t}, -1\right\rangle\right]$$

$$= \frac{1}{(e^{2t} + 1)^2}\, \left\langle \sqrt{2}e^t\left(1 - e^{2t}\right), 2e^{2t}, 2e^{2t}\right\rangle$$

Then

$$|\mathbf{T}'(t)| = \frac{1}{(e^{2t} + 1)^2}\sqrt{2e^{2t}(1 - 2e^{2t} + e^{4t}) + 4e^{4t} + 4e^{4t}} = \frac{1}{(e^{2t} + 1)^2}\sqrt{2e^{2t}(1 + 2e^{2t} + e^{4t})}$$

$$= \frac{1}{(e^{2t} + 1)^2}\sqrt{2e^{2t}\left(1 + e^{2t}\right)^2} = \frac{\sqrt{2}e^t(1 + e^{2t})}{(e^{2t} + 1)^2} = \frac{\sqrt{2}\, e^t}{e^{2t} + 1}$$

Therefore

$$\mathbf{N}(t) = \frac{\mathbf{T}'(t)}{|\mathbf{T}'(t)|} = \frac{e^{2t}+1}{\sqrt{2}\,e^t} \frac{1}{(e^{2t}+1)^2} \left\langle \sqrt{2}\,e^t(1-e^{2t}), 2e^{2t}, 2e^{2t} \right\rangle$$

$$= \frac{1}{\sqrt{2}\,e^t(e^{2t}+1)} \left\langle \sqrt{2}\,e^t(1-e^{2t}), 2e^{2t}, 2e^{2t} \right\rangle = \frac{1}{e^{2t}+1}\left\langle 1-e^{2t}, \sqrt{2}\,e^t, \sqrt{2}\,e^t \right\rangle$$

(b) $\kappa(t) = \dfrac{|\mathbf{T}'(t)|}{|\mathbf{r}'(t)|} = \dfrac{\sqrt{2}\,e^t}{e^{2t}+1} \cdot \dfrac{1}{e^t+e^{-t}} = \dfrac{\sqrt{2}\,e^t}{e^{3t}+2e^t+e^{-t}} = \dfrac{\sqrt{2}\,e^{2t}}{e^{4t}+2e^{2t}+1} = \dfrac{\sqrt{2}\,e^{2t}}{(e^{2t}+1)^2}.$

15. $\mathbf{r}'(t) = 2t\,\mathbf{i} + \mathbf{k}$, $\mathbf{r}''(t) = 2\,\mathbf{i}$, $|\mathbf{r}'(t)| = \sqrt{(2t)^2 + 0^2 + 1^2} = \sqrt{4t^2+1}$, $\mathbf{r}'(t) \times \mathbf{r}''(t) = 2\,\mathbf{j}$, $|\mathbf{r}'(t) \times \mathbf{r}''(t)| = 2$.

Then $\kappa(t) = \dfrac{|\mathbf{r}'(t) \times \mathbf{r}''(t)|}{|\mathbf{r}'(t)|^3} = \dfrac{2}{\left(\sqrt{4t^2+1}\right)^3} = \dfrac{2}{(4t^2+1)^{3/2}}.$

17. $\mathbf{r}'(t) = 3\,\mathbf{i} + 4\cos t\,\mathbf{j} - 4\sin t\,\mathbf{k}$, $\mathbf{r}''(t) = -4\sin t\,\mathbf{j} - 4\cos t\,\mathbf{k}$, $|\mathbf{r}'(t)| = \sqrt{9 + 16\cos^2 t + 16\sin^2 t} = \sqrt{9+16} = 5$,

$\mathbf{r}'(t) \times \mathbf{r}''(t) = -16\,\mathbf{i} + 12\cos t\,\mathbf{j} - 12\sin t\,\mathbf{k}$, $|\mathbf{r}'(t) \times \mathbf{r}''(t)| = \sqrt{256 + 144\cos^2 t + 144\sin^2 t} = \sqrt{400} = 20$.

Then $\kappa(t) = \dfrac{|\mathbf{r}'(t) \times \mathbf{r}''(t)|}{|\mathbf{r}'(t)|^3} = \dfrac{20}{5^3} = \dfrac{4}{25}.$

19. $\mathbf{r}'(t) = \langle 1, 2t, 3t^2 \rangle$. The point $(1, 1, 1)$ corresponds to $t = 1$, and

$\mathbf{r}'(1) = \langle 1, 2, 3 \rangle \;\Rightarrow\; |\mathbf{r}'(1)| = \sqrt{1+4+9} = \sqrt{14}$. $\mathbf{r}''(t) = \langle 0, 2, 6t \rangle \;\Rightarrow\; \mathbf{r}''(1) = \langle 0, 2, 6 \rangle$.

$\mathbf{r}'(1) \times \mathbf{r}''(1) = \langle 6, -6, 2 \rangle$, so $|\mathbf{r}'(1) \times \mathbf{r}''(1)| = \sqrt{36+36+4} = \sqrt{76}$. Then

$\kappa(1) = \dfrac{|\mathbf{r}'(1) \times \mathbf{r}''(1)|}{|\mathbf{r}'(1)|^3} = \dfrac{\sqrt{76}}{\sqrt{14}^3} = \dfrac{1}{7}\sqrt{\dfrac{19}{14}}.$

21. $f(x) = xe^x$, $f'(x) = xe^x + e^x$, $f''(x) = xe^x + 2e^x$,

$\kappa(x) = \dfrac{|f''(x)|}{[1 + (f'(x))^2]^{3/2}} = \dfrac{|xe^x + 2e^x|}{[1 + (xe^x + e^x)^2]^{3/2}} = \dfrac{|x+2|\,e^x}{[1 + (xe^x + e^x)^2]^{3/2}}$

23. $f(x) = 4x^{5/2}$, $f'(x) = 10x^{3/2}$, $f''(x) = 15x^{1/2}$, $\kappa(x) = \dfrac{|f''(x)|}{[1 + (f'(x))^2]^{3/2}} = \dfrac{\left|15x^{1/2}\right|}{[1 + (10x^{3/2})^2]^{3/2}} = \dfrac{15\sqrt{x}}{(1 + 100x^3)^{3/2}}$

25. Since $y' = y'' = e^x$, the curvature is $\kappa(x) = \dfrac{|y''(x)|}{[1 + (y'(x))^2]^{3/2}} = \dfrac{e^x}{(1 + e^{2x})^{3/2}} = e^x(1 + e^{2x})^{-3/2}$.

To find the maximum curvature, we first find the critical numbers of $\kappa(x)$:

$\kappa'(x) = e^x(1 + e^{2x})^{-3/2} + e^x\left(-\frac{3}{2}\right)(1 + e^{2x})^{-5/2}(2e^{2x}) = e^x\dfrac{1 + e^{2x} - 3e^{2x}}{(1 + e^{2x})^{5/2}} = e^x\dfrac{1 - 2e^{2x}}{(1 + e^{2x})^{5/2}}.$

$\kappa'(x) = 0$ when $1 - 2e^{2x} = 0$, so $e^{2x} = \frac{1}{2}$ or $x = -\frac{1}{2}\ln 2$. And since $1 - 2e^{2x} > 0$ for $x < -\frac{1}{2}\ln 2$ and $1 - 2e^{2x} < 0$ for

$x > -\frac{1}{2}\ln 2$, the maximum curvature is attained at the point $\left(-\frac{1}{2}\ln 2, e^{(-\ln 2)/2}\right) = \left(-\frac{1}{2}\ln 2, \frac{1}{\sqrt{2}}\right)$. Since

$\lim\limits_{x \to \infty} e^x(1 + e^{2x})^{-3/2} = 0$, $\kappa(x)$ approaches 0 as $x \to \infty$.

27. (a) C appears to be changing direction more quickly at P than Q, so we would expect the curvature to be greater at P.

(b) First we sketch approximate osculating circles at P and Q.

Using the axes scale as a guide, we measure the radius of the

osculating circle at P to be approximately 0.8 units, thus

$\rho = \dfrac{1}{\kappa} \;\Rightarrow\; \kappa = \dfrac{1}{\rho} \approx \dfrac{1}{0.8} \approx 1.3$. Similarly, we estimate

the radius of the osculating circle at Q to be 1.4 units, so

$\kappa = \dfrac{1}{\rho} \approx \dfrac{1}{1.4} \approx 0.7$.

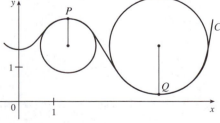

29. $y = x^{-2}$ $\Rightarrow$ $y' = -2x^{-3}$, $y'' = 6x^{-4}$, and $\kappa(x) = \dfrac{|y''|}{\left[1 + (y')^2\right]^{3/2}} = \dfrac{\left|6x^{-4}\right|}{\left[1 + (-2x^{-3})^2\right]^{3/2}} = \dfrac{6}{x^4 \left(1 + 4x^{-6}\right)^{3/2}}$. The

appearance of the two humps in this graph is perhaps a little surprising, but it is explained by the fact that $y = x^{-2}$ increases

asymptotically at the origin from both directions, and so its graph has very little bend there. (Note that $\kappa(0)$ is undefined.)

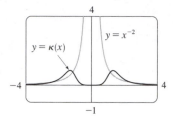

31. Notice that the curve b has two inflection points at which the graph appears almost straight. We would expect the curvature to

be 0 or nearly 0 at these values, but the curve a isn't near 0 there. Thus, a must be the graph of $y = f(x)$ rather than the graph

of curvature, and b is the graph of $y = \kappa(x)$.

33. Using a CAS, we find (after simplifying)

$$\kappa(t) = \frac{6\sqrt{4\cos^2 t - 12\cos t + 13}}{(17 - 12\cos t)^{3/2}}.$$ (To compute cross

products in Maple, use the `Linalg` package and the

`crossprod(a,b)` command; in Mathematica, use

`Cross[a,b]`.) Curvature is largest at integer multiples of 2π.

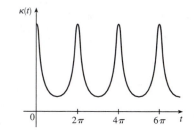

35. $x = e^t \cos t$ $\Rightarrow$ $\dot{x} = e^t(\cos t - \sin t)$ $\Rightarrow$ $\ddot{x} = e^t(-\sin t - \cos t) + e^t(\cos t - \sin t) = -2e^t \sin t$,

$y = e^t \sin t$ $\Rightarrow$ $\dot{y} = e^t(\cos t + \sin t)$ $\Rightarrow$ $\ddot{y} = e^t(-\sin t + \cos t) + e^t(\cos t + \sin t) = 2e^t \cos t$. Then

$$\kappa(t) = \frac{|\dot{x}\ddot{y} - \dot{y}\ddot{x}|}{(\dot{x}^2 + \dot{y}^2)^{3/2}} = \frac{\left|e^t(\cos t - \sin t)(2e^t \cos t) - e^t(\cos t + \sin t)(-2e^t \sin t)\right|}{\left([e^t(\cos t - \sin t)]^2 + [e^t(\cos t + \sin t)]^2\right)^{3/2}}$$

$$= \frac{\left|2e^{2t}(\cos^2 t - \sin t \cos t + \sin t \cos t + \sin^2 t)\right|}{\left[e^{2t}(\cos^2 t - 2\cos t \sin t + \sin^2 t + \cos^2 t + 2\cos t \sin t + \sin^2 t)\right]^{3/2}}$$

$$= \frac{\left|2e^{2t}(1)\right|}{\left[e^{2t}(1 + 1)\right]^{3/2}} = \frac{2e^{2t}}{e^{3t}(2)^{3/2}} = \frac{1}{\sqrt{2}\,e^t}$$

37. $\left(1, \frac{2}{3}, 1\right)$ corresponds to $t = 1$. $\mathbf{T}(t) = \dfrac{\mathbf{r}'(t)}{|\mathbf{r}'(t)|} = \dfrac{\langle 2t, 2t^2, 1 \rangle}{\sqrt{4t^2 + 4t^4 + 1}} = \dfrac{\langle 2t, 2t^2, 1 \rangle}{2t^2 + 1}$, so $\mathbf{T}(1) = \left\langle \frac{2}{3}, \frac{2}{3}, \frac{1}{3} \right\rangle$.

$\mathbf{T}'(t) = -4t(2t^2 + 1)^{-2} \langle 2t, 2t^2, 1 \rangle + (2t^2 + 1)^{-1} \langle 2, 4t, 0 \rangle$ [by Theorem 10.2.3 #3]

     $= (2t^2 + 1)^{-2} \langle -8t^2 + 4t^2 + 2, -8t^3 + 8t^3 + 4t, -4t \rangle = 2(2t^2 + 1)^{-2} \langle 1 - 2t^2, 2t, -2t \rangle$

$\mathbf{N}(t) = \dfrac{\mathbf{T}'(t)}{|\mathbf{T}'(t)|} = \dfrac{2(2t^2 + 1)^{-2} \langle 1 - 2t^2, 2t, -2t \rangle}{2(2t^2 + 1)^{-2}\sqrt{(1 - 2t^2)^2 + (2t)^2 + (-2t)^2}} = \dfrac{\langle 1 - 2t^2, 2t, -2t \rangle}{\sqrt{1 - 4t^2 + 4t^4 + 8t^2}} = \dfrac{\langle 1 - 2t^2, 2t, -2t \rangle}{1 + 2t^2}$

$\mathbf{N}(1) = \left\langle -\frac{1}{3}, \frac{2}{3}, -\frac{2}{3} \right\rangle$ and $\mathbf{B}(1) = \mathbf{T}(1) \times \mathbf{N}(1) = \left\langle -\frac{4}{9} - \frac{2}{9}, -\left(-\frac{4}{9} + \frac{1}{9}\right), \frac{4}{9} + \frac{2}{9} \right\rangle = \left\langle -\frac{2}{3}, \frac{1}{3}, \frac{2}{3} \right\rangle$.

39. $(0, \pi, -2)$ corresponds to $t = \pi$. $\mathbf{r}(t) = \langle 2 \sin 3t, t, 2 \cos 3t \rangle$ $\Rightarrow$

$$\mathbf{T}(t) = \frac{\mathbf{r}'(t)}{|\mathbf{r}'(t)|} = \frac{\langle 6 \cos 3t, 1, -6 \sin 3t \rangle}{\sqrt{36 \cos^2 3t + 1 + 36 \sin^2 3t}} = \tfrac{1}{\sqrt{37}} \langle 6 \cos 3t, 1, -6 \sin 3t \rangle.$$

$\mathbf{T}(\pi) = \tfrac{1}{\sqrt{37}} \langle -6, 1, 0 \rangle$ is a normal vector for the normal plane, and so $\langle -6, 1, 0 \rangle$ is also normal. Thus an equation for the

plane is $-6(x - 0) + 1(y - \pi) + 0(z + 2) = 0$ or $y - 6x = \pi$.

$$\mathbf{T}'(t) = \tfrac{1}{\sqrt{37}} \langle -18 \sin 3t, 0, -18 \cos 3t \rangle \quad \Rightarrow \quad |\mathbf{T}'(t)| = \frac{\sqrt{18^2 \sin^2 3t + 18^2 \cos^2 3t}}{\sqrt{37}} = \frac{18}{\sqrt{37}} \quad \Rightarrow$$

$\mathbf{N}(t) = \dfrac{\mathbf{T}'(t)}{|\mathbf{T}'(t)|} = \langle -\sin 3t, 0, -\cos 3t \rangle$. So $\mathbf{N}(\pi) = \langle 0, 0, 1 \rangle$ and $\mathbf{B}(\pi) = \tfrac{1}{\sqrt{37}} \langle -6, 1, 0 \rangle \times \langle 0, 0, 1 \rangle = \tfrac{1}{\sqrt{37}} \langle 1, 6, 0 \rangle$.

Since $\mathbf{B}(\pi)$ is a normal to the osculating plane, so is $\langle 1, 6, 0 \rangle$ and an equation for the plane is

$1(x - 0) + 6(y - \pi) + 0(z + 2) = 0$ or $x + 6y = 6\pi$.

41. The ellipse is given by the parametric equations $x = 2 \cos t$, $y = 3 \sin t$, so using the result from Exercise 34,

$$\kappa(t) = \frac{|\dot{x} \ddot{y} - \ddot{x} \dot{y}|}{(\dot{x}^2 + \dot{y}^2)^{3/2}} = \frac{|(-2 \sin t)(-3 \sin t) - (3 \cos t)(-2 \cos t)|}{(4 \sin^2 t + 9 \cos^2 t)^{3/2}} = \frac{6}{(4 \sin^2 t + 9 \cos^2 t)^{3/2}}$$

At $(2, 0)$, $t = 0$. Now $\kappa(0) = \frac{6}{27} = \frac{2}{9}$, so the radius of the osculating

circle is $1/\kappa(0) = \frac{9}{2}$ and its center is $\left(-\frac{5}{2}, 0\right)$. Its equation is therefore

$\left(x + \frac{5}{2}\right)^2 + y^2 = \frac{81}{4}$. At $(0, 3)$, $t = \frac{\pi}{2}$, and $\kappa\left(\frac{\pi}{2}\right) = \frac{6}{8} = \frac{3}{4}$. So the

radius of the osculating circle is $\frac{4}{3}$ and its center is $\left(0, \frac{5}{3}\right)$. Hence its

equation is $x^2 + \left(y - \frac{5}{3}\right)^2 = \frac{16}{9}$.

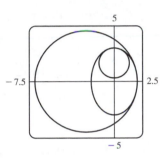

43. The tangent vector is normal to the normal plane, and the vector $\langle 6, 6, -8 \rangle$ is normal to the given plane. But $\mathbf{T}(t) \parallel \mathbf{r}'(t)$ and

$\langle 6, 6, -8 \rangle \parallel \langle 3, 3, -4 \rangle$, so we need to find t such that $\mathbf{r}'(t) \parallel \langle 3, 3, -4 \rangle$. $\mathbf{r}(t) = \langle t^3, 3t, t^4 \rangle$ $\Rightarrow$

$\mathbf{r}'(t) = \langle 3t^2, 3, 4t^3 \rangle \parallel \langle 3, 3, -4 \rangle$ when $t = -1$. So the planes are parallel at the point $\mathbf{r}(-1) = (-1, -3, 1)$.

45. $\kappa = \left| \dfrac{d\mathbf{T}}{ds} \right| = \left| \dfrac{d\mathbf{T}/dt}{ds/dt} \right| = \dfrac{|d\mathbf{T}/dt|}{ds/dt}$ and $\mathbf{N} = \dfrac{d\mathbf{T}/dt}{|d\mathbf{T}/dt|}$, so $\kappa \mathbf{N} = \dfrac{\left| \dfrac{d\mathbf{T}}{dt} \right| \dfrac{d\mathbf{T}}{dt}}{\left| \dfrac{d\mathbf{T}}{dt} \right| \dfrac{ds}{dt}} = \dfrac{d\mathbf{T}/dt}{ds/dt} = \dfrac{d\mathbf{T}}{ds}$ by the Chain Rule.

47. (a) $|\mathbf{B}| = 1 \Rightarrow \mathbf{B} \cdot \mathbf{B} = 1 \Rightarrow \dfrac{d}{ds}(\mathbf{B} \cdot \mathbf{B}) = 0 \Rightarrow 2 \dfrac{d\mathbf{B}}{ds} \cdot \mathbf{B} = 0 \Rightarrow \dfrac{d\mathbf{B}}{ds} \perp \mathbf{B}$

(b) $\mathbf{B} = \mathbf{T} \times \mathbf{N} \Rightarrow$

$$\frac{d\mathbf{B}}{ds} = \frac{d}{ds}(\mathbf{T} \times \mathbf{N}) = \frac{d}{dt}(\mathbf{T} \times \mathbf{N}) \frac{1}{ds/dt} = \frac{d}{dt}(\mathbf{T} \times \mathbf{N}) \frac{1}{|\mathbf{r}'(t)|}$$

$$= [(\mathbf{T}' \times \mathbf{N}) + (\mathbf{T} \times \mathbf{N}')] \frac{1}{|\mathbf{r}'(t)|} = \left[\left(\mathbf{T}' \times \frac{\mathbf{T}'}{|\mathbf{T}'|} \right) + (\mathbf{T} \times \mathbf{N}') \right] \frac{1}{|\mathbf{r}'(t)|} = \frac{\mathbf{T} \times \mathbf{N}'}{|\mathbf{r}'(t)|}$$

$$\Rightarrow \frac{d\mathbf{B}}{ds} \perp \mathbf{T}$$

(c) $\mathbf{B} = \mathbf{T} \times \mathbf{N} \Rightarrow \mathbf{T} \perp \mathbf{N}$, $\mathbf{B} \perp \mathbf{T}$ and $\mathbf{B} \perp \mathbf{N}$. So $\mathbf{B}$, $\mathbf{T}$ and $\mathbf{N}$ form an orthogonal set of vectors in the

three-dimensional space $\mathbb{R}^3$. From parts (a) and (b), $d\mathbf{B}/ds$ is perpendicular to both $\mathbf{B}$ and $\mathbf{T}$, so $d\mathbf{B}/ds$ is parallel to $\mathbf{N}$.

Therefore, $d\mathbf{B}/ds = -\tau(s)\mathbf{N}$, where $\tau(s)$ is a scalar.

(d) Since $\mathbf{B} = \mathbf{T} \times \mathbf{N}$, $\mathbf{T} \perp \mathbf{N}$ and both $\mathbf{T}$ and $\mathbf{N}$ are unit vectors, $\mathbf{B}$ is a unit vector mutually perpendicular to both $\mathbf{T}$ and $\mathbf{N}$. For a plane curve, $\mathbf{T}$ and $\mathbf{N}$ always lie in the plane of the curve, so that $\mathbf{B}$ is a constant unit vector always perpendicular to the plane. Thus $d\mathbf{B}/ds = \mathbf{0}$, but $d\mathbf{B}/ds = -\tau(s)\mathbf{N}$ and $\mathbf{N} \neq \mathbf{0}$, so $\tau(s) = 0$.

49. (a) $\mathbf{r}' = s'\,\mathbf{T} \;\Rightarrow\; \mathbf{r}'' = s''\,\mathbf{T} + s'\,\mathbf{T}' = s''\,\mathbf{T} + s'\,\dfrac{d\mathbf{T}}{ds}\,s' = s''\,\mathbf{T} + \kappa(s')^2\,\mathbf{N}$ by the first Serret-Frenet formula.

(b) Using part (a), we have

$$
\begin{aligned}
\mathbf{r}' \times \mathbf{r}'' &= (s'\,\mathbf{T}) \times [s''\,\mathbf{T} + \kappa(s')^2\,\mathbf{N}] \\
&= [(s'\,\mathbf{T}) \times (s''\,\mathbf{T})] + [(s'\mathbf{T}) \times (\kappa(s')^2\,\mathbf{N})] \qquad \text{[by Property 3 of the cross product]} \\
&= (s's'')(\mathbf{T} \times \mathbf{T}) + \kappa(s')^3(\mathbf{T} \times \mathbf{N}) = \mathbf{0} + \kappa(s')^3\,\mathbf{B} = \kappa(s')^3\,\mathbf{B}
\end{aligned}
$$

(c) Using part (a), we have

$$
\begin{aligned}
\mathbf{r}''' &= [s''\,\mathbf{T} + \kappa(s')^2\,\mathbf{N}]' = s'''\,\mathbf{T} + s''\,\mathbf{T}' + \kappa'(s')^2\,\mathbf{N} + 2\kappa s's''\,\mathbf{N} + \kappa(s')^2\,\mathbf{N}' \\
&= s'''\,\mathbf{T} + s''\,\dfrac{d\mathbf{T}}{ds}\,s' + \kappa'(s')^2\,\mathbf{N} + 2\kappa s's''\,\mathbf{N} + \kappa(s')^2\,\dfrac{d\mathbf{N}}{ds}\,s' \\
&= s'''\,\mathbf{T} + s''s'\kappa\,\mathbf{N} + \kappa'(s')^2\,\mathbf{N} + 2\kappa s's''\,\mathbf{N} + \kappa(s')^3(-\kappa\,\mathbf{T} + \tau\,\mathbf{B}) \qquad \text{[by the second formula]} \\
&= [s''' - \kappa^2(s')^3]\,\mathbf{T} + [3\kappa s's'' + \kappa'(s')^2]\,\mathbf{N} + \kappa\tau(s')^3\,\mathbf{B}
\end{aligned}
$$

(d) Using parts (b) and (c) and the facts that $\mathbf{B} \cdot \mathbf{T} = 0$, $\mathbf{B} \cdot \mathbf{N} = 0$, and $\mathbf{B} \cdot \mathbf{B} = 1$, we get

$$
\frac{(\mathbf{r}' \times \mathbf{r}'') \cdot \mathbf{r}'''}{|\mathbf{r}' \times \mathbf{r}''|^2} = \frac{\kappa(s')^3\,\mathbf{B} \cdot \{[s''' - \kappa^2(s')^3]\,\mathbf{T} + [3\kappa s's'' + \kappa'(s')^2]\,\mathbf{N} + \kappa\tau(s')^3\,\mathbf{B}\}}{|\kappa(s')^3\,\mathbf{B}|^2} = \frac{\kappa(s')^3\kappa\tau(s')^3}{[\kappa(s')^3]^2} = \tau
$$

51. For one helix, the vector equation is $\mathbf{r}(t) = \langle 10\cos t, 10\sin t, 34t/(2\pi) \rangle$ (measuring in angstroms), because the radius of each helix is 10 angstroms, and z increases by 34 angstroms for each increase of 2π in t. Using the arc length formula, letting t go from 0 to $2.9 \times 10^8 \times 2\pi$, we find the approximate length of each helix to be

$$
L = \int_0^{2.9 \times 10^8 \times 2\pi} |\mathbf{r}'(t)|\, dt = \int_0^{2.9 \times 10^8 \times 2\pi} \sqrt{(-10\sin t)^2 + (10\cos t)^2 + \left(\tfrac{34}{2\pi}\right)^2}\, dt
$$

$$
= \sqrt{100 + \left(\tfrac{34}{2\pi}\right)^2}\; t \,\Bigg]_0^{2.9 \times 10^8 \times 2\pi} = 2.9 \times 10^8 \times 2\pi \sqrt{100 + \left(\tfrac{34}{2\pi}\right)^2}
$$

$$
\approx 2.07 \times 10^{10}\ \text{Å} \;\text{— more than two meters!}
$$

10.4 Motion in Space: Velocity and Acceleration

1. (a) If $\mathbf{r}(t) = x(t)\,\mathbf{i} + y(t)\,\mathbf{j} + z(t)\,\mathbf{k}$ is the position vector of the particle at time t, then the average velocity over the time
interval $[0, 1]$ is

$$\mathbf{v}_{\text{ave}} = \frac{\mathbf{r}(1) - \mathbf{r}(0)}{1 - 0} = \frac{(4.5\,\mathbf{i} + 6.0\,\mathbf{j} + 3.0\,\mathbf{k}) - (2.7\,\mathbf{i} + 9.8\,\mathbf{j} + 3.7\,\mathbf{k})}{1} = 1.8\,\mathbf{i} - 3.8\,\mathbf{j} - 0.7\,\mathbf{k}.$$ Similarly, over the other

intervals we have

$$[0.5, 1]: \quad \mathbf{v}_{\text{ave}} = \frac{\mathbf{r}(1) - \mathbf{r}(0.5)}{1 - 0.5} = \frac{(4.5\,\mathbf{i} + 6.0\,\mathbf{j} + 3.0\,\mathbf{k}) - (3.5\,\mathbf{i} + 7.2\,\mathbf{j} + 3.3\,\mathbf{k})}{0.5} = 2.0\,\mathbf{i} - 2.4\,\mathbf{j} - 0.6\,\mathbf{k}$$

$$[1, 2]: \quad \mathbf{v}_{\text{ave}} = \frac{\mathbf{r}(2) - \mathbf{r}(1)}{2 - 1} = \frac{(7.3\,\mathbf{i} + 7.8\,\mathbf{j} + 2.7\,\mathbf{k}) - (4.5\,\mathbf{i} + 6.0\,\mathbf{j} + 3.0\,\mathbf{k})}{1} = 2.8\,\mathbf{i} + 1.8\,\mathbf{j} - 0.3\,\mathbf{k}$$

$$[1, 1.5]: \quad \mathbf{v}_{\text{ave}} = \frac{\mathbf{r}(1.5) - \mathbf{r}(1)}{1.5 - 1} = \frac{(5.9\,\mathbf{i} + 6.4\,\mathbf{j} + 2.8\,\mathbf{k}) - (4.5\,\mathbf{i} + 6.0\,\mathbf{j} + 3.0\,\mathbf{k})}{0.5} = 2.8\,\mathbf{i} + 0.8\,\mathbf{j} - 0.4\,\mathbf{k}$$

(b) We can estimate the velocity at $t = 1$ by averaging the average velocities over the time intervals $[0.5, 1]$ and $[1, 1.5]$:

$\mathbf{v}(1) \approx \frac{1}{2}[(2\,\mathbf{i} - 2.4\,\mathbf{j} - 0.6\,\mathbf{k}) + (2.8\,\mathbf{i} + 0.8\,\mathbf{j} - 0.4\,\mathbf{k})] = 2.4\,\mathbf{i} - 0.8\,\mathbf{j} - 0.5\,\mathbf{k}.$ Then the speed is

$|\mathbf{v}(1)| \approx \sqrt{(2.4)^2 + (-0.8)^2 + (-0.5)^2} \approx 2.58.$

3. $\mathbf{r}(t) = \left\langle -\frac{1}{2}t^2, t \right\rangle \quad \Rightarrow$ At $t = 2$:

$\mathbf{v}(t) = \mathbf{r}'(t) = \langle -t, 1 \rangle,$ $\mathbf{v}(2) = \langle -2, 1 \rangle$

$\mathbf{a}(t) = \mathbf{r}''(t) = \langle -1, 0 \rangle,$ $\mathbf{a}(2) = \langle -1, 0 \rangle$

$|\mathbf{v}(t)| = \sqrt{t^2 + 1}$

5. $\mathbf{r}(t) = 3\cos t\,\mathbf{i} + 2\sin t\,\mathbf{j} \quad \Rightarrow$

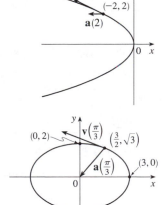

$\mathbf{v}(t) = -3\sin t\,\mathbf{i} + 2\cos t\,\mathbf{j}, \; \mathbf{v}\left(\frac{\pi}{3}\right) = -\frac{3\sqrt{3}}{2}\,\mathbf{i} + \mathbf{j}$

$\mathbf{a}(t) = -3\cos t\,\mathbf{i} - 2\sin t\,\mathbf{j}, \; \mathbf{a}\left(\frac{\pi}{3}\right) = -\frac{3}{2}\,\mathbf{i} - \sqrt{3}\,\mathbf{j}$

$|\mathbf{v}(t)| = \sqrt{9\sin^2 t + 4\cos^2 t} = \sqrt{4 + 5\sin^2 t}$

Notice that $x^2/9 + y^2/4 = \sin^2 t + \cos^2 t = 1$, so the path is an

ellipse.

7. $\mathbf{r}(t) = t\,\mathbf{i} + t^2\,\mathbf{j} + 2\,\mathbf{k} \quad \Rightarrow$

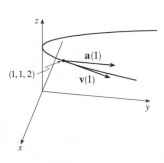

$\mathbf{v}(t) = \mathbf{i} + 2t\,\mathbf{j}, \; \mathbf{v}(1) = \mathbf{i} + 2\,\mathbf{j}$

$\mathbf{a}(t) = 2\,\mathbf{j}, \; \mathbf{a}(1) = 2\,\mathbf{j}$

$|\mathbf{v}(t)| = \sqrt{1 + 4t^2}$

Here $x = t$, $y = t^2 \quad \Rightarrow \quad y = x^2$ and $z = 2$, so the path of the

particle is a parabola in the plane $z = 2$.

9. $\mathbf{r}(t) = \left\langle t^2 + 1, t^3, t^2 - 1 \right\rangle \quad \Rightarrow \quad \mathbf{v}(t) = \mathbf{r}'(t) = \left\langle 2t, 3t^2, 2t \right\rangle, \; \mathbf{a}(t) = \mathbf{v}'(t) = \left\langle 2, 6t, 2 \right\rangle,$

$|\mathbf{v}(t)| = \sqrt{(2t)^2 + (3t^2)^2 + (2t)^2} = \sqrt{9t^4 + 8t^2} = |t|\sqrt{9t^2 + 8}.$

11. $r(t) = \sqrt{2}\,t\,i + e^t\,j + e^{-t}\,k \Rightarrow v(t) = r'(t) = \sqrt{2}\,i + e^t\,j - e^{-t}\,k,\ a(t) = v'(t) = e^t\,j + e^{-t}\,k,$

$|v(t)| = \sqrt{2 + e^{2t} + e^{-2t}} = \sqrt{(e^t + e^{-t})^2} = e^t + e^{-t}.$

13. $a(t) = i + 2j \Rightarrow v(t) = \int a(t)\,dt = \int (i + 2j)\,dt = t\,i + 2t\,j + C$ and $k = v(0) = C$, so $C = k$ and

$v(t) = t\,i + 2t\,j + k.\ \ r(t) = \int v(t)\,dt = \int (t\,i + 2t\,j + k)\,dt = \frac{1}{2}t^2\,i + t^2\,j + t\,k + D.$ But $i = r(0) = D$, so $D = i$

and $r(t) = \left(\frac{1}{2}t^2 + 1\right)i + t^2\,j + t\,k.$

15. (a) $a(t) = 2t\,i + \sin t\,j + \cos 2t\,k \Rightarrow$

$v(t) = \int (2t\,i + \sin t\,j + \cos 2t\,k)\,dt = t^2\,i - \cos t\,j + \frac{1}{2}\sin 2t\,k + C$

and $i = v(0) = -j + C$, so $C = i + j$ and

$v(t) = (t^2 + 1)\,i + (1 - \cos t)\,j + \frac{1}{2}\sin 2t\,k.$

$r(t) = \int [(t^2 + 1)\,i + (1 - \cos t)\,j + \frac{1}{2}\sin 2t\,k]\,dt$

$= \left(\frac{1}{3}t^3 + t\right)i + (t - \sin t)\,j - \frac{1}{4}\cos 2t\,k + D$

(b)

But $j = r(0) = -\frac{1}{4}k + D$, so $D = j + \frac{1}{4}k$ and $r(t) = \left(\frac{1}{3}t^3 + t\right)i + (t - \sin t + 1)j + \left(\frac{1}{4} - \frac{1}{4}\cos 2t\right)k.$

17. $r(t) = \langle t^2, 5t, t^2 - 16t \rangle \Rightarrow v(t) = \langle 2t, 5, 2t - 16 \rangle,\ |v(t)| = \sqrt{4t^2 + 25 + 4t^2 - 64t + 256} = \sqrt{8t^2 - 64t + 281}$ and

$\frac{d}{dt}|v(t)| = \frac{1}{2}(8t^2 - 64t + 281)^{-1/2}(16t - 64).$ This is zero if and only if the numerator is zero, that is, $16t - 64 = 0$ or

$t = 4.$ Since $\frac{d}{dt}|v(t)| < 0$ for $t < 4$ and $\frac{d}{dt}|v(t)| > 0$ for $t > 4$, the minimum speed of $\sqrt{153}$ is attained at $t = 4$ units of

time.

19. $|F(t)| = 20$ N in the direction of the positive z-axis, so $F(t) = 20\,k$. Also $m = 4$ kg, $r(0) = 0$ and $v(0) = i - j$. Since

$20k = F(t) = 4\,a(t),\ a(t) = 5\,k.$ Then $v(t) = 5t\,k + c_1$ where $c_1 = i - j$ so $v(t) = i - j + 5t\,k$ and the speed is

$|v(t)| = \sqrt{1 + 1 + 25t^2} = \sqrt{25t^2 + 2}.$ Also $r(t) = t\,i - t\,j + \frac{5}{2}t^2\,k + c_2$ and $0 = r(0)$, so $c_2 = 0$ and

$r(t) = t\,i - t\,j + \frac{5}{2}t^2\,k.$

21. $|v(0)| = 500$ m/s and since the angle of elevation is $30\,°$, the direction of the velocity is $\frac{1}{2}\left(\sqrt{3}\,i + j\right).$ Thus

$v(0) = 250\left(\sqrt{3}\,i + j\right)$ and if we set up the axes so the projectile starts at the origin, then $r(0) = 0$. Ignoring air resistance, the

only force is that due to gravity, so $F(t) = -mg\,j$ where $g \approx 9.8$ m/s^2. Thus $a(t) = -g\,j$ and $v(t) = -gt\,j + c_1$. But

$250\left(\sqrt{3}\,i + j\right) = v(0) = c_1$, so $v(t) = 250\sqrt{3}\,i + (250 - gt)\,j$ and $r(t) = 250\sqrt{3}\,t\,i + \left(250t - \frac{1}{2}gt^2\right)j + c_2$ where

$0 = r(0) = c_2.$ Thus $r(t) = 250\sqrt{3}\,t\,i + \left(250t - \frac{1}{2}gt^2\right)j.$

(a) Setting $250t - \frac{1}{2}gt^2 = 0$ gives $t = 0$ or $t = \frac{500}{g} \approx 51.0$ s. So the range is $250\sqrt{3} \cdot \frac{500}{g} \approx 22$ km.

(b) $0 = \frac{d}{dt}\left(250t - \frac{1}{2}gt^2\right) = 250 - gt$ implies that the maximum height is attained when $t = 250/g \approx 25.5$ s. Thus, the

maximum height is $(250)(250/g) - g(250/g)^2\frac{1}{2} = (250)^2/(2g) \approx 3.2$ km.

(c) From part (a), impact occurs at $t = 500/g \approx 51.0.$ Thus, the velocity at impact is

$v(500/g) = 250\sqrt{3}\,i + [250 - g(500/g)]\,j = 250\sqrt{3}\,i - 250\,j$ and the speed is $|v(500/g)| = 250\sqrt{3 + 1} = 500$ m/s.

23. As in Example 5, $r(t) = (v_0 \cos 45\,°)t\,i + \left[(v_0 \sin 45\,°)t - \frac{1}{2}gt^2\right]j = \frac{1}{2}\left[v_0\sqrt{2}\,t\,i + \left(v_0\sqrt{2}\,t - gt^2\right)j\right].$ Then the ball lands

at $t = \frac{v_0\sqrt{2}}{g}$ s. Now since it lands 90 m away, $90 = \frac{1}{2}v_0\sqrt{2}\,\frac{v_0\sqrt{2}}{g}$ or $v_0^2 = 90g$ and the initial velocity is

$v_0 = \sqrt{90g} \approx 30$ m/s.

25. Let α be the angle of elevation. Then $v_0 = 150$ m/s and from Example 5, the horizontal distance traveled by the projectile is

$$d = \frac{v_0^2 \sin 2\alpha}{g}. \text{ Thus } \frac{150^2 \sin 2\alpha}{g} = 800 \quad \Rightarrow \quad \sin 2\alpha = \frac{800g}{150^2} \approx 0.3484 \quad \Rightarrow \quad 2\alpha \approx 20.4\,° \text{ or } 180 - 20.4 = 159.6\,°.$$

Two angles of elevation then are $\alpha \approx 10.2\,°$ and $\alpha \approx 79.8\,°$.

27. Place the catapult at the origin and assume the catapult is 100 meters from the city, so the city lies between $(100, 0)$ and

$(600, 0)$. The initial speed is $v_0 = 80$ m/s and let θ be the angle the catapult is set at. As in Example 5, the trajectory of the

catapulted rock is given by $\mathbf{r}(t) = (80 \cos \theta)t\,\mathbf{i} + \left[(80 \sin \theta)t - 4.9t^2\right]\mathbf{j}$. The top of the near city wall is at $(100, 15)$ which

the rock will hit when $(80 \cos \theta)\, t = 100 \quad \Rightarrow \quad t = \dfrac{5}{4 \cos \theta}$ and $(80 \sin \theta)t - 4.9t^2 = 15 \quad \Rightarrow$

$$80 \sin \theta \cdot \frac{5}{4 \cos \theta} - 4.9\left(\frac{5}{4 \cos \theta}\right)^2 = 15 \quad \Rightarrow \quad 100 \tan \theta - 7.65625 \sec^2 \theta = 15. \text{ Replacing } \sec^2 \theta \text{ with } \tan^2 \theta + 1 \text{ gives}$$

$7.65625 \tan^2 \theta - 100 \tan \theta + 22.62625 = 0$. Using the quadratic formula, we have $\tan \theta \approx 0.230324, 12.8309 \quad \Rightarrow$

$\theta \approx 13.0\,°, 85.5\,°$. So for $13.0\,° < \theta < 85.5\,°$, the rock will land beyond the near city wall. The base of the far wall is located

at $(600, 0)$ which the rock hits if $(80 \cos \theta)t = 600 \quad \Rightarrow \quad t = \dfrac{15}{2 \cos \theta}$ and $(80 \sin \theta)t - 4.9t^2 = 0 \quad \Rightarrow$

$$80 \sin \theta \cdot \frac{15}{2 \cos \theta} - 4.9\left(\frac{15}{2 \cos \theta}\right)^2 = 0 \quad \Rightarrow \quad 600 \tan \theta - 275.625 \sec^2 \theta = 0 \quad \Rightarrow$$

$275.625 \tan^2 \theta - 600 \tan \theta + 275.625 = 0$. Solutions are $\tan \theta \approx 0.658678, 1.51819 \quad \Rightarrow \quad \theta \approx 33.4\,°, 56.6\,°$. Thus the

rock lands beyond the enclosed city ground for $33.4\,° < \theta < 56.6\,°$, and the angles that allow the rock to land on city ground

are $13.0\,° < \theta < 33.4\,°, 56.6\,° < \theta < 85.5\,°$. If you consider that the rock can hit the far wall and bounce back into the city,

we calculate the angles that cause the rock to hit the top of the wall at $(600, 15)$: $(80 \cos \theta)t = 600 \quad \Rightarrow \quad t = \dfrac{15}{2 \cos \theta}$ and

$(80 \sin \theta)t - 4.9t^2 = 15 \quad \Rightarrow \quad 600 \tan \theta - 275.625 \sec^2 \theta = 15 \quad \Rightarrow \quad 275.625 \tan^2 \theta - 600 \tan \theta + 290.625 = 0.$

Solutions are $\tan \theta \approx 0.727506, 1.44936 \quad \Rightarrow \quad \theta \approx 36.0\,°, 55.4\,°$, so the catapult should be set with angle θ where

$13.0\,° < \theta < 36.0\,°, 55.4\,° < \theta < 85.5\,°$.

29. (a) After t seconds, the boat will be $5t$ meters west of point A. The

velocity of the water at that location is $\frac{3}{400}(5t)(40 - 5t)\,\mathbf{j}$. The

velocity of the boat in still water is $5\,\mathbf{i}$, so the resultant velocity of the

boat is $\mathbf{v}(t) = 5\,\mathbf{i} + \frac{3}{400}(5t)(40 - 5t)\,\mathbf{j} = 5\,\mathbf{i} + \left(\frac{3}{2}t - \frac{3}{16}t^2\right)\mathbf{j}$.

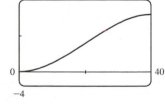

Integrating, we obtain $\mathbf{r}(t) = 5t\,\mathbf{i} + \left(\frac{3}{4}t^2 - \frac{1}{16}t^3\right)\mathbf{j} + \mathbf{C}$.

If we place the origin at A (and consider $\mathbf{j}$ to coincide with the northern direction) then $\mathbf{r}(0) = \mathbf{0} \quad \Rightarrow \quad \mathbf{C} = \mathbf{0}$ and we

have $\mathbf{r}(t) = 5t\,\mathbf{i} + \left(\frac{3}{4}t^2 - \frac{1}{16}t^3\right)\mathbf{j}$. The boat reaches the east bank after 8 s, and it is located at

$\mathbf{r}(8) = 5(8)\mathbf{i} + \left(\frac{3}{4}(8)^2 - \frac{1}{16}(8)^3\right)\mathbf{j} = 40\,\mathbf{i} + 16\,\mathbf{j}$. Thus the boat is 16 m downstream.

(b) Let α be the angle north of east that the boat heads. Then the velocity of the boat in still water is given by

$5(\cos \alpha)\,\mathbf{i} + 5(\sin \alpha)\,\mathbf{j}$. At t seconds, the boat is $5(\cos \alpha)t$ meters from the west bank, at which point the velocity of the

water is $\frac{3}{400}[5(\cos \alpha)t][40 - 5(\cos \alpha)t]\,\mathbf{j}$. The resultant velocity of the boat is given by

$$\mathbf{v}(t) = 5(\cos\alpha)\,\mathbf{i} + \left[5\sin\alpha + \tfrac{3}{400}(5t\cos\alpha)(40 - 5t\cos\alpha)\right]\mathbf{j}$$

$$= (5\cos\alpha)\,\mathbf{i} + \left(5\sin\alpha + \tfrac{3}{2}t\cos\alpha - \tfrac{3}{16}t^2\cos^2\alpha\right)\mathbf{j}$$

Integrating, $\mathbf{r}(t) = (5t\cos\alpha)\,\mathbf{i} + \left(5t\sin\alpha + \tfrac{3}{4}t^2\cos\alpha - \tfrac{1}{16}t^3\cos^2\alpha\right)\mathbf{j}$ (where we have again placed

the origin at A). The boat will reach the east bank when $5t\cos\alpha = 40 \;\Rightarrow\; t = \dfrac{40}{5\cos\alpha} = \dfrac{8}{\cos\alpha}$.

In order to land at point $B\,(40, 0)$ we need $5t\sin\alpha + \tfrac{3}{4}t^2\cos\alpha - \tfrac{1}{16}t^3\cos^2\alpha = 0 \;\Rightarrow$

$$5\left(\frac{8}{\cos\alpha}\right)\sin\alpha + \frac{3}{4}\left(\frac{8}{\cos\alpha}\right)^2\cos\alpha - \frac{1}{16}\left(\frac{8}{\cos\alpha}\right)^3\cos^2\alpha = 0 \;\Rightarrow\; \frac{1}{\cos\alpha}(40\sin\alpha + 48 - 32) = 0 \;\Rightarrow$$

$40\sin\alpha + 16 = 0 \;\Rightarrow\; \sin\alpha = -\tfrac{2}{5}$. Thus $\alpha = \sin^{-1}\left(-\tfrac{2}{5}\right) \approx -23.6\,°$, so the boat should head 23.6° south of

east (upstream).

The path does seem realistic. The boat initially heads upstream to

counteract the effect of the current. Near the center of the river, the

current is stronger and the boat is pushed downstream. When the

boat nears the eastern bank, the current is slower and the boat is

able to progress upstream to arrive at point B.

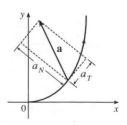

31. $\mathbf{r}(t) = (3t - t^3)\,\mathbf{i} + 3t^2\,\mathbf{j} \;\Rightarrow\; \mathbf{r}'(t) = (3 - 3t^2)\,\mathbf{i} + 6t\,\mathbf{j}$,

$|\mathbf{r}'(t)| = \sqrt{(3 - 3t^2)^2 + (6t)^2} = \sqrt{9 + 18t^2 + 9t^4} = \sqrt{(3 - 3t^2)^2} = 3 + 3t^2$,

$\mathbf{r}''(t) = -6t\,\mathbf{i} + 6\,\mathbf{j}$, $\mathbf{r}'(t) \times \mathbf{r}''(t) = (18 + 18t^2)\,\mathbf{k}$. Then Equation 9 gives

$$a_T = \frac{\mathbf{r}'(t)\cdot\mathbf{r}''(t)}{|\mathbf{r}'(t)|} = \frac{(3 - 3t^2)(-6t) + (6t)(6)}{3 + 3t^2} = \frac{18t + 18t^3}{3 + 3t^2} = \frac{18t(1 + t^2)}{3(1 + t^2)} = 6t \quad \left[\text{or by Equation 8,}\right.$$

$a_T = v' = \dfrac{d}{dt}\left[3 + 3t^2\right] = 6t\Bigr]$ and Equation 10 gives $a_N = \dfrac{|\mathbf{r}'(t) \times \mathbf{r}''(t)|}{|\mathbf{r}'(t)|} = \dfrac{18 + 18t^2}{3 + 3t^2} = \dfrac{18(1 + t^2)}{3(1 + t^2)} = 6.$

33. $\mathbf{r}(t) = \cos t\,\mathbf{i} + \sin t\,\mathbf{j} + t\,\mathbf{k} \;\Rightarrow\; \mathbf{r}'(t) = -\sin t\,\mathbf{i} + \cos t\,\mathbf{j} + \mathbf{k}$, $|\mathbf{r}'(t)| = \sqrt{\sin^2 t + \cos^2 t + 1} = \sqrt{2}$,

$\mathbf{r}''(t) = -\cos t\,\mathbf{i} - \sin t\,\mathbf{j}$, $\mathbf{r}'(t) \times \mathbf{r}''(t) = \sin t\,\mathbf{i} - \cos t\,\mathbf{j} + \mathbf{k}$. Then $a_T = \dfrac{\mathbf{r}'(t)\cdot\mathbf{r}''(t)}{|\mathbf{r}'(t)|} = \dfrac{\sin t\cos t - \sin t\cos t}{\sqrt{2}} = 0$

and $a_N = \dfrac{|\mathbf{r}'(t) \times \mathbf{r}''(t)|}{|\mathbf{r}'(t)|} = \dfrac{\sqrt{\sin^2 t + \cos^2 t + 1}}{\sqrt{2}} = \dfrac{\sqrt{2}}{\sqrt{2}} = 1$.

35. The tangential component of $\mathbf{a}$ is the length of the projection of $\mathbf{a}$ onto $\mathbf{T}$, so we

sketch the scalar projection of $\mathbf{a}$ in the tangential direction to the curve and

estimate its length to be 4.5 (using the fact that $\mathbf{a}$ has length 10 as a guide).

Similarly, the normal component of $\mathbf{a}$ is the length of the projection of $\mathbf{a}$ onto $\mathbf{N}$,

so we sketch the scalar projection of $\mathbf{a}$ in the normal direction to the curve and

estimate its length to be 9.0. Thus $a_T \approx 4.5$ cm/s^2 and $a_N \approx 9.0$ cm/s^2.

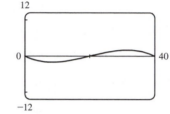

37. If the engines are turned off at time t, then the spacecraft will continue to travel in the direction of $\mathbf{v}(t)$, so we need a t such

that for some scalar $s > 0$, $\mathbf{r}(t) + s\,\mathbf{v}(t) = \langle 6, 4, 9 \rangle$. $\mathbf{v}(t) = \mathbf{r}'(t) = \mathbf{i} + \dfrac{1}{t}\mathbf{j} + \dfrac{8t}{(t^2 + 1)^2}\mathbf{k} \quad \Rightarrow$

$$\mathbf{r}(t) + s\,\mathbf{v}(t) = \left\langle 3 + t + s, 2 + \ln t + \dfrac{s}{t}, 7 - \dfrac{4}{t^2 + 1} + \dfrac{8st}{(t^2 + 1)^2} \right\rangle \quad \Rightarrow \quad 3 + t + s = 6 \quad \Rightarrow \quad s = 3 - t,$$

so $7 - \dfrac{4}{t^2 + 1} + \dfrac{8(3 - t)t}{(t^2 + 1)^2} = 9 \iff \dfrac{24t - 12t^2 - 4}{(t^2 + 1)^2} = 2 \iff t^4 + 8t^2 - 12t + 3 = 0$. It is easily seen that $t = 1$ is a

root of this polynomial. Also $2 + \ln 1 + \dfrac{3 - 1}{1} = 4$, so $t = 1$ is the desired solution.

10.5   Parametric Surfaces

1. $\mathbf{r}(u, v) = (u + v)\,\mathbf{i} + (3 - v)\,\mathbf{j} + (1 + 4u + 5v)\,\mathbf{k} = \langle 0, 3, 1 \rangle + u\,\langle 1, 0, 4 \rangle + v\,\langle 1, -1, 5 \rangle$. From Example 3, we recognize

this as a vector equation of a plane through the point $(0, 3, 1)$ and containing vectors $\mathbf{a} = \langle 1, 0, 4 \rangle$ and $\mathbf{b} = \langle 1, -1, 5 \rangle$. If we

wish to find a more conventional equation for the plane, a normal vector to the plane is $\mathbf{a} \times \mathbf{b} = \begin{vmatrix} \mathbf{i} & \mathbf{j} & \mathbf{k} \\ 1 & 0 & 4 \\ 1 & -1 & 5 \end{vmatrix} = 4\mathbf{i} - \mathbf{j} - \mathbf{k}$

and an equation of the plane is $4(x - 0) - (y - 3) - (z - 1) = 0$ or $4x - y - z = -4$.

3. $\mathbf{r}(s, t) = \langle s, t, t^2 - s^2 \rangle$, so the corresponding parametric equations for the surface are $x = s$, $y = t$, $z = t^2 - s^2$. For any

point (x, y, z) on the surface, we have $z = y^2 - x^2$. With no restrictions on the parameters, the surface is $z = y^2 - x^2$, which

we recognize as a hyperbolic paraboloid.

5. $\mathbf{r}(u, v) = \langle u^2 + 1, v^3 + 1, u + v \rangle$, $-1 \le u \le 1$, $-1 \le v \le 1$.

The surface has parametric equations $x = u^2 + 1$, $y = v^3 + 1$, $z = u + v$,

$-1 \le u \le 1$, $-1 \le v \le 1$. If we keep u constant at u_0, $x = u_0^2 + 1$, a

constant, so the corresponding grid curves must be the curves parallel to

the yz-plane. If v is constant, we have $y = v_0^3 + 1$, a constant, so these

grid curves are the curves parallel to the xz-plane.

7. $\mathbf{r}(u, v) = \langle \cos^3 u \cos^3 v, \sin^3 u \cos^3 v, \sin^3 v \rangle$.

The surface has parametric equations $x = \cos^3 u \cos^3 v$,

$y = \sin^3 u \cos^3 v$, $z = \sin^3 v$, $0 \le u \le \pi$, $0 \le v \le 2\pi$. Note that if

$v = v_0$ is constant then $z = \sin^3 v_0$ is constant, so the corresponding grid

curves must be the curves parallel to the xy-plane. The vertically oriented

grid curves, then, correspond to $u = u_0$ being held constant, giving

$x = \cos^3 u_0 \cos^3 v$, $y = \sin^3 u_0 \cos^3 v$, $z = \sin^3 v$. These curves lie in

vertical planes that contain the z-axis.

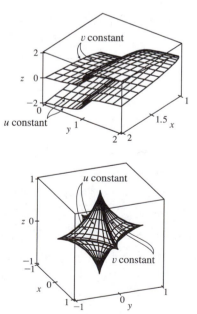

9. $x = \cos u \sin 2v$, $y = \sin u \sin 2v$, $z = \sin v$.

The complete graph of the surface is given by the parametric domain

$0 \le u \le \pi, 0 \le v \le 2\pi$. Note that if $v = v_0$ is constant, the parametric

equations become $x = \cos u \sin 2v_0$, $y = \sin u \sin 2v_0$, $z = \sin v_0$

which represent a circle of radius $\sin 2v_0$ in the plane $z = \sin v_0$. So the

circular grid curves we see lying horizontally are the grid curves which

have v constant. The vertical grid curves, then, correspond to $u = u_0$

being held constant, giving $x = \cos u_0 \sin 2v$ and $y = \sin u_0 \sin 2v$

with $z = \sin v$ which has a "figure-eight" shape.

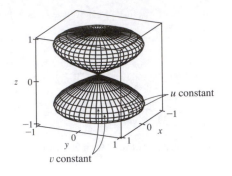

11. $\mathbf{r}(u, v) = \cos v\, \mathbf{i} + \sin v\, \mathbf{j} + u\, \mathbf{k}$. The parametric equations for the surface are $x = \cos v$, $y = \sin v$, $z = u$. Then $x^2 + y^2 = \cos^2 v + \sin^2 v = 1$ and $z = u$ with no restriction on u, so we have a circular cylinder, graph IV. The grid curves with u constant are the horizontal circles we see in the plane $z = u$. If v is constant, both x and y are constant with z free to vary, so the corresponding grid curves are the lines on the cylinder parallel to the z-axis.

13. $\mathbf{r}(u, v) = u \cos v\, \mathbf{i} + u \sin v\, \mathbf{j} + v\, \mathbf{k}$. The parametric equations for the surface are $x = u \cos v$, $y = u \sin v$, $z = v$. We look at the grid curves first; if we fix v, then x and y parametrize a straight line in the plane $z = v$ which intersects the z-axis. If u is held constant, the projection onto the xy-plane is circular; with $z = v$, each grid curve is a helix. The surface is a spiraling ramp, graph I.

15. $x = (u - \sin u) \cos v$, $y = (1 - \cos u) \sin v$, $z = u$. If u is held constant, x and y give an equation of an ellipse in the plane $z = u$, thus the grid curves are horizontally oriented ellipses. Note that when $u = 0$, the "ellipse" is the single point $(0, 0, 0)$, and when $u = \pi$, we have $y = 0$ while x ranges from $-\pi$ to π, a line segment parallel to the x-axis in the plane $z = \pi$. This is the upper "seam" we see in graph II. When v is held constant, $z = u$ is free to vary, so the corresponding grid curves are the curves we see running up and down along the surface.

17. From Example 3, parametric equations for the plane through the point $(1, 2, -3)$ that contains the vectors $\mathbf{a} = \langle 1, 1, -1 \rangle$ and $\mathbf{b} = \langle 1, -1, 1 \rangle$ are $x = 1 + u(1) + v(1) = 1 + u + v$, $y = 2 + u(1) + v(-1) = 2 + u - v$, $z = -3 + u(-1) + v(1) = -3 - u + v$.

19. Solving the equation for y gives $y^2 = 1 - x^2 + z^2 \quad \Rightarrow \quad y = \sqrt{1 - x^2 + z^2}$. (We choose the positive root since we want the part of the hyperboloid that corresponds to $y \ge 0$.) If we let x and z be the parameters, parametric equations are $x = x$, $z = z$, $y = \sqrt{1 - x^2 + z^2}$.

21. Since the cone intersects the sphere in the circle $x^2 + y^2 = 2$, $z = \sqrt{2}$ and we want the portion of the sphere above this, we can parametrize the surface as $x = x$, $y = y$, $z = \sqrt{4 - x^2 - y^2}$ where $x^2 + y^2 \le 2$.

Alternate solution: Using spherical coordinates, $x = 2 \sin \phi \cos \theta$, $y = 2 \sin \phi \sin \theta$, $z = 2 \cos \phi$ where $0 \le \phi \le \frac{\pi}{4}$ and $0 \le \theta \le 2\pi$.

23. Parametric equations are $x = x$, $y = 4 \cos \theta$, $z = 4 \sin \theta$, $0 \le x \le 5$, $0 \le \theta \le 2\pi$.

25. The surface appears to be a portion of a circular cylinder of radius 3 with axis the x-axis. An equation of the cylinder is $y^2 + z^2 = 9$, and we can impose the restrictions $0 \le x \le 5$, $y \le 0$ to obtain the portion shown.

To graph the surface on a CAS, we can use parametric equations $x = u$, $y = 3 \cos v$, $z = 3 \sin v$ with the parameter domain $0 \le u \le 5$, $\frac{\pi}{2} \le v \le \frac{3\pi}{2}$. Alternatively, we can regard x and z as parameters. Then parametric equations are $x = x$, $z = z$, $y = -\sqrt{9 - z^2}$, where $0 \le x \le 5$ and $-3 \le z \le 3$.

27. Using Equations 3, we have the parametrization $x = x$, $y = e^{-x} \cos \theta$, $z = e^{-x} \sin \theta$, $0 \le x \le 3$, $0 \le \theta \le 2\pi$.

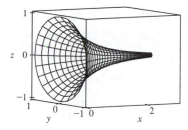

29. (a) $x = a \sin u \cos v$, $y = b \sin u \sin v$, $z = c \cos u$ $\Rightarrow$

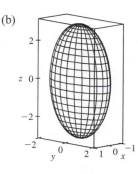

(b)

$$\frac{x^2}{a^2} + \frac{y^2}{b^2} + \frac{z^2}{c^2} = (\sin u \cos v)^2 + (\sin u \sin v)^2 + (\cos u)^2$$

$$= \sin^2 u + \cos^2 u = 1$$

and since the ranges of u and v are sufficient to generate the entire graph,

the parametric equations represent an ellipsoid.

31. (a) Replacing $\cos u$ by $\sin u$ and $\sin u$ by $\cos u$ gives parametric equations

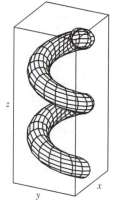

$x = (2 + \sin v) \sin u$, $y = (2 + \sin v) \cos u$, $z = u + \cos v$. From the graph, it

appears that the direction of the spiral is reversed. We can verify this observation

by noting that the projection of the spiral grid curves onto the xy-plane, given by

$x = (2 + \sin v) \sin u$, $y = (2 + \sin v) \cos u$, $z = 0$, draws a circle in the

clockwise direction for each value of v. The original equations, on the other hand,

give circular projections drawn in the counterclockwise direction. The equation for

z is identical in both surfaces, so as z increases, these grid curves spiral up in

opposite directions for the two surfaces.

(b) Replacing $\cos u$ by $\cos 2u$ and $\sin u$ by $\sin 2u$ gives parametric equations

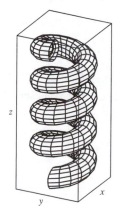

$x = (2 + \sin v) \cos 2u$, $y = (2 + \sin v) \sin 2u$, $z = u + \cos v$. From the graph, it

appears that the number of coils in the surface doubles within the same parametric

domain. We can verify this observation by noting that the projection of the spiral

grid curves onto the xy-plane, given by $x = (2 + \sin v) \cos 2u$,

$y = (2 + \sin v) \sin 2u$, $z = 0$ (where v is constant), complete circular revolutions

for $0 \le u \le \pi$ while the original surface requires $0 \le u \le 2\pi$ for a complete

revolution. Thus, the new surface winds around twice as fast as the original

surface, and since the equation for z is identical in both surfaces, we observe twice

as many circular coils in the same z-interval.

10 Review

1. A vector function is a function whose domain is a set of real numbers and whose range is a set of vectors. To find the derivative or integral, we can differentiate or integrate each component of the vector function.

2. The tip of the moving vector $\mathbf{r}(t)$ of a continuous vector function traces out a space curve.

3. (a) A curve represented by the vector function $\mathbf{r}(t)$ is smooth if $\mathbf{r}'(t)$ is continuous and $\mathbf{r}'(t) \neq \mathbf{0}$ on its parametric domain (except possibly at the endpoints).

   (b) The tangent vector to a smooth curve at a point P with position vector $\mathbf{r}(t)$ is the vector $\mathbf{r}'(t)$. The tangent line at P is the line through P parallel to the tangent vector $\mathbf{r}'(t)$. The unit tangent vector is $\mathbf{T}(t) = \dfrac{\mathbf{r}'(t)}{|\mathbf{r}'(t)|}$.

4. (a)–(f) See Theorem 10.2.3.

5. Use Formula 10.3.2, or equivalently 10.3.3.

6. (a) The curvature of a curve is $\kappa = \left| \dfrac{d\mathbf{T}}{ds} \right|$ where $\mathbf{T}$ is the unit tangent vector.

   (b) $\kappa(t) = \left| \dfrac{\mathbf{T}'(t)}{\mathbf{r}'(t)} \right|$ (c) $\kappa(t) = \dfrac{|\mathbf{r}'(t) \times \mathbf{r}''(t)|}{|\mathbf{r}'(t)|^3}$ (d) $\kappa(x) = \dfrac{|f''(x)|}{[1 + (f'(x))^2]^{3/2}}$

7. (a) The unit normal vector: $\mathbf{N}(t) = \dfrac{\mathbf{T}'(t)}{|\mathbf{T}'(t)|}$. The binormal vector: $\mathbf{B}(t) = \mathbf{T}(t) \times \mathbf{N}(t)$.

   (b) See the discussion at the bottom of page 713.

8. (a) If $\mathbf{r}(t)$ is the position vector of the particle on the space curve, the velocity $\mathbf{v}(t) = \mathbf{r}'(t)$, the speed is given by $|\mathbf{v}(t)|$, and the acceleration $\mathbf{a}(t) = \mathbf{v}'(t) = \mathbf{r}''(t)$.

   (b) $\mathbf{a} = a_T \mathbf{T} + a_N \mathbf{N}$ where $a_T = v'$ and $a_N = \kappa v^2$.

9. See the statement of Kepler's Laws on page 722.

10. See the discussion on pages 728 and 729.

1. True. If we reparametrize the curve by replacing $u = t^3$, we have $\mathbf{r}(u) = u\,\mathbf{i} + 2u\,\mathbf{j} + 3u\,\mathbf{k}$, which is a line through the origin with direction vector $\mathbf{i} + 2\,\mathbf{j} + 3\,\mathbf{k}$.

3. False. $\mathbf{r}'(t) = \langle -\sin t, 2t, 4t^3 \rangle$, and since $\mathbf{r}'(0) = \langle 0, 0, 0 \rangle = \mathbf{0}$, the curve is not smooth.

5. False. By Formula 5 of Theorem 10.2.3, $\dfrac{d}{dt}\,[\mathbf{u}(t) \times \mathbf{v}(t)] = \mathbf{u}'(t) \times \mathbf{v}(t) + \mathbf{u}(t) \times \mathbf{v}'(t)$.

7. False. κ is the magnitude of the rate of change of the unit tangent vector $\mathbf{T}$ with respect to arc length s, not with respect to t.

9. True. See the discussion at the bottom of page 713.

1. (a) The corresponding parametric equations for the curve are $x = t$, $y = \cos \pi t$, $z = \sin \pi t$. Since $y^2 + z^2 = 1$, the curve is contained in a circular cylinder with axis the x-axis. Since $x = t$, the curve is a helix.

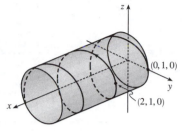

(b) $\mathbf{r}(t) = t\,\mathbf{i} + \cos \pi t\,\mathbf{j} + \sin \pi t\,\mathbf{k} \quad \Rightarrow \quad \mathbf{r}'(t) = \mathbf{i} - \pi \sin \pi t\,\mathbf{j} + \pi \cos \pi t\,\mathbf{k} \quad \Rightarrow \quad \mathbf{r}''(t) = -\pi^2 \cos \pi t\,\mathbf{j} - \pi^2 \sin \pi t\,\mathbf{k}$

3. The projection of the curve C of intersection onto the xy-plane is the circle $x^2 + y^2 = 16, z = 0$. So we can write $x = 4\cos t, y = 4\sin t, 0 \le t \le 2\pi$. From the equation of the plane, we have $z = 5 - x = 5 - 4\cos t$, so parametric equations for C are $x = 4\cos t, y = 4\sin t, z = 5 - 4\cos t, 0 \le t \le 2\pi$, and the corresponding vector function is $\mathbf{r}(t) = 4\cos t\,\mathbf{i} + 4\sin t\,\mathbf{j} + (5 - 4\cos t)\,\mathbf{k}, 0 \le t \le 2\pi$.

5. $\displaystyle\int_0^1 (t^2\,\mathbf{i} + t\cos \pi t\,\mathbf{j} + \sin \pi t\,\mathbf{k})\, dt = \left(\int_0^1 t^2\, dt\right)\mathbf{i} + \left(\int_0^1 t\cos \pi t\, dt\right)\mathbf{j} + \left(\int_0^1 \sin \pi t\, dt\right)\mathbf{k}$

$\qquad = \left[\tfrac{1}{3}t^3\right]_0^1 \mathbf{i} + \left(\tfrac{t}{\pi}\sin \pi t\right]_0^1 - \int_0^1 \tfrac{1}{\pi}\sin \pi t\, dt\right)\mathbf{j} + \left[-\tfrac{1}{\pi}\cos \pi t\right]_0^1 \mathbf{k}$

$\qquad = \tfrac{1}{3}\mathbf{i} + \left[\tfrac{1}{\pi^2}\cos \pi t\right]_0^1 \mathbf{j} + \tfrac{2}{\pi}\mathbf{k} = \tfrac{1}{3}\mathbf{i} - \tfrac{2}{\pi^2}\mathbf{j} + \tfrac{2}{\pi}\mathbf{k}$

where we integrated by parts in the y-component.

7. $\mathbf{r}(t) = \langle t^2, t^3, t^4 \rangle \quad \Rightarrow \quad \mathbf{r}'(t) = \langle 2t, 3t^2, 4t^3 \rangle \quad \Rightarrow \quad |\mathbf{r}'(t)| = \sqrt{4t^2 + 9t^4 + 16t^6}$ and

$L = \displaystyle\int_0^3 |\mathbf{r}'(t)|\, dt = \int_0^3 \sqrt{4t^2 + 9t^4 + 16t^6}\, dt$. Using Simpson's Rule with $f(t) = \sqrt{4t^2 + 9t^4 + 16t^6}$ and $n = 6$ we

have $\Delta t = \frac{3-0}{6} = \frac{1}{2}$ and

$L \approx \tfrac{\Delta t}{3}\left[f(0) + 4f\left(\tfrac{1}{2}\right) + 2f(1) + 4f\left(\tfrac{3}{2}\right) + 2f(2) + 4f\left(\tfrac{5}{2}\right) + f(3)\right]$

$\qquad = \tfrac{1}{6}\left[\sqrt{0+0+0} + 4\cdot\sqrt{4\left(\tfrac{1}{2}\right)^2 + 9\left(\tfrac{1}{2}\right)^4 + 16\left(\tfrac{1}{2}\right)^6} + 2\cdot\sqrt{4(1)^2 + 9(1)^4 + 16(1)^6}\right.$

$\qquad\qquad + 4\cdot\sqrt{4\left(\tfrac{3}{2}\right)^2 + 9\left(\tfrac{3}{2}\right)^4 + 16\left(\tfrac{3}{2}\right)^6} + 2\cdot\sqrt{4(2)^2 + 9(2)^4 + 16(2)^6}$

$\qquad\qquad \left. + 4\cdot\sqrt{4\left(\tfrac{5}{2}\right)^2 + 9\left(\tfrac{5}{2}\right)^4 + 16\left(\tfrac{5}{2}\right)^6} + \sqrt{4(3)^2 + 9(3)^4 + 16(3)^6}\right]$

$\qquad \approx 86.631$

9. The angle of intersection of the two curves, θ, is the angle between their respective tangents at the point of intersection. For both curves the point $(1, 0, 0)$ occurs when $t = 0$. $\mathbf{r}_1'(t) = -\sin t\,\mathbf{i} + \cos t\,\mathbf{j} + \mathbf{k} \quad \Rightarrow$ $\mathbf{r}_1'(0) = \mathbf{j} + \mathbf{k}$ and $\mathbf{r}_2'(t) = \mathbf{i} + 2t\,\mathbf{j} + 3t^2\,\mathbf{k} \quad \Rightarrow \quad \mathbf{r}_2'(0) = \mathbf{i}$. $\mathbf{r}_1'(0) \cdot \mathbf{r}_2'(0) = (\mathbf{j} + \mathbf{k}) \cdot \mathbf{i} = 0$. Therefore, the curves intersect in a right angle, that is, $\theta = \frac{\pi}{2}$.

11. (a) $\mathbf{T}(t) = \dfrac{\mathbf{r}'(t)}{|\mathbf{r}'(t)|} = \dfrac{\langle t^2, t, 1 \rangle}{|\langle t^2, t, 1 \rangle|} = \dfrac{\langle t^2, t, 1 \rangle}{\sqrt{t^4 + t^2 + 1}}$

(b) $\mathbf{T}'(t) = -\frac{1}{2}(t^4 + t^2 + 1)^{-3/2}(4t^3 + 2t)\langle t^2, t, 1 \rangle + (t^4 + t^2 + 1)^{-1/2}\langle 2t, 1, 0 \rangle$

$$= \frac{-2t^3 - t}{(t^4 + t^2 + 1)^{3/2}}\langle t^2, t, 1 \rangle + \frac{1}{(t^4 + t^2 + 1)^{1/2}}\langle 2t, 1, 0 \rangle$$

$$= \frac{\langle -2t^5 - t^3, -2t^4 - t^2, -2t^3 - t \rangle + \langle 2t^5 + 2t^3 + 2t, t^4 + t^2 + 1, 0 \rangle}{(t^4 + t^2 + 1)^{3/2}}$$

$$= \frac{\langle 2t, -t^4 + 1, -2t^3 - t \rangle}{(t^4 + t^2 + 1)^{3/2}}$$

$$|\mathbf{T}'(t)| = \frac{\sqrt{4t^2 + t^8 - 2t^4 + 1 + 4t^6 + 4t^4 + t^2}}{(t^4 + t^2 + 1)^{3/2}} = \frac{\sqrt{t^8 + 4t^6 + 2t^4 + 5t^2}}{(t^4 + t^2 + 1)^{3/2}}, \text{ and } \mathbf{N}(t) = \frac{\langle 2t, 1 - t^4, -2t^3 - t \rangle}{\sqrt{t^8 + 4t^6 + 2t^4 + 5t^2}}.$$

(c) $\kappa(t) = \dfrac{|\mathbf{T}'(t)|}{|\mathbf{r}'(t)|} = \dfrac{\sqrt{t^8 + 4t^6 + 2t^4 + 5t^2}}{(t^4 + t^2 + 1)^2}$

13. $y' = 4x^3$, $y'' = 12x^2$ and $\kappa(x) = \dfrac{|y''|}{[1 + (y')^2]^{3/2}} = \dfrac{|12x^2|}{(1 + 16x^6)^{3/2}}$, so $\kappa(1) = \dfrac{12}{17^{3/2}}$.

15. $\mathbf{r}(t) = \langle \sin 2t, t, \cos 2t \rangle$ $\Rightarrow$ $\mathbf{r}'(t) = \langle 2\cos 2t, 1, -2\sin 2t \rangle$ $\Rightarrow$ $\mathbf{T}(t) = \frac{1}{\sqrt{5}}\langle 2\cos 2t, 1, -2\sin 2t \rangle$ $\Rightarrow$

$\mathbf{T}'(t) = \frac{1}{\sqrt{5}}\langle -4\sin 2t, 0, -4\cos 2t \rangle$ $\Rightarrow$ $\mathbf{N}(t) = \langle -\sin 2t, 0, -\cos 2t \rangle$. So $\mathbf{N} = \mathbf{N}(\pi) = \langle 0, 0, -1 \rangle$ and

$\mathbf{B} = \mathbf{T} \times \mathbf{N} = \frac{1}{\sqrt{5}}\langle -1, 2, 0 \rangle$. So a normal to the osculating plane is $\langle -1, 2, 0 \rangle$ and an equation is

$-1(x - 0) + 2(y - \pi) + 0(z - 1) = 0$ or $x - 2y + 2\pi = 0$.

17. $\mathbf{r}(t) = t\ln t\,\mathbf{i} + t\,\mathbf{j} + e^{-t}\,\mathbf{k}$, $\mathbf{v}(t) = \mathbf{r}'(t) = (1 + \ln t)\,\mathbf{i} + \mathbf{j} - e^{-t}\,\mathbf{k}$,

$|\mathbf{v}(t)| = \sqrt{(1 + \ln t)^2 + 1^2 + (-e^{-t})^2} = \sqrt{2 + 2\ln t + (\ln t)^2 + e^{-2t}}$, $\mathbf{a}(t) = \mathbf{v}'(t) = \frac{1}{t}\mathbf{i} + e^{-t}\,\mathbf{k}$.

19. We set up the axes so that the shot leaves the athlete's hand 7 ft above the origin. Then we are given $\mathbf{r}(0) = 7\mathbf{j}$,

$|\mathbf{v}(0)| = 43$ ft/s, and $\mathbf{v}(0)$ has direction given by a $45\,°$ angle of elevation. Then a unit vector in the direction of $\mathbf{v}(0)$ is

$\frac{1}{\sqrt{2}}(\mathbf{i} + \mathbf{j})$ $\Rightarrow$ $\mathbf{v}(0) = \frac{43}{\sqrt{2}}(\mathbf{i} + \mathbf{j})$. Assuming air resistance is negligible, the only external force is due to gravity, so as in

Example 10.4.5 we have $\mathbf{a} = -g\mathbf{j}$ where here $g \approx 32$ ft/s^2. Since $\mathbf{v}'(t) = \mathbf{a}(t)$, we integrate, giving $\mathbf{v}(t) = -gt\mathbf{j} + \mathbf{C}$

where $\mathbf{C} = \mathbf{v}(0) = \frac{43}{\sqrt{2}}(\mathbf{i} + \mathbf{j})$ $\Rightarrow$ $\mathbf{v}(t) = \frac{43}{\sqrt{2}}\mathbf{i} + \left(\frac{43}{\sqrt{2}} - gt\right)\mathbf{j}$. Since $\mathbf{r}'(t) = \mathbf{v}(t)$ we integrate again, so

$\mathbf{r}(t) = \frac{43}{\sqrt{2}}t\,\mathbf{i} + \left(\frac{43}{\sqrt{2}}t - \frac{1}{2}gt^2\right)\mathbf{j} + \mathbf{D}$. But $\mathbf{D} = \mathbf{r}(0) = 7\mathbf{j}$ $\Rightarrow$ $\mathbf{r}(t) = \frac{43}{\sqrt{2}}t\,\mathbf{i} + \left(\frac{43}{\sqrt{2}}t - \frac{1}{2}gt^2 + 7\right)\mathbf{j}$.

(a) At 2 seconds, the shot is at $\mathbf{r}(2) = \frac{43}{\sqrt{2}}(2)\,\mathbf{i} + \left(\frac{43}{\sqrt{2}}(2) - \frac{1}{2}g(2)^2 + 7\right)\mathbf{j} \approx 60.8\,\mathbf{i} + 3.8\,\mathbf{j}$, so the shot is about 3.8 ft above

the ground, at a horizontal distance of 60.8 ft from the athlete.

(b) The shot reaches its maximum height when the vertical component of velocity is 0: $\frac{43}{\sqrt{2}} - gt = 0$ $\Rightarrow$

$t = \dfrac{43}{\sqrt{2}\,g} \approx 0.95$ s. Then $\mathbf{r}(0.95) \approx 28.9\,\mathbf{i} + 21.4\,\mathbf{j}$, so the maximum height is approximately 21.4 ft.

(c) The shot hits the ground when the vertical component of $\mathbf{r}(t)$ is 0, so $\frac{43}{\sqrt{2}}t - \frac{1}{2}gt^2 + 7 = 0$ $\Rightarrow$ $-16t^2 + \frac{43}{\sqrt{2}}t + 7 = 0$

$\Rightarrow$ $t \approx 2.11$ s. $\mathbf{r}(2.11) \approx 64.2\,\mathbf{i} - 0.08\,\mathbf{j}$, thus the shot lands approximately 64.2 ft from the athlete.

21. From Example 4 in Section 10.5, a parametric representation of the sphere $x^2 + y^2 + z^2 = 4$ is $x = 2\sin\phi\cos\theta$,

$y = 2\sin\phi\sin\theta$, $z = 2\cos\phi$ with $0 \le \theta \le 2\pi$ and $0 \le \phi \le \pi$. We can restrict the surface to that portion between the planes

$z = 1$ and $z = -1$ by restricting $-1 \le z \le 1$ $\Rightarrow$ $-1 \le 2\cos\phi \le 1$ $\Rightarrow$ $\frac{\pi}{3} \le \phi \le \frac{2\pi}{3}$.

23. By the Fundamental Theorem of Calculus, $\mathbf{r}'(t) = \langle \sin(\pi t^2/2), \cos(\pi t^2/2) \rangle$, $|\mathbf{r}'(t)| = 1$ and so $\mathbf{T}(t) = \mathbf{r}'(t)$. Thus

$\mathbf{T}'(t) = \pi t \langle \sin(\pi t^2/2), \cos(\pi t^2/2) \rangle$ and the curvature is $\kappa = |\mathbf{T}'(t)| = \sqrt{(\pi t)^2(1)} = \pi\,|t|$.

☐ FOCUS ON PROBLEM SOLVING

1. (a) $\mathbf{r}(t) = R\cos\omega t\,\mathbf{i} + R\sin\omega t\,\mathbf{j} \Rightarrow \mathbf{v} = \mathbf{r}'(t) = -\omega R\sin\omega t\,\mathbf{i} + \omega R\cos\omega t\,\mathbf{j}$, so $\mathbf{r} = R(\cos\omega t\,\mathbf{i} + \sin\omega t\,\mathbf{j})$ and

$\mathbf{v} = \omega R(-\sin\omega t\,\mathbf{i} + \cos\omega t\,\mathbf{j})$. $\mathbf{v} \cdot \mathbf{r} = \omega R^2(-\cos\omega t\sin\omega t + \sin\omega t\cos\omega t) = 0$, so $\mathbf{v} \perp \mathbf{r}$. Since $\mathbf{r}$ points along a

radius of the circle, and $\mathbf{v} \perp \mathbf{r}$, $\mathbf{v}$ is tangent to the circle. Because it is a velocity vector, $\mathbf{v}$ points in the direction of motion.

(b) In (a), we wrote $\mathbf{v}$ in the form $\omega R\,\mathbf{u}$, where $\mathbf{u}$ is the unit vector $-\sin\omega t\,\mathbf{i} + \cos\omega t\,\mathbf{j}$. Clearly $|\mathbf{v}| = \omega R\,|\mathbf{u}| = \omega R$. At

speed ωR, the particle completes one revolution, a distance $2\pi R$, in time $T = \dfrac{2\pi R}{\omega R} = \dfrac{2\pi}{\omega}$.

(c) $\mathbf{a} = \dfrac{d\mathbf{v}}{dt} = -\omega^2 R\cos\omega t\,\mathbf{i} - \omega^2 R\sin\omega t\,\mathbf{j} = -\omega^2 R(\cos\omega t\,\mathbf{i} + \sin\omega t\,\mathbf{j})$, so $\mathbf{a} = -\omega^2\mathbf{r}$. This shows that $\mathbf{a}$ is proportional

to $\mathbf{r}$ and points in the opposite direction (toward the origin). Also, $|\mathbf{a}| = \omega^2\,|\mathbf{r}| = \omega^2 R$.

(d) By Newton's Second Law (see Section 10.4), $\mathbf{F} = m\mathbf{a}$, so $|\mathbf{F}| = m\,|\mathbf{a}| = mR\omega^2 = \dfrac{m\,(\omega R)^2}{R} = \dfrac{m\,|\mathbf{v}|^2}{R}$.

3. (a) The projectile reaches maximum height when $0 = \dfrac{dy}{dt} = \dfrac{d}{dt}\left[(v_0\sin\alpha)t - \tfrac{1}{2}gt^2\right] = v_0\sin\alpha - gt$; that is, when

$t = \dfrac{v_0\sin\alpha}{g}$ and $y = (v_0\sin\alpha)\left(\dfrac{v_0\sin\alpha}{g}\right) - \dfrac{1}{2}g\left(\dfrac{v_0\sin\alpha}{g}\right)^2 = \dfrac{v_0^2\sin^2\alpha}{2g}$. This is the maximum height attained when

the projectile is fired with an angle of elevation α. This maximum height is largest when $\alpha = \dfrac{\pi}{2}$. In that case, $\sin\alpha = 1$

and the maximum height is $\dfrac{v_0^2}{2g}$.

(b) Let $R = v_0^2/g$. We are asked to consider the parabola $x^2 + 2Ry - R^2 = 0$ which can be rewritten as $y = -\dfrac{1}{2R}x^2 + \dfrac{R}{2}$.

The points on or inside this parabola are those for which $-R \le x \le R$ and $0 \le y \le \dfrac{-1}{2R}x^2 + \dfrac{R}{2}$. When the projectile is

fired at angle of elevation α, the points (x, y) along its path satisfy the relations $x = (v_0\cos\alpha)\,t$ and

$y = (v_0\sin\alpha)t - \tfrac{1}{2}gt^2$, where $0 \le t \le (2v_0\sin\alpha)/g$ (as in Example 10.4.5). Thus

$|x| \le \left|v_0\cos\alpha\left(\dfrac{2v_0\sin\alpha}{g}\right)\right| = \left|\dfrac{v_0^2}{g}\sin 2\alpha\right| \le \left|\dfrac{v_0^2}{g}\right| = |R|$. This shows that $-R \le x \le R$.

For t in the specified range, we also have $y = t\left(v_0\sin\alpha - \tfrac{1}{2}gt\right) = \tfrac{1}{2}gt\left(\dfrac{2v_0\sin\alpha}{g} - t\right) \ge 0$ and

$y = (v_0\sin\alpha)\dfrac{x}{v_0\cos\alpha} - \dfrac{g}{2}\left(\dfrac{x}{v_0\cos\alpha}\right)^2 = (\tan\alpha)x - \dfrac{g}{2v_0^2\cos^2\alpha}x^2 = -\dfrac{1}{2R\cos^2\alpha}x^2 + (\tan\alpha)x$. Thus

$$y - \left(\dfrac{-1}{2R}x^2 + \dfrac{R}{2}\right) = \dfrac{-1}{2R\cos^2\alpha}x^2 + \dfrac{1}{2R}x^2 + (\tan\alpha)x - \dfrac{R}{2}$$

$$= \dfrac{x^2}{2R}\left(1 - \dfrac{1}{\cos^2\alpha}\right) + (\tan\alpha)x - \dfrac{R}{2} = \dfrac{x^2(1 - \sec^2\alpha) + 2R(\tan\alpha)x - R^2}{2R}$$

$$= \dfrac{-(\tan^2\alpha)x^2 + 2R(\tan\alpha)x - R^2}{2R} = \dfrac{-[(\tan\alpha)x - R]^2}{2R} \le 0$$

[continued]

99

We have shown that every target that can be hit by the projectile lies on or inside the parabola $y = -\dfrac{1}{2R} x^2 + \dfrac{R}{2}$.

Now let (a, b) be any point on or inside the parabola $y = -\dfrac{1}{2R} x^2 + \dfrac{R}{2}$. Then $-R \le a \le R$ and $0 \le b \le -\dfrac{1}{2R} a^2 + \dfrac{R}{2}$.

We seek an angle α such that (a, b) lies in the path of the projectile; that is, we wish to find an angle α such that

$b = -\dfrac{1}{2R\cos^2\alpha} a^2 + (\tan\alpha)\, a$ or equivalently $b = \dfrac{-1}{2R}(\tan^2\alpha + 1)a^2 + (\tan\alpha)\, a$. Rearranging this equation we get

$\dfrac{a^2}{2R} \tan^2\alpha - a\tan\alpha + \left(\dfrac{a^2}{2R} + b\right) = 0$ or $a^2(\tan\alpha)^2 - 2aR(\tan\alpha) + (a^2 + 2bR) = 0$ $(*)$. This quadratic equation

for $\tan\alpha$ has real solutions exactly when the discriminant is nonnegative. Now $B^2 - 4AC \ge 0$ $\Leftrightarrow$

$(-2aR)^2 - 4a^2(a^2 + 2bR) \ge 0$ $\Leftrightarrow$ $4a^2(R^2 - a^2 - 2bR) \ge 0$ $\Leftrightarrow$ $-a^2 - 2bR + R^2 \ge 0$ $\Leftrightarrow$

$b \le \dfrac{1}{2R}(R^2 - a^2)$ $\Leftrightarrow$ $b \le \dfrac{-1}{2R} a^2 + \dfrac{R}{2}$. This condition is satisfied since (a, b) is on or inside the parabola

$y = -\dfrac{1}{2R} x^2 + \dfrac{R}{2}$. It follows that (a, b) lies in the path of the projectile when $\tan\alpha$ satisfies $(*)$, that is, when

$\tan\alpha = \dfrac{2aR \pm \sqrt{4a^2(R^2 - a^2 - 2bR)}}{2a^2} = \dfrac{R \pm \sqrt{R^2 - 2bR - a^2}}{a}$.

(c)

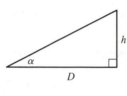

If the gun is pointed at a target with height h at a distance D downrange, then

$\tan\alpha = h/D$. When the projectile reaches a distance D downrange (remember

we are assuming that it doesn't hit the ground first), we have $D = x = (v_0\cos\alpha)t$,

so $t = \dfrac{D}{v_0\cos\alpha}$ and $y = (v_0\sin\alpha)t - \tfrac{1}{2}gt^2 = D\tan\alpha - \dfrac{gD^2}{2v_0^2\cos^2\alpha}$.

Meanwhile, the target, whose x-coordinate is also D, has fallen from height h to height

$h - \tfrac{1}{2}gt^2 = D\tan\alpha - \dfrac{gD^2}{2v_0^2\cos^2\alpha}$. Thus the projectile hits the target.

5. (a) $\mathbf{a} = -g\,\mathbf{j}$ $\Rightarrow$ $\mathbf{v} = \mathbf{v_0} - gt\,\mathbf{j} = 2\,\mathbf{i} - gt\,\mathbf{j}$ $\Rightarrow$ $\mathbf{s} = \mathbf{s_0} + 2t\,\mathbf{i} - \tfrac{1}{2}gt^2\,\mathbf{j} = 3.5\,\mathbf{j} + 2t\,\mathbf{i} - \tfrac{1}{2}gt^2\,\mathbf{j}$ $\Rightarrow$

$\mathbf{s} = 2t\,\mathbf{i} + \left(3.5 - \tfrac{1}{2}gt^2\right)\mathbf{j}$. Therefore $y = 0$ when $t = \sqrt{7/g}$ seconds. At that instant, the ball is $2\sqrt{7/g} \approx 0.94$ ft to the

right of the table top. Its coordinates (relative to an origin on the floor directly under the table's edge) are $(0.94, 0)$. At

impact, the velocity is $\mathbf{v} = 2\,\mathbf{i} - \sqrt{7g}\,\mathbf{j}$, so the speed is $|\mathbf{v}| = \sqrt{4 + 7g} \approx 15$ ft/s.

(b) The slope of the curve when $t = \sqrt{\dfrac{7}{g}}$ is $\dfrac{dy}{dx} = \dfrac{dy/dt}{dx/dt} = \dfrac{-gt}{2} = \dfrac{-g\sqrt{7/g}}{2} = \dfrac{-\sqrt{7g}}{2}$. Thus $\cot\theta = \dfrac{\sqrt{7g}}{2}$

and $\theta \approx 7.6°$.

(c) From (a), $|\mathbf{v}| = \sqrt{4 + 7g}$. So the ball rebounds with speed $0.8\sqrt{4 + 7g} \approx 12.08$ ft/s at angle of inclination

$90° - \theta \approx 82.3886°$. By Example 10.4.5, the horizontal distance traveled between bounces is $d = \dfrac{v_0^2\sin 2\alpha}{g}$, where

$v_0 \approx 12.08$ ft/s and $\alpha \approx 82.3886°$. Therefore, $d \approx 1.197$ ft. So the ball strikes the floor at about

$2\sqrt{7/g} + 1.197 \approx 2.13$ ft to the right of the table's edge.

7. The trajectory of the projectile is given by $\mathbf{r}(t) = (v\cos\alpha)\,t\,\mathbf{i} + \left[(v\sin\alpha)\,t - \frac{1}{2}gt^2\right]\mathbf{j}$, so
$\mathbf{v}(t) = \mathbf{r}'(t) = v\cos\alpha\,\mathbf{i} + (v\sin\alpha - gt)\,\mathbf{j}$ and

$$|\mathbf{v}(t)| = \sqrt{(v\cos\alpha)^2 + (v\sin\alpha - gt)^2} = \sqrt{v^2 - (2vg\sin\alpha)\,t + g^2t^2}$$

$$= \sqrt{g^2\left(t^2 - \frac{2v}{g}(\sin\alpha)\,t + \frac{v^2}{g^2}\right)} = g\sqrt{\left(t - \frac{v}{g}\sin\alpha\right)^2 + \frac{v^2}{g^2} - \frac{v^2}{g^2}\sin^2\alpha}$$

$$= g\sqrt{\left(t - \frac{v}{g}\sin\alpha\right)^2 + \frac{v^2}{g^2}\cos^2\alpha}$$

The projectile hits the ground when $(v\sin\alpha)\,t - \frac{1}{2}gt^2 = 0 \;\Rightarrow\; t = \frac{2v}{g}\sin\alpha$, so the distance travelled by the projectile is

$$L(\alpha) = \int_0^{(2v/g)\sin\alpha} |\mathbf{v}(t)|\,dt = \int_0^{(2v/g)\sin\alpha} g\sqrt{\left(t - \frac{v}{g}\sin\alpha\right)^2 + \frac{v^2}{g^2}\cos^2\alpha}\,dt$$

$$= g\left[\frac{t - (v/g)\sin\alpha}{2}\sqrt{\left(t - \frac{v}{g}\sin\alpha\right)^2 + \left(\frac{v}{g}\cos\alpha\right)^2}\right.$$

$$\left. + \frac{[(v/g)\cos\alpha]^2}{2}\ln\left(t - \frac{v}{g}\sin\alpha + \sqrt{\left(t - \frac{v}{g}\sin\alpha\right)^2 + \left(\frac{v}{g}\cos\alpha\right)^2}\right)\right]_0^{(2v/g)\sin\alpha}$$

[using Formula 21 in the Table of Integrals]

$$= \frac{g}{2}\left[\frac{v}{g}\sin\alpha\sqrt{\left(\frac{v}{g}\sin\alpha\right)^2 + \left(\frac{v}{g}\cos\alpha\right)^2} + \left(\frac{v}{g}\cos\alpha\right)^2\ln\left(\frac{v}{g}\sin\alpha + \sqrt{\left(\frac{v}{g}\sin\alpha\right)^2 + \left(\frac{v}{g}\cos\alpha\right)^2}\right)\right.$$

$$\left. + \frac{v}{g}\sin\alpha\sqrt{\left(\frac{v}{g}\sin\alpha\right)^2 + \left(\frac{v}{g}\cos\alpha\right)^2} - \left(\frac{v}{g}\cos\alpha\right)^2\ln\left(-\frac{v}{g}\sin\alpha + \sqrt{\left(\frac{v}{g}\sin\alpha\right)^2 + \left(\frac{v}{g}\cos\alpha\right)^2}\right)\right]$$

$$= \frac{g}{2}\left[\frac{v}{g}\sin\alpha\cdot\frac{v}{g} + \frac{v^2}{g^2}\cos^2\alpha\ln\left(\frac{v}{g}\sin\alpha + \frac{v}{g}\right) + \frac{v}{g}\sin\alpha\cdot\frac{v}{g} - \frac{v^2}{g^2}\cos^2\alpha\ln\left(-\frac{v}{g}\sin\alpha + \frac{v}{g}\right)\right]$$

$$= \frac{v^2}{g}\sin\alpha + \frac{v^2}{2g}\cos^2\alpha\ln\left(\frac{(v/g)\sin\alpha + v/g}{-(v/g)\sin\alpha + v/g}\right) = \frac{v^2}{g}\sin\alpha + \frac{v^2}{2g}\cos^2\alpha\ln\left(\frac{1 + \sin\alpha}{1 - \sin\alpha}\right)$$

We want to maximize $L(\alpha)$ for $0 \le \alpha \le \pi/2$.

$$L'(\alpha) = \frac{v^2}{g}\cos\alpha + \frac{v^2}{2g}\left[\cos^2\alpha\cdot\frac{1 - \sin\alpha}{1 + \sin\alpha}\cdot\frac{2\cos\alpha}{(1 - \sin\alpha)^2} - 2\cos\alpha\sin\alpha\ln\left(\frac{1 + \sin\alpha}{1 - \sin\alpha}\right)\right]$$

$$= \frac{v^2}{g}\cos\alpha + \frac{v^2}{2g}\left[\cos^2\alpha\cdot\frac{2}{\cos\alpha} - 2\cos\alpha\sin\alpha\ln\left(\frac{1 + \sin\alpha}{1 - \sin\alpha}\right)\right]$$

$$= \frac{v^2}{g}\cos\alpha + \frac{v^2}{g}\cos\alpha\left[1 - \sin\alpha\ln\left(\frac{1 + \sin\alpha}{1 - \sin\alpha}\right)\right] = \frac{v^2}{g}\cos\alpha\left[2 - \sin\alpha\ln\left(\frac{1 + \sin\alpha}{1 - \sin\alpha}\right)\right]$$

$L(\alpha)$ has critical points for $0 < \alpha < \pi/2$ when $L'(\alpha) = 0 \;\Rightarrow\; 2 - \sin\alpha\ln\left(\frac{1+\sin\alpha}{1-\sin\alpha}\right) = 0$ (since $\cos\alpha \ne 0$).

Solving by graphing (or using a CAS) gives $\alpha \approx 0.9855$. Compare values at the critical point and the endpoints:

$L(0) = 0$, $L(\pi/2) = v^2/g$, and $L(0.9855) \approx 1.20v^2/g$. Thus the distance travelled by the projectile is maximized

for $\alpha \approx 0.9855$ or $\approx 56°$.

11 ☐ PARTIAL DERIVATIVES

11.1 Functions of Several Variables

1. (a) From Table 1, $f(-15, 40) = -27$, which means that if the temperature is $-15°C$ and the wind speed is 40 km/h, then the air would feel equivalent to approximately $-27°C$ without wind.

(b) The question is asking: when the temperature is $-20°C$, what wind speed gives a wind-chill index of $-30°C$? From Table 1, the speed is 20 km/h.

(c) The question is asking: when the wind speed is 20 km/h, what temperature gives a wind-chill index of $-49°C$? From Table 1, the temperature is $-35°C$.

(d) The function $W = f(-5, v)$ means that we fix T at -5 and allow v to vary, resulting in a function of one variable. In other words, the function gives wind-chill index values for different wind speeds when the temperature is $-5°C$. From Table 1 (look at the row corresponding to $T = -5$), the function decreases and appears to approach a constant value as v increases.

(e) The function $W = f(T, 50)$ means that we fix v at 50 and allow T to vary, again giving a function of one variable. In other words, the function gives wind-chill index values for different temperatures when the wind speed is 50 km/h . From Table 1 (look at the column corresponding to $v = 50$), the function increases almost linearly as T increases.

3. If the amounts of labor and capital are both doubled, we replace L, K in the function with $2L, 2K$, giving

$$P(2L, 2K) = 1.01(2L)^{0.75}(2K)^{0.25} = 1.01(2^{0.75})(2^{0.25})L^{0.75}K^{0.25} = (2^1)1.01L^{0.75}K^{0.25} = 2P(L, K)$$

Thus, the production is doubled. It is also true for the general case $P(L, K) = bL^{\alpha}K^{1-\alpha}$:

$$P(2L, 2K) = b(2L)^{\alpha}(2K)^{1-\alpha} = b(2^{\alpha})(2^{1-\alpha})L^{\alpha}K^{1-\alpha} = (2^{\alpha+1-\alpha})bL^{\alpha}K^{1-\alpha} = 2P(L, K).$$

5. $\ln(9 - x^2 - 9y^2)$ is defined only when $9 - x^2 - 9y^2 > 0$, or $\frac{1}{9}x^2 + y^2 < 1$. So the domain of f is $\left\{(x, y) \mid \frac{1}{9}x^2 + y^2 < 1\right\}$, the interior of an ellipse.

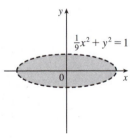

7. (a) $f(2, -1, 6) = e^{\sqrt{6-2^2-(-1)^2}} = e^{\sqrt{1}} = e.$

(b) $e^{\sqrt{z-x^2-y^2}}$ is defined when $z - x^2 - y^2 \geq 0 \Rightarrow z \geq x^2 + y^2$. Thus the domain of f is $\left\{(x, y, z) \mid z \geq x^2 + y^2\right\}$.

(c) Since $\sqrt{z - x^2 - y^2} \geq 0$, we have $e^{\sqrt{z-x^2-y^2}} \geq 1$. Thus the range of f is $[1, \infty)$.

9. The point $(-3, 3)$ lies between the level curves with z-values 50 and 60. Since the point is a little closer to the level curve with $z = 60$, we estimate that $f(-3, 3) \approx 56$. The point $(3, -2)$ appears to be just about halfway between the level curves with z-values 30 and 40, so we estimate $f(3, -2) \approx 35$. The graph rises as we approach the origin, gradually from above, steeply from below.

11. Near A, the level curves are very close together, indicating that the terrain is quite steep. At B, the level curves are much farther apart, so we would expect the terrain to be much less steep than near A, perhaps almost flat.

13.

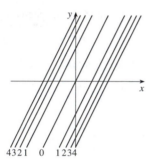

15. The level curves are $(y - 2x)^2 = k$ or $y = 2x \pm \sqrt{k}$, $k \geq 0$, a family of pairs of parallel lines.

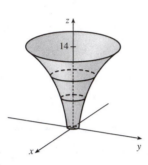

17. The level curves are $y - \ln x = k$ or $y = \ln x + k$.

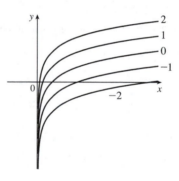

19. The level curves are $ye^x = k$ or $y = ke^{-x}$, a family of exponential curves.

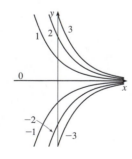

21. The level curves are $\sqrt{y^2 - x^2} = k$ or $y^2 - x^2 = k^2$, $k \geq 0$. When $k = 0$ the level curve is the pair of lines $y = \pm x$. For $k > 0$, the level curves are hyperbolas with axis the y-axis.

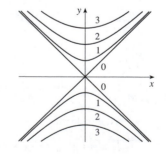

23. The contour map consists of the level curves $k = x^2 + 9y^2$, a family
of ellipses with major axis the x-axis. (Or, if $k = 0$, the origin.)
The graph of $f(x, y)$ is the surface $z = x^2 + 9y^2$, an elliptic
paraboloid.

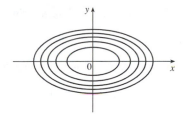

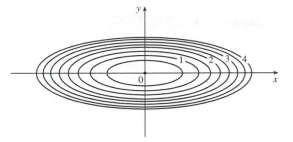

If we visualize lifting each ellipse $k = x^2 + 9y^2$ of the contour map
to the plane $z = k$, we have horizontal traces that indicate the shape
of the graph of f.

25. The isothermals are given by $k = 100/\left(1 + x^2 + 2y^2\right)$ or
$x^2 + 2y^2 = (100 - k)/k$ $(0 < k \le 100)$, a family of ellipses.

27. $f(x, y) = e^x \cos y$

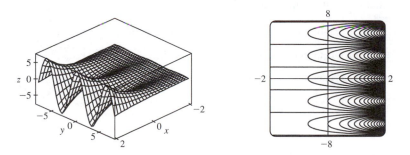

Traces parallel to the yz-plane (such as the left-front trace in the first graph above) are cosine curves. The amplitudes of these
curves decrease as x decreases.

29. $f(x, y) = xy^2 - x^3$

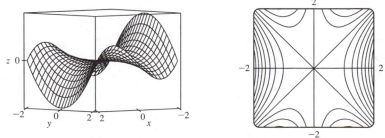

The traces parallel to the yz-plane (such as the left-front trace in the graph above) are parabolas; those parallel to the xz-plane
(such as the right-front trace) are cubic curves. The surface is called a monkey saddle because a monkey sitting on the surface
near the origin has places for both legs and tail to rest.

31. (a) C   **(b)** II

Reasons: This function is periodic in both x and y, and the function is the same when x is interchanged with y, so its graph is symmetric about the plane $y = x$. In addition, the function is 0 along the x- and y-axes. These conditions are satisfied only by C and II.

33. (a) F   **(b)** I

Reasons: This function is periodic in both x and y but is constant along the lines $y = x + k$, a condition satisfied only by F and I.

35. (a) B   **(b)** VI

Reasons: This function is 0 along the lines $x = \pm 1$ and $y = \pm 1$. The only contour map in which this could occur is VI. Also note that the trace in the xz-plane is the parabola $z = 1 - x^2$ and the trace in the yz-plane is the parabola $z = 1 - y^2$, so the graph is B.

37. $k = x + 3y + 5z$ is a family of parallel planes with normal vector $\langle 1, 3, 5 \rangle$.

39. $k = x^2 - y^2 + z^2$ are the equations of the level surfaces. For $k = 0$, the surface is a right circular cone with vertex the origin and axis the y-axis. For $k > 0$, we have a family of hyperboloids of one sheet with axis the y-axis. For $k < 0$, we have a family of hyperboloids of two sheets with axis the y-axis.

41. (a) The graph of g is the graph of f shifted upward 2 units.

(b) The graph of g is the graph of f stretched vertically by a factor of 2.

(c) The graph of g is the graph of f reflected about the xy-plane.

(d) The graph of $g(x, y) = -f(x, y) + 2$ is the graph of f reflected about the xy-plane and then shifted upward 2 units.

43. $f(x, y) = e^{cx^2 + y^2}$. First, if $c = 0$, the graph is the cylindrical surface $z = e^{y^2}$ (whose level curves are parallel lines). When $c > 0$, the vertical trace above the y-axis remains fixed while the sides of the surface in the x-direction "curl" upward, giving the graph a shape resembling an elliptic paraboloid. The level curves of the surface are ellipses centered at the origin.

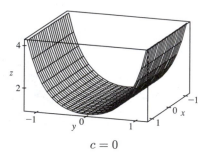

$c = 0$

For $0 < c < 1$, the ellipses have major axis the x-axis and the eccentricity increases as $c \to 0$.

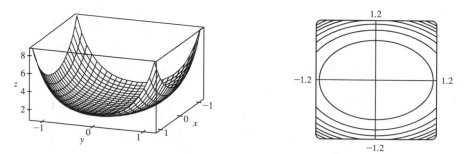

$c = 0.5$ (level curves in increments of 1)

For $c = 1$ the level curves are circles centered at the origin.

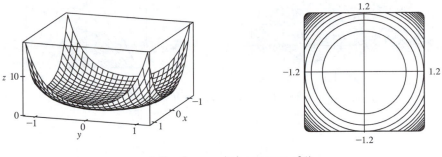

$c = 1$ (level curves in increments of 1)

When $c > 1$, the level curves are ellipses with major axis the y-axis, and the eccentricity increases as c increases.

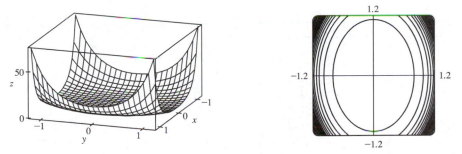

$c = 2$ (level curves in increments of 4)

For values of $c < 0$, the sides of the surface in the x-direction curl downward and approach the xy-plane (while the vertical trace $x = 0$ remains fixed), giving a saddle-shaped appearance to the graph near the point $(0, 0, 1)$. The level curves consist of a family of hyperbolas. As c decreases, the surface becomes flatter in the x-direction and the surface's approach to the curve in the trace $x = 0$ becomes steeper, as the graphs demonstrate.

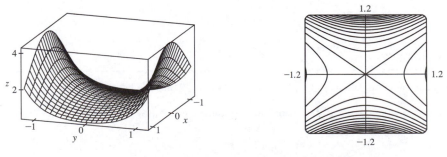

$c = -0.5$ (level curves in increments of 0.25)

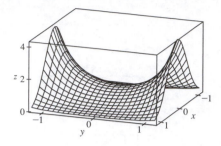 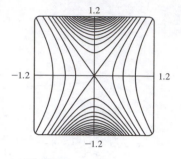

$c = -2$ (level curves in increments of 0.25)

45. (a) $P = bL^\alpha K^{1-\alpha} \implies \dfrac{P}{K} = bL^\alpha K^{-\alpha} \implies \dfrac{P}{K} = b\left(\dfrac{L}{K}\right)^\alpha \implies \ln\dfrac{P}{K} = \ln\left(b\left(\dfrac{L}{K}\right)^\alpha\right) \implies$

$\ln\dfrac{P}{K} = \ln b + \alpha \ln\left(\dfrac{L}{K}\right)$

(b) We list the values for $\ln(L/K)$ and $\ln(P/K)$ for the years 1899–1922. (Historically, these values were rounded to 2 decimal places.)

Year	$x = \ln(L/K)$	$y = \ln(P/K)$		Year	$x = \ln(L/K)$	$y = \ln(P/K)$
1899	0	0		1911	−0.38	−0.34
1900	−0.02	−0.06		1912	−0.38	−0.24
1901	−0.04	−0.02		1913	−0.41	−0.25
1902	−0.04	0		1914	−0.47	−0.37
1903	−0.07	−0.05		1915	−0.53	−0.34
1904	−0.13	−0.12		1916	−0.49	−0.28
1905	−0.18	−0.04		1917	−0.53	−0.39
1906	−0.20	−0.07		1918	−0.60	−0.50
1907	−0.23	−0.15		1919	−0.68	−0.57
1908	−0.41	−0.38		1920	−0.74	−0.57
1909	−0.33	−0.24		1921	−1.05	−0.85
1910	−0.35	−0.27		1922	−0.98	−0.59

After entering the (x, y) pairs into a calculator or CAS, the resulting least squares regression line through the points is approximately $y = 0.75136x + 0.01053$, which we round to $y = 0.75x + 0.01$.

(c) Comparing the regression line from part (b) to the equation $y = \ln b + \alpha x$ with $x = \ln(L/K)$ and $y = \ln(P/K)$, we have $\alpha = 0.75$ and $\ln b = 0.01 \implies b = e^{0.01} \approx 1.01$. Thus, the Cobb-Douglas production function is $P = bL^\alpha K^{1-\alpha} = 1.01L^{0.75}K^{0.25}$.

11.2 Limits and Continuity

1. In general, we can't say anything about $f(3, 1)$! $\displaystyle\lim_{(x,y)\to(3,1)} f(x, y) = 6$ means that the values of $f(x, y)$ approach 6 as (x, y) approaches, but is not equal to, $(3, 1)$. If f is continuous, we know that $\displaystyle\lim_{(x,y)\to(a,b)} f(x, y) = f(a, b)$, so $\displaystyle\lim_{(x,y)\to(3,1)} f(x, y) = f(3, 1) = 6$.

3. We make a table of values of $f(x, y) = \dfrac{x^2 y^3 + x^3 y^2 - 5}{2 - xy}$ for a set of (x, y) points near the origin.

x \ y	−0.2	−0.1	−0.05	0	0.05	0.1	0.2
−0.2	−2.551	−2.525	−2.513	−2.500	−2.488	−2.475	−2.451
−0.1	−2.525	−2.513	−2.506	−2.500	−2.494	−2.488	2.475
−0.05	−2.513	−2.506	−2.503	−2.500	−2.497	−2.494	−2.488
0	−2.500	−2.500	−2.500		−2.500	−2.500	−2.500
0.05	−2.488	−2.494	−2.497	−2.500	−2.503	−2.506	−2.513
0.1	−2.475	−2.488	−2.494	−2.500	−2.506	−2.513	−2.525
0.2	−2.451	−2.475	−2.488	−2.500	−2.513	−2.525	−2.551

As the table shows, the values of $f(x, y)$ seem to approach -2.5 as (x, y) approaches the origin from a variety of different directions. This suggests that $\lim\limits_{(x,y) \to (0,0)} f(x, y) = -2.5$.

Since f is a rational function, it is continuous on its domain. f is defined at $(0, 0)$, so we can use direct substitution to establish that $\lim\limits_{(x,y) \to (0,0)} f(x, y) = \dfrac{0^2 0^3 + 0^3 0^2 - 5}{2 - 0 \cdot 0} = -\dfrac{5}{2}$, verifying our guess.

5. $f(x, y) = x^5 + 4x^3 y - 5xy^2$ is a polynomial, and hence continuous, so
$$\lim_{(x,y) \to (5,-2)} f(x, y) = f(5, -2) = 5^5 + 4(5)^3(-2) - 5(5)(-2)^2 = 2025.$$

7. $f(x, y) = y^4 / (x^4 + 3y^4)$. First approach $(0, 0)$ along the x-axis. Then $f(x, 0) = 0/x^4 = 0$ for $x \neq 0$, so $f(x, y) \to 0$. Now approach $(0, 0)$ along the y-axis. Then for $y \neq 0$, $f(0, y) = y^4/3y^4 = 1/3$, so $f(x, y) \to 1/3$. Since f has two different limits along two different lines, the limit does not exist.

9. $f(x, y) = (xy \cos y)/(3x^2 + y^2)$. On the x-axis, $f(x, 0) = 0$ for $x \neq 0$, so $f(x, y) \to 0$ as $(x, y) \to (0, 0)$ along the x-axis. Approaching $(0, 0)$ along the line $y = x$, $f(x, x) = (x^2 \cos x)/4x^2 = \frac{1}{4} \cos x$ for $x \neq 0$, so $f(x, y) \to \frac{1}{4}$ along this line. Thus the limit does not exist.

11. $f(x, y) = \dfrac{xy}{\sqrt{x^2 + y^2}}$. We can see that the limit along any line through $(0, 0)$ is 0, as well as along other paths through $(0, 0)$ such as $x = y^2$ and $y = x^2$. So we suspect that the limit exists and equals 0; we use the Squeeze Theorem to prove our assertion. $0 \leq \left| \dfrac{xy}{\sqrt{x^2 + y^2}} \right| \leq |x|$ since $|y| \leq \sqrt{x^2 + y^2}$, and $|x| \to 0$ as $(x, y) \to (0, 0)$. So $\lim\limits_{(x,y) \to (0,0)} f(x, y) = 0$.

13. Let $f(x, y) = \dfrac{2x^2 y}{x^4 + y^2}$. Then $f(x, 0) = 0$ for $x \neq 0$, so $f(x, y) \to 0$ as $(x, y) \to (0, 0)$ along the x-axis. But $f(x, x^2) = \dfrac{2x^4}{2x^4} = 1$ for $x \neq 0$, so $f(x, y) \to 1$ as $(x, y) \to (0, 0)$ along the parabola $y = x^2$. Thus the limit doesn't exist.

15. $\lim\limits_{(x,y) \to (0,0)} \dfrac{x^2 + y^2}{\sqrt{x^2 + y^2 + 1} - 1} = \lim\limits_{(x,y) \to (0,0)} \dfrac{x^2 + y^2}{\sqrt{x^2 + y^2 + 1} - 1} \cdot \dfrac{\sqrt{x^2 + y^2 + 1} + 1}{\sqrt{x^2 + y^2 + 1} + 1}$

$$= \lim_{(x,y) \to (0,0)} \dfrac{(x^2 + y^2)\left(\sqrt{x^2 + y^2 + 1} + 1\right)}{x^2 + y^2} = \lim_{(x,y) \to (0,0)} \left(\sqrt{x^2 + y^2 + 1} + 1\right) = 2$$

17. $f(x, y, z) = \dfrac{xy + yz^2 + xz^2}{x^2 + y^2 + z^4}$. Then $f(x, 0, 0) = 0/x^2 = 0$ for $x \neq 0$, so as $(x, y, z) \to (0, 0, 0)$ along the x-axis, $f(x, y, z) \to 0$. But $f(x, x, 0) = x^2/(2x^2) = \frac{1}{2}$ for $x \neq 0$, so as $(x, y, z) \to (0, 0, 0)$ along the line $y = x$, $z = 0$, $f(x, y, z) \to \frac{1}{2}$. Thus the limit doesn't exist.

19.

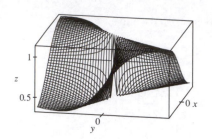

From the ridges on the graph, we see that as $(x, y) \rightarrow (0, 0)$ along the lines under the two ridges, $f(x, y)$ approaches different values. So the limit does not exist.

21. $h(x, y) = g(f(x, y)) = (2x + 3y - 6)^2 + \sqrt{2x + 3y - 6}$. Since f is a polynomial, it is continuous on $\mathbb{R}^2$ and g is continuous on its domain $\{t \mid t \geq 0\}$. Thus h is continuous on its domain

$D = \{(x, y) \mid 2x + 3y - 6 \geq 0\} = \{(x, y) \mid y \geq -\frac{2}{3}x + 2\}$, which consists of all points on or above the line $y = -\frac{2}{3}x + 2$.

23.

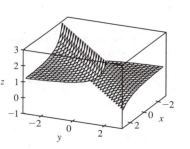

From the graph, it appears that f is discontinuous along the line $y = x$. If we consider $f(x, y) = e^{1/(x-y)}$ as a composition of functions, $g(x, y) = 1/(x - y)$ is a rational function and therefore continuous except where $x - y = 0 \Rightarrow y = x$. Since the function $h(t) = e^t$ is continuous everywhere, the composition $h(g(x, y)) = e^{1/(x-y)} = f(x, y)$ is continuous except along the line $y = x$, as we suspected.

25. The functions $\sin(xy)$ and $e^x - y^2$ are continuous everywhere, so $F(x, y) = \dfrac{\sin(xy)}{e^x - y^2}$ is continuous except where

$e^x - y^2 = 0 \Rightarrow y^2 = e^x \Rightarrow y = \pm\sqrt{e^x} = \pm e^{\frac{1}{2}x}$. Thus F is continuous on its domain $\{(x, y) \mid y \neq \pm e^{x/2}\}$.

27. $G(x, y) = \ln(x^2 + y^2 - 4) = g(f(x, y))$ where $f(x, y) = x^2 + y^2 - 4$, continuous on $\mathbb{R}^2$, and $g(t) = \ln t$, continuous on

its domain $\{t \mid t > 0\}$. Thus G is continuous on its domain $\{(x, y) \mid x^2 + y^2 - 4 > 0\} = \{(x, y) \mid x^2 + y^2 > 4\}$, the exterior of the circle $x^2 + y^2 = 4$.

29. $\sqrt{y}$ is continuous on its domain $\{y \mid y \geq 0\}$ and $x^2 - y^2 + z^2$ is continuous everywhere, so $f(x, y, z) = \dfrac{\sqrt{y}}{x^2 - y^2 + z^2}$ is

continuous for $y \geq 0$ and $x^2 - y^2 + z^2 \neq 0 \Rightarrow y^2 \neq x^2 + z^2$, that is, $\{(x, y, z) \mid y \geq 0, y \neq \sqrt{x^2 + z^2}\}$.

31. $f(x, y) = \begin{cases} \dfrac{x^2 y^3}{2x^2 + y^2} & \text{if } (x, y) \neq (0, 0) \\ 1 & \text{if } (x, y) = (0, 0) \end{cases}$ The first piece of f is a rational function defined everywhere except at the

origin, so f is continuous on $\mathbb{R}^2$ except possibly at the origin. Since $x^2 \leq 2x^2 + y^2$, we have $\left|x^2 y^3/(2x^2 + y^2)\right| \leq |y^3|$. We

know that $\left|y^3\right| \rightarrow 0$ as $(x, y) \rightarrow (0, 0)$. So, by the Squeeze Theorem, $\displaystyle\lim_{(x,y)\to(0,0)} f(x, y) = \lim_{(x,y)\to(0,0)} \dfrac{x^2 y^3}{2x^2 + y^2} = 0$. But

$f(0, 0) = 1$, so f is discontinuous at $(0, 0)$. Therefore, f is continuous on the set $\{(x, y) \mid (x, y) \neq (0, 0)\}$.

33. $\displaystyle\lim_{(x,y)\to(0,0)} \frac{x^3 + y^3}{x^2 + y^2} = \lim_{r\to 0^+} \frac{(r\cos\theta)^3 + (r\sin\theta)^3}{r^2} = \lim_{r\to 0^+} (r\cos^3\theta + r\sin^3\theta) = 0$

35. $\displaystyle\lim_{(x,y,z)\to(0,0,0)} \frac{xyz}{x^2 + y^2 + z^2} = \lim_{\rho\to 0^+} \frac{(\rho\sin\phi\cos\theta)(\rho\sin\phi\sin\theta)(\rho\cos\phi)}{\rho^2} = \lim_{\rho\to 0^+} (\rho\sin^2\phi\cos\phi\sin\theta\cos\theta) = 0$

11.3 Partial Derivatives

1. (a) $\partial T/\partial x$ represents the rate of change of T when we fix y and t and consider T as a function of the single variable x, which describes how quickly the temperature changes when longitude changes but latitude and time are constant. $\partial T/\partial y$ represents the rate of change of T when we fix x and t and consider T as a function of y, which describes how quickly the temperature changes when latitude changes but longitude and time are constant. $\partial T/\partial t$ represents the rate of change of T when we fix x and y and consider T as a function of t, which describes how quickly the temperature changes over time for a constant longitude and latitude.

(b) $f_x(158, 21, 9)$ represents the rate of change of temperature at longitude $158°$W, latitude $21°$N at 9:00 A.M. when only longitude varies. Since the air is warmer to the west than to the east, increasing longitude results in an increased air temperature, so we would expect $f_x(158, 21, 9)$ to be positive. $f_y(158, 21, 9)$ represents the rate of change of temperature at the same time and location when only latitude varies. Since the air is warmer to the south and cooler to the north, increasing latitude results in a decreased air temperature, so we would expect $f_y(158, 21, 9)$ to be negative. $f_t(158, 21, 9)$ represents the rate of change of temperature at the same time and location when only time varies. Since typically air temperature increases from the morning to the afternoon as the sun warms it, we would expect $f_t(158, 21, 9)$ to be positive.

3. (a) By Definition 4, $f_T(-15, 30) = \lim\limits_{h \to 0} \dfrac{f(-15+h, 30) - f(-15, 30)}{h}$, which we can approximate by considering $h = 5$ and $h = -5$ and using the values given in the table:

$f_T(-15, 30) \approx \dfrac{f(-10, 30) - f(-15, 30)}{5} = \dfrac{-20 - (-26)}{5} = \dfrac{6}{5} = 1.2$,

$f_T(-15, 30) \approx \dfrac{f(-20, 30) - f(-15, 30)}{-5} = \dfrac{-33 - (-26)}{-5} = \dfrac{-7}{-5} = 1.4$. Averaging these values, we estimate

$f_T(-15, 30)$ to be approximately 1.3. Thus, when the actual temperature is $-15°$C and the wind speed is 30 km/h, the apparent temperature rises by about $1.3°$C for every degree that the actual temperature rises.

Similarly, $f_v(-15, 30) = \lim\limits_{h \to 0} \dfrac{f(-15, 30+h) - f(-15, 30)}{h}$ which we can approximate by considering $h = 10$ and

$h = -10$: $f_v(-15, 30) \approx \dfrac{f(-15, 40) - f(-15, 30)}{10} = \dfrac{-27 - (-26)}{10} = \dfrac{-1}{10} = -0.1$,

$f_v(-15, 30) \approx \dfrac{f(-15, 20) - f(-15, 30)}{-10} = \dfrac{-24 - (-26)}{-10} = \dfrac{2}{-10} = -0.2$. Averaging these values, we estimate

$f_v(-15, 30)$ to be approximately -0.15. Thus, when the actual temperature is $-15°$C and the wind speed is 30 km/h, the apparent temperature decreases by about $0.15°$C for every km/h that the wind speed increases.

(b) For a fixed wind speed v, the values of the wind-chill index W increase as temperature T increases (look at a column of the table), so $\dfrac{\partial W}{\partial T}$ is positive. For a fixed temperature T, the values of W decrease (or remain constant) as v increases (look at a row of the table), so $\dfrac{\partial W}{\partial v}$ is negative (or perhaps 0).

(c) For fixed values of T, the function values $f(T, v)$ appear to become constant (or nearly constant) as v increases, so the corresponding rate of change is 0 or near 0 as v increases. This suggests that $\lim\limits_{v \to \infty} (\partial W/\partial v) = 0$.

5. (a) If we start at $(1, 2)$ and move in the positive x-direction, the graph of f increases. Thus $f_x(1, 2)$ is positive.

(b) If we start at $(1, 2)$ and move in the positive y-direction, the graph of f decreases. Thus $f_y(1, 2)$ is negative.

7. First of all, if we start at the point $(3, -3)$ and move in the positive y-direction, we see that both b and c decrease, while a increases. Both b and c have a low point at about $(3, -1.5)$, while a is 0 at this point. So a is definitely the graph of f_y, and one of b and c is the graph of f. To see which is which, we start at the point $(-3, -1.5)$ and move in the positive x-direction. b traces out a line with negative slope, while c traces out a parabola opening downward. This tells us that b is the x-derivative of c. So c is the graph of f, b is the graph of f_x, and a is the graph of f_y.

9. $f(x, y) = 16 - 4x^2 - y^2 \Rightarrow f_x(x, y) = -8x$ and $f_y(x, y) = -2y \Rightarrow f_x(1, 2) = -8$ and $f_y(1, 2) = -4$. The graph of f is the paraboloid $z = 16 - 4x^2 - y^2$ and the vertical plane $y = 2$ intersects it in the parabola $z = 12 - 4x^2$, $y = 2$ (the curve C_1 in the first figure).

The slope of the tangent line to this parabola at $(1, 2, 8)$ is $f_x(1, 2) = -8$. Similarly the plane $x = 1$ intersects the paraboloid in the parabola $z = 12 - y^2$, $x = 1$ (the curve C_2 in the second figure) and the slope of the tangent line at $(1, 2, 8)$ is $f_y(1, 2) = -4$.

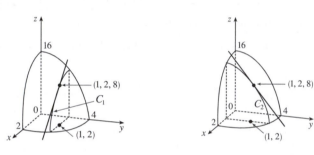

11. $f(x, y) = x^2 + y^2 + x^2y \Rightarrow f_x = 2x + 2xy$, $f_y = 2y + x^2$

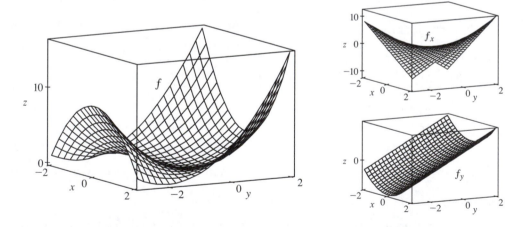

Note that the traces of f in planes parallel to the xz-plane are parabolas which open downward for $y < -1$ and upward for $y > -1$, and the traces of f_x in these planes are straight lines, which have negative slopes for $y < -1$ and positive slopes for $y > -1$. The traces of f in planes parallel to the yz-plane are parabolas which always open upward, and the traces of f_y in these planes are straight lines with positive slopes.

13. $f(x, y) = 3x - 2y^4 \Rightarrow f_x(x, y) = 3 - 0 = 3, f_y(x, y) = 0 - 8y^3 = -8y^3$

15. $z = xe^{3y} \Rightarrow \dfrac{\partial z}{\partial x} = e^{3y}, \dfrac{\partial z}{\partial y} = 3xe^{3y}$

17. $f(x, y) = \dfrac{x - y}{x + y} \Rightarrow f_x(x, y) = \dfrac{(1)(x + y) - (x - y)(1)}{(x + y)^2} = \dfrac{2y}{(x + y)^2}$,

$f_y(x, y) = \dfrac{(-1)(x + y) - (x - y)(1)}{(x + y)^2} = -\dfrac{2x}{(x + y)^2}$

19. $w = \sin \alpha \cos \beta \Rightarrow \dfrac{\partial w}{\partial \alpha} = \cos \alpha \cos \beta, \dfrac{\partial w}{\partial \beta} = -\sin \alpha \sin \beta$

21. $f(r, s) = r \ln(r^2 + s^2) \Rightarrow f_r(r, s) = r \cdot \dfrac{2r}{r^2 + s^2} + \ln(r^2 + s^2) \cdot 1 = \dfrac{2r^2}{r^2 + s^2} + \ln(r^2 + s^2)$,

$f_s(r, s) = r \cdot \dfrac{2s}{r^2 + s^2} + 0 = \dfrac{2rs}{r^2 + s^2}$

23. $u = te^{w/t} \Rightarrow \dfrac{\partial u}{\partial t} = t \cdot e^{w/t}(-wt^{-2}) + e^{w/t} \cdot 1 = e^{w/t} - \dfrac{w}{t}e^{w/t} = e^{w/t}\left(1 - \dfrac{w}{t}\right), \dfrac{\partial u}{\partial w} = te^{w/t} \cdot \dfrac{1}{t} = e^{w/t}$

25. $f(x, y, z) = xy^2z^3 + 3yz \Rightarrow f_x(x, y, z) = y^2z^3, f_y(x, y, z) = 2xyz^3 + 3z, f_z(x, y, z) = 3xy^2z^2 + 3y$

27. $w = \ln(x + 2y + 3z) \Rightarrow \dfrac{\partial w}{\partial x} = \dfrac{1}{x + 2y + 3z}, \dfrac{\partial w}{\partial y} = \dfrac{2}{x + 2y + 3z}, \dfrac{\partial w}{\partial z} = \dfrac{3}{x + 2y + 3z}$

29. $u = xe^{-t} \sin \theta \Rightarrow \dfrac{\partial u}{\partial x} = e^{-t} \sin \theta, \dfrac{\partial u}{\partial t} = -xe^{-t} \sin \theta, \dfrac{\partial u}{\partial \theta} = xe^{-t} \cos \theta$

31. $f(x, y, z, t) = xyz^2 \tan(yt) \Rightarrow f_x(x, y, z, t) = yz^2 \tan(yt)$,

$f_y(x, y, z, t) = xyz^2 \cdot \sec^2(yt) \cdot t + xz^2 \tan(yt) = xyz^2t \sec^2(yt) + xz^2 \tan(yt)$,

$f_z(x, y, z, t) = 2xyz \tan(yt), f_t(x, y, z, t) = xyz^2 \sec^2(yt) \cdot y = xy^2z^2 \sec^2(yt)$.

33. $u = \sqrt{x_1^2 + x_2^2 + \cdots + x_n^2}$. For each $i = 1, \ldots, n, u_{x_i} = \frac{1}{2}\left(x_1^2 + x_2^2 + \cdots + x_n^2\right)^{-1/2}(2x_i) = \dfrac{x_i}{\sqrt{x_1^2 + x_2^2 + \cdots + x_n^2}}$.

35. $f(x, y) = \sqrt{x^2 + y^2} \Rightarrow f_x(x, y) = \frac{1}{2}(x^2 + y^2)^{-1/2}(2x) = \dfrac{x}{\sqrt{x^2 + y^2}}$, so $f_x(3, 4) = \dfrac{3}{\sqrt{3^2 + 4^2}} = \dfrac{3}{5}$.

37. $f(x, y, z) = \dfrac{x}{y + z} = x(y + z)^{-1} \Rightarrow f_z(x, y, z) = x(-1)(y + z)^{-2} = -\dfrac{x}{(y + z)^2}$, so

$f_z(3, 2, 1) = -\dfrac{3}{(2 + 1)^2} = -\dfrac{1}{3}$.

39. $f(x, y) = xy^2 - x^3y \Rightarrow$

$f_x(x, y) = \lim_{h \to 0} \dfrac{f(x + h, y) - f(x, y)}{h} = \lim_{h \to 0} \dfrac{(x + h)y^2 - (x + h)^3y - (xy^2 - x^3y)}{h}$

$= \lim_{h \to 0} \dfrac{h(y^2 - 3x^2y - 3xyh - yh^2)}{h} = \lim_{h \to 0}(y^2 - 3x^2y - 3xyh - yh^2) = y^2 - 3x^2y$

$f_y(x, y) = \lim_{h \to 0} \dfrac{f(x, y + h) - f(x, y)}{h} = \lim_{h \to 0} \dfrac{x(y + h)^2 - x^3(y + h) - (xy^2 - x^3y)}{h}$

$= \lim_{h \to 0} \dfrac{h(2xy + xh - x^3)}{h} = \lim_{h \to 0}(2xy + xh - x^3) = 2xy - x^3$

41. $x^2 + y^2 + z^2 = 3xyz$ ⇒ $\dfrac{\partial}{\partial x}(x^2 + y^2 + z^2) = \dfrac{\partial}{\partial x}(3xyz)$ ⇒ $2x + 0 + 2z\dfrac{\partial z}{\partial x} = 3y\left(x\dfrac{\partial z}{\partial x} + z \cdot 1\right)$ ⇔

$2z\dfrac{\partial z}{\partial x} - 3xy\dfrac{\partial z}{\partial x} = 3yz - 2x$ ⇔ $(2z - 3xy)\dfrac{\partial z}{\partial x} = 3yz - 2x$, so $\dfrac{\partial z}{\partial x} = \dfrac{3yz - 2x}{2z - 3xy}$.

$\dfrac{\partial}{\partial y}(x^2 + y^2 + z^2) = \dfrac{\partial}{\partial y}(3xyz)$ ⇒ $0 + 2y + 2z\dfrac{\partial z}{\partial y} = 3x\left(y\dfrac{\partial z}{\partial y} + z \cdot 1\right)$ ⇔ $2z\dfrac{\partial z}{\partial y} - 3xy\dfrac{\partial z}{\partial y} = 3xz - 2y$ ⇔

$(2z - 3xy)\dfrac{\partial z}{\partial y} = 3xz - 2y$, so $\dfrac{\partial z}{\partial y} = \dfrac{3xz - 2y}{2z - 3xy}$.

43. $x - z = \arctan(yz)$ ⇒ $\dfrac{\partial}{\partial x}(x - z) = \dfrac{\partial}{\partial x}(\arctan(yz))$ ⇒ $1 - \dfrac{\partial z}{\partial x} = \dfrac{1}{1 + (yz)^2} \cdot y\dfrac{\partial z}{\partial x}$ ⇔

$1 = \left(\dfrac{y}{1 + y^2 z^2} + 1\right)\dfrac{\partial z}{\partial x}$ ⇔ $1 = \left(\dfrac{y + 1 + y^2 z^2}{1 + y^2 z^2}\right)\dfrac{\partial z}{\partial x}$, so $\dfrac{\partial z}{\partial x} = \dfrac{1 + y^2 z^2}{1 + y + y^2 z^2}$.

$\dfrac{\partial}{\partial y}(x - z) = \dfrac{\partial}{\partial y}(\arctan(yz))$ ⇒ $0 - \dfrac{\partial z}{\partial y} = \dfrac{1}{1 + (yz)^2} \cdot \left(y\dfrac{\partial z}{\partial y} + z \cdot 1\right)$ ⇔

$-\dfrac{z}{1 + y^2 z^2} = \left(\dfrac{y}{1 + y^2 z^2} + 1\right)\dfrac{\partial z}{\partial y}$ ⇔ $-\dfrac{z}{1 + y^2 z^2} = \left(\dfrac{y + 1 + y^2 z^2}{1 + y^2 z^2}\right)\dfrac{\partial z}{\partial y}$ ⇔ $\dfrac{\partial z}{\partial y} = -\dfrac{z}{1 + y + y^2 z^2}$.

45. (a) $z = f(x) + g(y)$ ⇒ $\dfrac{\partial z}{\partial x} = f'(x)$, $\dfrac{\partial z}{\partial y} = g'(y)$

(b) $z = f(x + y)$. Let $u = x + y$. Then $\dfrac{\partial z}{\partial x} = \dfrac{df}{du}\dfrac{\partial u}{\partial x} = \dfrac{df}{du}(1) = f'(u) = f'(x + y)$,

$\dfrac{\partial z}{\partial y} = \dfrac{df}{du}\dfrac{\partial u}{\partial y} = \dfrac{df}{du}(1) = f'(u) = f'(x + y)$.

47. $f(x, y) = x^4 - 3x^2 y^3$ ⇒ $f_x(x, y) = 4x^3 - 6xy^3$, $f_y(x, y) = -9x^2 y^2$. Then $f_{xx}(x, y) = 12x^2 - 6y^3$,

$f_{xy}(x, y) = -18xy^2$, $f_{yx}(x, y) = -18xy^2$, and $f_{yy}(x, y) = -18x^2 y$.

49. $z = \dfrac{x}{x + y} = x(x + y)^{-1}$ ⇒ $z_x = \dfrac{1(x + y) - 1(x)}{(x + y)^2} = \dfrac{y}{(x + y)^2}$, $z_y = x(-1)(x + y)^{-2} = -\dfrac{x}{(x + y)^2}$. Then

$z_{xx} = y(-2)(x + y)^{-3} = -\dfrac{2y}{(x + y)^3}$, $z_{xy} = \dfrac{1(x + y)^2 - y(2)(x + y)}{[(x + y)^2]^2} = \dfrac{x + y - 2y}{(x + y)^3} = \dfrac{x - y}{(x + y)^3}$,

$z_{yx} = -\dfrac{1(x + y)^2 - x(2)(x + y)}{[(x + y)^2]^2} = -\dfrac{-x^2 + xy + y^2}{(x + y)^2} = \dfrac{(x + y)(x - y)}{(x + y)^2} = \dfrac{x - y}{(x + y)^3}$, and

$z_{yy} = -x(-2)(x + y)^{-3} = \dfrac{2x}{(x + y)^3}$.

51. $u = e^{-s}\sin t$ ⇒ $u_s = -e^{-s}\sin t$, $u_t = e^{-s}\cos t$. Then $u_{ss} = e^{-s}\sin t$, $u_{st} = -e^{-s}\cos t$, $u_{ts} = -e^{-s}\cos t$,

and $u_{tt} = -e^{-s}\sin t$.

53. $u = x\sin(x + 2y)$ ⇒ $u_x = x \cdot \cos(x + 2y)(1) + \sin(x + 2y) \cdot 1 = x\cos(x + 2y) + \sin(x + 2y)$,

$u_{xy} = x(-\sin(x + 2y)(2)) + \cos(x + 2y)(2) = 2\cos(x + 2y) - 2x\sin(x + 2y)$

and $u_y = x\cos(x + 2y)(2) = 2x\cos(x + 2y)$,

$u_{yx} = 2x \cdot (-\sin(x + 2y)(1)) + \cos(x + 2y) \cdot 2 = 2\cos(x + 2y) - 2x\sin(x + 2y)$. Thus $u_{xy} = u_{yx}$.

55. $f(x, y) = 3xy^4 + x^3 y^2$ ⇒ $f_x = 3y^4 + 3x^2 y^2$, $f_{xx} = 6xy^2$, $f_{xxy} = 12xy$ and $f_y = 12xy^3 + 2x^3 y$,

$f_{yy} = 36xy^2 + 2x^3$, $f_{yyy} = 72xy$.

57. $f(x, y, z) = \cos(4x + 3y + 2z)$ ⟹

$f_x = -\sin(4x + 3y + 2z)(4) = -4\sin(4x + 3y + 2z)$,

$f_{xy} = -4\cos(4x + 3y + 2z)(3) = -12\cos(4x + 3y + 2z)$,

$f_{xyz} = -12(-\sin(4x + 3y + 2z))(2) = 24\sin(4x + 3y + 2z)$ and

$f_y = -\sin(4x + 3y + 2z)(3) = -3\sin(4x + 3y + 2z)$,

$f_{yz} = -3\cos(4x + 3y + 2z)(2) = -6\cos(4x + 3y + 2z)$,

$f_{yzz} = -6(-\sin(4x + 3y + 2z))(2) = 12\sin(4x + 3y + 2z)$.

59. $u = e^{r\theta}\sin\theta$ ⟹ $\dfrac{\partial u}{\partial \theta} = e^{r\theta}\cos\theta + \sin\theta \cdot e^{r\theta}(r) = e^{r\theta}(\cos\theta + r\sin\theta)$,

$\dfrac{\partial^2 u}{\partial r\, \partial \theta} = e^{r\theta}(\sin\theta) + (\cos\theta + r\sin\theta)\,e^{r\theta}(\theta) = e^{r\theta}(\sin\theta + \theta\cos\theta + r\theta\sin\theta)$,

$\dfrac{\partial^3 u}{\partial r^2\, \partial \theta} = e^{r\theta}(\theta\sin\theta) + (\sin\theta + \theta\cos\theta + r\theta\sin\theta) \cdot e^{r\theta}(\theta) = \theta e^{r\theta}(2\sin\theta + \theta\cos\theta + r\theta\sin\theta)$.

61. By Definition 4, $f_x(3, 2) = \lim\limits_{h \to 0} \dfrac{f(3 + h, 2) - f(3, 2)}{h}$ which we can approximate by considering $h = 0.5$

and $h = -0.5$: $f_x(3, 2) \approx \dfrac{f(3.5, 2) - f(3, 2)}{0.5} = \dfrac{22.4 - 17.5}{0.5} = 9.8$,

$f_x(3, 2) \approx \dfrac{f(2.5, 2) - f(3, 2)}{-0.5} = \dfrac{10.2 - 17.5}{-0.5} = 14.6$. Averaging these values, we estimate $f_x(3, 2)$ to be

approximately 12.2. Similarly, $f_x(3, 2.2) = \lim\limits_{h \to 0} \dfrac{f(3 + h, 2.2) - f(3, 2.2)}{h}$ which we can approximate by

considering $h = 0.5$ and $h = -0.5$: $f_x(3, 2.2) \approx \dfrac{f(3.5, 2.2) - f(3, 2.2)}{0.5} = \dfrac{26.1 - 15.9}{0.5} = 20.4$,

$f_x(3, 2.2) \approx \dfrac{f(2.5, 2.2) - f(3, 2.2)}{-0.5} = \dfrac{9.3 - 15.9}{-0.5} = 13.2$. Averaging these values, we have $f_x(3, 2.2) \approx 16.8$.

To estimate $f_{xy}(3, 2)$, we first need an estimate for $f_x(3, 1.8)$: $f_x(3, 1.8) \approx \dfrac{f(3.5, 1.8) - f(3, 1.8)}{0.5} = \dfrac{20.0 - 18.1}{0.5} = 3.8$,

$f_x(3, 1.8) \approx \dfrac{f(2.5, 1.8) - f(3, 1.8)}{-0.5} = \dfrac{12.5 - 18.1}{-0.5} = 11.2$. Averaging these values, we get $f_x(3, 1.8) \approx 7.5$.

Now $f_{xy}(x, y) = \dfrac{\partial}{\partial y}[f_x(x, y)]$ and $f_x(x, y)$ is itself a function of 2 variables, so Definition 4 says that

$f_{xy}(x, y) = \dfrac{\partial}{\partial y}[f_x(x, y)] = \lim\limits_{h \to 0} \dfrac{f_x(x, y + h) - f_x(x, y)}{h}$ ⟹ $f_{xy}(3, 2) = \lim\limits_{h \to 0} \dfrac{f_x(3, 2 + h) - f_x(3, 2)}{h}$.

We can estimate this value using our previous work with $h = 0.2$ and $h = -0.2$:

$f_{xy}(3, 2) \approx \dfrac{f_x(3, 2.2) - f_x(3, 2)}{0.2} = \dfrac{16.8 - 12.2}{0.2} = 23$, $f_{xy}(3, 2) \approx \dfrac{f_x(3, 1.8) - f_x(3, 2)}{-0.2} = \dfrac{7.5 - 12.2}{-0.2} = 23.5$.

Averaging these values, we estimate $f_{xy}(3, 2)$ to be approximately 23.25.

63. $u = e^{-\alpha^2 k^2 t}\sin kx$ ⟹ $u_x = ke^{-\alpha^2 k^2 t}\cos kx$, $u_{xx} = -k^2 e^{-\alpha^2 k^2 t}\sin kx$, and $u_t = -\alpha^2 k^2 e^{-\alpha^2 k^2 t}\sin kx$.

Thus $\alpha^2 u_{xx} = u_t$.

65. $u = \dfrac{1}{\sqrt{x^2 + y^2 + z^2}} \Rightarrow u_x = \left(-\frac{1}{2}\right)(x^2 + y^2 + z^2)^{-3/2}(2x) = -x(x^2 + y^2 + z^2)^{-3/2}$ and

$u_{xx} = -(x^2 + y^2 + z^2)^{-3/2} - x\left(-\frac{3}{2}\right)(x^2 + y^2 + z^2)^{-5/2}(2x) = \dfrac{2x^2 - y^2 - z^2}{(x^2 + y^2 + z^2)^{5/2}}$.

By symmetry, $u_{yy} = \dfrac{2y^2 - x^2 - z^2}{(x^2 + y^2 + z^2)^{5/2}}$ and $u_{zz} = \dfrac{2z^2 - x^2 - y^2}{(x^2 + y^2 + z^2)^{5/2}}$.

Thus $u_{xx} + u_{yy} + u_{zz} = \dfrac{2x^2 - y^2 - z^2 + 2y^2 - x^2 - z^2 + 2z^2 - x^2 - y^2}{(x^2 + y^2 + z^2)^{5/2}} = 0$.

67. Let $v = x + at$, $w = x - at$. Then $u_t = \dfrac{\partial[f(v) + g(w)]}{\partial t} = \dfrac{df(v)}{dv}\dfrac{\partial v}{\partial t} + \dfrac{dg(w)}{dw}\dfrac{\partial w}{\partial t} = af'(v) - ag'(w)$ and

$u_{tt} = \dfrac{\partial[af'(v) - ag'(w)]}{\partial t} = a[af''(v) + ag''(w)] = a^2[f''(v) + g''(w)]$. Similarly, by using the Chain Rule we have

$u_x = f'(v) + g'(w)$ and $u_{xx} = f''(v) + g''(w)$. Thus $u_{tt} = a^2 u_{xx}$.

69. If we fix $K = K_0$, $P(L, K_0)$ is a function of a single variable L, and $\dfrac{dP}{dL} = \alpha \dfrac{P}{L}$ is a separable differential equation. Then

$\dfrac{dP}{P} = \alpha \dfrac{dL}{L} \Rightarrow \displaystyle\int \dfrac{dP}{P} = \int \alpha \dfrac{dL}{L} \Rightarrow \ln|P| = \alpha \ln|L| + C(K_0)$, where $C(K_0)$ can depend on K_0. Then

$|P| = e^{\alpha \ln|L| + C(K_0)}$, and since $P > 0$ and $L > 0$, we have $P = e^{\alpha \ln L} e^{C(K_0)} = e^{C(K_0)} e^{\ln L^\alpha} = C_1(K_0) L^\alpha$ where

$C_1(K_0) = e^{C(K_0)}$.

71. By the Chain Rule, taking the partial derivative of both sides with respect to R_1 gives

$\dfrac{\partial R^{-1}}{\partial R} \dfrac{\partial R}{\partial R_1} = \dfrac{\partial\left[(1/R_1) + (1/R_2) + (1/R_3)\right]}{\partial R_1}$ or $-R^{-2} \dfrac{\partial R}{\partial R_1} = -R_1^{-2}$. Thus $\dfrac{\partial R}{\partial R_1} = \dfrac{R^2}{R_1^2}$.

73. By Exercise 72, $PV = mRT \Rightarrow P = \dfrac{mRT}{V}$, so $\dfrac{\partial P}{\partial T} = \dfrac{mR}{V}$. Also, $PV = mRT \Rightarrow V = \dfrac{mRT}{P}$

and $\dfrac{\partial V}{\partial T} = \dfrac{mR}{P}$. Since $T = \dfrac{PV}{mR}$, we have $T \dfrac{\partial P}{\partial T} \dfrac{\partial V}{\partial T} = \dfrac{PV}{mR} \cdot \dfrac{mR}{V} \cdot \dfrac{mR}{P} = mR$.

75. $\dfrac{\partial K}{\partial m} = \frac{1}{2}v^2$, $\dfrac{\partial K}{\partial v} = mv$, $\dfrac{\partial^2 K}{\partial v^2} = m$. Thus $\dfrac{\partial K}{\partial m} \cdot \dfrac{\partial^2 K}{\partial v^2} = \frac{1}{2}v^2 m = K$.

77. $f_x(x, y) = x + 4y \Rightarrow f_{xy}(x, y) = 4$ and $f_y(x, y) = 3x - y \Rightarrow f_{yx}(x, y) = 3$. Since f_{xy} and f_{yx} are continuous

everywhere but $f_{xy}(x, y) \neq f_{yx}(x, y)$, Clairaut's Theorem implies that such a function $f(x, y)$ does not exist.

79. By the geometry of partial derivatives, the slope of the tangent line is $f_x(1, 2)$. By implicit differentiation of

$4x^2 + 2y^2 + z^2 = 16$, we get $8x + 2z\,(\partial z/\partial x) = 0 \Rightarrow \partial z/\partial x = -4x/z$, so when $x = 1$ and $z = 2$ we have

$\partial z/\partial x = -2$. So the slope is $f_x(1, 2) = -2$. Thus the tangent line is given by $z - 2 = -2(x - 1)$, $y = 2$. Taking the

parameter to be $t = x - 1$, we can write parametric equations for this line: $x = 1 + t$, $y = 2$, $z = 2 - 2t$.

81. Let $g(x) = f(x, 0) = x(x^2)^{-3/2}e^0 = x\,|x|^{-3}$. But we are using the point $(1, 0)$, so near $(1, 0)$, $g(x) = x^{-2}$. Then

$g'(x) = -2x^{-3}$ and $g'(1) = -2$, so using (1) we have $f_x(1, 0) = g'(1) = -2$.

83. (a)

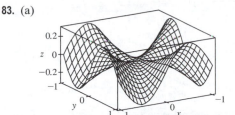

(b) For $(x, y) \neq (0, 0)$, $f_x(x, y) = \dfrac{(3x^2 y - y^3)(x^2 + y^2) - (x^3 y - xy^3)(2x)}{(x^2 + y^2)^2} = \dfrac{x^4 y + 4x^2 y^3 - y^5}{(x^2 + y^2)^2}$, and by symmetry

$f_y(x, y) = \dfrac{x^5 - 4x^3 y^2 - xy^4}{(x^2 + y^2)^2}$.

(c) $f_x(0, 0) = \lim\limits_{h \to 0} \dfrac{f(h, 0) - f(0, 0)}{h} = \lim\limits_{h \to 0} \dfrac{(0/h^2) - 0}{h} = 0$ and $f_y(0, 0) = \lim\limits_{h \to 0} \dfrac{f(0, h) - f(0, 0)}{h} = 0$.

(d) By (3), $f_{xy}(0, 0) = \dfrac{\partial f_x}{\partial y} = \lim\limits_{h \to 0} \dfrac{f_x(0, h) - f_x(0, 0)}{h} = \lim\limits_{h \to 0} \dfrac{(-h^5 - 0)/h^4}{h} = -1$ while by (2),

$f_{yx}(0, 0) = \dfrac{\partial f_y}{\partial x} = \lim\limits_{h \to 0} \dfrac{f_y(h, 0) - f_y(0, 0)}{h} = \lim\limits_{h \to 0} \dfrac{h^5/h^4}{h} = 1$.

(e) For $(x, y) \neq (0, 0)$, we use a CAS to compute

$$f_{xy}(x, y) = \dfrac{x^6 + 9x^4 y^2 - 9x^2 y^4 - y^6}{(x^2 + y^2)^3}.$$

Now as $(x, y) \to (0, 0)$ along the x-axis, $f_{xy}(x, y) \to 1$ while as $(x, y) \to (0, 0)$ along the y-axis, $f_{xy}(x, y) \to -1$. Thus f_{xy} isn't continuous at $(0, 0)$ and Clairaut's Theorem doesn't apply, so there is no contradiction. The graphs of f_{xy} and f_{yx} are identical except at the origin, where we observe the discontinuity.

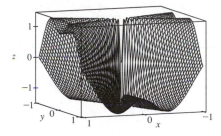

11.4 Tangent Planes and Linear Approximations

1. $z = f(x, y) = 4x^2 - y^2 + 2y \Rightarrow f_x(x, y) = 8x$, $f_y(x, y) = -2y + 2$, so $f_x(-1, 2) = -8$, $f_y(-1, 2) = -2$. By Equation 2, an equation of the tangent plane is $z - 4 = f_x(-1, 2)[x - (-1)] + f_y(-1, 2)(y - 2) \Rightarrow$
$z - 4 = -8(x + 1) - 2(y - 2)$ or $z = -8x - 2y$.

3. $z = f(x, y) = y \cos(x - y) \Rightarrow f_x = y(-\sin(x - y)(1)) = -y \sin(x - y)$,
$f_y = y(-\sin(x - y)(-1)) + \cos(x - y) = y \sin(x - y) + \cos(x - y)$, so $f_x(2, 2) = -2 \sin(0) = 0$,
$f_y(2, 2) = 2 \sin(0) + \cos(0) = 1$ and an equation of the tangent plane is $z - 2 = 0(x - 2) + 1(y - 2)$ or $z = y$.

5. $z = f(x, y) = x^2 + xy + 3y^2$, so $f_x(x, y) = 2x + y \Rightarrow f_x(1, 1) = 3$, $f_y(x, y) = x + 6y \Rightarrow f_y(1, 1) = 7$ and an equation of the tangent plane is $z - 5 = 3(x - 1) + 7(y - 1)$ or $z = 3x + 7y - 5$. After zooming in, the surface and the tangent plane become almost indistinguishable. (Here, the tangent plane is below the surface.) If we zoom in farther, the surface and the tangent plane will appear to coincide.

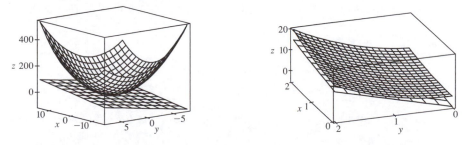

7. $f(x, y) = \dfrac{xy \sin(x - y)}{1 + x^2 + y^2}$. A CAS gives $f_x(x, y) = \dfrac{y \sin(x - y) + xy \cos(x - y)}{1 + x^2 + y^2} - \dfrac{2x^2 y \sin(x - y)}{(1 + x^2 + y^2)^2}$ and

$f_y(x, y) = \dfrac{x \sin(x - y) - xy \cos(x - y)}{1 + x^2 + y^2} - \dfrac{2xy^2 \sin(x - y)}{(1 + x^2 + y^2)^2}$. We use the CAS to evaluate these at $(1, 1)$, and then

substitute the results into Equation 2 to compute an equation of the tangent plane: $z = \frac{1}{3}x - \frac{1}{3}y$. The surface and tangent

plane are shown in the first graph below. After zooming in, the surface and the tangent plane become almost indistinguishable,

as shown in the second graph. (Here, the tangent plane is shown with fewer traces than the surface.) If we zoom in farther, the

surface and the tangent plane will appear to coincide.

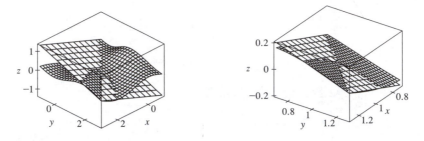

9. $f(x, y) = x\sqrt{y}$. The partial derivatives are $f_x(x, y) = \sqrt{y}$ and $f_y(x, y) = \dfrac{x}{2\sqrt{y}}$, so $f_x(1, 4) = 2$ and $f_y(1, 4) = \frac{1}{4}$. Both f_x

and f_y are continuous functions for $y > 0$, so by Theorem 8, f is differentiable at $(1, 4)$. By Equation 3, the linearization of f

at $(1, 4)$ is given by $L(x, y) = f(1, 4) + f_x(1, 4)(x - 1) + f_y(1, 4)(y - 4) = 2 + 2(x - 1) + \frac{1}{4}(y - 4) = 2x + \frac{1}{4}y - 1$.

11. $f(x, y) = \tan^{-1}(x + 2y)$. The partial derivatives are $f_x(x, y) = \dfrac{1}{1 + (x + 2y)^2}$ and $f_y(x, y) = \dfrac{2}{1 + (x + 2y)^2}$, so

$f_x(1, 0) = \frac{1}{2}$ and $f_y(1, 0) = 1$. Both f_x and f_y are continuous functions, so f is differentiable at $(1, 0)$, and the linearization

of f at $(1, 0)$ is $L(x, y) = f(1, 0) + f_x(1, 0)(x - 1) + f_y(1, 0)(y - 0) = \frac{\pi}{4} + \frac{1}{2}(x - 1) + 1(y) = \frac{1}{2}x + y + \frac{\pi}{4} - \frac{1}{2}$.

13. $f(x, y) = \sqrt{20 - x^2 - 7y^2}$ $\Rightarrow$ $f_x(x, y) = -\dfrac{x}{\sqrt{20 - x^2 - 7y^2}}$ and $f_y(x, y) = -\dfrac{7y}{\sqrt{20 - x^2 - 7y^2}}$,

so $f_x(2, 1) = -\frac{2}{3}$ and $f_y(2, 1) = -\frac{7}{3}$. Then the linear approximation of f at $(2, 1)$ is given by

$f(x, y) \approx f(2, 1) + f_x(2, 1)(x - 2) + f_y(2, 1)(y - 1) = 3 - \frac{2}{3}(x - 2) - \frac{7}{3}(y - 1) = -\frac{2}{3}x - \frac{7}{3}y + \frac{20}{3}$.

Thus $f(1.95, 1.08) \approx -\frac{2}{3}(1.95) - \frac{7}{3}(1.08) + \frac{20}{3} = 2.84\overline{6}$.

15. $f(x, y, z) = \sqrt{x^2 + y^2 + z^2}$ $\Rightarrow$ $f_x(x, y, z) = \dfrac{x}{\sqrt{x^2 + y^2 + z^2}}$, $f_y(x, y, z) = \dfrac{y}{\sqrt{x^2 + y^2 + z^2}}$, and

$f_z(x, y, z) = \dfrac{z}{\sqrt{x^2 + y^2 + z^2}}$, so $f_x(3, 2, 6) = \frac{3}{7}$, $f_y(3, 2, 6) = \frac{2}{7}$, and $f_z(3, 2, 6) = \frac{6}{7}$. Then the linear approximation of f

at $(3, 2, 6)$ is given by

$$f(x, y, z) \approx f(3, 2, 6) + f_x(3, 2, 6)(x - 3) + f_y(3, 2, 6)(y - 2) + f_z(3, 2, 6)(z - 6)$$
$$= 7 + \frac{3}{7}(x - 3) + \frac{2}{7}(y - 2) + \frac{6}{7}(z - 6) = \frac{3}{7}x + \frac{2}{7}y + \frac{6}{7}z$$

Thus $\sqrt{(3.02)^2 + (1.97)^2 + (5.99)^2} = f(3.02, 1.97, 5.99) \approx \frac{3}{7}(3.02) + \frac{2}{7}(1.97) + \frac{6}{7}(5.99) \approx 6.9914$.

17. From the table, $f(94, 80) = 127$. To estimate $f_T(94, 80)$ and $f_H(94, 80)$ we follow the procedure used in Section 11.3. Since

$f_T(94, 80) = \lim_{h \to 0} \dfrac{f(94 + h, 80) - f(94, 80)}{h}$, we approximate this quantity with $h = \pm 2$ and use the values given in the

table: $f_T(94, 80) \approx \dfrac{f(96, 80) - f(94, 80)}{2} = \dfrac{135 - 127}{2} = 4$, $f_T(94, 80) \approx \dfrac{f(92, 80) - f(94, 80)}{-2} = \dfrac{119 - 127}{-2} = 4$.

Averaging these values gives $f_T(94, 80) \approx 4$. Similarly, $f_H(94, 80) = \lim_{h \to 0} \dfrac{f(94, 80 + h) - f(94, 80)}{h}$, so we use $h = \pm 5$:

$f_H(94, 80) \approx \dfrac{f(94, 85) - f(94, 80)}{5} = \dfrac{132 - 127}{5} = 1$, $f_H(94, 80) \approx \dfrac{f(94, 75) - f(94, 80)}{-5} = \dfrac{122 - 127}{-5} = 1$.

Averaging these values gives $f_H(94, 80) \approx 1$. The linear approximation, then, is

$$f(T, H) \approx f(94, 80) + f_T(94, 80)(T - 94) + f_H(94, 80)(H - 80)$$
$$\approx 127 + 4(T - 94) + 1(H - 80)$$

Thus when $T = 95$ and $H = 78$, $f(95, 78) \approx 127 + 4(95 - 94) + 1(78 - 80) = 129$, so we estimate the heat index to be approximately $129°$F.

19. $z = x^3 \ln(y^2) \Rightarrow dz = \dfrac{\partial z}{\partial x} dx + \dfrac{\partial z}{\partial y} dy = 3x^2 \ln(y^2)\, dx + x^3 \cdot \dfrac{1}{y^2}(2y)\, dy = 3x^2 \ln(y^2)\, dx + \dfrac{2x^3}{y}\, dy$.

21. $R = \alpha\beta^2 \cos\gamma \Rightarrow dR = \dfrac{\partial R}{\partial \alpha} d\alpha + \dfrac{\partial R}{\partial \beta} d\beta + \dfrac{\partial R}{\partial \gamma} d\gamma = \beta^2 \cos\gamma\, d\alpha + 2\alpha\beta \cos\gamma\, d\beta - \alpha\beta^2 \sin\gamma\, d\gamma$

23. $dx = \Delta x = 0.05$, $dy = \Delta y = 0.1$, $z = 5x^2 + y^2$, $z_x = 10x$, $z_y = 2y$. Thus when $x = 1$ and $y = 2$,
$dz = z_x(1, 2)\, dx + z_y(1, 2)\, dy = (10)(0.05) + (4)(0.1) = 0.9$ while
$\Delta z = f(1.05, 2.1) - f(1, 2) = 5(1.05)^2 + (2.1)^2 - 5 - 4 = 0.9225$.

25. $dA = \dfrac{\partial A}{\partial x} dx + \dfrac{\partial A}{\partial y} dy = y\, dx + x\, dy$ and $|\Delta x| \le 0.1$, $|\Delta y| \le 0.1$. We use $dx = 0.1$, $dy = 0.1$ with $x = 30$, $y = 24$; then

the maximum error in the area is about $dA = 24(0.1) + 30(0.1) = 5.4$ cm^2.

27. The volume of a can is $V = \pi r^2 h$ and $\Delta V \approx dV$ is an estimate of the amount of tin. Here $dV = 2\pi rh\, dr + \pi r^2\, dh$, so put $dr = 0.04$, $dh = 0.08$ (0.04 on top, 0.04 on bottom) and then $\Delta V \approx dV = 2\pi(48)(0.04) + \pi(16)(0.08) \approx 16.08$ cm^3. Thus the amount of tin is about 16 cm^3.

29. The maximum errors in measurement are 2%, so $\left|\dfrac{\Delta w}{w}\right| \le 0.02$, $\left|\dfrac{\Delta h}{h}\right| \le 0.02$. The percentage error in the calculated surface area is

$$\dfrac{\Delta S}{S} \approx \dfrac{dS}{S} = \dfrac{0.1091(0.425w^{0.425-1})h^{0.725}\, dw + 0.1091w^{0.425}(0.725h^{0.725-1})\, dh}{0.1091w^{0.425}h^{0.725}}$$

$$= 0.425\dfrac{dw}{w} + 0.725\dfrac{dh}{h}$$

To estimate the maximum percentage error, we use $\dfrac{dw}{w} = \left|\dfrac{\Delta w}{w}\right| = 0.02$ and

$\dfrac{dh}{h} = \left|\dfrac{\Delta h}{h}\right| = 0.02 \Rightarrow \dfrac{dS}{S} = 0.425\,(0.02) + 0.725\,(0.02) = 0.023$. Thus the maximum percentage error is approximately 2.3%.

31. First we find $\dfrac{\partial R}{\partial R_1}$ implicitly by taking partial derivatives of both sides with respect to R_1:

$$\frac{\partial}{\partial R_1}\left[\frac{1}{R}\right] = \frac{\partial\left[(1/R_1) + (1/R_2) + (1/R_3)\right]}{\partial R_1} \Rightarrow -R^{-2}\frac{\partial R}{\partial R_1} = -R_1^{-2} \Rightarrow \frac{\partial R}{\partial R_1} = \frac{R^2}{R_1^2}. \text{ Then by symmetry,}$$

$\dfrac{\partial R}{\partial R_2} = \dfrac{R^2}{R_2^2}, \dfrac{\partial R}{\partial R_3} = \dfrac{R^2}{R_3^2}$. When $R_1 = 25$, $R_2 = 40$ and $R_3 = 50$, $\dfrac{1}{R} = \dfrac{17}{200} \Leftrightarrow R = \dfrac{200}{17}$ ohms. Since the possible

error for each R_i is 0.5%, the maximum error of R is attained by setting $\Delta R_i = 0.005R_i$. So

$$\Delta R \approx dR = \frac{\partial R}{\partial R_1}\Delta R_1 + \frac{\partial R}{\partial R_2}\Delta R_2 + \frac{\partial R}{\partial R_3}\Delta R_3 = (0.005)R^2\left[\frac{1}{R_1} + \frac{1}{R_2} + \frac{1}{R_3}\right] = (0.005)R = \tfrac{1}{17} \approx 0.059 \text{ ohms.}$$

33. $\mathbf{r}(u,v) = (u+v)\,\mathbf{i} + 3u^2\,\mathbf{j} + (u-v)\,\mathbf{k}$.

$\mathbf{r}_u = \mathbf{i} + 6u\,\mathbf{j} + \mathbf{k}$ and $\mathbf{r}_v = \mathbf{i} - \mathbf{k}$, so

$\mathbf{r}_u \times \mathbf{r}_v = -6u\,\mathbf{i} + 2\,\mathbf{j} - 6u\,\mathbf{k}$. Since the point $(2,3,0)$

corresponds to $u = 1$, $v = 1$, a normal vector to the surface at

$(2,3,0)$ is $-6\,\mathbf{i} + 2\,\mathbf{j} - 6\,\mathbf{k}$, and an equation of the tangent plane is

$-6x + 2y - 6z = -6$ or $3x - y + 3z = 3$.

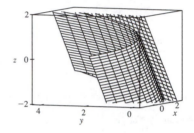

35. $\mathbf{r}(u,v) = u^2\,\mathbf{i} + 2u\sin v\,\mathbf{j} + u\cos v\,\mathbf{k} \Rightarrow \mathbf{r}(1,0) = (1,0,1)$.

$\mathbf{r}_u = 2u\,\mathbf{i} + 2\sin v\,\mathbf{j} + \cos v\,\mathbf{k}$ and $\mathbf{r}_v = 2u\cos v\,\mathbf{j} - u\sin v\,\mathbf{k}$,

so a normal vector to the surface at the point $(1,0,1)$ is

$\mathbf{r}_u(1,0) \times \mathbf{r}_v(1,0) = (2\,\mathbf{i} + \mathbf{k}) \times (2\,\mathbf{j}) = -2\,\mathbf{i} + 4\,\mathbf{k}$.

Thus an equation of the tangent plane at $(1,0,1)$ is

$-2(x-1) + 0(y-0) + 4(z-1) = 0$ or $-x + 2z = 1$.

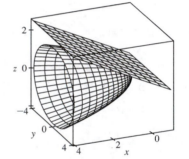

37. $\mathbf{r}(u,v) = u\,\mathbf{i} + \ln(uv)\,\mathbf{j} + v\,\mathbf{k} \Rightarrow \mathbf{r}_u(u,v) = \mathbf{i} + \frac{1}{u}\,\mathbf{j}$,

$\mathbf{r}_v(u,v) = \frac{1}{v}\,\mathbf{j} + \mathbf{k}$. $\mathbf{r}(1,1) = \mathbf{i} + \mathbf{k}$, so the point corresponding to

$u = 1$, $v = 1$ is $(1,0,1)$. A normal vector for the tangent plane is

$\mathbf{r}_u(1,1) \times \mathbf{r}_v(1,1) = (\mathbf{i}+\mathbf{j}) \times (\mathbf{j}+\mathbf{k}) = \mathbf{i} - \mathbf{j} + \mathbf{k}$, so an equation

of the tangent plane is $(x-1) - (y-0) + (z-1) = 0$

or $x - y + z = 2$.

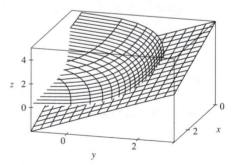

39. $\Delta z = f(a+\Delta x, b+\Delta y) - f(a,b) = (a+\Delta x)^2 + (b+\Delta y)^2 - (a^2 + b^2)$

$\quad = a^2 + 2a\,\Delta x + (\Delta x)^2 + b^2 + 2b\,\Delta y + (\Delta y)^2 - a^2 - b^2 = 2a\,\Delta x + (\Delta x)^2 + 2b\,\Delta y + (\Delta y)^2$

But $f_x(a,b) = 2a$ and $f_y(a,b) = 2b$ and so $\Delta z = f_x(a,b)\,\Delta x + f_y(a,b)\,\Delta y + \Delta x\,\Delta x + \Delta y\,\Delta y$, which is Definition 7

with $\varepsilon_1 = \Delta x$ and $\varepsilon_2 = \Delta y$. Hence f is differentiable.

41. To show that f is continuous at (a, b) we need to show that $\lim\limits_{(x,y) \to (a,b)} f(x,y) = f(a,b)$ or

equivalently $\lim\limits_{(\Delta x, \Delta y) \to (0,0)} f(a + \Delta x, b + \Delta y) = f(a,b)$. Since f is differentiable at (a,b),

$f(a + \Delta x, b + \Delta y) - f(a,b) = \Delta z = f_x(a,b)\,\Delta x + f_y(a,b)\,\Delta y + \varepsilon_1\,\Delta x + \varepsilon_2\,\Delta y$, where ε_1 and $\varepsilon_2 \to 0$ as

$(\Delta x, \Delta y) \to (0,0)$. Thus $f(a + \Delta x, b + \Delta y) = f(a,b) + f_x(a,b)\,\Delta x + f_y(a,b)\,\Delta y + \varepsilon_1\,\Delta x + \varepsilon_2\,\Delta y$. Taking the limit of

both sides as $(\Delta x, \Delta y) \to (0,0)$ gives $\lim\limits_{(\Delta x, \Delta y) \to (0,0)} f(a + \Delta x, b + \Delta y) = f(a,b)$. Thus f is continuous at (a,b).

11.5   The Chain Rule

1. $z = \sin x \cos y$, $x = \pi t$, $y = \sqrt{t}$ $\Rightarrow$

$$\frac{dz}{dt} = \frac{\partial z}{\partial x}\frac{dx}{dt} + \frac{\partial z}{\partial y}\frac{dy}{dt} = \cos x \cos y \cdot \pi + \sin x\,(-\sin y) \cdot \tfrac{1}{2}t^{-1/2} = \pi \cos x \cos y - \frac{1}{2\sqrt{t}}\sin x \sin y$$

3. $w = xe^{y/z}$, $x = t^2$, $y = 1 - t$, $z = 1 + 2t$ $\Rightarrow$

$$\frac{dw}{dt} = \frac{\partial w}{\partial x}\frac{dx}{dt} + \frac{\partial w}{\partial y}\frac{dy}{dt} + \frac{\partial w}{\partial z}\frac{dz}{dt} = e^{y/z} \cdot 2t + xe^{y/z}\left(\frac{1}{z}\right) \cdot (-1) + xe^{y/z}\left(-\frac{y}{z^2}\right) \cdot 2 = e^{y/z}\left(2t - \frac{x}{z} - \frac{2xy}{z^2}\right)$$

5. $z = x^2 + xy + y^2$, $x = s + t$, $y = st$ $\Rightarrow$

$$\frac{\partial z}{\partial s} = \frac{\partial z}{\partial x}\frac{\partial x}{\partial s} + \frac{\partial z}{\partial y}\frac{\partial y}{\partial s} = (2x + y)(1) + (x + 2y)(t) = 2x + y + xt + 2yt$$

$$\frac{\partial z}{\partial t} = \frac{\partial z}{\partial x}\frac{\partial x}{\partial t} + \frac{\partial z}{\partial y}\frac{\partial y}{\partial t} = (2x + y)(1) + (x + 2y)(s) = 2x + y + xs + 2ys$$

7. $z = e^r \cos\theta$, $r = st$, $\theta = \sqrt{s^2 + t^2}$ $\Rightarrow$

$$\frac{\partial z}{\partial s} = \frac{\partial z}{\partial r}\frac{\partial r}{\partial s} + \frac{\partial z}{\partial \theta}\frac{\partial \theta}{\partial s} = e^r \cos\theta \cdot t + e^r(-\sin\theta) \cdot \tfrac{1}{2}(s^2 + t^2)^{-1/2}(2s)$$

$$= te^r \cos\theta - e^r \sin\theta \cdot \frac{s}{\sqrt{s^2 + t^2}} = e^r\left(t\cos\theta - \frac{s}{\sqrt{s^2 + t^2}}\sin\theta\right)$$

$$\frac{\partial z}{\partial t} = \frac{\partial z}{\partial r}\frac{\partial r}{\partial t} + \frac{\partial z}{\partial \theta}\frac{\partial \theta}{\partial t} = e^r \cos\theta \cdot s + e^r(-\sin\theta) \cdot \tfrac{1}{2}(s^2 + t^2)^{-1/2}(2t)$$

$$= se^r \cos\theta - e^r \sin\theta \cdot \frac{t}{\sqrt{s^2 + t^2}} = e^r\left(s\cos\theta - \frac{t}{\sqrt{s^2 + t^2}}\sin\theta\right)$$

9. When $t = 3$, $x = g(3) = 2$ and $y = h(3) = 7$. By the Chain Rule (2),

$$\frac{dz}{dt} = \frac{\partial f}{\partial x}\frac{dx}{dt} + \frac{\partial f}{\partial y}\frac{dy}{dt} = f_x(2,7)g'(3) + f_y(2,7)\,h'(3) = (6)(5) + (-8)(-4) = 62.$$

11. $g(u,v) = f(x(u,v), y(u,v))$ where $x = e^u + \sin v$, $y = e^u + \cos v$ $\Rightarrow$ $\dfrac{\partial x}{\partial u} = e^u$, $\dfrac{\partial x}{\partial v} = \cos v$, $\dfrac{\partial y}{\partial u} = e^u$,

$\dfrac{\partial y}{\partial v} = -\sin v$. By the Chain Rule (3), $\dfrac{\partial g}{\partial u} = \dfrac{\partial f}{\partial x}\dfrac{\partial x}{\partial u} + \dfrac{\partial f}{\partial y}\dfrac{\partial y}{\partial u}$. Then

$$g_u(0,0) = f_x(x(0,0), y(0,0))\,x_u(0,0) + f_y(x(0,0), y(0,0))\,y_u(0,0)$$

$$= f_x(1,2)(e^0) + f_y(1,2)(e^0) = 2(1) + 5(1) = 7$$

Similarly $\dfrac{\partial g}{\partial v} = \dfrac{\partial f}{\partial x}\dfrac{\partial x}{\partial v} + \dfrac{\partial f}{\partial y}\dfrac{\partial y}{\partial v}$. Then

$$g_v(0,0) = f_x(x(0,0), y(0,0))\,x_v(0,0) + f_y(x(0,0), y(0,0))\,y_v(0,0)$$

$$= f_x(1,2)(\cos 0) + f_y(1,2)(-\sin 0) = 2(1) + 5(0) = 2$$

13.

$$u = f(x, y),\ x = x(r, s, t),\ y = y(r, s, t)\ \Rightarrow$$

$$\frac{\partial u}{\partial r} = \frac{\partial u}{\partial x}\frac{\partial x}{\partial r} + \frac{\partial u}{\partial y}\frac{\partial y}{\partial r},\ \frac{\partial u}{\partial s} = \frac{\partial u}{\partial x}\frac{\partial x}{\partial s} + \frac{\partial u}{\partial y}\frac{\partial y}{\partial s},$$

$$\frac{\partial u}{\partial t} = \frac{\partial u}{\partial x}\frac{\partial x}{\partial t} + \frac{\partial u}{\partial y}\frac{\partial y}{\partial t}$$

15.

$$v = f(p, q, r),\ p = p(x, y, z),\ q = q(x, y, z),\ r = r(x, y, z)\ \Rightarrow$$

$$\frac{\partial v}{\partial x} = \frac{\partial v}{\partial p}\frac{\partial p}{\partial x} + \frac{\partial v}{\partial q}\frac{\partial q}{\partial x} + \frac{\partial v}{\partial r}\frac{\partial r}{\partial x},\ \frac{\partial v}{\partial y} = \frac{\partial v}{\partial p}\frac{\partial p}{\partial y} + \frac{\partial v}{\partial q}\frac{\partial q}{\partial y} + \frac{\partial v}{\partial r}\frac{\partial r}{\partial y},$$

$$\frac{\partial v}{\partial z} = \frac{\partial v}{\partial p}\frac{\partial p}{\partial z} + \frac{\partial v}{\partial q}\frac{\partial q}{\partial z} + \frac{\partial v}{\partial r}\frac{\partial r}{\partial z}$$

17. $z = x^2 + xy^3,\ x = uv^2 + w^3,\ y = u + ve^w\ \Rightarrow\ \dfrac{\partial z}{\partial u} = \dfrac{\partial z}{\partial x}\dfrac{\partial x}{\partial u} + \dfrac{\partial z}{\partial y}\dfrac{\partial y}{\partial u} = (2x + y^3)(v^2) + (3xy^2)(1),$

$\dfrac{\partial z}{\partial v} = \dfrac{\partial z}{\partial x}\dfrac{\partial x}{\partial v} + \dfrac{\partial z}{\partial y}\dfrac{\partial y}{\partial v} = (2x + y^3)(2uv) + (3xy^2)(e^w),\ \dfrac{\partial z}{\partial w} = \dfrac{\partial z}{\partial x}\dfrac{\partial x}{\partial w} + \dfrac{\partial z}{\partial y}\dfrac{\partial y}{\partial w} = (2x + y^3)(3w^2) + (3xy^2)(ve^w).$

When $u = 2$, $v = 1$, and $w = 0$, we have $x = 2$, $y = 3$, so $\dfrac{\partial z}{\partial u} = (31)(1) + (54)(1) = 85$, $\dfrac{\partial z}{\partial v} = (31)(4) + (54)(1) = 178$,

$\dfrac{\partial z}{\partial w} = (31)(0) + (54)(1) = 54.$

19. $R = \ln(u^2 + v^2 + w^2),\ u = x + 2y,\ v = 2x - y,\ w = 2xy\ \Rightarrow$

$$\frac{\partial R}{\partial x} = \frac{\partial R}{\partial u}\frac{\partial u}{\partial x} + \frac{\partial R}{\partial v}\frac{\partial v}{\partial x} + \frac{\partial R}{\partial w}\frac{\partial w}{\partial x} = \frac{2u}{u^2 + v^2 + w^2}(1) + \frac{2v}{u^2 + v^2 + w^2}(2) + \frac{2w}{u^2 + v^2 + w^2}(2y)$$

$$= \frac{2u + 4v + 4wy}{u^2 + v^2 + w^2},$$

$$\frac{\partial R}{\partial y} = \frac{\partial R}{\partial u}\frac{\partial u}{\partial y} + \frac{\partial R}{\partial v}\frac{\partial v}{\partial y} + \frac{\partial R}{\partial w}\frac{\partial w}{\partial y} = \frac{2u}{u^2 + v^2 + w^2}(2) + \frac{2v}{u^2 + v^2 + w^2}(-1) + \frac{2w}{u^2 + v^2 + w^2}(2x)$$

$$= \frac{4u - 2v + 4wx}{u^2 + v^2 + w^2}.$$

When $x = y = 1$ we have $u = 3$, $v = 1$, and $w = 2$, so $\dfrac{\partial R}{\partial x} = \dfrac{9}{7}$ and $\dfrac{\partial R}{\partial y} = \dfrac{9}{7}$.

21. $u = x^2 + yz,\ x = pr\cos\theta,\ y = pr\sin\theta,\ z = p + r\ \Rightarrow$

$$\frac{\partial u}{\partial p} = \frac{\partial u}{\partial x}\frac{\partial x}{\partial p} + \frac{\partial u}{\partial y}\frac{\partial y}{\partial p} + \frac{\partial u}{\partial z}\frac{\partial z}{\partial p} = (2x)(r\cos\theta) + (z)(r\sin\theta) + (y)(1) = 2xr\cos\theta + zr\sin\theta + y,$$

$$\frac{\partial u}{\partial r} = \frac{\partial u}{\partial x}\frac{\partial x}{\partial r} + \frac{\partial u}{\partial y}\frac{\partial y}{\partial r} + \frac{\partial u}{\partial z}\frac{\partial z}{\partial r} = (2x)(p\cos\theta) + (z)(p\sin\theta) + (y)(1) = 2xp\cos\theta + zp\sin\theta + y,$$

$$\frac{\partial u}{\partial\theta} = \frac{\partial u}{\partial x}\frac{\partial x}{\partial\theta} + \frac{\partial u}{\partial y}\frac{\partial y}{\partial\theta} + \frac{\partial u}{\partial z}\frac{\partial z}{\partial\theta} = (2x)(-pr\sin\theta) + (z)(pr\cos\theta) + (y)(0) = -2xpr\sin\theta + zpr\cos\theta.$$

When $p = 2$, $r = 3$, and $\theta = 0$ we have $x = 6$, $y = 0$, and $z = 5$, so $\dfrac{\partial u}{\partial p} = 36$, $\dfrac{\partial u}{\partial r} = 24$, and $\dfrac{\partial u}{\partial\theta} = 30$.

23. $\sqrt{xy} = 1 + x^2y$, so let $F(x, y) = (xy)^{1/2} - 1 - x^2y = 0$. Then by Equation 6

$$\frac{dy}{dx} = -\frac{F_x}{F_y} = -\frac{\frac{1}{2}(xy)^{-1/2}(y) - 2xy}{\frac{1}{2}(xy)^{-1/2}(x) - x^2} = -\frac{y - 4xy\sqrt{xy}}{x - 2x^2\sqrt{xy}} = \frac{4(xy)^{3/2} - y}{x - 2x^2\sqrt{xy}}.$$

25. $x^2 + y^2 + z^2 = 3xyz$, so let $F(x, y, z) = x^2 + y^2 + z^2 - 3xyz = 0$. Then by Equations 7

$$\frac{\partial z}{\partial x} = -\frac{F_x}{F_z} = -\frac{2x - 3yz}{2z - 3xy} = \frac{3yz - 2x}{2z - 3xy}\quad\text{and}\quad\frac{\partial z}{\partial y} = -\frac{F_y}{F_z} = -\frac{2y - 3xz}{2z - 3xy} = \frac{3xz - 2y}{2z - 3xy}.$$

27. $x - z = \arctan(yz)$, so let $F(x, y, z) = x - z - \arctan(yz) = 0$. Then

$$\frac{\partial z}{\partial x} = -\frac{F_x}{F_z} = -\frac{1}{-1 - \dfrac{1}{1 + (yz)^2}(y)} = \frac{1 + y^2 z^2}{1 + y + y^2 z^2} \text{ and}$$

$$\frac{\partial z}{\partial y} = -\frac{F_y}{F_z} = -\frac{-\dfrac{1}{1 + (yz)^2}(z)}{-1 - \dfrac{1}{1 + (yz)^2}(y)} = -\frac{\dfrac{z}{1 + y^2 z^2}}{\dfrac{1 + y^2 z^2 + y}{1 + y^2 z^2}} = -\frac{z}{1 + y + y^2 z^2}.$$

29. Since x and y are each functions of t, $T(x, y)$ is a function of t, so by the Chain Rule, $\dfrac{dT}{dt} = \dfrac{\partial T}{\partial x}\dfrac{dx}{dt} + \dfrac{\partial T}{\partial y}\dfrac{dy}{dt}$. After

3 seconds, $x = \sqrt{1 + t} = \sqrt{1 + 3} = 2$, $y = 2 + \frac{1}{3}t = 2 + \frac{1}{3}(3) = 3$, $\dfrac{dx}{dt} = \dfrac{1}{2\sqrt{1+t}} = \dfrac{1}{2\sqrt{1+3}} = \dfrac{1}{4}$, and $\dfrac{dy}{dt} = \dfrac{1}{3}$.

Then $\dfrac{dT}{dt} = T_x(2, 3)\dfrac{dx}{dt} + T_y(2, 3)\dfrac{dy}{dt} = 4\left(\frac{1}{4}\right) + 3\left(\frac{1}{3}\right) = 2$. Thus the temperature is rising at a rate of $2°\text{C/s}$.

31. $C = 1449.2 + 4.6T - 0.055T^2 + 0.00029T^3 + 0.016D$, so $\dfrac{\partial C}{\partial T} = 4.6 - 0.11T + 0.00087T^2$ and $\dfrac{\partial C}{\partial D} = 0.016$.

According to the graph, the diver is experiencing a temperature of approximately $12.5°\text{C}$ at $t = 20$ minutes, so

$\dfrac{\partial C}{\partial T} = 4.6 - 0.11(12.5) + 0.00087(12.5)^2 \approx 3.36$. By sketching tangent lines at $t = 20$ to the graphs given, we estimate

$\dfrac{dD}{dt} \approx \dfrac{1}{2}$ and $\dfrac{dT}{dt} \approx -\dfrac{1}{10}$. Then, by the Chain Rule, $\dfrac{dC}{dt} = \dfrac{\partial C}{\partial T}\dfrac{dT}{dt} + \dfrac{\partial C}{\partial D}\dfrac{dD}{dt} \approx (3.36)\left(-\frac{1}{10}\right) + (0.016)\left(\frac{1}{2}\right) \approx -0.33$.

Thus the speed of sound experienced by the diver is decreasing at a rate of approximately 0.33 m/s per minute.

33. (a) $V = \ell w h$, so by the Chain Rule,

$$\frac{dV}{dt} = \frac{\partial V}{\partial \ell}\frac{d\ell}{dt} + \frac{\partial V}{\partial w}\frac{dw}{dt} + \frac{\partial V}{\partial h}\frac{dh}{dt} = wh\frac{d\ell}{dt} + \ell h\frac{dw}{dt} + \ell w\frac{dh}{dt}$$

$$= 2 \cdot 2 \cdot 2 + 1 \cdot 2 \cdot 2 + 1 \cdot 2 \cdot (-3) = 6 \text{ m}^3/\text{s}$$

(b) $S = 2(\ell w + \ell h + wh)$, so by the Chain Rule,

$$\frac{dS}{dt} = \frac{\partial S}{\partial \ell}\frac{d\ell}{dt} + \frac{\partial S}{\partial w}\frac{dw}{dt} + \frac{\partial S}{\partial h}\frac{dh}{dt} = 2(w + h)\frac{d\ell}{dt} + 2(\ell + h)\frac{dw}{dt} + 2(\ell + w)\frac{dh}{dt}$$

$$= 2(2 + 2)2 + 2(1 + 2)2 + 2(1 + 2)(-3) = 10 \text{ m}^2/\text{s}$$

(c) $L^2 = \ell^2 + w^2 + h^2 \;\Rightarrow\; 2L\dfrac{dL}{dt} = 2\ell\dfrac{d\ell}{dt} + 2w\dfrac{dw}{dt} + 2h\dfrac{dh}{dt} = 2(1)(2) + 2(2)(2) + 2(2)(-3) = 0 \;\Rightarrow$

$dL/dt = 0 \text{ m/s}$.

35. $\dfrac{dP}{dt} = 0.05$, $\dfrac{dT}{dt} = 0.15$, $V = 8.31\dfrac{T}{P}$ and $\dfrac{dV}{dt} = \dfrac{8.31}{P}\dfrac{dT}{dt} - 8.31\dfrac{T}{P^2}\dfrac{dP}{dt}$. Thus when $P = 20$ and $T = 320$,

$\dfrac{dV}{dt} = 8.31\left[\dfrac{0.15}{20} - \dfrac{(0.05)(320)}{400}\right] \approx -0.27 \text{ L/s}$.

37. (a) By the Chain Rule, $\dfrac{\partial z}{\partial r} = \dfrac{\partial z}{\partial x}\cos\theta + \dfrac{\partial z}{\partial y}\sin\theta$, $\dfrac{\partial z}{\partial \theta} = \dfrac{\partial z}{\partial x}(-r\sin\theta) + \dfrac{\partial z}{\partial y}r\cos\theta$.

(b) $\left(\dfrac{\partial z}{\partial r}\right)^2 = \left(\dfrac{\partial z}{\partial x}\right)^2\cos^2\theta + 2\dfrac{\partial z}{\partial x}\dfrac{\partial z}{\partial y}\cos\theta\sin\theta + \left(\dfrac{\partial z}{\partial y}\right)^2\sin^2\theta$,

$\left(\dfrac{\partial z}{\partial \theta}\right)^2 = \left(\dfrac{\partial z}{\partial x}\right)^2 r^2\sin^2\theta - 2\dfrac{\partial z}{\partial x}\dfrac{\partial z}{\partial y}r^2\cos\theta\sin\theta + \left(\dfrac{\partial z}{\partial y}\right)^2 r^2\cos^2\theta$. Thus

$\left(\dfrac{\partial z}{\partial r}\right)^2 + \dfrac{1}{r^2}\left(\dfrac{\partial z}{\partial \theta}\right)^2 = \left[\left(\dfrac{\partial z}{\partial x}\right)^2 + \left(\dfrac{\partial z}{\partial y}\right)^2\right](\cos^2\theta + \sin^2\theta) = \left(\dfrac{\partial z}{\partial x}\right)^2 + \left(\dfrac{\partial z}{\partial y}\right)^2$.

39. Let $u = x - y$. Then $\dfrac{\partial z}{\partial x} = \dfrac{dz}{du}\dfrac{\partial u}{\partial x} = \dfrac{dz}{du}$ and $\dfrac{\partial z}{\partial y} = \dfrac{dz}{du}(-1)$. Thus $\dfrac{\partial z}{\partial x} + \dfrac{\partial z}{\partial y} = 0$.

41. Let $u = x + at$, $v = x - at$. Then $z = f(u) + g(v)$, so $\partial z/\partial u = f'(u)$ and $\partial z/\partial v = g'(v)$.

Thus $\dfrac{\partial z}{\partial t} = \dfrac{\partial z}{\partial u}\dfrac{\partial u}{\partial t} + \dfrac{\partial z}{\partial v}\dfrac{\partial v}{\partial t} = af'(u) - ag'(v)$ and

$$\frac{\partial^2 z}{\partial t^2} = a\frac{\partial}{\partial t}[f'(u) - g'(v)] = a\left(\frac{df'(u)}{du}\frac{\partial u}{\partial t} - \frac{dg'(v)}{dv}\frac{\partial v}{\partial t}\right) = a^2 f''(u) + a^2 g''(v).$$

Similarly $\dfrac{\partial z}{\partial x} = f'(u) + g'(v)$ and $\dfrac{\partial^2 z}{\partial x^2} = f''(u) + g''(v)$. Thus $\dfrac{\partial^2 z}{\partial t^2} = a^2 \dfrac{\partial^2 z}{\partial x^2}$.

43. $\dfrac{\partial z}{\partial s} = \dfrac{\partial z}{\partial x}2s + \dfrac{\partial z}{\partial y}2r$. Then

$$\frac{\partial^2 z}{\partial r\,\partial s} = \frac{\partial}{\partial r}\left(\frac{\partial z}{\partial x}2s\right) + \frac{\partial}{\partial r}\left(\frac{\partial z}{\partial y}2r\right)$$

$$= \frac{\partial^2 z}{\partial x^2}\frac{\partial x}{\partial r}2s + \frac{\partial}{\partial y}\left(\frac{\partial z}{\partial x}\right)\frac{\partial y}{\partial r}2s + \frac{\partial z}{\partial x}\frac{\partial}{\partial r}2s + \frac{\partial^2 z}{\partial y^2}\frac{\partial y}{\partial r}2r + \frac{\partial}{\partial x}\left(\frac{\partial z}{\partial y}\right)\frac{\partial x}{\partial r}2r + \frac{\partial z}{\partial y}2$$

$$= 4rs\frac{\partial^2 z}{\partial x^2} + \frac{\partial^2 z}{\partial y\,\partial x}4s^2 + 0 + 4rs\frac{\partial^2 z}{\partial y^2} + \frac{\partial^2 z}{\partial x\,\partial y}4r^2 + 2\frac{\partial z}{\partial y}$$

By the continuity of the partials, $\dfrac{\partial^2 z}{\partial r\partial s} = 4rs\dfrac{\partial^2 z}{\partial x^2} + 4rs\dfrac{\partial^2 z}{\partial y^2} + (4r^2 + 4s^2)\dfrac{\partial^2 z}{\partial x\,\partial y} + 2\dfrac{\partial z}{\partial y}$.

45. $\dfrac{\partial z}{\partial r} = \dfrac{\partial z}{\partial x}\cos\theta + \dfrac{\partial z}{\partial y}\sin\theta$ and $\dfrac{\partial z}{\partial \theta} = -\dfrac{\partial z}{\partial x}r\sin\theta + \dfrac{\partial z}{\partial y}r\cos\theta$. Then

$$\frac{\partial^2 z}{\partial r^2} = \cos\theta\left(\frac{\partial^2 z}{\partial x^2}\cos\theta + \frac{\partial^2 z}{\partial y\,\partial x}\sin\theta\right) + \sin\theta\left(\frac{\partial^2 z}{\partial y^2}\sin\theta + \frac{\partial^2 z}{\partial x\,\partial y}\cos\theta\right)$$

$$= \cos^2\theta\frac{\partial^2 z}{\partial x^2} + 2\cos\theta\sin\theta\frac{\partial^2 z}{\partial x\,\partial y} + \sin^2\theta\frac{\partial^2 z}{\partial y^2}$$

and

$$\frac{\partial^2 z}{\partial \theta^2} = -r\cos\theta\frac{\partial z}{\partial x} + (-r\sin\theta)\left(\frac{\partial^2 z}{\partial x^2}(-r\sin\theta) + \frac{\partial^2 z}{\partial y\,\partial x}r\cos\theta\right)$$

$$-r\sin\theta\frac{\partial z}{\partial y} + r\cos\theta\left(\frac{\partial^2 z}{\partial y^2}r\cos\theta + \frac{\partial^2 z}{\partial x\,\partial y}(-r\sin\theta)\right)$$

$$= -r\cos\theta\frac{\partial z}{\partial x} - r\sin\theta\frac{\partial z}{\partial y} + r^2\sin^2\theta\frac{\partial^2 z}{\partial x^2} - 2r^2\cos\theta\sin\theta\frac{\partial^2 z}{\partial x\,\partial y} + r^2\cos^2\theta\frac{\partial^2 z}{\partial y^2}$$

Thus

$$\frac{\partial^2 z}{\partial r^2} + \frac{1}{r^2}\frac{\partial^2 z}{\partial \theta^2} + \frac{1}{r}\frac{\partial z}{\partial r} = (\cos^2\theta + \sin^2\theta)\frac{\partial^2 z}{\partial x^2} + (\sin^2\theta + \cos^2\theta)\frac{\partial^2 z}{\partial y^2} - \frac{1}{r}\cos\theta\frac{\partial z}{\partial x}$$

$$-\frac{1}{r}\sin\theta\frac{\partial z}{\partial y} + \frac{1}{r}\left(\cos\theta\frac{\partial z}{\partial x} + \sin\theta\frac{\partial z}{\partial y}\right)$$

$$= \frac{\partial^2 z}{\partial x^2} + \frac{\partial^2 z}{\partial y^2} \text{ as desired.}$$

47. $F(x, y, z) = 0$ is assumed to define z as a function of x and y, that is, $z = f(x, y)$. So by (7), $\dfrac{\partial z}{\partial x} = -\dfrac{F_x}{F_z}$ since $F_z \neq 0$.

Similarly, it is assumed that $F(x, y, z) = 0$ defines x as a function of y and z, that is $x = h(x, z)$. Then $F(h(y, z), y, z) = 0$
and by the Chain Rule, $F_x\dfrac{\partial x}{\partial y} + F_y\dfrac{\partial y}{\partial y} + F_z\dfrac{\partial z}{\partial y} = 0$. But $\dfrac{\partial z}{\partial y} = 0$ and $\dfrac{\partial y}{\partial y} = 1$, so $F_x\dfrac{\partial x}{\partial y} + F_y = 0 \;\Rightarrow\; \dfrac{\partial x}{\partial y} = -\dfrac{F_y}{F_x}$.

A similar calculation shows that $\dfrac{\partial y}{\partial z} = -\dfrac{F_z}{F_y}$. Thus $\dfrac{\partial z}{\partial x}\dfrac{\partial x}{\partial y}\dfrac{\partial y}{\partial z} = \left(-\dfrac{F_x}{F_z}\right)\left(-\dfrac{F_y}{F_x}\right)\left(-\dfrac{F_z}{F_y}\right) = -1$.

11.6   Directional Derivatives and the Gradient Vector

1. First we draw a line passing through Raleigh and the eye of the hurricane. We can approximate the directional derivative at Raleigh in the direction of the eye of the hurricane by the average rate of change of pressure between the points where this line intersects the contour lines closest to Raleigh. In the direction of the eye of the hurricane, the pressure changes from 996 millibars to 992 millibars. We estimate the distance between these two points to be approximately 40 miles, so the rate of change of pressure in the direction given is approximately $\frac{992 - 996}{40} = -0.1$ millibar/mi.

3. $D_{\mathbf{u}} f(-20, 30) = \nabla f(-20, 30) \cdot \mathbf{u} = f_T(-20, 30)\left(\frac{1}{\sqrt{2}}\right) + f_v(-20, 30)\left(\frac{1}{\sqrt{2}}\right)$.

$f_T(-20, 30) = \lim\limits_{h \to 0} \dfrac{f(-20 + h, 30) - f(-20, 30)}{h}$, so we can approximate $f_T(-20, 30)$ by considering $h = \pm 5$ and using

the values given in the table: $f_T(-20, 30) \approx \dfrac{f(-15, 30) - f(-20, 30)}{5} = \dfrac{-26 - (-33)}{5} = 1.4$,

$f_T(-20, 30) \approx \dfrac{f(-25, 30) - f(-20, 30)}{-5} = \dfrac{-39 - (-33)}{-5} = 1.2$. Averaging these values gives $f_T(-20, 30) \approx 1.3$.

Similarly, $f_v(-20, 30) = \lim\limits_{h \to 0} \dfrac{f(-20, 30 + h) - f(-20, 30)}{h}$, so we can approximate $f_v(-20, 30)$ with $h = \pm 10$:

$f_v(-20, 30) \approx \dfrac{f(-20, 40) - f(-20, 30)}{10} = \dfrac{-34 - (-33)}{10} = -0.1$,

$f_v(-20, 30) \approx \dfrac{f(-20, 20) - f(-20, 30)}{-10} = \dfrac{-30 - (-33)}{-10} = -0.3$. Averaging these values gives $f_v(-20, 30) \approx -0.2$.

Then $D_{\mathbf{u}} f(-20, 30) \approx 1.3\left(\frac{1}{\sqrt{2}}\right) + (-0.2)\left(\frac{1}{\sqrt{2}}\right) \approx 0.778$.

5. $f(x, y) = \sqrt{5x - 4y} \quad \Rightarrow \quad f_x(x, y) = \frac{1}{2}(5x - 4y)^{-1/2}(5) = \dfrac{5}{2\sqrt{5x - 4y}}$ and

$f_y(x, y) = \frac{1}{2}(5x - 4y)^{-1/2}(-4) = -\dfrac{2}{\sqrt{5x - 4y}}$. If $\mathbf{u}$ is a unit vector in the direction of $\theta = -\frac{\pi}{6}$, then from Equation 6,

$D_{\mathbf{u}} f(4, 1) = f_x(4, 1) \cos\left(-\frac{\pi}{6}\right) + f_y(4, 1) \sin\left(-\frac{\pi}{6}\right) = \frac{5}{8} \cdot \frac{\sqrt{3}}{2} + \left(-\frac{1}{2}\right)\left(-\frac{1}{2}\right) = \frac{5\sqrt{3}}{16} + \frac{1}{4}$.

7. $f(x, y) = 5xy^2 - 4x^3 y$

(a) $\nabla f(x, y) = \langle f_x(x, y), f_y(x, y) \rangle = \langle 5y^2 - 12x^2 y, 10xy - 4x^3 \rangle$

(b) $\nabla f(1, 2) = \langle 5(2)^2 - 12(1)^2(2), 10(1)(2) - 4(1)^3 \rangle = \langle -4, 16 \rangle$

(c) By Equation 9, $D_{\mathbf{u}} f(1, 2) = \nabla f(1, 2) \cdot \mathbf{u} = \langle -4, 16 \rangle \cdot \langle \frac{5}{13}, \frac{12}{13} \rangle = (-4)\left(\frac{5}{13}\right) + (16)\left(\frac{12}{13}\right) = \frac{172}{13}$.

9. $f(x, y, z) = xe^{2yz}$

(a) $\nabla f(x, y, z) = \langle f_x(x, y, z), f_y(x, y, z), f_z(x, y, z) \rangle = \langle e^{2yz}, 2xze^{2yz}, 2xye^{2yz} \rangle$

(b) $\nabla f(3, 0, 2) = \langle 1, 12, 0 \rangle$

(c) By Equation 14, $D_{\mathbf{u}} f(3, 0, 2) = \nabla f(3, 0, 2) \cdot \mathbf{u} = \langle 1, 12, 0 \rangle \cdot \langle \frac{2}{3}, -\frac{2}{3}, \frac{1}{3} \rangle = \frac{2}{3} - \frac{24}{3} + 0 = -\frac{22}{3}$.

11. $f(x, y) = 1 + 2x\sqrt{y} \quad \Rightarrow \quad \nabla f(x, y) = \left\langle 2\sqrt{y}, 2x \cdot \frac{1}{2}y^{-1/2} \right\rangle = \langle 2\sqrt{y}, x/\sqrt{y} \rangle$, $\nabla f(3, 4) = \langle 4, \frac{3}{2} \rangle$, and a unit vector in

the direction of $\mathbf{v}$ is $\mathbf{u} = \dfrac{1}{\sqrt{4^2 + (-3)^2}}\langle 4, -3 \rangle = \langle \frac{4}{5}, -\frac{3}{5} \rangle$, so $D_{\mathbf{u}} f(3, 4) = \nabla f(3, 4) \cdot \mathbf{u} = \langle 4, \frac{3}{2} \rangle \cdot \langle \frac{4}{5}, -\frac{3}{5} \rangle = \frac{23}{10}$.

13. $g(s,t) = s^2 e^t \Rightarrow \nabla g(s,t) = 2se^t \mathbf{i} + s^2 e^t \mathbf{j}$, $\nabla g(2,0) = 4\mathbf{i} + 4\mathbf{j}$, and a unit vector in the direction of $\mathbf{v}$ is

$\mathbf{u} = \frac{1}{\sqrt{2}}(\mathbf{i} + \mathbf{j})$, so $D_{\mathbf{u}}\, g(2,0) = \nabla g(2,0) \cdot \mathbf{u} = (4\mathbf{i} + 4\mathbf{j}) \cdot \frac{1}{\sqrt{2}}(\mathbf{i} + \mathbf{j}) = \frac{8}{\sqrt{2}} = 4\sqrt{2}$.

15. $g(x,y,z) = (x + 2y + 3z)^{3/2} \Rightarrow$

$\nabla g(x,y,z) = \left\langle \frac{3}{2}(x+2y+3z)^{1/2}(1), \frac{3}{2}(x+2y+3z)^{1/2}(2), \frac{3}{2}(x+2y+3z)^{1/2}(3) \right\rangle$

$\qquad = \left\langle \frac{3}{2}\sqrt{x+2y+3z}, 3\sqrt{x+2y+3z}, \frac{9}{2}\sqrt{x+2y+3z} \right\rangle$, $\nabla g(1,1,2) = \left\langle \frac{9}{2}, 9, \frac{27}{2} \right\rangle$,

and a unit vector in the direction of $\mathbf{v} = 2\mathbf{j} - \mathbf{k}$ is $\mathbf{u} = \frac{2}{\sqrt{5}}\mathbf{j} - \frac{1}{\sqrt{5}}\mathbf{k}$, so

$D_{\mathbf{u}}\, g(1,1,2) = \left\langle \frac{9}{2}, 9, \frac{27}{2} \right\rangle \cdot \left\langle 0, \frac{2}{\sqrt{5}}, -\frac{1}{\sqrt{5}} \right\rangle = \frac{18}{\sqrt{5}} - \frac{27}{2\sqrt{5}} = \frac{9}{2\sqrt{5}}$.

17. $f(x,y) = \sqrt{xy} \Rightarrow \nabla f(x,y) = \left\langle \frac{1}{2}(xy)^{-1/2}(y), \frac{1}{2}(xy)^{-1/2}(x) \right\rangle = \left\langle \frac{y}{2\sqrt{xy}}, \frac{x}{2\sqrt{xy}} \right\rangle$, so $\nabla f(2,8) = \left\langle 1, \frac{1}{4} \right\rangle$.

The unit vector in the direction of $\overrightarrow{PQ} = \langle 5 - 2, 4 - 8 \rangle = \langle 3, -4 \rangle$ is $\mathbf{u} = \left\langle \frac{3}{5}, -\frac{4}{5} \right\rangle$, so

$D_{\mathbf{u}}\, f(2,8) = \nabla f(2,8) \cdot \mathbf{u} = \left\langle 1, \frac{1}{4} \right\rangle \cdot \left\langle \frac{3}{5}, -\frac{4}{5} \right\rangle = \frac{2}{5}$.

19. $f(x,y) = y^2/x = y^2 x^{-1} \Rightarrow \nabla f(x,y) = \left\langle -y^2 x^{-2}, 2yx^{-1} \right\rangle = \left\langle -y^2/x^2, 2y/x \right\rangle$.

$\nabla f(2,4) = \langle -4, 4 \rangle$, or equivalently $\langle -1, 1 \rangle$, is the direction of maximum rate of change, and the maximum rate

is $|\nabla f(2,4)| = \sqrt{16 + 16} = 4\sqrt{2}$.

21. $f(x,y,z) = \ln(xy^2z^3) \Rightarrow \nabla f(x,y,z) = \left\langle \frac{y^2 z^3}{xy^2 z^3}, \frac{2xyz^3}{xy^2 z^3}, \frac{3xy^2 z^2}{xy^2 z^3} \right\rangle = \left\langle \frac{1}{x}, \frac{2}{y}, \frac{3}{z} \right\rangle$.

$\nabla f(1,-2,-3) = \langle 1, -1, -1 \rangle$ is the direction of maximum rate of change and the maximum rate is $|\nabla f(1,-2,-3)| = \sqrt{3}$.

23. (a) As in the proof of Theorem 15, $D_{\mathbf{u}}\, f = |\nabla f| \cos\theta$. Since the minimum value of $\cos\theta$ is -1 occurring when $\theta = \pi$, the minimum value of $D_{\mathbf{u}}\, f$ is $-|\nabla f|$ occurring when $\theta = \pi$, that is when $\mathbf{u}$ is in the opposite direction of ∇f (assuming $\nabla f \neq \mathbf{0}$).

(b) $f(x,y) = x^4 y - x^2 y^3 \Rightarrow \nabla f(x,y) = \left\langle 4x^3 y - 2xy^3, x^4 - 3x^2 y^2 \right\rangle$, so f decreases fastest at the point $(2,-3)$ in the direction $-\nabla f(2,-3) = -\langle 12, -92 \rangle = \langle -12, 92 \rangle$.

25. The direction of fastest change is $\nabla f(x,y) = (2x - 2)\mathbf{i} + (2y - 4)\mathbf{j}$, so we need to find all points (x,y) where $\nabla f(x,y)$ is

parallel to $\mathbf{i} + \mathbf{j} \Leftrightarrow (2x - 2)\mathbf{i} + (2y - 4)\mathbf{j} = k(\mathbf{i} + \mathbf{j}) \Leftrightarrow k = 2x - 2$ and $k = 2y - 4$. Then $2x - 2 = 2y - 4 \Rightarrow$

$y = x + 1$, so the direction of fastest change is $\mathbf{i} + \mathbf{j}$ at all points on the line $y = x + 1$.

27. $T = \dfrac{k}{\sqrt{x^2 + y^2 + z^2}}$ and $120 = T(1,2,2) = \dfrac{k}{3}$ so $k = 360$.

(a) $\mathbf{u} = \dfrac{\langle 1, -1, 1 \rangle}{\sqrt{3}}$,

$D_{\mathbf{u}}T(1,2,2) = \nabla T(1,2,2) \cdot \mathbf{u} = \left[-360(x^2 + y^2 + z^2)^{-3/2}\langle x,y,z \rangle \right]_{(1,2,2)} \cdot \mathbf{u} = -\frac{40}{3}\langle 1,2,2 \rangle \cdot \frac{1}{\sqrt{3}}\langle 1,-1,1 \rangle = -\frac{40}{3\sqrt{3}}$

(b) From (a), $\nabla T = -360(x^2 + y^2 + z^2)^{-3/2}\langle x,y,z \rangle$, and since $\langle x,y,z \rangle$ is the position vector of the point (x,y,z), the vector $-\langle x,y,z \rangle$, and thus ∇T, always points toward the origin.

29. $\nabla V(x, y, z) = \langle 10x - 3y + yz, xz - 3x, xy \rangle$, $\nabla V(3, 4, 5) = \langle 38, 6, 12 \rangle$

(a) $D_{\mathbf{u}} V(3, 4, 5) = \langle 38, 6, 12 \rangle \cdot \frac{1}{\sqrt{3}} \langle 1, 1, -1 \rangle = \frac{32}{\sqrt{3}}$

(b) $\nabla V(3, 4, 5) = \langle 38, 6, 12 \rangle$ or equivalently $\langle 19, 3, 6 \rangle$.

(c) $|\nabla V(3, 4, 5)| = \sqrt{38^2 + 6^2 + 12^2} = \sqrt{1624} = 2\sqrt{406}$

31. A unit vector in the direction of $\overrightarrow{AB}$ is $\mathbf{i}$ and a unit vector in the direction of $\overrightarrow{AC}$ is $\mathbf{j}$. Thus $D_{\overrightarrow{AB}} f(1, 3) = f_x(1, 3) = 3$ and

$D_{\overrightarrow{AC}} f(1, 3) = f_y(1, 3) = 26$. Therefore $\nabla f(1, 3) = \langle f_x(1, 3), f_y(1, 3) \rangle = \langle 3, 26 \rangle$, and by definition,

$D_{\overrightarrow{AD}} f(1, 3) = \nabla f \cdot \mathbf{u}$ where $\mathbf{u}$ is a unit vector in the direction of $\overrightarrow{AD}$, which is $\langle \frac{5}{13}, \frac{12}{13} \rangle$. Therefore,

$D_{\overrightarrow{AD}} f(1, 3) = \langle 3, 26 \rangle \cdot \langle \frac{5}{13}, \frac{12}{13} \rangle = 3 \cdot \frac{5}{13} + 26 \cdot \frac{12}{13} = \frac{327}{13}$.

33. (a) $\nabla(au + bv) = \left\langle \dfrac{\partial(au + bv)}{\partial x}, \dfrac{\partial(au + bv)}{\partial y} \right\rangle = \left\langle a\dfrac{\partial u}{\partial x} + b\dfrac{\partial v}{\partial x}, a\dfrac{\partial u}{\partial y} + b\dfrac{\partial v}{\partial y} \right\rangle = a\left\langle \dfrac{\partial u}{\partial x}, \dfrac{\partial u}{\partial y} \right\rangle + b\left\langle \dfrac{\partial v}{\partial x}, \dfrac{\partial v}{\partial y} \right\rangle$

$= a\nabla u + b\nabla v$

(b) $\nabla(uv) = \left\langle v\dfrac{\partial u}{\partial x} + u\dfrac{\partial v}{\partial x}, v\dfrac{\partial u}{\partial y} + u\dfrac{\partial v}{\partial y} \right\rangle = v\left\langle \dfrac{\partial u}{\partial x}, \dfrac{\partial u}{\partial y} \right\rangle + u\left\langle \dfrac{\partial v}{\partial x}, \dfrac{\partial v}{\partial y} \right\rangle = v\nabla u + u\nabla v$

(c) $\nabla\left(\dfrac{u}{v}\right) = \left\langle \dfrac{v\dfrac{\partial u}{\partial x} - u\dfrac{\partial v}{\partial x}}{v^2}, \dfrac{v\dfrac{\partial u}{\partial y} - u\dfrac{\partial v}{\partial y}}{v^2} \right\rangle = \dfrac{v\left\langle \dfrac{\partial u}{\partial x}, \dfrac{\partial u}{\partial y} \right\rangle - u\left\langle \dfrac{\partial v}{\partial x}, \dfrac{\partial v}{\partial y} \right\rangle}{v^2} = \dfrac{v\nabla u - u\nabla v}{v^2}$

(d) $\nabla u^n = \left\langle \dfrac{\partial(u^n)}{\partial x}, \dfrac{\partial(u^n)}{\partial y} \right\rangle = \left\langle nu^{n-1}\dfrac{\partial u}{\partial x}, nu^{n-1}\dfrac{\partial u}{\partial y} \right\rangle = nu^{n-1}\nabla u$.

35. Let $F(x, y, z) = x^2 - 2y^2 + z^2 + yz$. Then $x^2 - 2y^2 + z^2 + yz = 2$ is a level surface of F

and $\nabla F(x, y, z) = \langle 2x, -4y + z, 2z + y \rangle$.

(a) $\nabla F(2, 1, -1) = \langle 4, -5, -1 \rangle$ is a normal vector for the tangent plane at $(2, 1, -1)$, so an equation of the tangent plane is

$4(x - 2) - 5(y - 1) - 1(z + 1) = 0$ or $4x - 5y - z = 4$.

(b) The normal line has direction $\langle 4, -5, -1 \rangle$, so parametric equations are $x = 2 + 4t$, $y = 1 - 5t$, $z = -1 - t$, and

symmetric equations are $\dfrac{x - 2}{4} = \dfrac{y - 1}{-5} = \dfrac{z + 1}{-1}$.

37. $F(x, y, z) = -z + xe^y \cos z \Rightarrow \nabla F(x, y, z) = \langle e^y \cos z, xe^y \cos z, -1 - xe^y \sin z \rangle$, $\nabla F(1, 0, 0) = \langle 1, 1, -1 \rangle$

(a) $1(x - 1) + 1(y - 0) - 1(z - 0) = 0$ or $x + y - z = 1$

(b) $x - 1 = y = -z$

39. $F(x, y, z) = xy + yz + zx$, $\nabla F(x, y, z) = \langle y + z, x + z, y + x \rangle$, $\nabla F(1, 1, 1) = \langle 2, 2, 2 \rangle$, so an equation of the tangent

plane is $2x + 2y + 2z = 6$ or $x + y + z = 3$, and the normal line is given by $x - 1 = y - 1 = z - 1$ or $x = y = z$.

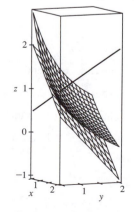

41. $\nabla f(x, y) = \langle 2x, 8y \rangle$, $\nabla f(2, 1) = \langle 4, 8 \rangle$. The tangent line has

equation $\nabla f(2, 1) \cdot \langle x - 2, y - 1 \rangle = 0$ $\Rightarrow$

$4(x - 2) + 8(y - 1) = 0$, which simplifies to $x + 2y = 4$.

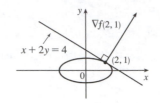

43. $\nabla F(x_0, y_0, z_0) = \left\langle \dfrac{2x_0}{a^2}, \dfrac{2y_0}{b^2}, \dfrac{2z_0}{c^2} \right\rangle$. Thus an equation of the tangent plane at (x_0, y_0, z_0) is

$\dfrac{2x_0}{a^2} x + \dfrac{2y_0}{b^2} y + \dfrac{2z_0}{c^2} z = 2 \left(\dfrac{x_0^2}{a^2} + \dfrac{y_0^2}{b^2} + \dfrac{z_0^2}{c^2} \right) = 2(1) = 2$ since (x_0, y_0, z_0) is a point on the ellipsoid. Hence

$\dfrac{x_0}{a^2} x + \dfrac{y_0}{b^2} y + \dfrac{z_0}{c^2} z = 1$ is an equation of the tangent plane.

45. $\nabla f(x_0, y_0, z_0) = \langle 2x_0, -2y_0, 4z_0 \rangle$ and the given line has direction numbers $2, 4, 6$, so $\langle 2x_0, -2y_0, 4z_0 \rangle = k \langle 2, 4, 6 \rangle$ or

$x_0 = k$, $y_0 = -2k$ and $z_0 = \frac{3}{2}k$. But $x_0^2 - y_0^2 + 2z_0^2 = 1$ or $\left(1 - 4 + \frac{9}{2} \right) k^2 = 1$, so $k = \pm \sqrt{\frac{2}{3}} = \pm \frac{\sqrt{6}}{3}$ and there are two

such points: $\left(\pm \frac{\sqrt{6}}{3}, \mp \frac{2\sqrt{6}}{3}, \pm \frac{\sqrt{6}}{2} \right)$.

47. Let (x_0, y_0, z_0) be a point on the surface. Then an equation of the tangent plane at the point is

$\dfrac{x}{2\sqrt{x_0}} + \dfrac{y}{2\sqrt{y_0}} + \dfrac{z}{2\sqrt{z_0}} = \dfrac{\sqrt{x_0} + \sqrt{y_0} + \sqrt{z_0}}{2}$. But $\sqrt{x_0} + \sqrt{y_0} + \sqrt{z_0} = \sqrt{c}$, so the equation is

$\dfrac{x}{\sqrt{x_0}} + \dfrac{y}{\sqrt{y_0}} + \dfrac{z}{\sqrt{z_0}} = \sqrt{c}$. The x-, y-, and z-intercepts are $\sqrt{cx_0}$, $\sqrt{cy_0}$ and $\sqrt{cz_0}$ respectively. (The x-intercept is found by

setting $y = z = 0$ and solving the resulting equation for x, and the y- and z-intercepts are found similarly.) So the sum of the

intercepts is $\sqrt{c}\left(\sqrt{x_0} + \sqrt{y_0} + \sqrt{z_0} \right) = c$, a constant.

49. If $f(x, y, z) = z - x^2 - y^2$ and $g(x, y, z) = 4x^2 + y^2 + z^2$, then the tangent line is perpendicular to both ∇f and ∇g at

$(-1, 1, 2)$. The vector $\mathbf{v} = \nabla f \times \nabla g$ will therefore be parallel to the tangent line. We have: $\nabla f(x, y, z) = \langle -2x, -2y, 1 \rangle$

$\Rightarrow$ $\nabla f(-1, 1, 2) = \langle 2, -2, 1 \rangle$, and $\nabla g(x, y, z) = \langle 8x, 2y, 2z \rangle$ $\Rightarrow$ $\nabla g(-1, 1, 2) = \langle -8, 2, 4 \rangle$. Hence

$\mathbf{v} = \nabla f \times \nabla g = \begin{vmatrix} \mathbf{i} & \mathbf{j} & \mathbf{k} \\ 2 & -2 & 1 \\ -8 & 2 & 4 \end{vmatrix} = -10\,\mathbf{i} - 16\,\mathbf{j} - 12\,\mathbf{k}$. Parametric equations are: $x = -1 - 10t$, $y = 1 - 16t$, $z = 2 - 12t$.

51. (a) The direction of the normal line of F is given by ∇F, and that of G by ∇G. Assuming that

$\nabla F \neq 0 \neq \nabla G$, the two normal lines are perpendicular at P if $\nabla F \cdot \nabla G = 0$ at P $\Leftrightarrow$

$\langle \partial F / \partial x, \partial F / \partial y, \partial F / \partial z \rangle \cdot \langle \partial G / \partial x, \partial G / \partial y, \partial G / \partial z \rangle = 0$ at P $\Leftrightarrow$ $F_x G_x + F_y G_y + F_z G_z = 0$ at P.

(b) Here $F = x^2 + y^2 - z^2$ and $G = x^2 + y^2 + z^2 - r^2$, so

$\nabla F \cdot \nabla G = \langle 2x, 2y, -2z \rangle \cdot \langle 2x, 2y, 2z \rangle = 4x^2 + 4y^2 - 4z^2 = 4F = 0$, since the point $\langle x, y, z \rangle$ lies on the graph of

$F = 0$. To see that this is true without using calculus, note that $G = 0$ is the equation of a sphere centered at the origin and

$F = 0$ is the equation of a right circular cone with vertex at the origin (which is generated by lines through the origin). At

any point of intersection, the sphere's normal line (which passes through the origin) lies on the cone, and thus is

perpendicular to the cone's normal line. So the surfaces with equations $F = 0$ and $G = 0$ are everywhere orthogonal.

53. Let $\mathbf{u} = \langle a, b \rangle$ and $\mathbf{v} = \langle c, d \rangle$. Then we know that at the given point, $D_{\mathbf{u}} f = \nabla f \cdot \mathbf{u} = a f_x + b f_y$ and $D_{\mathbf{v}} f = \nabla f \cdot \mathbf{v} = c f_x + d f_y$. But these are just two linear equations in the two unknowns f_x and f_y, and since $\mathbf{u}$ and $\mathbf{v}$ are not parallel, we can solve the equations to find $\nabla f = \langle f_x, f_y \rangle$ at the given point. In fact,

$$\nabla f = \left\langle \frac{d\, D_{\mathbf{u}} f - b\, D_{\mathbf{v}} f}{ad - bc}, \frac{a\, D_{\mathbf{v}} f - c\, D_{\mathbf{u}} f}{ad - bc} \right\rangle.$$

11.7   Maximum and Minimum Values

1. (a) First we compute $D(1, 1) = f_{xx}(1, 1)\, f_{yy}(1, 1) - [f_{xy}(1, 1)]^2 = (4)(2) - (1)^2 = 7$. Since $D(1, 1) > 0$ and $f_{xx}(1, 1) > 0$, f has a local minimum at $(1, 1)$ by the Second Derivatives Test.

(b) $D(1, 1) = f_{xx}(1, 1)\, f_{yy}(1, 1) - [f_{xy}(1, 1)]^2 = (4)(2) - (3)^2 = -1$. Since $D(1, 1) < 0$, f has a saddle point at $(1, 1)$ by the Second Derivatives Test.

3. In the figure, a point at approximately $(1, 1)$ is enclosed by level curves which are oval in shape and indicate that as we move away from the point in any direction the values of f are increasing. Hence we would expect a local minimum at or near $(1, 1)$. The level curves near $(0, 0)$ resemble hyperbolas, and as we move away from the origin, the values of f increase in some directions and decrease in others, so we would expect to find a saddle point there.

To verify our predictions, we have $f(x, y) = 4 + x^3 + y^3 - 3xy \;\Rightarrow\; f_x(x, y) = 3x^2 - 3y, f_y(x, y) = 3y^2 - 3x$. We have critical points where these partial derivatives are equal to 0: $3x^2 - 3y = 0, 3y^2 - 3x = 0$. Substituting $y = x^2$ from the first equation into the second equation gives $3(x^2)^2 - 3x = 0 \;\Rightarrow\; 3x(x^3 - 1) = 0 \;\Rightarrow\; x = 0$ or $x = 1$. Then we have two critical points, $(0, 0)$ and $(1, 1)$. The second partial derivatives are $f_{xx}(x, y) = 6x, f_{xy}(x, y) = -3$, and $f_{yy}(x, y) = 6y$, so $D(x, y) = f_{xx}(x, y)\, f_{yy}(x, y) - [f_{xy}(x, y)]^2 = (6x)(6y) - (-3)^2 = 36xy - 9$. Then $D(0, 0) = 36(0)(0) - 9 = -9$, and $D(1, 1) = 36(1)(1) - 9 = 27$. Since $D(0, 0) < 0$, f has a saddle point at $(0, 0)$ by the Second Derivatives Test. Since $D(1, 1) > 0$ and $f_{xx}(1, 1) > 0$, f has a local minimum at $(1, 1)$.

5. $f(x, y) = 9 - 2x + 4y - x^2 - 4y^2 \;\Rightarrow\; f_x = -2 - 2x, f_y = 4 - 8y$, $f_{xx} = -2, f_{xy} = 0, f_{yy} = -8$. Then $f_x = 0$ and $f_y = 0$ imply $x = -1$ and $y = \frac{1}{2}$, and the only critical point is $\left(-1, \frac{1}{2}\right)$.

$D(x, y) = f_{xx} f_{yy} - (f_{xy})^2 = (-2)(-8) - 0^2 = 16$, and since $D\left(-1, \frac{1}{2}\right) = 16 > 0$ and $f_{xx}\left(-1, \frac{1}{2}\right) = -2 < 0$, $f\left(-1, \frac{1}{2}\right) = 11$ is a local maximum by the Second Derivatives Test.

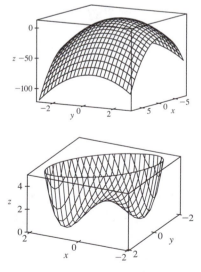

7. $f(x, y) = x^4 + y^4 - 4xy + 2 \;\Rightarrow\; f_x = 4x^3 - 4y, f_y = 4y^3 - 4x$, $f_{xx} = 12x^2, f_{xy} = -4, f_{yy} = 12y^2$. Then $f_x = 0$ implies $y = x^3$, and substitution into $f_y = 0 \;\Rightarrow\; x = y^3$ gives $x^9 - x = 0 \;\Rightarrow\; x(x^8 - 1) = 0 \;\Rightarrow\; x = 0$ or $x = \pm 1$. Thus the critical points are $(0, 0)$, $(1, 1)$, and $(-1, -1)$. Now $D(0, 0) = 0 \cdot 0 - (-4)^2 = -16 < 0$, so $(0, 0)$ is a saddle point. $D(1, 1) = (12)(12) - (-4)^2 > 0$ and $f_{xx}(1, 1) = 12 > 0$, so $f(1, 1) = 0$ is a local minimum. $D(-1, -1) = (12)(12) - (-4)^2 > 0$ and $f_{xx} = (-1, -1) = 12 > 0$, so $f(-1, -1) = 0$ is also a local minimum.

9. $f(x, y) = (1 + xy)(x + y) = x + y + x^2y + xy^2 \implies$

$f_x = 1 + 2xy + y^2$, $f_y = 1 + x^2 + 2xy$, $f_{xx} = 2y$, $f_{xy} = 2x + 2y$,

$f_{yy} = 2x$. Then $f_x = 0$ implies $1 + 2xy + y^2 = 0$ and $f_y = 0$ implies

$1 + x^2 + 2xy = 0$. Subtracting the second equation from the first gives

$y^2 - x^2 = 0 \implies y = \pm x$, but if $y = x$ then $1 + 2xy + y^2 = 0 \implies$

$1 + 3x^2 = 0$ which has no real solution. If $y = -x$ then

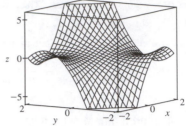

$1 + 2xy + y^2 = 0 \implies 1 - x^2 = 0 \implies x = \pm 1$, so critical points

are $(1, -1)$ and $(-1, 1)$. $D(1, -1) = (-2)(2) - 0 < 0$ and $D(-1, 1) = (2)(-2) - 0 < 0$, so $(-1, 1)$ and $(1, -1)$ are

saddle points.

11. $f(x, y) = e^x \cos y \implies f_x = e^x \cos y$, $f_y = -e^x \sin y$. Now $f_x = 0$

implies $\cos y = 0$ or $y = \frac{\pi}{2} + n\pi$ for n an integer. But $\sin\left(\frac{\pi}{2} + n\pi\right) \neq 0$,

so there are no critical points.

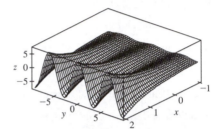

13. $f(x, y) = x \sin y \implies f_x = \sin y$, $f_y = x \cos y$, $f_{xx} = 0$, $f_{yy} = -x$

$\sin y$ and $f_{xy} = \cos y$. Then $f_x = 0$ if and only if $y = n\pi$, n an integer,

and substituting into $f_y = 0$ requires $x = 0$ for each of these y-values. Thus

the critical points are $(0, n\pi)$, n an integer. But

$D(0, n\pi) = -\cos^2(n\pi) < 0$ so each critical point is a saddle point.

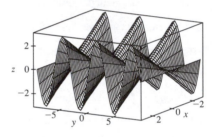

15. $f(x, y) = (x^2 + y^2)e^{y^2 - x^2} \implies f_x = (x^2 + y^2)e^{y^2 - x^2}(-2x) + 2xe^{y^2 - x^2} = 2xe^{y^2 - x^2}(1 - x^2 - y^2)$,

$f_y = (x^2 + y^2)e^{y^2 - x^2}(2y) + 2ye^{y^2 - x^2} = 2ye^{y^2 - x^2}(1 + x^2 + y^2)$,

$f_{xx} = 2xe^{y^2 - x^2}(-2x) + (1 - x^2 - y^2)\left(2x\left(-2xe^{y^2 - x^2}\right) + 2e^{y^2 - x^2}\right) = 2e^{y^2 - x^2}((1 - x^2 - y^2)(1 - 2x^2) - 2x^2)$,

$f_{xy} = 2xe^{y^2 - x^2}(-2y) + 2x(2y)e^{y^2 - x^2}(1 - x^2 - y^2) = -4xye^{y^2 - x^2}(x^2 + y^2)$,

$f_{yy} = 2ye^{y^2 - x^2}(2y) + (1 + x^2 + y^2)\left(2y\left(2ye^{y^2 - x^2}\right) + 2e^{y^2 - x^2}\right) = 2e^{y^2 - x^2}((1 + x^2 + y^2)(1 + 2y^2) + 2y^2)$.

$f_y = 0$ implies $y = 0$, and substituting into $f_x = 0$ gives

$2xe^{-x^2}(1 - x^2) = 0 \implies x = 0$ or $x = \pm 1$.

Thus the critical points are $(0, 0)$ and $(\pm 1, 0)$. $D(0, 0) = (2)(2) - 0 > 0$

and $f_{xx}(0, 0) = 2 > 0$, so $f(0, 0) = 0$ is a local minimum.

$D(\pm 1, 0) = (-4e^{-1})(4e^{-1}) - 0 < 0$ so $(\pm 1, 0)$ are saddle points.

17. $f(x, y) = 3x^2 y + y^3 - 3x^2 - 3y^2 + 2$

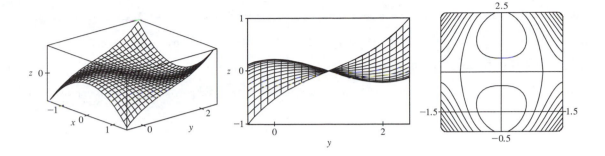

From the graphs, it appears that f has a local maximum $f(0, 0) \approx 2$ and a local minimum $f(0, 2) \approx -2$. There appear to be saddle points near $(\pm 1, 1)$.

$f_x = 6xy - 6x$, $f_y = 3x^2 + 3y^2 - 6y$. Then $f_x = 0$ implies $x = 0$ or $y = 1$ and when $x = 0$, $f_y = 0$ implies $y = 0$ or $y = 2$; when $y = 1$, $f_y = 0$ implies $x^2 = 1$ or $x = \pm 1$. Thus the critical points are $(0, 0)$, $(0, 2)$, $(\pm 1, 1)$. Now $f_{xx} = 6y - 6$, $f_{yy} = 6y - 6$ and $f_{xy} = 6x$, so $D(0, 0) = D(0, 2) = 36 > 0$ while $D(\pm 1, 1) = -36 < 0$ and $f_{xx}(0, 0) = -6$, $f_{xx}(0, 2) = 6$. Hence $(\pm 1, 1)$ are saddle points while $f(0, 0) = 2$ is a local maximum and $f(0, 2) = -2$ is a local minimum.

19. $f(x, y) = \sin x + \sin y + \sin(x + y)$, $0 \le x \le 2\pi$, $0 \le y \le 2\pi$

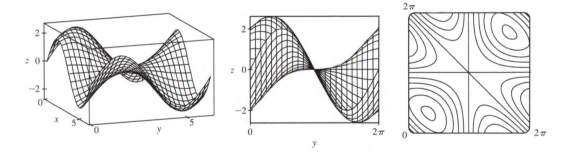

From the graphs it appears that f has a local maximum at about $(1, 1)$ with value approximately 2.6, a local minimum at about $(5, 5)$ with value approximately -2.6, and a saddle point at about $(3, 3)$.

$f_x = \cos x + \cos(x + y)$, $f_y = \cos y + \cos(x + y)$, $f_{xx} = -\sin x - \sin(x + y)$, $f_{yy} = -\sin y - \sin(x + y)$, $f_{xy} = -\sin(x + y)$. Setting $f_x = 0$ and $f_y = 0$ and subtracting gives $\cos x - \cos y = 0$ or $\cos x = \cos y$. Thus $x = y$ or $x = 2\pi - y$. If $x = y$, $f_x = 0$ becomes $\cos x + \cos 2x = 0$ or $2\cos^2 x + \cos x - 1 = 0$, a quadratic in $\cos x$. Thus $\cos x = -1$ or $\frac{1}{2}$ and $x = \pi$, $\frac{\pi}{3}$, or $\frac{5\pi}{3}$, yielding the critical points (π, π), $\left(\frac{\pi}{3}, \frac{\pi}{3}\right)$ and $\left(\frac{5\pi}{3}, \frac{5\pi}{3}\right)$. Similarly if $x = 2\pi - y$, $f_x = 0$ becomes $(\cos x) + 1 = 0$ and the resulting critical point is (π, π). Now $D(x, y) = \sin x \sin y + \sin x \sin(x + y) + \sin y \sin(x + y)$. So $D(\pi, \pi) = 0$ and the Second Derivatives Test doesn't apply. $D\left(\frac{\pi}{3}, \frac{\pi}{3}\right) = \frac{9}{4} > 0$ and $f_{xx}\left(\frac{\pi}{3}, \frac{\pi}{3}\right) < 0$ so $f\left(\frac{\pi}{3}, \frac{\pi}{3}\right) = \frac{3\sqrt{3}}{2}$ is a local maximum while $D\left(\frac{5\pi}{3}, \frac{5\pi}{3}\right) = \frac{9}{4} > 0$ and $f_{xx}\left(\frac{5\pi}{3}, \frac{5\pi}{3}\right) > 0$, so $f\left(\frac{5\pi}{3}, \frac{5\pi}{3}\right) = -\frac{3\sqrt{3}}{2}$ is a local minimum.

21. $f(x, y) = x^4 - 5x^2 + y^2 + 3x + 2 \implies f_x(x, y) = 4x^3 - 10x + 3$ and $f_y(x, y) = 2y$. $f_y = 0 \implies y = 0$, and the graph

of f_x shows that the roots of $f_x = 0$ are approximately $x = -1.714, 0.312$ and 1.402. (Alternatively, we could have used a

calculator or a CAS to find these roots.) So to three decimal places, the critical points are $(-1.714, 0)$, $(1.402, 0)$, and

$(0.312, 0)$. Now since $f_{xx} = 12x^2 - 10$, $f_{xy} = 0$, $f_{yy} = 2$, and $D = 24x^2 - 20$, we have $D(-1.714, 0) > 0$,

$f_{xx}(-1.714, 0) > 0$, $D(1.402, 0) > 0$, $f_{xx}(1.402, 0) > 0$, and $D(0.312, 0) < 0$. Therefore $f(-1.714, 0) \approx -9.200$ and

$f(1.402, 0) \approx 0.242$ are local minima, and $(0.312, 0)$ is a saddle point. The lowest point on the graph is approximately

$(-1.714, 0, -9.200)$.

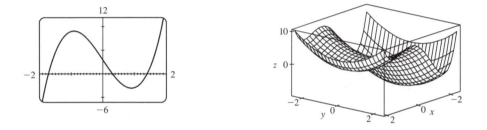

23. $f(x, y) = 2x + 4x^2 - y^2 + 2xy^2 - x^4 - y^4 \implies f_x(x, y) = 2 + 8x + 2y^2 - 4x^3$, $f_y(x, y) = -2y + 4xy - 4y^3$. Now

$f_y = 0 \iff 2y(2y^2 - 2x + 1) = 0 \iff y = 0$ or $y^2 = x - \frac{1}{2}$. The first of these implies that $f_x = -4x^3 + 8x + 2$, and

the second implies that $f_x = 2 + 8x + 2\left(x - \frac{1}{2}\right) - 4x^3 = -4x^3 + 10x + 1$. From the graphs, we see that the first possibility

for f_x has roots at approximately $-1.267, -0.259$, and 1.526, and the second has a root at approximately 1.629 (the negative

roots do not give critical points, since $y^2 = x - \frac{1}{2}$ must be positive). So to three decimal places, f has critical points at

$(-1.267, 0)$, $(-0.259, 0)$, $(1.526, 0)$, and $(1.629, \pm 1.063)$. Now since $f_{xx} = 8 - 12x^2$, $f_{xy} = 4y$, $f_{yy} = 4x - 12y^2$, and

$D = (8 - 12x^2)(4x - 12y^2) - 16y^2$, we have $D(-1.267, 0) > 0$, $f_{xx}(-1.267, 0) > 0$, $D(-0.259, 0) < 0$,

$D(1.526, 0) < 0$, $D(1.629, \pm 1.063) > 0$, and $f_{xx}(1.629, \pm 1.063) < 0$. Therefore, to three decimal places,

$f(-1.267, 0) \approx 1.310$ and $f(1.629, \pm 1.063) \approx 8.105$ are local maxima, and $(-0.259, 0)$ and $(1.526, 0)$ are saddle points.

The highest points on the graph are approximately $(1.629, \pm 1.063, 8.105)$.

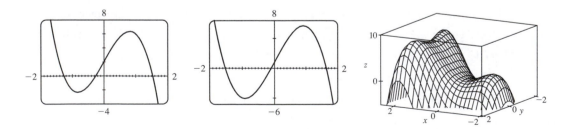

25. Since f is a polynomial it is continuous on D, so an absolute maximum and minimum exist. Here $f_x = 4$, $f_y = -5$ so there

are no critical points inside D. Thus the absolute extrema must both occur on the boundary. Along L_1, $x = 0$ and

$f(0, y) = 1 - 5y$ for $0 \le y \le 3$, a decreasing function in y, so the maximum value is $f(0, 0) = 1$ and the minimum value is

$f(0, 3) = -14$. Along L_2, $y = 0$ and $f(x, 0) = 1 + 4x$ for $0 \le x \le 2$, an increasing

function in x, so the minimum value is $f(0, 0) = 1$ and the maximum value

is $f(2, 0) = 9$. Along L_3, $y = -\frac{3}{2}x + 3$ and $f\left(x, -\frac{3}{2}x + 3\right) = \frac{23}{2}x - 14$

for $0 \le x \le 2$, an increasing function in x, so the minimum value is

$f(0, 3) = -14$ and the maximum value is $f(2, 0) = 9$. Thus the absolue

maximum of f on D is $f(2, 0) = 9$ and the absolute minimum is

$f(0, 3) = -14$.

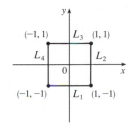

27. $f_x(x, y) = 2x + 2xy$, $f_y(x, y) = 2y + x^2$, and setting $f_x = f_y = 0$

gives $(0, 0)$ as the only critical point in D, with $f(0, 0) = 4$.

On L_1: $y = -1$, $f(x, -1) = 5$, a constant.

On L_2: $x = 1$, $f(1, y) = y^2 + y + 5$, a quadratic in y which attains its maximum

at $(1, 1)$, $f(1, 1) = 7$ and its minimum at $\left(1, -\frac{1}{2}\right)$, $f\left(1, -\frac{1}{2}\right) = \frac{19}{4}$.

On L_3: $f(x, 1) = 2x^2 + 5$ which attains its maximum at $(-1, 1)$ and $(1, 1)$ with

$f(\pm 1, 1) = 7$ and its minimum at $(0, 1)$, $f(0, 1) = 5$.

On L_4: $f(-1, y) = y^2 + y + 5$ with maximum at $(-1, 1)$, $f(-1, 1) = 7$ and minimum at $\left(-1, -\frac{1}{2}\right)$, $f\left(-1, -\frac{1}{2}\right) = \frac{19}{4}$.

Thus the absolute maximum is attained at both $(\pm 1, 1)$ with $f(\pm 1, 1) = 7$ and the absolute minimum on D is attained at

$(0, 0)$ with $f(0, 0) = 4$.

29. $f(x, y) = x^4 + y^4 - 4xy + 2$ is a polynomial and hence continuous on D,

so it has an absolute maximum and minimum on D. In Exercise 7, we found

the critical points of f; only $(1, 1)$ with $f(1, 1) = 0$ is inside D. On L_1:

$y = 0$, $f(x, 0) = x^4 + 2$, $0 \le x \le 3$, a polynomial in x which attains its

maximum at $x = 3$, $f(3, 0) = 83$, and its minimum at $x = 0$, $f(0, 0) = 2$.

On L_2: $x = 3$, $f(3, y) = y^4 - 12y + 83$, $0 \le y \le 2$, a polynomial in y

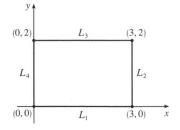

which attains its minimum at $y = \sqrt[3]{3}$, $f(3, \sqrt[3]{3}) = 83 - 9\sqrt[3]{3} \approx 70.0$, and its maximum at $y = 0$, $f(3, 0) = 83$. On L_3:

$y = 2$, $f(x, 2) = x^4 - 8x + 18$, $0 \le x \le 3$, a polynomial in x which attains its minimum at $x = \sqrt[3]{2}$,

$f(\sqrt[3]{2}, 2) = 18 - 6\sqrt[3]{2} \approx 10.4$, and its maximum at $x = 3$, $f(3, 2) = 75$. On L_4: $x = 0$, $f(0, y) = y^4 + 2$, $0 \le y \le 2$, a

polynomial in y which attains its maximum at $y = 2$, $f(0, 2) = 18$, and its minimum at $y = 0$, $f(0, 0) = 2$. Thus the absolute

maximum of f on D is $f(3, 0) = 83$ and the absolute minimum is $f(1, 1) = 0$.

31. $f(x, y) = -(x^2 - 1)^2 - (x^2 y - x - 1)^2 \Rightarrow f_x(x, y) = -2(x^2 - 1)(2x) - 2(x^2 y - x - 1)(2xy - 1)$ and

$f_y(x, y) = -2(x^2 y - x - 1)x^2$. Setting $f_y(x, y) = 0$ gives either $x = 0$ or $x^2 y - x - 1 = 0$. There are no critical points for

$x = 0$, since $f_x(0, y) = -2$, so we set $x^2 y - x - 1 = 0 \Leftrightarrow y = \dfrac{x + 1}{x^2}$ $(x \neq 0)$, so

$f_x\left(x, \dfrac{x+1}{x^2}\right) = -2(x^2 - 1)(2x) - 2\left(x^2 \dfrac{x+1}{x^2} - x - 1\right)\left(2x \dfrac{x+1}{x^2} - 1\right) = -4x(x^2 - 1)$. Therefore

$f_x(x, y) = f_y(x, y) = 0$ at the points $(1, 2)$ and $(-1, 0)$. To classify these critical points, we calculate

$f_{xx}(x, y) = -12x^2 - 12x^2 y^2 + 12xy + 4y + 2,\ f_{yy}(x, y) = -2x^4$,

and $f_{xy}(x, y) = -8x^3 y + 6x^2 + 4x$. In order to use the Second Derivatives

Test we calculate

$D(-1, 0) = f_{xx}(-1, 0)\,f_{yy}(-1, 0) - [f_{xy}(-1, 0)]^2 = 16 > 0$,

$f_{xx}(-1, 0) = -10 < 0,\ D(1, 2) = 16 > 0$, and $f_{xx}(1, 2) = -26 < 0$, so

both $(-1, 0)$ and $(1, 2)$ give local maxima.

33. Let d be the distance from $(2, 1, -1)$ to any point (x, y, z) on the plane $x + y - z = 1$, so

$d = \sqrt{(x-2)^2 + (y-1)^2 + (z+1)^2}$ where $z = x + y - 1$, and we minimize

$d^2 = f(x, y) = (x-2)^2 + (y-1)^2 + (x+y)^2$. Then $f_x(x, y) = 2(x-2) + 2(x+y) = 4x + 2y - 4$,

$f_y(x, y) = 2(y-1) + 2(x+y) = 2x + 4y - 2$. Solving $4x + 2y - 4 = 0$ and $2x + 4y - 2 = 0$ simultaneously gives $x = 1$,

$y = 0$. An absolute minimum exists (since there is a minimum distance from the point to the plane) and it must occur at a

critical point, so the shortest distance occurs for $x = 1$, $y = 0$ for which $d = \sqrt{(1-2)^2 + (0-1)^2 + (1+0)^2} = \sqrt{3}$.

35. Let d be the distance from the point $(4, 2, 0)$ to any point (x, y, z) on the cone, so $d = \sqrt{(x-4)^2 + (y-2)^2 + z^2}$ where

$z^2 = x^2 + y^2$, and we minimize $d^2 = (x-4)^2 + (y-2)^2 + x^2 + y^2 = f(x, y)$. Then

$f_x(x, y) = 2(x-4) + 2x = 4x - 8,\ f_y(x, y) = 2(y-2) + 2y = 4y - 4$, and the critical points occur when

$f_x = 0 \Rightarrow x = 2,\ f_y = 0 \Rightarrow y = 1$. Thus the only critical point is $(2, 1)$. An absolute minimum exists (since there

is a minimum distance from the cone to the point) which must occur at a critical point, so the points on the cone closest to

$(4, 2, 0)$ are $\left(2, 1, \pm\sqrt{5}\right)$.

37. $x + y + z = 100$, so maximize $f(x, y) = xy(100 - x - y)$. $f_x = 100y - 2xy - y^2,\ f_y = 100x - x^2 - 2xy,\ f_{xx} = -2y$,

$f_{yy} = -2x,\ f_{xy} = 100 - 2x - 2y$. Then $f_x = 0$ implies $y = 0$ or $y = 100 - 2x$. Substituting $y = 0$ into $f_y = 0$ gives

$x = 0$ or $x = 100$ and substituting $y = 100 - 2x$ into $f_y = 0$ gives $3x^2 - 100x = 0$ so $x = 0$ or $\frac{100}{3}$. Thus the critical points

are $(0, 0)$, $(100, 0)$, $(0, 100)$ and $\left(\frac{100}{3}, \frac{100}{3}\right)$.

$D(0, 0) = D(100, 0) = D(0, 100) = -10,000$ while $D\left(\frac{100}{3}, \frac{100}{3}\right) = \frac{10,000}{3}$ and $f_{xx}\left(\frac{100}{3}, \frac{100}{3}\right) = -\frac{200}{3} < 0$. Thus $(0, 0)$,

$(100, 0)$ and $(0, 100)$ are saddle points whereas $f\left(\frac{100}{3}, \frac{100}{3}\right)$ is a local maximum. Thus the numbers are $x = y = z = \frac{100}{3}$.

39. Maximize $f(x, y) = xy(36 - 9x^2 - 36y^2)^{1/2}/2$ with (x, y, z) in first octant. Then

$$f_x = \frac{y(36 - 9x^2 - 36y^2)^{1/2}}{2} + \frac{-9x^2y(36 - 9x^2 - 36y^2)^{-1/2}}{2} = \frac{(36y - 18x^2y - 36y^3)}{2(36 - 9x^2 - 36y^2)^{1/2}} \text{ and}$$

$f_y = \dfrac{36x - 9x^3 - 72xy^2}{2(36 - 9x^2 - 36y^2)^{1/2}}$. Setting $f_x = 0$ gives $y = 0$ or $y^2 = \dfrac{2 - x^2}{2}$ but $y > 0$, so only the latter solution applies.

Substituting this y into $f_y = 0$ gives $x^2 = \frac{4}{3}$ or $x = \frac{2}{\sqrt{3}}$, $y = \frac{1}{\sqrt{3}}$ and then $z^2 = (36 - 12 - 12)/4 = 3$. The fact

that this gives a maximum volume follows from the geometry. This maximum volume is

$$V = (2x)(2y)(2z) = 8\left(\tfrac{2}{\sqrt{3}}\right)\left(\tfrac{1}{\sqrt{3}}\right)(\sqrt{3}) = \tfrac{16}{\sqrt{3}}.$$

41. Maximize $f(x, y) = \dfrac{xy}{3}(6 - x - 2y)$, then the maximum volume is $V = xyz$. $f_x = \frac{1}{3}(6y - 2xy - y^2) = \frac{1}{3}y(6 - 2x - 2y)$

and $f_y = \frac{1}{3}x(6 - x - 4y)$. Setting $f_x = 0$ and $f_y = 0$ gives the critical point $(2, 1)$ which geometrically must yield a

maximum. Thus the volume of the largest such box is $V = (2)(1)\left(\frac{2}{3}\right) = \frac{4}{3}$.

43. Let the dimensions be x, y, and z; then $4x + 4y + 4z = c$ and the volume is

$$V = xyz = xy(\tfrac{1}{4}c - x - y) = \tfrac{1}{4}cxy - x^2y - xy^2, x > 0, y > 0. \text{ Then } V_x = \tfrac{1}{4}cy - 2xy - y^2 \text{ and } V_y = \tfrac{1}{4}cx - x^2 - 2xy,$$

so $V_x = 0 = V_y$ when $2x + y = \frac{1}{4}c$ and $x + 2y = \frac{1}{4}c$. Solving, we get $x = \frac{1}{12}c$, $y = \frac{1}{12}c$ and $z = \frac{1}{4}c - x - y = \frac{1}{12}c$. From

the geometrical nature of the problem, this critical point must give an absolute maximum. Thus the box is a cube with edge

length $\frac{1}{12}c$.

45. Let the dimensions be x, y and z, then minimize $xy + 2(xz + yz)$ if $xyz = 32{,}000$ m³. Then

$f(x, y) = xy + [64{,}000(x + y)/xy] = xy + 64{,}000(x^{-1} + y^{-1})$, $f_x = y - 64{,}000x^{-2}$, $f_y = x - 64{,}000y^{-2}$. And $f_x = 0$

implies $y = 64{,}000/x^2$; substituting into $f_y = 0$ implies $x^3 = 64{,}000$ or $x = 40$ and then $y = 40$. Now

$D(x, y) = [(2)(64{,}000)]^2 x^{-3}y^{-3} - 1 > 0$ for $(40, 40)$ and $f_{xx}(40, 40) > 0$ so this is indeed a minimum. Thus the

dimensions of the box are $x = y = 40$ cm, $z = 20$ cm.

47. Let x, y, z be the dimensions of the rectangular box. Then the volume of the box is xyz and

$L = \sqrt{x^2 + y^2 + z^2} \;\Rightarrow\; L^2 = x^2 + y^2 + z^2 \;\Rightarrow\; z = \sqrt{L^2 - x^2 - y^2}$. Substituting, we have volume
$V(x, y) = xy\sqrt{L^2 - x^2 - y^2}$, $x, y > 0$.

$$V_x = xy \cdot \tfrac{1}{2}(L^2 - x^2 - y^2)^{-1/2}(-2x) + y\sqrt{L^2 - x^2 - y^2} = y\sqrt{L^2 - x^2 - y^2} - \frac{x^2y}{\sqrt{L^2 - x^2 - y^2}},$$

$$V_y = x\sqrt{L^2 - x^2 - y^2} - \frac{xy^2}{\sqrt{L^2 - x^2 - y^2}},$$

$V_x = 0$ implies $y(L^2 - x^2 - y^2) = x^2y \;\Rightarrow\; y(L^2 - 2x^2 - y^2) = 0 \;\Rightarrow\; 2x^2 + y^2 = L^2$ (since $y > 0$), and $V_y = 0$

implies $x(L^2 - x^2 - y^2) = xy^2 \;\Rightarrow\; x(L^2 - x^2 - 2y^2) = 0 \;\Rightarrow\; x^2 + 2y^2 = L^2$ (since $x > 0$). Substituting

$y^2 = L^2 - 2x^2$ into $x^2 + 2y^2 = L^2$ gives $x^2 + 2L^2 - 4x^2 = L^2 \;\Rightarrow\; 3x^2 = L^2 \;\Rightarrow\; x = L/\sqrt{3}$ (since $x > 0$) and then

$y = \sqrt{L^2 - 2(L/\sqrt{3})^2} = L/\sqrt{3}$. So the only critical point is $(L/\sqrt{3}, L/\sqrt{3})$ which, from the geometrical nature

of the problem, must give an absolute maximum. Thus the maximum volume is

$$V(L/\sqrt{3}, L/\sqrt{3}) = (L/\sqrt{3})^2 \sqrt{L^2 - (L/\sqrt{3})^2 - (L/\sqrt{3})^2} = L^3/(3\sqrt{3}) \text{ cubic units.}$$

49. Note that here the variables are m and b, and $f(m, b) = \sum\limits_{i=1}^{n} [y_i - (mx_i + b)]^2$. Then $f_m = \sum\limits_{i=1}^{n} -2x_i[y_i - (mx_i + b)] = 0$

implies $\sum\limits_{i=1}^{n} (x_i y_i - mx_i^2 - bx_i) = 0$ or $\sum\limits_{i=1}^{n} x_i y_i = m \sum\limits_{i=1}^{n} x_i^2 + b \sum\limits_{i=1}^{n} x_i$ and $f_b = \sum\limits_{i=1}^{n} -2[y_i - (mx_i + b)] = 0$ implies

$\sum\limits_{i=1}^{n} y_i = m \sum\limits_{i=1}^{n} x_i + \sum\limits_{i=1}^{n} b = m \left(\sum\limits_{i=1}^{n} x_i \right) + nb$. Thus we have the two desired equations. Now

$f_{mm} = \sum\limits_{i=1}^{n} 2x_i^2$, $f_{bb} = \sum\limits_{i=1}^{n} 2 = 2n$ and $f_{mb} = \sum\limits_{i=1}^{n} 2x_i$. And $f_{mm}(m, b) > 0$ always and

$D(m, b) = 4n \left(\sum\limits_{i=1}^{n} x_i^2 \right) - 4 \left(\sum\limits_{i=1}^{n} x_i \right)^2 = 4 \left[n \left(\sum\limits_{i=1}^{n} x_i^2 \right) - \left(\sum\limits_{i=1}^{n} x_i \right)^2 \right] > 0$ always so the solutions of these two

equations do indeed minimize $\sum\limits_{i=1}^{n} d_i^2$.

11.8 Lagrange Multipliers

1. At the extreme values of f, the level curves of f just touch the curve $g(x, y) = 8$ with a common tangent line. (See Figure 1

and the accompanying discussion.) We can observe several such occurrences on the contour map, but the level curve

$f(x, y) = c$ with the largest value of c which still intersects the curve $g(x, y) = 8$ is approximately $c = 59$, and the smallest

value of c corresponding to a level curve which intersects $g(x, y) = 8$ appears to be $c = 30$. Thus we estimate the maximum

value of f subject to the constraint $g(x, y) = 8$ to be about 59 and the minimum to be 30.

3. $f(x, y) = x^2 + y^2$, $g(x, y) = xy = 1$, and $\nabla f = \lambda \nabla g$ ⇒ $\langle 2x, 2y \rangle = \langle \lambda y, \lambda x \rangle$, so $2x = \lambda y$, $2y = \lambda x$, and $xy = 1$.

From the last equation, $x \neq 0$ and $y \neq 0$, so $2x = \lambda y$ ⇒ $\lambda = 2x/y$. Substituting, we have

$2y = (2x/y) x$ ⇒ $y^2 = x^2$ ⇒ $y = \pm x$. But $xy = 1$, so $x = y = \pm 1$ and the possible points for the extreme values

of f are $(1, 1)$ and $(-1, -1)$. Here there is no maximum value, since the constraint $xy = 1$ allows x or y to become arbitrarily

large, and hence $f(x, y) = x^2 + y^2$ can be made arbitrarily large. The minimum value is $f(1, 1) = f(-1, -1) = 2$.

5. $f(x, y) = x^2 y$, $g(x, y) = x^2 + 2y^2 = 6$ ⇒ $\nabla f = \langle 2xy, x^2 \rangle$, $\lambda \nabla g = \langle 2\lambda x, 4\lambda y \rangle$. Then $2xy = 2\lambda x$ implies $x = 0$ or

$\lambda = y$. If $x = 0$, then $x^2 = 4\lambda y$ implies $\lambda = 0$ or $y = 0$. However, if $y = 0$ then $g(x, y) = 0$, a contradiction. So $\lambda = 0$ and

then $g(x, y) = 6$ ⇒ $y = \pm\sqrt{3}$. If $\lambda = y$, then $x^2 = 4\lambda y$ implies $x^2 = 4y^2$, and so $g(x, y) = 6$ ⇒ $4y^2 + 2y^2 = 6$

⇒ $y^2 = 1$ ⇒ $y = \pm 1$. Thus f has possible extreme values at the points $(0, \pm\sqrt{3})$, $(\pm 2, 1)$, and $(\pm 2, -1)$. After

evaluating f at these points, we find the maximum value to be $f(\pm 2, 1) = 4$ and the minimum to be $f(\pm 2, -1) = -4$.

7. $f(x, y, z) = 2x + 6y + 10z$, $g(x, y, z) = x^2 + y^2 + z^2 = 35$ ⇒ $\nabla f = \langle 2, 6, 10 \rangle$, $\lambda \nabla g = \langle 2\lambda x, 2\lambda y, 2\lambda z \rangle$. Then

$2\lambda x = 2$, $2\lambda y = 6$, $2\lambda z = 10$ imply $x = \dfrac{1}{\lambda}$, $y = \dfrac{3}{\lambda}$, and $z = \dfrac{5}{\lambda}$. But $35 = x^2 + y^2 + z^2 = \left(\dfrac{1}{\lambda} \right)^2 + \left(\dfrac{3}{\lambda} \right)^2 + \left(\dfrac{5}{\lambda} \right)^2$ ⇒

$35 = \dfrac{35}{\lambda^2}$ ⇒ $\lambda = \pm 1$, so f has possible extreme values at the points $(1, 3, 5)$, $(-1, -3, -5)$. The maximum value of f on

$x^2 + y^2 + z^2 = 35$ is $f(1, 3, 5) = 70$, and the minimum is $f(-1, -3, -5) = -70$.

9. $f(x, y, z) = xyz, g(x, y, z) = x^2 + 2y^2 + 3z^2 = 6 \Rightarrow \nabla f = \langle yz, xz, xy \rangle, \lambda \nabla g = \langle 2\lambda x, 4\lambda y, 6\lambda z \rangle$. Then $\nabla f = \lambda \nabla g$

implies $\lambda = (yz)/(2x) = (xz)/(4y) = (xy)/(6z)$ or $x^2 = 2y^2$ and $z^2 = \frac{2}{3}y^2$. Thus $x^2 + 2y^2 + 3z^2 = 6$ implies $6y^2 = 6$

or $y = \pm 1$. Then the possible points are $\left(\sqrt{2}, \pm 1, \sqrt{\frac{2}{3}}\right), \left(\sqrt{2}, \pm 1, -\sqrt{\frac{2}{3}}\right), \left(-\sqrt{2}, \pm 1, \sqrt{\frac{2}{3}}\right), \left(-\sqrt{2}, \pm 1, -\sqrt{\frac{2}{3}}\right)$. The

maximum value of f on the ellipsoid is $\frac{2}{\sqrt{3}}$, occurring when all coordinates are positive or exactly two are negative and the

minimum is $-\frac{2}{\sqrt{3}}$ occurring when 1 or 3 of the coordinates are negative.

11. $f(x, y, z) = x^2 + y^2 + z^2, g(x, y, z) = x^4 + y^4 + z^4 = 1 \Rightarrow \nabla f = \langle 2x, 2y, 2z \rangle, \lambda \nabla g = \langle 4\lambda x^3, 4\lambda y^3, 4\lambda z^3 \rangle$.

Case 1: If $x \neq 0, y \neq 0$ and $z \neq 0$, then $\nabla f = \lambda \nabla g$ implies $\lambda = 1/(2x^2) = 1/(2y^2) = 1/(2z^2)$ or $x^2 = y^2 = z^2$ and

$3x^4 = 1$ or $x = \pm\frac{1}{\sqrt[4]{3}}$ giving the points $\left(\pm\frac{1}{\sqrt[4]{3}}, \frac{1}{\sqrt[4]{3}}, \frac{1}{\sqrt[4]{3}}\right), \left(\pm\frac{1}{\sqrt[4]{3}}, -\frac{1}{\sqrt[4]{3}}, \frac{1}{\sqrt[4]{3}}\right), \left(\pm\frac{1}{\sqrt[4]{3}}, \frac{1}{\sqrt[4]{3}}, -\frac{1}{\sqrt[4]{3}}\right), \left(\pm\frac{1}{\sqrt[4]{3}}, -\frac{1}{\sqrt[4]{3}}, -\frac{1}{\sqrt[4]{3}}\right)$

all with an f-value of $\sqrt{3}$.

Case 2: If one of the variables equals zero and the other two are not zero, then the squares of the two nonzero coordinates are

equal with common value $\frac{1}{\sqrt{2}}$ and corresponding f value of $\sqrt{2}$.

Case 3: If exactly two of the variables are zero, then the third variable has value ± 1 with the corresponding f value of 1. Thus

on $x^4 + y^4 + z^4 = 1$, the maximum value of f is $\sqrt{3}$ and the minimum value is 1.

13. $f(x, y, z, t) = x + y + z + t, g(x, y, z, t) = x^2 + y^2 + z^2 + t^2 = 1 \Rightarrow \langle 1, 1, 1, 1 \rangle = \langle 2\lambda x, 2\lambda y, 2\lambda z, 2\lambda t \rangle$, so

$\lambda = 1/(2x) = 1/(2y) = 1/(2z) = 1/(2t)$ and $x = y = z = t$. But $x^2 + y^2 + z^2 + t^2 = 1$, so the possible points are

$\left(\pm\frac{1}{2}, \pm\frac{1}{2}, \pm\frac{1}{2}, \pm\frac{1}{2}\right)$. Thus the maximum value of f is $f\left(\frac{1}{2}, \frac{1}{2}, \frac{1}{2}, \frac{1}{2}\right) = 2$ and the minimum value is

$f\left(-\frac{1}{2}, -\frac{1}{2}, -\frac{1}{2}, -\frac{1}{2}\right) = -2$.

15. $f(x, y, z) = x + 2y, g(x, y, z) = x + y + z = 1, h(x, y, z) = y^2 + z^2 = 4 \Rightarrow \nabla f = \langle 1, 2, 0 \rangle, \lambda \nabla g = \langle \lambda, \lambda, \lambda \rangle$ and

$\mu \nabla h = \langle 0, 2\mu y, 2\mu z \rangle$. Then $1 = \lambda, 2 = \lambda + 2\mu y$ and $0 = \lambda + 2\mu z$ so $\mu y = \frac{1}{2} = -\mu z$ or $y = 1/(2\mu), z = -1/(2\mu)$. Thus

$x + y + z = 1$ implies $x = 1$ and $y^2 + z^2 = 4$ implies $\mu = \pm\frac{1}{2\sqrt{2}}$. Then the possible points are $\left(1, \pm\sqrt{2}, \mp\sqrt{2}\right)$ and the

maximum value is $f\left(1, \sqrt{2}, -\sqrt{2}\right) = 1 + 2\sqrt{2}$ and the minimum value is $f\left(1, -\sqrt{2}, \sqrt{2}\right) = 1 - 2\sqrt{2}$.

17. $f(x, y, z) = yz + xy, g(x, y, z) = xy = 1, h(x, y, z) = y^2 + z^2 = 1 \Rightarrow \nabla f = \langle y, x + z, y \rangle, \lambda \nabla g = \langle \lambda y, \lambda x, 0 \rangle$,

$\mu \nabla h = \langle 0, 2\mu y, 2\mu z \rangle$. Then $y = \lambda y$ implies $\lambda = 1$ [$y \neq 0$ since $g(x, y, z) = 1$], $x + z = \lambda x + 2\mu y$ and $y = 2\mu z$. Thus

$\mu = z/(2y) = y/(2y)$ or $y^2 = z^2$, and so $y^2 + z^2 = 1$ implies $y = \pm\frac{1}{\sqrt{2}}, z = \pm\frac{1}{\sqrt{2}}$. Then $xy = 1$ implies $x = \pm\sqrt{2}$ and

the possible points are $\left(\pm\sqrt{2}, \pm\frac{1}{\sqrt{2}}, \frac{1}{\sqrt{2}}\right), \left(\pm\sqrt{2}, \pm\frac{1}{\sqrt{2}}, -\frac{1}{\sqrt{2}}\right)$. Hence the maximum of f subject to the constraints is

$f\left(\pm\sqrt{2}, \pm\frac{1}{\sqrt{2}}, \pm\frac{1}{\sqrt{2}}\right) = \frac{3}{2}$ and the minimum is $f\left(\pm\sqrt{2}, \pm\frac{1}{\sqrt{2}}, \mp\frac{1}{\sqrt{2}}\right) = \frac{1}{2}$.

Note: Since $xy = 1$ is one of the constraints we could have solved the problem by solving $f(y, z) = yz + 1$ subject to

$y^2 + z^2 = 1$.

19. $f(x, y) = e^{-xy}$. For the interior of the region, we find the critical points: $f_x = -ye^{-xy}$, $f_y = -xe^{-xy}$, so the only critical

point is $(0, 0)$, and $f(0, 0) = 1$. For the boundary, we use Lagrange multipliers. $g(x, y) = x^2 + 4y^2 = 1$ $\Rightarrow$

$\lambda \nabla g = \langle 2\lambda x, 8\lambda y \rangle$, so setting $\nabla f = \lambda \nabla g$ we get $-ye^{-xy} = 2\lambda x$ and $-xe^{-xy} = 8\lambda y$. The first of these gives

$e^{-xy} = -2\lambda x/y$, and then the second gives $-x(-2\lambda x/y) = 8\lambda y$ $\Rightarrow$ $x^2 = 4y^2$. Solving this last equation with the

constraint $x^2 + 4y^2 = 1$ gives $x = \pm\frac{1}{\sqrt{2}}$ and $y = \pm\frac{1}{2\sqrt{2}}$. Now $f\left(\pm\frac{1}{\sqrt{2}}, \mp\frac{1}{2\sqrt{2}}\right) = e^{1/4} \approx 1.284$ and

$f\left(\pm\frac{1}{\sqrt{2}}, \pm\frac{1}{2\sqrt{2}}\right) = e^{-1/4} \approx 0.779$. The former are the maxima on the region and the latter are the minima.

21. $P(L, K) = bL^\alpha K^{1-\alpha}$, $g(L, K) = mL + nK = p$ $\Rightarrow$ $\nabla P = \langle \alpha bL^{\alpha-1}K^{1-\alpha}, (1-\alpha)bL^\alpha K^{-\alpha}\rangle$, $\lambda \nabla g = \langle \lambda m, \lambda n \rangle$.

Then $\alpha b(K/L)^{1-\alpha} = \lambda m$ and $(1-\alpha)b(L/K)^\alpha = \lambda n$ and $mL + nK = p$, so $\alpha b(K/L)^{1-\alpha}/m = (1-\alpha)b(L/K)^\alpha/n$ or

$n\alpha/[m(1-\alpha)] = (L/K)^\alpha(L/K)^{1-\alpha}$ or $L = Kn\alpha/[m(1-\alpha)]$. Substituting into $mL + nK = p$ gives $K = (1-\alpha)p/n$

and $L = \alpha p/m$ for the maximum production.

23. Let the sides of the rectangle be x and y. Then $f(x, y) = xy$, $g(x, y) = 2x + 2y = p$ $\Rightarrow$ $\nabla f(x, y) = \langle y, x \rangle$,

$\lambda \nabla g = \langle 2\lambda, 2\lambda \rangle$. Then $\lambda = \frac{1}{2}y = \frac{1}{2}x$ implies $x = y$ and the rectangle with maximum area is a square with side length $\frac{1}{4}p$.

25. Let $f(x, y, z) = d^2 = (x-2)^2 + (y-1)^2 + (z+1)^2$, then we want to minimize f subject to the constraint

$g(x, y, z) = x + y - z = 1$. $\nabla f = \lambda \nabla g$ $\Rightarrow$ $\langle 2(x-2), 2(y-1), 2(z+1) \rangle = \lambda \langle 1, 1, -1 \rangle$, so $x = (\lambda+4)/2$,

$y = (\lambda+2)/2$, $z = -(\lambda+2)/2$. Substituting into the constraint equation gives $\dfrac{\lambda+4}{2} + \dfrac{\lambda+2}{2} + \dfrac{\lambda+2}{2} = 1$ $\Rightarrow$

$3\lambda + 8 = 2$ $\Rightarrow$ $\lambda = -2$, so $x = 1$, $y = 0$, and $z = 0$. This must correspond to a minimum, so the shortest distance is

$d = \sqrt{(1-2)^2 + (0-1)^2 + (0+1)^2} = \sqrt{3}$.

27. Let $f(x, y, z) = d^2 = (x-4)^2 + (y-2)^2 + z^2$. Then we want to minimize f subject to the constraint

$g(x, y, z) = x^2 + y^2 - z^2 = 0$. $\nabla f = \lambda \nabla g$ $\Rightarrow$ $\langle 2(x-4), 2(y-2), 2z \rangle = \langle 2\lambda x, 2\lambda y, -2\lambda z \rangle$, so $x - 4 = \lambda x$,

$y - 2 = \lambda y$, and $z = -\lambda z$. From the last equation we have $z + \lambda z = 0$ $\Rightarrow$ $z(1 + \lambda) = 0$, so either $z = 0$ or $\lambda = -1$.

But from the constraint equation we have $z = 0$ $\Rightarrow$ $x^2 + y^2 = 0$ $\Rightarrow$ $x = y = 0$ which is not possible from the first

two equations. So $\lambda = -1$ and $x - 4 = \lambda x$ $\Rightarrow$ $x = 2$, $y - 2 = \lambda y$ $\Rightarrow$ $y = 1$, and

$x^2 + y^2 - z^2 = 0$ $\Rightarrow$ $4 + 1 - z^2 = 0$ $\Rightarrow$ $z = \pm\sqrt{5}$. This must correspond to a minimum, so the points on the cone

closest to $(4, 2, 0)$ are $\left(2, 1, \pm\sqrt{5}\right)$.

29. $f(x, y, z) = xyz$, $g(x, y, z) = x + y + z = 100$ $\Rightarrow$ $\nabla f = \langle yz, xz, xy \rangle = \lambda \nabla g = \langle \lambda, \lambda, \lambda \rangle$. Then $\lambda = yz = xz = xy$

implies $x = y = z = \frac{100}{3}$.

31. If the dimensions are $2x$, $2y$ and $2z$, then $f(x, y, z) = 8xyz$ and $g(x, y, z) = 9x^2 + 36y^2 + 4z^2 = 36$ $\Rightarrow$

$\nabla f = \langle 8yz, 8xz, 8xy \rangle = \lambda \nabla g = \langle 18\lambda x, 72\lambda y, 8\lambda z \rangle$. Thus $18\lambda x = 8yz$, $72\lambda y = 8xz$, $8\lambda z = 8xy$ so $x^2 = 4y^2$, $z^2 = 9y^2$

and $36y^2 + 36y^2 + 36y^2 = 36$ or $y = \frac{1}{\sqrt{3}}$ ($y > 0$). Thus the volume of the largest such box is $8\left(\frac{1}{\sqrt{3}}\right)\left(\frac{2}{\sqrt{3}}\right)\left(\frac{3}{\sqrt{3}}\right) = 16/\sqrt{3}$.

33. $f(x, y, z) = xyz$, $g(x, y, z) = x + 2y + 3z = 6$ $\Rightarrow$ $\nabla f = \langle yz, xz, xy \rangle = \lambda \nabla g = \langle \lambda, 2\lambda, 3\lambda \rangle$.

Then $\lambda = yz = \frac{1}{2}xz = \frac{1}{3}xy$ implies $x = 2y$, $z = \frac{2}{3}y$. But $2y + 2y + 2y = 6$ so $y = 1$, $x = 2$, $z = \frac{2}{3}$ and the volume

is $V = \frac{4}{3}$.

35. $f(x, y, z) = xyz$, $g(x, y, z) = 4(x + y + z) = c$ $\Rightarrow$ $\nabla f = \langle yz, xz, xy \rangle$, $\lambda \nabla g = \langle 4\lambda, 4\lambda, 4\lambda \rangle$. Thus

$4\lambda = yz = xz = xy$ or $x = y = z = \frac{1}{12}c$ are the dimensions giving the maximum volume.

37. If the dimensions of the box are given by x, y, and z, then we need to find the maximum value of $f(x, y, z) = xyz$

$(x, y, z > 0)$ subject to the constraint $L = \sqrt{x^2 + y^2 + z^2}$ or $g(x, y, z) = x^2 + y^2 + z^2 = L^2$. $\nabla f = \lambda \nabla g$ $\Rightarrow$

$\langle yz, xz, xy \rangle = \lambda \langle 2x, 2y, 2z \rangle$, so $yz = 2\lambda x$ $\Rightarrow$ $\lambda = \dfrac{yz}{2x}$, $xz = 2\lambda y$ $\Rightarrow$ $\lambda = \dfrac{xz}{2y}$, and $xy = 2\lambda z$ $\Rightarrow$ $\lambda = \dfrac{xy}{2z}$.

Thus $\lambda = \dfrac{yz}{2x} = \dfrac{xz}{2y}$ $\Rightarrow$ $x^2 = y^2$ [since $z \neq 0$] $\Rightarrow$ $x = y$ and $\lambda = \dfrac{yz}{2x} = \dfrac{xy}{2z}$ $\Rightarrow$ $x = z$ [since $y \neq 0$].

Substituting into the constraint equation gives $x^2 + x^2 + x^2 = L^2$ $\Rightarrow$ $x^2 = L^2/3$ $\Rightarrow$ $x = L/\sqrt{3} = y = z$ and the

maximum volume is $\left(L/\sqrt{3}\right)^3 = L^3/\left(3\sqrt{3}\right)$.

39. We need to find the extreme values of $f(x, y, z) = x^2 + y^2 + z^2$ subject to the two constraints $g(x, y, z) = x + y + 2z = 2$

and $h(x, y, z) = x^2 + y^2 - z = 0$. $\nabla f = \langle 2x, 2y, 2z \rangle$, $\lambda \nabla g = \langle \lambda, \lambda, 2\lambda \rangle$ and $\mu \nabla h = \langle 2\mu x, 2\mu y, -\mu \rangle$. Thus we need

(1) $2x = \lambda + 2\mu x$, (2) $2y = \lambda + 2\mu y$, (3) $2z = 2\lambda - \mu$, (4) $x + y + 2z = 2$, and (5) $x^2 + y^2 - z = 0$. From (1) and (2),

$2(x - y) = 2\mu(x - y)$, so if $x \neq y$, $\mu = 1$. Putting this in (3) gives $2z = 2\lambda - 1$ or $\lambda = z + \frac{1}{2}$, but putting $\mu = 1$ into (1)

says $\lambda = 0$. Hence $z + \frac{1}{2} = 0$ or $z = -\frac{1}{2}$. Then (4) and (5) become $x + y - 3 = 0$ and $x^2 + y^2 + \frac{1}{2} = 0$. The last equation

cannot be true, so this case gives no solution. So we must have $x = y$. Then (4) and (5) become $2x + 2z = 2$ and $2x^2 - z = 0$

which imply $z = 1 - x$ and $z = 2x^2$. Thus $2x^2 = 1 - x$ or $2x^2 + x - 1 = (2x - 1)(x + 1) = 0$ so $x = \frac{1}{2}$ or $x = -1$. The

two points to check are $\left(\frac{1}{2}, \frac{1}{2}, \frac{1}{2}\right)$ and $(-1, -1, 2)$: $f\left(\frac{1}{2}, \frac{1}{2}, \frac{1}{2}\right) = \frac{3}{4}$ and $f(-1, -1, 2) = 6$. Thus $\left(\frac{1}{2}, \frac{1}{2}, \frac{1}{2}\right)$ is the point on the

ellipse nearest the origin and $(-1, -1, 2)$ is the one farthest from the origin.

41. $f(x, y, z) = ye^{x-z}$, $g(x, y, z) = 9x^2 + 4y^2 + 36z^2 = 36$, $h(x, y, z) = xy + yz = 1$. $\nabla f = \lambda \nabla g + \mu \nabla h$ $\Rightarrow$

$\left\langle ye^{x-z}, e^{x-z}, -ye^{x-z} \right\rangle = \lambda \langle 18x, 8y, 72z \rangle + \mu \langle y, x + z, y \rangle$, so $ye^{x-z} = 18\lambda x + \mu y$, $e^{x-z} = 8\lambda y + \mu(x + z)$,

$-ye^{x-z} = 72\lambda z + \mu y$, $9x^2 + 4y^2 + 36z^2 = 36$, $xy + yz = 1$. Using a CAS to solve these 5 equations simultaneously for x,

y, z, λ, and μ (in Maple, use the `allvalues` command), we get 4 real-valued solutions:

$$x \approx 0.222444, \quad y \approx -2.157012, \quad z \approx -0.686049, \quad \lambda \approx -0.200401, \quad \mu \approx 2.108584$$

$$x \approx -1.951921, \quad y \approx -0.545867, \quad z \approx 0.119973, \quad \lambda \approx 0.003141, \quad \mu \approx -0.076238$$

$$x \approx 0.155142, \quad y \approx 0.904622, \quad z \approx 0.950293, \quad \lambda \approx -0.012447, \quad \mu \approx 0.489938$$

$$x \approx 1.138731, \quad y \approx 1.768057, \quad z \approx -0.573138, \quad \lambda \approx 0.317141, \quad \mu \approx 1.862675$$

Substituting these values into f gives $f(0.222444, -2.157012, -0.686049) \approx -5.3506$,

$f(-1.951921, -0.545867, 0.119973) \approx -0.0688$, $f(0.155142, 0.904622, 0.950293) \approx 0.4084$,

$f(1.138731, 1.768057, -0.573138) \approx 9.7938$. Thus the maximum is approximately 9.7938, and the minimum is

approximately -5.3506.

43. (a) We wish to maximize $f(x_1, x_2, \ldots, x_n) = \sqrt[n]{x_1 x_2 \cdots x_n}$ subject to

$g(x_1, x_2, \ldots, x_n) = x_1 + x_2 + \cdots + x_n = c$ and $x_i > 0$.

$$\nabla f = \left\langle \frac{1}{n}(x_1 x_2 \cdots x_n)^{\frac{1}{n}-1}(x_2 \cdots x_n),\ \frac{1}{n}(x_1 x_2 \cdots x_n)^{\frac{1}{n}-1}(x_1 x_3 \cdots x_n),\ \ldots,\right.$$

$$\left. \frac{1}{n}(x_1 x_2 \cdots x_n)^{\frac{1}{n}-1}(x_1 \cdots x_{n-1}) \right\rangle$$

and $\lambda \nabla g = \langle \lambda, \lambda, \ldots, \lambda \rangle$, so we need to solve the system of equations

$$\frac{1}{n}(x_1 x_2 \cdots x_n)^{\frac{1}{n}-1}(x_2 \cdots x_n) = \lambda \quad \Rightarrow \quad x_1^{1/n} x_2^{1/n} \cdots x_n^{1/n} = n\lambda x_1$$

$$\frac{1}{n}(x_1 x_2 \cdots x_n)^{\frac{1}{n}-1}(x_1 x_3 \cdots x_n) = \lambda \quad \Rightarrow \quad x_1^{1/n} x_2^{1/n} \cdots x_n^{1/n} = n\lambda x_2$$

$$\vdots$$

$$\frac{1}{n}(x_1 x_2 \cdots x_n)^{\frac{1}{n}-1}(x_1 \cdots x_{n-1}) = \lambda \quad \Rightarrow \quad x_1^{1/n} x_2^{1/n} \cdots x_n^{1/n} = n\lambda x_n$$

This implies $n\lambda x_1 = n\lambda x_2 = \cdots = n\lambda x_n$. Note $\lambda \neq 0$, otherwise we can't have all $x_i > 0$. Thus $x_1 = x_2 = \cdots = x_n$.

But $x_1 + x_2 + \cdots + x_n = c \quad \Rightarrow \quad nx_1 = c \quad \Rightarrow \quad x_1 = \dfrac{c}{n} = x_2 = x_3 = \cdots = x_n$. Then the only point where f can

have an extreme value is $\left(\dfrac{c}{n}, \dfrac{c}{n}, \ldots, \dfrac{c}{n}\right)$. Since we can choose values for $(x_1, x_2, \ldots, x_n)$ that make f as close to

zero (but not equal) as we like, f has no minimum value. Thus the maximum value is

$$f\left(\frac{c}{n}, \frac{c}{n}, \ldots, \frac{c}{n}\right) = \sqrt[n]{\frac{c}{n} \cdot \frac{c}{n} \cdots \frac{c}{n}} = \frac{c}{n}.$$

(b) From part (a), $\dfrac{c}{n}$ is the maximum value of f. Thus $f(x_1, x_2, \ldots, x_n) = \sqrt[n]{x_1 x_2 \cdots x_n} \leq \dfrac{c}{n}$. But

$x_1 + x_2 + \cdots + x_n = c$, so $\sqrt[n]{x_1 x_2 \cdots x_n} \leq \dfrac{x_1 + x_2 + \cdots + x_n}{n}$. These two means are equal when f attains its

maximum value $\dfrac{c}{n}$, but this can occur only at the point $\left(\dfrac{c}{n}, \dfrac{c}{n}, \ldots, \dfrac{c}{n}\right)$ we found in part (a). So the means are equal only

when $x_1 = x_2 = x_3 = \cdots = x_n = \dfrac{c}{n}$.

11 Review

CONCEPT CHECK

1. (a) A function f of two variables is a rule that assigns to each ordered pair (x, y) of real numbers in its domain a unique real number denoted by $f(x, y)$.

(b) One way to visualize a function of two variables is by graphing it, resulting in the surface $z = f(x, y)$. Another method for visualizing a function of two variables is a contour map. The contour map consists of level curves of the function which are horizontal traces of the graph of the function projected onto the xy-plane.

2. A function f of three variables is a rule that assigns to each ordered triple (x, y, z) in its domain a unique real number $f(x, y, z)$. We can visualize a function of three variables by examining its level surfaces $f(x, y, z) = k$, where k is a constant.

3. $\lim\limits_{(x,y)\to(a,b)} f(x, y) = L$ means the values of $f(x, y)$ approach the number L as the point (x, y) approaches the point (a, b) along any path that is within the domain of f. We can show that a limit at a point does not exist by finding two different paths approaching the point along which $f(x, y)$ has different limits.

4. (a) See Definition 11.2.3.

(b) If f is continuous on $\mathbb{R}^2$, its graph will appear as a surface without holes or breaks.

5. (a) See (2) and (3) in Section 11.3.

  (b) See "Interpretations of Partial Derivatives" on page 759.

  (c) To find f_x, regard y as a constant and differentiate $f(x, y)$ with respect to x. To find f_y, regard x as a constant and differentiate $f(x, y)$ with respect to y.

6. See the statement of Clairaut's Theorem on page 763.

7. (a) See (2) in Section 11.4.

  (b) See (19) and the preceding discussion in Section 11.6.

  (c) See "Tangent Planes to Parametric Surfaces" on page 777.

8. See (3) and (4) and the accompanying discussion in Section 11.4. We can interpret the linearization of f at (a, b) geometrically as the linear function whose graph is the tangent plane to the graph of f at (a, b). Thus it is the linear function which best approximates f near (a, b).

9. (a) See Definition 11.4.7.

  (b) Use Theorem 11.4.8.

10. See (10) and the associated discussion in Section 11.4.

11. See (2) and (3) in Section 11.5.

12. See (7) and the preceding discussion in Section 11.5.

13. (a) See Definition 11.6.2. We can interpret it as the rate of change of f at (x_0, y_0) in the direction of $\mathbf{u}$. Geometrically, if P is the point $(x_0, y_0, f(x_0, y_0))$ on the graph of f and C is the curve of intersection of the graph of f with the vertical plane that passes through P in the direction $\mathbf{u}$, the directional derivative of f at (x_0, y_0) in the direction of $\mathbf{u}$ is the slope of the tangent line to C at P. (See Figure 5 in Section 11.6.)

  (b) See Theorem 11.6.3.

14. (a) See (8) and (13) in Section 11.6.

  (b) $D_{\mathbf{u}} f(x, y) = \nabla f(x, y) \cdot \mathbf{u}$ or $D_{\mathbf{u}} f(x, y, z) = \nabla f(x, y, z) \cdot \mathbf{u}$

  (c) The gradient vector of a function points in the direction of maximum rate of increase of the function. On a graph of the function, the gradient points in the direction of steepest ascent.

15. (a) f has a local maximum at (a, b) if $f(x, y) \le f(a, b)$ when (x, y) is near (a, b).

  (b) f has an absolute maximum at (a, b) if $f(x, y) \le f(a, b)$ for all points (x, y) in the domain of f.

  (c) f has a local minimum at (a, b) if $f(x, y) \ge f(a, b)$ when (x, y) is near (a, b).

  (d) f has an absolute minimum at (a, b) if $f(x, y) \ge f(a, b)$ for all points (x, y) in the domain of f.

  (e) f has a saddle point at (a, b) if $f(a, b)$ is a local maximum in one direction but a local minimum in another.

16. (a) By Theorem 11.7.2, if f has a local maximum at (a, b) and the first-order partial derivatives of f exist there, then $f_x(a, b) = 0$ and $f_y(a, b) = 0$.

  (b) A critical point of f is a point (a, b) such that $f_x(a, b) = 0$ and $f_y(a, b) = 0$ or one of these partial derivatives does not exist.

17. See (3) in Section 11.7.

18. (a) See Figure 11 and the accompanying discussion in Section 11.7.

  (b) See Theorem 11.7.8.

  (c) See the procedure outlined in (9) in Section 11.7.

19. See the discussion beginning on page 813; see "Two Constraints" on page 817.

TRUE-FALSE QUIZ

1. True. $f_y(a, b) = \lim\limits_{h \to 0} \dfrac{f(a, b+h) - f(a, b)}{h}$ from Equation 11.3.3. Let $h = y - b$. As $h \to 0$, $y \to b$. Then by substituting, we get $f_y(a, b) = \lim\limits_{y \to b} \dfrac{f(a, y) - f(a, b)}{y - b}$.

3. False. $f_{xy} = \dfrac{\partial^2 f}{\partial y \, \partial x}$.

5. False. See Example 11.2.3.

7. True. If f has a local minimum and f is differentiable at (a, b) then by Theorem 11.7.2, $f_x(a, b) = 0$ and $f_y(a, b) = 0$, so $\nabla f(a, b) = \langle f_x(a, b), f_y(a, b) \rangle = \langle 0, 0 \rangle = \mathbf{0}$.

9. False. $\nabla f(x, y) = \langle 0, 1/y \rangle$.

11. True. $\nabla f = \langle \cos x, \cos y \rangle$, so $|\nabla f| = \sqrt{\cos^2 x + \cos^2 y}$. But $|\cos \theta| \leq 1$, so $|\nabla f| \leq \sqrt{2}$. Now $D_{\mathbf{u}} f(x, y) = \nabla f \cdot \mathbf{u} = |\nabla f| \, |\mathbf{u}| \cos \theta$, but $\mathbf{u}$ is a unit vector, so $|D_{\mathbf{u}} f(x, y)| \leq \sqrt{2} \cdot 1 \cdot 1 = \sqrt{2}$.

EXERCISES

1. $\ln(x + y + 1)$ is defined only when

$x + y + 1 > 0 \quad \Rightarrow \quad y > -x - 1$, so the domain of

f is $\{(x, y) \mid y > -x - 1\}$, all those points above the

line $y = -x - 1$.

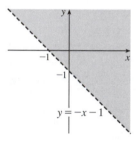

3. $z = f(x, y) = 1 - y^2$, a parabolic cylinder.

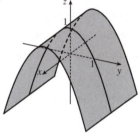

5. The level curves are $\sqrt{4x^2 + y^2} = k$ or

$4x^2 + y^2 = k^2$, $k \geq 0$, a family of ellipses.

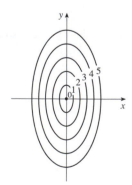

7.

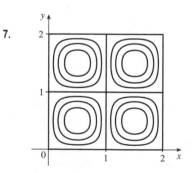

9. f is a rational function, so it is continuous on its domain. Since f is defined at $(1, 1)$, we use direct substitution to evaluate the

limit: $\displaystyle\lim_{(x,y)\to(1,1)} \frac{2xy}{x^2 + 2y^2} = \frac{2(1)(1)}{1^2 + 2(1)^2} = \frac{2}{3}$.

11. (a) $T_x(6, 4) = \displaystyle\lim_{h\to0} \frac{T(6+h, 4) - T(6, 4)}{h}$, so we can approximate $T_x(6, 4)$ by considering $h = \pm 2$ and

using the values given in the table: $T_x(6, 4) \approx \dfrac{T(8, 4) - T(6, 4)}{2} = \dfrac{86 - 80}{2} = 3$,

$T_x(6, 4) \approx \dfrac{T(4, 4) - T(6, 4)}{-2} = \dfrac{72 - 80}{-2} = 4$. Averaging these values, we estimate $T_x(6, 4)$ to be approximately

$3.5°\text{C/m}$. Similarly, $T_y(6, 4) = \displaystyle\lim_{h\to0} \frac{T(6, 4+h) - T(6, 4)}{h}$, which we can approximate with $h = \pm 2$:

$T_y(6, 4) \approx \dfrac{T(6, 6) - T(6, 4)}{2} = \dfrac{75 - 80}{2} = -2.5$, $T_y(6, 4) \approx \dfrac{T(6, 2) - T(6, 4)}{-2} = \dfrac{87 - 80}{-2} = -3.5$. Averaging these

values, we estimate $T_y(6, 4)$ to be approximately $-3.0°\text{C/m}$.

(b) Here $\mathbf{u} = \left\langle \frac{1}{\sqrt{2}}, \frac{1}{\sqrt{2}} \right\rangle$, so by Equation 11.6.9, $D_{\mathbf{u}}\, T(6, 4) = \nabla T(6, 4) \cdot \mathbf{u} = T_x(6, 4)\frac{1}{\sqrt{2}} + T_y(6, 4)\frac{1}{\sqrt{2}}$. Using our

estimates from part (a), we have $D_{\mathbf{u}}\, T(6, 4) \approx (3.5)\frac{1}{\sqrt{2}} + (-3.0)\frac{1}{\sqrt{2}} = \frac{1}{2\sqrt{2}} \approx 0.35$. This means that as we move

through the point $(6, 4)$ in the direction of $\mathbf{u}$, the temperature increases at a rate of approximately $0.35°\text{C/m}$.

Alternatively, we can use Definition 11.6.2: $D_{\mathbf{u}}\, T(6, 4) = \displaystyle\lim_{h\to0} \dfrac{T\left(6 + h\,\frac{1}{\sqrt{2}}, 4 + h\,\frac{1}{\sqrt{2}}\right) - T(6, 4)}{h}$,

which we can estimate with $h = \pm 2\sqrt{2}$. Then $D_{\mathbf{u}}\, T(6, 4) \approx \dfrac{T(8, 6) - T(6, 4)}{2\sqrt{2}} = \dfrac{80 - 80}{2\sqrt{2}} = 0$,

$D_{\mathbf{u}}\, T(6, 4) \approx \dfrac{T(4, 2) - T(6, 4)}{-2\sqrt{2}} = \dfrac{74 - 80}{-2\sqrt{2}} = \dfrac{3}{\sqrt{2}}$. Averaging these values, we have $D_{\mathbf{u}}\, T(6, 4) \approx \frac{3}{2\sqrt{2}} \approx 1.1°\text{C/m}$.

(c) $T_{xy}(x, y) = \dfrac{\partial}{\partial y}[T_x(x, y)] = \displaystyle\lim_{h\to0} \dfrac{T_x(x, y+h) - T_x(x, y)}{h}$, so $T_{xy}(6, 4) = \displaystyle\lim_{h\to0} \dfrac{T_x(6, 4+h) - T_x(6, 4)}{h}$ which we can

estimate with $h = \pm 2$. We have $T_x(6, 4) \approx 3.5$ from part (a), but we will also need values for $T_x(6, 6)$ and $T_x(6, 2)$. If we

use $h = \pm 2$ and the values given in the table, we have

$T_x(6, 6) \approx \dfrac{T(8, 6) - T(6, 6)}{2} = \dfrac{80 - 75}{2} = 2.5$, $T_x(6, 6) \approx \dfrac{T(4, 6) - T(6, 6)}{-2} = \dfrac{68 - 75}{-2} = 3.5$.

Averaging these values, we estimate $T_x(6, 6) \approx 3.0$. Similarly,

$T_x(6, 2) \approx \dfrac{T(8, 2) - T_x(6, 2)}{2} = \dfrac{90 - 87}{2} = 1.5$, $T_x(6, 2) \approx \dfrac{T(4, 2) - T(6, 2)}{-2} = \dfrac{74 - 87}{-2} = 6.5$.

Averaging these values, we estimate $T_x(6, 2) \approx 4.0$. Finally, we estimate $T_{xy}(6, 4)$:

$T_{xy}(6, 4) \approx \dfrac{T_x(6, 6) - T_x(6, 4)}{2} = \dfrac{3.0 - 3.5}{2} = -0.25$, $T_{xy}(6, 4) \approx \dfrac{T_x(6, 2) - T_x(6, 4)}{-2} = \dfrac{4.0 - 3.5}{-2} = -0.25$.

Averaging these values, we have $T_{xy}(6, 4) \approx -0.25$.

13. $f(x, y) = \sqrt{2x + y^2} \quad\Rightarrow\quad f_x = \frac{1}{2}(2x + y^2)^{-1/2}(2) = \dfrac{1}{\sqrt{2x + y^2}}$, $f_y = \frac{1}{2}(2x + y^2)^{-1/2}(2y) = \dfrac{y}{\sqrt{2x + y^2}}$

15. $g(u, v) = u\tan^{-1} v \quad\Rightarrow\quad g_u = \tan^{-1} v$, $g_v = \dfrac{u}{1 + v^2}$

17. $T(p, q, r) = p\ln(q + e^r) \quad\Rightarrow\quad T_p = \ln(q + e^r)$, $T_q = \dfrac{p}{q + e^r}$, $T_r = \dfrac{pe^r}{q + e^r}$

19. $f(x, y) = 4x^3 - xy^2 \quad\Rightarrow\quad f_x = 12x^2 - y^2$, $f_y = -2xy$, $f_{xx} = 24x$, $f_{yy} = -2x$, and $f_{xy} = f_{yx} = -2y$.

21. $f(x, y, z) = x^k y^l z^m \quad\Rightarrow\quad f_x = kx^{k-1}y^l z^m$, $f_y = lx^k y^{l-1} z^m$, $f_z = mx^k y^l z^{m-1}$, $f_{xx} = k(k-1)x^{k-2}y^l z^m$,

$f_{yy} = l(l-1)x^k y^{l-2} z^m$, $f_{zz} = m(m-1)x^k y^l z^{m-2}$, $f_{xy} = f_{yx} = klx^{k-1}y^{l-1}z^m$, $f_{xz} = f_{zx} = kmx^{k-1}y^l z^{m-1}$, and

$f_{yz} = f_{zy} = lmx^k y^{l-1} z^{m-1}$.

23. $z = xy + xe^{y/x}$ $\Rightarrow$ $\dfrac{\partial z}{\partial x} = y - \dfrac{y}{x}e^{y/x} + e^{y/x}$, $\dfrac{\partial z}{\partial y} = x + e^{y/x}$ and

$$x\dfrac{\partial z}{\partial x} + y\dfrac{\partial z}{\partial y} = x\left(y - \dfrac{y}{x}e^{y/x} + e^{y/x}\right) + y\left(x + e^{y/x}\right)$$

$$= xy - ye^{y/x} + xe^{y/x} + xy + ye^{y/x} = xy + xy + xe^{y/x} = xy + z$$

25. (a) $z_x = 6x + 2$ $\Rightarrow$ $z_x(1, -2) = 8$ and $z_y = -2y$ $\Rightarrow$ $z_y(1, -2) = 4$, so an equation of the tangent plane is
$z - 1 = 8(x - 1) + 4(y + 2)$ or $z = 8x + 4y + 1$.

(b) A normal vector to the tangent plane (and the surface) at $(1, -2, 1)$ is $\langle 8, 4, -1 \rangle$. Then parametric equations for the normal
line there are $x = 1 + 8t$, $y = -2 + 4t$, $z = 1 - t$, and symmetric equations are $\dfrac{x-1}{8} = \dfrac{y+2}{4} = \dfrac{z-1}{-1}$.

27. (a) Let $F(x, y, z) = x^2 + 2y^2 - 3z^2$. Then $F_x = 2x$, $F_y = 4y$, $F_z = -6z$, so $F_x(2, -1, 1) = 4$, $F_y(2, -1, 1) = -4$,
$F_z(2, -1, 1) = -6$. From Equation 11.6.19, an equation of the tangent plane is $4(x - 2) - 4(y + 1) - 6(z - 1) = 0$ or
equivalently $2x - 2y - 3z = 3$.

(b) From Equations 11.6.20, symmetric equations for the normal line are $\dfrac{x-2}{4} = \dfrac{y+1}{-4} = \dfrac{z-1}{-6}$.

29. (a) $\mathbf{r}(u, v) = (u + v)\,\mathbf{i} + u^2\,\mathbf{j} + v^2\,\mathbf{k}$ and the point $(3, 4, 1)$ corresponds to $u = 2$, $v = 1$. Then $\mathbf{r}_u = \mathbf{i} + 2u\,\mathbf{j}$ $\Rightarrow$
$\mathbf{r}_u(2, 1) = \mathbf{i} + 4\,\mathbf{j}$ and $\mathbf{r}_v = \mathbf{i} + 2v\,\mathbf{k}$ $\Rightarrow$ $\mathbf{r}_v(2, 1) = \mathbf{i} + 2\,\mathbf{j}$. A normal vector to the surface at $(3, 4, 1)$ is
$\mathbf{r}_u \times \mathbf{r}_v = 8\,\mathbf{i} - 2\,\mathbf{j} - 4\,\mathbf{k}$, so an equation of the tangent plane there is $8(x - 3) - 2(y - 4) - 4(z - 1) = 0$ or equivalently
$4x - y - 2z = 6$.

(b) A direction vector for the normal line through $(3, 4, 1)$ is $8\,\mathbf{i} - 2\,\mathbf{j} - 4\,\mathbf{k}$, so a vector equation is
$\mathbf{r}(t) = (3\,\mathbf{i} + 4\,\mathbf{j} + \mathbf{k}) + t\,(8\,\mathbf{i} - 2\,\mathbf{j} - 4\,\mathbf{k})$, and the corresponding parametric equations are $x = 3 + 8t$, $y = 4 - 2t$,
$z = 1 - 4t$.

31. The hyperboloid is a level surface of the function $F(x, y, z) = x^2 + 4y^2 - z^2$, so a normal vector to the surface at (x_0, y_0, z_0)
is $\nabla F(x_0, y_0, z_0) = \langle 2x_0, 8y_0, -2z_0 \rangle$. A normal vector for the plane $2x + 2y + z = 5$ is $\langle 2, 2, 1 \rangle$. For the planes to be
parallel, we need the normal vectors to be parallel, so $\langle 2x_0, 8y_0, -2z_0 \rangle = k\,\langle 2, 2, 1 \rangle$, or $x_0 = k$, $y_0 = \frac{1}{4}k$, and $z_0 = -\frac{1}{2}k$.
But $x_0^2 + 4y_0^2 - z_0^2 = 4$ $\Rightarrow$ $k^2 + \frac{1}{4}k^2 - \frac{1}{4}k^2 = 4$ $\Rightarrow$ $k^2 = 4$ $\Rightarrow$ $k = \pm 2$. So there are two such points:
$\left(2, \frac{1}{2}, -1\right)$ and $\left(-2, -\frac{1}{2}, 1\right)$.

33. $f(x, y, z) = x^3\sqrt{y^2 + z^2}$ $\Rightarrow$ $f_x(x, y, z) = 3x^2\sqrt{y^2 + z^2}$, $f_y(x, y, z) = \dfrac{yx^3}{\sqrt{y^2 + z^2}}$, and $f_z(x, y, z) = \dfrac{zx^3}{\sqrt{y^2 + z^2}}$, so
$f(2, 3, 4) = 8(5) = 40$, $f_x(2, 3, 4) = 3(4)\sqrt{25} = 60$, $f_y(2, 3, 4) = \dfrac{3(8)}{\sqrt{25}} = \dfrac{24}{5}$, and $f_z(2, 3, 4) = \dfrac{4(8)}{\sqrt{25}} = \dfrac{32}{5}$. Then the
linear approximation of f at $(2, 3, 4)$ is

$$f(x, y, z) \approx f(2, 3, 4) + f_x(2, 3, 4)(x - 2) + f_y(2, 3, 4)(y - 3) + f_z(2, 3, 4)(z - 4)$$

$$= 40 + 60(x - 2) + \tfrac{24}{5}(y - 3) + \tfrac{32}{5}(z - 4) = 60x + \tfrac{24}{5}y + \tfrac{32}{5}z - 120$$

Then $(1.98)^3\sqrt{(3.01)^2 + (3.97)^2} = f(1.98, 3.01, 3.97) \approx 60(1.98) + \tfrac{24}{5}(3.01) + \tfrac{32}{5}(3.97) - 120 = 38.656$.

35. $\dfrac{du}{dp} = \dfrac{\partial u}{\partial x}\dfrac{dx}{dp} + \dfrac{\partial u}{\partial y}\dfrac{dy}{dp} + \dfrac{\partial u}{\partial z}\dfrac{dz}{dp} = 2xy^3(1 + 6p) + 3x^2y^2(pe^p + e^p) + 4z^3(p\cos p + \sin p)$

37. By the Chain Rule, $\dfrac{\partial z}{\partial s} = \dfrac{\partial z}{\partial x}\dfrac{\partial x}{\partial s} + \dfrac{\partial z}{\partial y}\dfrac{\partial y}{\partial s}$. When $s = 1$ and $t = 2$, $x = g(1, 2) = 3$ and $y = h(1, 2) = 6$, so

$\dfrac{\partial z}{\partial s} = f_x(3, 6)g_s(1, 2) + f_y(3, 6)h_s(1, 2) = (7)(-1) + (8)(-5) = -47$. Similarly, $\dfrac{\partial z}{\partial t} = \dfrac{\partial z}{\partial x}\dfrac{\partial x}{\partial t} + \dfrac{\partial z}{\partial y}\dfrac{\partial y}{\partial t}$, so

$\dfrac{\partial z}{\partial t} = f_x(3, 6)g_t(1, 2) + f_y(3, 6)h_t(1, 2) = (7)(4) + (8)(10) = 108$.

39. $\dfrac{\partial z}{\partial x} = 2xf'(x^2 - y^2), \dfrac{\partial z}{\partial y} = 1 - 2yf'(x^2 - y^2)$ $\left[\text{where } f' = \dfrac{df}{d(x^2 - y^2)}\right]$. Then

$y\dfrac{\partial z}{\partial x} + x\dfrac{\partial z}{\partial y} = 2xyf'(x^2 - y^2) + x - 2xyf'(x^2 - y^2) = x.$

41. $\dfrac{\partial z}{\partial x} = \dfrac{\partial z}{\partial u}y + \dfrac{\partial z}{\partial v}\dfrac{-y}{x^2}$ and

$$\dfrac{\partial^2 z}{\partial x^2} = y\dfrac{\partial}{\partial x}\left(\dfrac{\partial z}{\partial u}\right) + \dfrac{2y}{x^3}\dfrac{\partial z}{\partial v} + \dfrac{-y}{x^2}\dfrac{\partial}{\partial x}\left(\dfrac{\partial z}{\partial v}\right) = \dfrac{2y}{x^3}\dfrac{\partial z}{\partial v} + y\left(\dfrac{\partial^2 z}{\partial u^2}y + \dfrac{\partial^2 z}{\partial v\,\partial u}\dfrac{-y}{x^2}\right) + \dfrac{-y}{x^2}\left(\dfrac{\partial^2 z}{\partial v^2}\dfrac{-y}{x^2} + \dfrac{\partial^2 z}{\partial u\,\partial v}y\right)$$

$$= \dfrac{2y}{x^3}\dfrac{\partial z}{\partial v} + y^2\dfrac{\partial^2 z}{\partial u^2} - \dfrac{2y^2}{x^2}\dfrac{\partial^2 z}{\partial u\,\partial v} + \dfrac{y^2}{x^4}\dfrac{\partial^2 z}{\partial v^2}$$

Also $\dfrac{\partial z}{\partial y} = x\dfrac{\partial z}{\partial u} + \dfrac{1}{x}\dfrac{\partial z}{\partial v}$ and

$$\dfrac{\partial^2 z}{\partial y^2} = x\dfrac{\partial}{\partial y}\left(\dfrac{\partial z}{\partial u}\right) + \dfrac{1}{x}\dfrac{\partial}{\partial y}\left(\dfrac{\partial z}{\partial v}\right) = x\left(\dfrac{\partial^2 z}{\partial u^2}x + \dfrac{\partial^2 z}{\partial v\,\partial u}\dfrac{1}{x}\right) + \dfrac{1}{x}\left(\dfrac{\partial^2 z}{\partial v^2}\dfrac{1}{x} + \dfrac{\partial^2 z}{\partial u\,\partial v}x\right)$$

$$= x^2\dfrac{\partial^2 z}{\partial u^2} + 2\dfrac{\partial^2 z}{\partial u\,\partial v} + \dfrac{1}{x^2}\dfrac{\partial^2 z}{\partial v^2}$$

Thus

$$x^2\dfrac{\partial^2 z}{\partial x^2} - y^2\dfrac{\partial^2 z}{\partial y^2} = \dfrac{2y}{x}\dfrac{\partial z}{\partial v} + x^2 y^2\dfrac{\partial^2 z}{\partial u^2} - 2y^2\dfrac{\partial^2 z}{\partial u\,\partial v} + \dfrac{y^2}{x^2}\dfrac{\partial^2 z}{\partial v^2} - x^2 y^2\dfrac{\partial^2 z}{\partial u^2} - 2y^2\dfrac{\partial^2 z}{\partial u\,\partial v} - \dfrac{y^2}{x^2}\dfrac{\partial^2 z}{\partial v^2}$$

$$= \dfrac{2y}{x}\dfrac{\partial z}{\partial v} - 4y^2\dfrac{\partial^2 z}{\partial u\,\partial v} = 2v\dfrac{\partial z}{\partial v} - 4uv\dfrac{\partial^2 z}{\partial u\,\partial v}$$

since $y = xv = \dfrac{uv}{y}$ or $y^2 = uv$.

43. $\nabla f = \left\langle z^2\sqrt{y}\,e^{x\sqrt{y}}, \dfrac{xz^2 e^{x\sqrt{y}}}{2\sqrt{y}}, 2ze^{x\sqrt{y}}\right\rangle = ze^{x\sqrt{y}}\left\langle z\sqrt{y}, \dfrac{xz}{2\sqrt{y}}, 2\right\rangle$

45. $\nabla f = \langle 1/\sqrt{x}, -2y\rangle, \nabla f(1, 5) = \langle 1, -10\rangle, \mathbf{u} = \frac{1}{5}\langle 3, -4\rangle$. Then $D_{\mathbf{u}}f(1, 5) = \frac{43}{5}$.

47. $\nabla f = \langle 2xy, x^2 + 1/(2\sqrt{y})\rangle, |\nabla f(2, 1)| = |\langle 4, \frac{9}{2}\rangle|$. Thus the maximum rate of change of f at $(2, 1)$ is $\dfrac{\sqrt{145}}{2}$ in the direction $\langle 4, \frac{9}{2}\rangle$.

49. First we draw a line passing through Homestead and the eye of the hurricane. We can approximate the directional derivative at Homestead in the direction of the eye of the hurricane by the average rate of change of wind speed between the points where this line intersects the contour lines closest to Homestead. In the direction of the eye of the hurricane, the wind speed changes from 45 to 50 knots. We estimate the distance between these two points to be approximately 8 miles, so the rate of change of wind speed in the direction given is approximately $\frac{50 - 45}{8} = \frac{5}{8} = 0.625$ knot/mi.

51. $f(x, y) = x^2 - xy + y^2 + 9x - 6y + 10 \implies f_x = 2x - y + 9,$
$f_y = -x + 2y - 6, f_{xx} = 2 = f_{yy}, f_{xy} = -1$. Then $f_x = 0$ and
$f_y = 0$ imply $y = 1, x = -4$. Thus the only critical point is $(-4, 1)$
and $f_{xx}(-4, 1) > 0, D(-4, 1) = 3 > 0$, so $f(-4, 1) = -11$ is a
local minimum.

53. $f(x, y) = 3xy - x^2y - xy^2$ $\Rightarrow$ $f_x = 3y - 2xy - y^2$,

$f_y = 3x - x^2 - 2xy$, $f_{xx} = -2y$, $f_{yy} = -2x$, $f_{xy} = 3 - 2x - 2y$. Then

$f_x = 0$ implies $y(3 - 2x - y) = 0$ so $y = 0$ or $y = 3 - 2x$. Substituting

into $f_y = 0$ implies $x(3 - x) = 0$ or $3x(-1 + x) = 0$. Hence the critical

points are $(0, 0)$, $(3, 0)$, $(0, 3)$ and $(1, 1)$.

$D(0, 0) = D(3, 0) = D(0, 3) = -9 < 0$ so $(0, 0)$, $(3, 0)$, and $(0, 3)$ are

saddle points. $D(1, 1) = 3 > 0$ and $f_{xx}(1, 1) = -2 < 0$, so $f(1, 1) = 1$

is a local maximum.

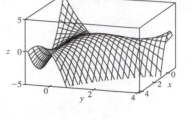

55. First solve inside D. Here $f_x = 4y^2 - 2xy^2 - y^3$, $f_y = 8xy - 2x^2y - 3xy^2$.

Then $f_x = 0$ implies $y = 0$ or $y = 4 - 2x$, but $y = 0$ isn't inside D. Substituting

$y = 4 - 2x$ into $f_y = 0$ implies $x = 0$, $x = 2$ or $x = 1$, but $x = 0$ isn't inside D,

and when $x = 2$, $y = 0$ but $(2, 0)$ isn't inside D. Thus the only critical point inside

D is $(1, 2)$ and $f(1, 2) = 4$. Secondly we consider the boundary of D.

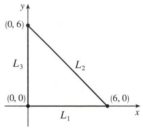

On L_1, $f(x, 0) = 0$ and so $f = 0$ on L_1. On L_2, $x = -y + 6$ and

$f(-y + 6, y) = y^2(6 - y)(-2) = -2(6y^2 - y^3)$ which has critical points

at $y = 0$ and $y = 4$. Then $f(6, 0) = 0$ while $f(2, 4) = -64$. On L_3, $f(0, y) = 0$, so $f = 0$ on L_3. Thus on D the absolute

maximum of f is $f(1, 2) = 4$ while the absolute minimum is $f(2, 4) = -64$.

57. $f(x, y) = x^3 - 3x + y^4 - 2y^2$

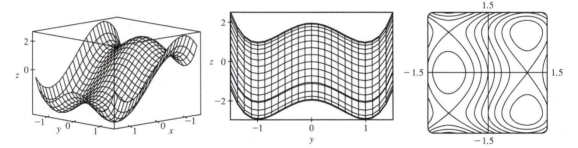

From the graphs, it appears that f has a local maximum $f(-1, 0) \approx 2$, local minima $f(1, \pm1) \approx -3$, and saddle points at

$(-1, \pm1)$ and $(1, 0)$.

To find the exact quantities, we calculate $f_x = 3x^2 - 3 = 0$ $\Leftrightarrow$ $x = \pm1$ and $f_y = 4y^3 - 4y = 0$ $\Leftrightarrow$

$y = 0, \pm1$, giving the critical points estimated above. Also $f_{xx} = 6x$, $f_{xy} = 0$, $f_{yy} = 12y^2 - 4$, so using the Second

Derivatives Test, $D(-1, 0) = 24 > 0$ and $f_{xx}(-1, 0) = -6 < 0$ indicating a local maximum $f(-1, 0) = 2$;

$D(1, \pm1) = 48 > 0$ and $f_{xx}(1, \pm1) = 6 > 0$ indicating local minima $f(1, \pm1) = -3$; and $D(-1, \pm1) = -48$ and

$D(1, 0) = -24$, indicating saddle points.

59. $f(x, y) = x^2y$, $g(x, y) = x^2 + y^2 = 1$ $\Rightarrow$ $\nabla f = \langle 2xy, x^2 \rangle = \lambda \nabla g = \langle 2\lambda x, 2\lambda y \rangle$. Then $2xy = 2\lambda x$ and $x^2 = 2\lambda y$

imply $\lambda = x^2/(2y)$ and $\lambda = y$ if $x \neq 0$ and $y \neq 0$. Hence $x^2 = 2y^2$. Then $x^2 + y^2 = 1$ implies $3y^2 = 1$ so $y = \pm\frac{1}{\sqrt{3}}$ and

$x = \pm\sqrt{\frac{2}{3}}$. [Note if $x = 0$ then $x^2 = 2\lambda y$ implies $y = 0$ and $f(0, 0) = 0$.] Thus the possible points are $\left(\pm\sqrt{\frac{2}{3}}, \pm\frac{1}{\sqrt{3}}\right)$ and

the absolute maxima are $f\left(\pm\sqrt{\frac{2}{3}}, \frac{1}{\sqrt{3}}\right) = \frac{2}{3\sqrt{3}}$ while the absolute minima are $f\left(\pm\sqrt{\frac{2}{3}}, -\frac{1}{\sqrt{3}}\right) = -\frac{2}{3\sqrt{3}}$.

61. $f(x, y, z) = xyz$, $g(x, y, z) = x^2 + y^2 + z^2 = 3$. $\nabla f = \lambda \nabla g \Rightarrow \langle yz, xz, xy \rangle = \lambda \langle 2x, 2y, 2z \rangle$. If any of x, y, or z is

zero, then $x = y = z = 0$ which contradicts $x^2 + y^2 + z^2 = 3$. Then $\lambda = \dfrac{yz}{2x} = \dfrac{xz}{2y} = \dfrac{xy}{2z} \Rightarrow 2y^2 z = 2x^2 z \Rightarrow$

$y^2 = x^2$, and similarly $2yz^2 = 2x^2 y \Rightarrow z^2 = x^2$. Substituting into the constraint equation gives $x^2 + x^2 + x^2 = 3 \Rightarrow$

$x^2 = 1 = y^2 = z^2$. Thus the possible points are $(1, 1, \pm 1)$, $(1, -1, \pm 1)$, $(-1, 1, \pm 1)$, $(-1, -1, \pm 1)$. The absolute maximum

is $f(1, 1, 1) = f(1, -1, -1) = f(-1, 1, -1) = f(-1, -1, 1) = 1$ and the absolute minimum is

$f(1, 1, -1) = f(1, -1, 1) = f(-1, 1, 1) = f(-1, -1, -1) = -1$.

63. $f(x, y, z) = x^2 + y^2 + z^2$, $g(x, y, z) = xy^2 z^3 = 2 \Rightarrow \nabla f = \langle 2x, 2y, 2z \rangle = \lambda \nabla g = \langle \lambda y^2 z^3, 2\lambda xyz^3, 3\lambda xy^2 z^2 \rangle$. Since

$xy^2 z^3 = 2$, $x \neq 0$, $y \neq 0$ and $z \neq 0$, so (1) $2x = \lambda y^2 z^3$, (2) $1 = \lambda xz^3$, (3) $2 = 3\lambda xy^2 z$. Then (2) and (3) imply

$\dfrac{1}{xz^3} = \dfrac{2}{3xy^2 z}$ or $y^2 = \frac{2}{3} z^2$ so $y = \pm z \sqrt{\frac{2}{3}}$. Similarly (1) and (3) imply $\dfrac{2x}{y^2 z^3} = \dfrac{2}{3xy^2 z}$ or $3x^2 = z^2$ so $x = \pm \frac{1}{\sqrt{3}} z$. But

$xy^2 z^3 = 2$ so x and z must have the same sign, that is, $x = \frac{1}{\sqrt{3}} z$. Thus $g(x, y, z) = 2$ implies $\frac{1}{\sqrt{3}} z \left(\frac{2}{3} z^2\right) z^3 = 2$ or

$z = \pm 3^{1/4}$ and the possible points are $(\pm 3^{-1/4}, 3^{-1/4}\sqrt{2}, \pm 3^{1/4})$, $(\pm 3^{-1/4}, -3^{-1/4}\sqrt{2}, \pm 3^{1/4})$. However at each of these

points f takes on the same value, $2\sqrt{3}$. But $(2, 1, 1)$ also satisfies $g(x, y, z) = 2$ and $f(2, 1, 1) = 6 > 2\sqrt{3}$. Thus f has an

absolute minimum value of $2\sqrt{3}$ and no absolute maximum subject to the constraint $xy^2 z^3 = 2$.

Alternate solution: $g(x, y, z) = xy^2 z^3 = 2$ implies $y^2 = \dfrac{2}{xz^3}$, so minimize $f(x, z) = x^2 + \dfrac{2}{xz^3} + z^2$. Then

$f_x = 2x - \dfrac{2}{x^2 z^3}$, $f_z = -\dfrac{6}{xz^4} + 2z$, $f_{xx} = 2 + \dfrac{4}{x^3 z^3}$, $f_{zz} = \dfrac{24}{xz^5} + 2$ and $f_{xz} = \dfrac{6}{x^2 z^4}$. Now $f_x = 0$ implies

$2x^3 z^3 - 2 = 0$ or $z = 1/x$. Substituting into $f_y = 0$ implies $-6x^3 + 2x^{-1} = 0$ or $x = \dfrac{1}{\sqrt[4]{3}}$, so the two critical points are

$\left(\pm \frac{1}{\sqrt[4]{3}}, \pm \sqrt[4]{3}\right)$. Then $D\left(\pm \frac{1}{\sqrt[4]{3}}, \pm \sqrt[4]{3}\right) = (2 + 4)\left(2 + \frac{24}{3}\right) - \left(\frac{6}{\sqrt{3}}\right)^2 > 0$ and $f_{xx}\left(\pm \frac{1}{\sqrt[4]{3}}, \pm \sqrt[4]{3}\right) = 6 > 0$, so each point

is a minimum. Finally, $y^2 = \dfrac{2}{xz^3}$, so the four points closest to the origin are $\left(\pm \frac{1}{\sqrt[4]{3}}, \frac{\sqrt{2}}{\sqrt[4]{3}}, \pm \sqrt[4]{3}\right)$, $\left(\pm \frac{1}{\sqrt[4]{3}}, -\frac{\sqrt{2}}{\sqrt[4]{3}}, \pm \sqrt[4]{3}\right)$.

65.

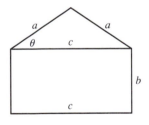

The area of the triangle is $\frac{1}{2} ca \sin \theta$ and the area of the rectangle is bc. Thus, the

area of the whole object is $f(a, b, c) = \frac{1}{2} ca \sin \theta + bc$. The perimeter of the object

is $g(a, b, c) = 2a + 2b + c = P$. To simplify $\sin \theta$ in terms of a, b, and c notice

that $a^2 \sin^2 \theta + \left(\frac{1}{2} c\right)^2 = a^2 \Rightarrow \sin \theta = \dfrac{1}{2a} \sqrt{4a^2 - c^2}$. Thus

$$f(a, b, c) = \frac{c}{4} \sqrt{4a^2 - c^2} + bc. \text{ (Instead of using } \theta, \text{ we could just have used the}$$

Pythagorean Theorem.) As a result, by Lagrange's method, we must find a, b, c, and λ by solving $\nabla f = \lambda \nabla g$ which gives the

following equations: (1) $ca(4a^2 - c^2)^{-1/2} = 2\lambda$, (2) $c = 2\lambda$, (3) $\frac{1}{4}(4a^2 - c^2)^{1/2} - \frac{1}{4} c^2 (4a^2 - c^2)^{-1/2} + b = \lambda$, and

(4) $2a + 2b + c = P$. From (2), $\lambda = \frac{1}{2} c$ and so (1) produces $ca(4a^2 - c^2)^{-1/2} = c \Rightarrow (4a^2 - c^2)^{1/2} = a \Rightarrow$

$4a^2 - c^2 = a^2 \Rightarrow$ (5) $c = \sqrt{3}\, a$. Similarly, since $(4a^2 - c^2)^{1/2} = a$ and $\lambda = \frac{1}{2} c$, (3) gives $\dfrac{a}{4} - \dfrac{c^2}{4a} + b = \dfrac{c}{2}$, so from

(5), $\dfrac{a}{4} - \dfrac{3a}{4} + b = \dfrac{\sqrt{3}\, a}{2} \Rightarrow -\dfrac{a}{2} - \dfrac{\sqrt{3}\, a}{2} = -b \Rightarrow$ (6) $b = \dfrac{a}{2}(1 + \sqrt{3})$. Substituting (5) and (6) into (4) we get:

$2a + a(1 + \sqrt{3}) + \sqrt{3}\, a = P \Rightarrow 3a + 2\sqrt{3}\, a = P \Rightarrow a = \dfrac{P}{3 + 2\sqrt{3}} = \dfrac{2\sqrt{3} - 3}{3} P$ and thus

$b = \dfrac{(2\sqrt{3} - 3)(1 + \sqrt{3})}{6} P = \dfrac{3 - \sqrt{3}}{6} P$ and $c = (2 - \sqrt{3})\, P$.

FOCUS ON PROBLEM SOLVING

1. The areas of the smaller rectangles are $A_1 = xy$, $A_2 = (L-x)y$,

   $A_3 = (L-x)(W-y)$, $A_4 = x(W-y)$. For $0 \leq x \leq L$,

   $0 \leq y \leq W$, let

   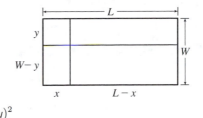

$$f(x,y) = A_1^2 + A_2^2 + A_3^2 + A_4^2$$

$$= x^2y^2 + (L-x)^2y^2 + (L-x)^2(W-y)^2 + x^2(W-y)^2$$

$$= [x^2 + (L-x)^2][y^2 + (W-y)^2]$$

Then we need to find the maximum and minimum values of $f(x,y)$. Here

$f_x(x,y) = [2x - 2(L-x)][y^2 + (W-y)^2] = 0 \Rightarrow 4x - 2L = 0$ or $x = \frac{1}{2}L$, and

$f_y(x,y) = [x^2 + (L-x)^2][2y - 2(W-y)] = 0 \Rightarrow 4y - 2W = 0$ or $y = W/2$. Also

$f_{xx} = 4[y^2 + (W-y)^2]$, $f_{yy} = 4[x^2 + (L-x)^2]$, and $f_{xy} = (4x - 2L)(4y - 2W)$. Then

$D = 16[y^2 + (W-y)^2][x^2 + (L-x)^2] - (4x - 2L)^2(4y - 2W)^2$. Thus when $x = \frac{1}{2}L$ and $y = \frac{1}{2}W$, $D > 0$ and

$f_{xx} = 2W^2 > 0$. Thus a minimum of f occurs at $\left(\frac{1}{2}L, \frac{1}{2}W\right)$ and this minimum value is $f\left(\frac{1}{2}L, \frac{1}{2}W\right) = \frac{1}{4}L^2W^2$. There are

no other critical points, so the maximum must occur on the boundary. Now along the width of the rectangle let

$g(y) = f(0,y) = f(L,y) = L^2[y^2 + (W-y)^2]$, $0 \leq y \leq W$. Then $g'(y) = L^2[2y - 2(W-y)] = 0 \Leftrightarrow y = \frac{1}{2}W$.

And $g\left(\frac{1}{2}\right) = \frac{1}{2}L^2W^2$. Checking the endpoints, we get $g(0) = g(W) = L^2W^2$. Along the length of the rectangle let

$h(x) = f(x,0) = f(x,W) = W^2[x^2 + (L-x)^2]$, $0 \leq x \leq L$. By symmetry $h'(x) = 0 \Leftrightarrow x = \frac{1}{2}L$ and

$h\left(\frac{1}{2}L\right) = \frac{1}{2}L^2W^2$. At the endpoints we have $h(0) = h(L) = L^2W^2$. Therefore L^2W^2 is the maximum value of f. This

maximum value of f occurs when the "cutting" lines correspond to sides of the rectangle.

3. (a) The area of a trapezoid is $\frac{1}{2}h(b_1 + b_2)$, where h is the height (the distance between the two parallel sides) and b_1, b_2 are

   the lengths of the bases (the parallel sides). From the figure in the text, we see that $h = x \sin\theta$, $b_1 = w - 2x$, and

   $b_2 = w - 2x + 2x \cos\theta$. Therefore the cross-sectional area of the rain gutter is

$$A(x,\theta) = \frac{1}{2}x \sin\theta \left[(w - 2x) + (w - 2x + 2x \cos\theta)\right]$$

$$= (x \sin\theta)(w - 2x + x \cos\theta)$$

$$= wx \sin\theta - 2x^2 \sin\theta + x^2 \sin\theta \cos\theta, \ 0 < x \leq \tfrac{1}{2}w, 0 < \theta \leq \tfrac{\pi}{2}$$

We look for the critical points of A: $\partial A/\partial x = w \sin\theta - 4x \sin\theta + 2x \sin\theta \cos\theta$ and

$\partial A/\partial\theta = wx \cos\theta - 2x^2 \cos\theta + x^2(\cos^2\theta - \sin^2\theta)$, so $\partial A/\partial x = 0 \Leftrightarrow \sin\theta(w - 4x + 2x \cos\theta) = 0 \Leftrightarrow$

$\cos\theta = \dfrac{4x - w}{2x} = 2 - \dfrac{w}{2x}$ $\left(0 < \theta \leq \tfrac{\pi}{2} \Rightarrow \sin\theta > 0\right)$. If, in addition, $\partial A/\partial\theta = 0$, then

$$0 = wx \cos\theta - 2x^2 \cos\theta + x^2(2\cos^2\theta - 1)$$

$$= wx\left(2 - \frac{w}{2x}\right) - 2x^2\left(2 - \frac{w}{2x}\right) + x^2\left[2\left(2 - \frac{w}{2x}\right)^2 - 1\right]$$

$$= 2wx - \tfrac{1}{2}w^2 - 4x^2 + wx + x^2\left[8 - \frac{4w}{x} + \frac{w^2}{2x^2} - 1\right] = -wx + 3x^2 = x(3x - w)$$

149

Since $x > 0$, we must have $x = \frac{1}{3}w$, in which case $\cos\theta = \frac{1}{2}$, so $\theta = \frac{\pi}{3}$, $\sin\theta = \frac{\sqrt{3}}{2}$, $k = \frac{\sqrt{3}}{6}w$, $b_1 = \frac{1}{3}w$, $b_2 = \frac{2}{3}w$,

and $A = \frac{\sqrt{3}}{12}w^2$. As in Example 11.7.6, we can argue from the physical nature of this problem that we have found a local maximum of A. Now checking the boundary of A, let

$g(\theta) = A(w/2, \theta) = \frac{1}{2}w^2\sin\theta - \frac{1}{2}w^2\sin\theta + \frac{1}{4}w^2\sin\theta\cos\theta = \frac{1}{8}w^2\sin 2\theta$, $0 < \theta \le \frac{\pi}{2}$. Clearly g is maximized when

$\sin 2\theta = 1$ in which case $A = \frac{1}{8}w^2$. Also along the line $\theta = \frac{\pi}{2}$, let $h(x) = A\left(x, \frac{\pi}{2}\right) = wx - 2x^2$, $0 < x < \frac{1}{2}w$ $\Rightarrow$

$h'(x) = w - 4x = 0$ $\Leftrightarrow$ $x = \frac{1}{4}w$, and $h\left(\frac{1}{4}w\right) = w\left(\frac{1}{4}w\right) - 2\left(\frac{1}{4}w\right)^2 = \frac{1}{8}w^2$. Since $\frac{1}{8}w^2 < \frac{\sqrt{3}}{12}w^2$, we conclude that the local maximum found earlier was an absolute maximum.

(b) If the metal were bent into a semi-circular gutter of radius r, we would have $w = \pi r$ and $A = \frac{1}{2}\pi r^2 = \frac{1}{2}\pi\left(\frac{w}{\pi}\right)^2 = \frac{w^2}{2\pi}$.

Since $\frac{w^2}{2\pi} > \frac{\sqrt{3}\,w^2}{12}$, it *would* be better to bend the metal into a gutter with a semicircular cross-section.

5. Let $g(x, y) = xf\left(\frac{y}{x}\right)$. Then $g_x(x, y) = f\left(\frac{y}{x}\right) + xf'\left(\frac{y}{x}\right)\left(-\frac{y}{x^2}\right) = f\left(\frac{y}{x}\right) - \frac{y}{x}f'\left(\frac{y}{x}\right)$ and

$g_y(x, y) = xf'\left(\frac{y}{x}\right)\left(\frac{1}{x}\right) = f'\left(\frac{y}{x}\right)$. Thus the tangent plane at (x_0, y_0, z_0) on the surface has equation

$z - x_0 f\left(\frac{y_0}{x_0}\right) = \left[f\left(\frac{y_0}{x_0}\right) - y_0 x_0^{-1}f'\left(\frac{y_0}{x_0}\right)\right](x - x_0) + f'\left(\frac{y_0}{x_0}\right)(y - y_0)$ $\Rightarrow$

$\left[f\left(\frac{y_0}{x_0}\right) - y_0 x_0^{-1}f'\left(\frac{y_0}{x_0}\right)\right]x + \left[f'\left(\frac{y_0}{x_0}\right)\right]y - z = 0$. But any plane whose equation is of the form $ax + by + cz = 0$

passes through the origin. Thus the origin is the common point of intersection.

7. (a) $x = r\cos\theta$, $y = r\sin\theta$, $z = z$. Then $\dfrac{\partial u}{\partial r} = \dfrac{\partial u}{\partial x}\dfrac{\partial x}{\partial r} + \dfrac{\partial u}{\partial y}\dfrac{\partial y}{\partial r} + \dfrac{\partial u}{\partial z}\dfrac{\partial z}{\partial r} = \dfrac{\partial u}{\partial x}\cos\theta + \dfrac{\partial u}{\partial y}\sin\theta$ and

$$\frac{\partial^2 u}{\partial r^2} = \cos\theta\left[\frac{\partial^2 u}{\partial x^2}\frac{\partial x}{\partial r} + \frac{\partial^2 u}{\partial y\,\partial x}\frac{\partial y}{\partial r} + \frac{\partial^2 u}{\partial z\,\partial x}\frac{\partial z}{\partial r}\right] + \sin\theta\left[\frac{\partial^2 u}{\partial y^2}\frac{\partial y}{\partial r} + \frac{\partial^2 u}{\partial x\,\partial y}\frac{\partial x}{\partial r} + \frac{\partial^2 u}{\partial z\,\partial y}\frac{\partial z}{\partial r}\right]$$

$$= \frac{\partial^2 u}{\partial x^2}\cos^2\theta + \frac{\partial^2 u}{\partial y^2}\sin^2\theta + 2\frac{\partial^2 u}{\partial y\,\partial x}\cos\theta\sin\theta$$

Similarly $\dfrac{\partial u}{\partial\theta} = -\dfrac{\partial u}{\partial x}r\sin\theta + \dfrac{\partial u}{\partial y}r\cos\theta$ and

$$\frac{\partial^2 u}{\partial\theta^2} = \frac{\partial^2 u}{\partial x^2}r^2\sin^2\theta + \frac{\partial^2 u}{\partial y^2}r^2\cos^2\theta - 2\frac{\partial^2 u}{\partial y\,\partial x}r^2\sin\theta\cos\theta - \frac{\partial u}{\partial x}r\cos\theta - \frac{\partial u}{\partial y}r\sin\theta.$$ So

$$\frac{\partial^2 u}{\partial r^2} + \frac{1}{r}\frac{\partial u}{\partial r} + \frac{1}{r^2}\frac{\partial^2 u}{\partial\theta^2} + \frac{\partial^2 u}{\partial z^2}$$

$$= \frac{\partial^2 u}{\partial x^2}\cos^2\theta + \frac{\partial^2 u}{\partial y^2}\sin^2\theta + 2\frac{\partial^2 u}{\partial y\,\partial x}\cos\theta\sin\theta + \frac{\partial u}{\partial x}\frac{\cos\theta}{r} + \frac{\partial u}{\partial y}\frac{\sin\theta}{r}$$

$$+ \frac{\partial^2 u}{\partial x^2}\sin^2\theta + \frac{\partial^2 u}{\partial y^2}\cos^2\theta - 2\frac{\partial^2 u}{\partial y\,\partial x}\sin\theta\cos\theta - \frac{\partial u}{\partial x}\frac{\cos\theta}{r} - \frac{\partial u}{\partial y}\frac{\sin\theta}{r} + \frac{\partial^2 u}{\partial z^2}$$

$$= \frac{\partial^2 u}{\partial x^2} + \frac{\partial^2 u}{\partial y^2} + \frac{\partial^2 u}{\partial z^2}$$

(b) $x = \rho \sin\phi \cos\theta$, $y = \rho \sin\phi \sin\theta$, $z = \rho \cos\phi$. Then

$$\frac{\partial u}{\partial \rho} = \frac{\partial u}{\partial x}\frac{\partial x}{\partial \rho} + \frac{\partial u}{\partial y}\frac{\partial y}{\partial \rho} + \frac{\partial u}{\partial z}\frac{\partial z}{\partial \rho} = \frac{\partial u}{\partial x}\sin\phi\cos\theta + \frac{\partial u}{\partial y}\sin\phi\sin\theta + \frac{\partial u}{\partial z}\cos\phi, \text{ and}$$

$$\frac{\partial^2 u}{\partial \rho^2} = \sin\phi\cos\theta\left[\frac{\partial^2 u}{\partial x^2}\frac{\partial x}{\partial \rho} + \frac{\partial^2 u}{\partial y\,\partial x}\frac{\partial y}{\partial \rho} + \frac{\partial^2 u}{\partial z\,\partial x}\frac{\partial z}{\partial \rho}\right]$$

$$+ \sin\phi\sin\theta\left[\frac{\partial^2 u}{\partial y^2}\frac{\partial y}{\partial \rho} + \frac{\partial^2 u}{\partial x\,\partial y}\frac{\partial x}{\partial \rho} + \frac{\partial^2 u}{\partial z\,\partial y}\frac{\partial z}{\partial \rho}\right]$$

$$+ \cos\phi\left[\frac{\partial^2 u}{\partial z^2}\frac{\partial z}{\partial \rho} + \frac{\partial^2 u}{\partial x\,\partial z}\frac{\partial x}{\partial \rho} + \frac{\partial^2 u}{\partial y\,\partial z}\frac{\partial y}{\partial \rho}\right]$$

$$= 2\frac{\partial^2 u}{\partial y\,\partial x}\sin^2\phi\sin\theta\cos\theta + 2\frac{\partial^2 u}{\partial z\,\partial x}\sin\phi\cos\phi\cos\theta + 2\frac{\partial^2 u}{\partial y\,\partial z}\sin\phi\cos\phi\sin\theta$$

$$+ \frac{\partial^2 u}{\partial x^2}\sin^2\phi\cos^2\theta + \frac{\partial^2 u}{\partial y^2}\sin^2\phi\sin^2\theta + \frac{\partial^2 u}{\partial z^2}\cos^2\phi$$

Similarly $\dfrac{\partial u}{\partial \phi} = \dfrac{\partial u}{\partial x}\rho\cos\phi\cos\theta + \dfrac{\partial u}{\partial y}\rho\cos\phi\sin\theta - \dfrac{\partial u}{\partial z}\rho\sin\phi$, and

$$\frac{\partial^2 u}{\partial \phi^2} = 2\frac{\partial^2 u}{\partial y\,\partial x}\rho^2\cos^2\phi\sin\theta\cos\theta - 2\frac{\partial^2 u}{\partial x\,\partial z}\rho^2\sin\phi\cos\phi\cos\theta$$

$$- 2\frac{\partial^2 u}{\partial y\,\partial z}\rho^2\sin\phi\cos\phi\sin\theta + \frac{\partial^2 u}{\partial x^2}\rho^2\cos^2\phi\cos^2\theta + \frac{\partial^2 u}{\partial y^2}\rho^2\cos^2\phi\sin^2\theta$$

$$+ \frac{\partial^2 u}{\partial z^2}\rho^2\sin^2\phi - \frac{\partial u}{\partial x}\rho\sin\phi\cos\theta - \frac{\partial u}{\partial y}\rho\sin\phi\sin\theta - \frac{\partial u}{\partial z}\rho\cos\phi$$

And $\dfrac{\partial u}{\partial \theta} = -\dfrac{\partial u}{\partial x}\rho\sin\phi\sin\theta + \dfrac{\partial u}{\partial y}\rho\sin\phi\cos\theta$, while

$$\frac{\partial^2 u}{\partial \theta^2} = -2\frac{\partial^2 u}{\partial y\,\partial x}\rho^2\sin^2\phi\cos\theta\sin\theta + \frac{\partial^2 u}{\partial x^2}\rho^2\sin^2\phi\sin^2\theta$$

$$+ \frac{\partial^2 u}{\partial y^2}\rho^2\sin^2\phi\cos^2\theta - \frac{\partial u}{\partial x}\rho\sin\phi\cos\theta - \frac{\partial u}{\partial y}\rho\sin\phi\sin\theta$$

Therefore

$$\frac{\partial^2 u}{\partial \rho^2} + \frac{2}{\rho}\frac{\partial u}{\partial \rho} + \frac{\cot\phi}{\rho^2}\frac{\partial u}{\partial \phi} + \frac{1}{\rho^2}\frac{\partial^2 u}{\partial \phi^2} + \frac{1}{\rho^2\sin^2\phi}\frac{\partial^2 u}{\partial \theta^2}$$

$$= \frac{\partial^2 u}{\partial x^2}\left[(\sin^2\phi\cos^2\theta) + (\cos^2\phi\cos^2\theta) + \sin^2\theta\right]$$

$$+ \frac{\partial^2 u}{\partial y^2}\left[(\sin^2\phi\sin^2\theta) + (\cos^2\phi\sin^2\theta) + \cos^2\theta\right] + \frac{\partial^2 u}{\partial z^2}\left[\cos^2\phi + \sin^2\phi\right]$$

$$+ \frac{\partial u}{\partial x}\left[\frac{2\sin^2\phi\cos\theta + \cos^2\phi\cos\theta - \sin^2\phi\cos\theta - \cos\theta}{\rho\sin\phi}\right]$$

$$+ \frac{\partial u}{\partial y}\left[\frac{2\sin^2\phi\sin\theta + \cos^2\phi\sin\theta - \sin^2\phi\sin\theta - \sin\theta}{\rho\sin\phi}\right]$$

But $2\sin^2\phi\cos\theta + \cos^2\phi\cos\theta - \sin^2\phi\cos\theta - \cos\theta = (\sin^2\phi + \cos^2\phi - 1)\cos\theta = 0$ and similarly the coefficient of $\partial u/\partial y$ is 0. Also $\sin^2\phi\cos^2\theta + \cos^2\phi\cos^2\theta + \sin^2\theta = \cos^2\theta(\sin^2\phi + \cos^2\phi) + \sin^2\theta = 1$, and similarly the coefficient of $\partial^2 u/\partial y^2$ is 1. So Laplace's Equation in spherical coordinates is as stated.

9. Since we are minimizing the area of the ellipse, and the circle lies above the x-axis, the ellipse will intersect the circle for only one value of y. This y-value must satisfy both the equation of the circle and the equation of the ellipse. Now

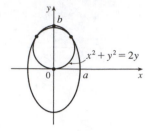

$\dfrac{x^2}{a^2} + \dfrac{y^2}{b^2} = 1 \;\Rightarrow\; x^2 = \dfrac{a^2}{b^2}\left(b^2 - y^2\right)$. Substituting into the equation of the

circle gives $\dfrac{a^2}{b^2}\left(b^2 - y^2\right) + y^2 - 2y = 0 \;\Rightarrow\; \left(\dfrac{b^2 - a^2}{b^2}\right)y^2 - 2y + a^2 = 0$.

In order for there to be only one solution to this quadratic equation, the discriminant must be 0, so $4 - 4a^2\,\dfrac{b^2 - a^2}{b^2} = 0 \;\Rightarrow$

$b^2 - a^2b^2 + a^4 = 0$. The area of the ellipse is $A(a, b) = \pi ab$, and we minimize this function subject to the constraint

$g(a, b) = b^2 - a^2b^2 + a^4 = 0$.

Now $\nabla A = \lambda \nabla g \;\Leftrightarrow\; \pi b = \lambda(4a^3 - 2ab^2),\; \pi a = \lambda(2b - 2ba^2) \;\Rightarrow\; \text{(1)}\; \lambda = \dfrac{\pi b}{2a(2a^2 - b^2)}$,

$\text{(2)}\; \lambda = \dfrac{\pi a}{2b(1 - a^2)}$, $\text{(3)}\; b^2 - a^2b^2 + a^4 = 0$. Comparing (1) and (2) gives $\dfrac{\pi b}{2a(2a^2 - b^2)} = \dfrac{\pi a}{2b(1 - a^2)} \;\Rightarrow$

$2\pi b^2 = 4\pi a^4 \;\Leftrightarrow\; a^2 = \dfrac{1}{\sqrt{2}}\,b$. Substitute this into (3) to get $b = \dfrac{3}{\sqrt{2}} \;\Rightarrow\; a = \sqrt{\dfrac{3}{2}}$.

12 □ MULTIPLE INTEGRALS

12.1 Double Integrals over Rectangles

1. (a) The subrectangles are shown in the figure.

The surface is the graph of $f(x, y) = xy$ and $\Delta A = 4$, so we estimate

$$V \approx \sum_{i=1}^{3} \sum_{j=1}^{2} f(x_i, y_j) \Delta A$$

$$= f(2, 2) \Delta A + f(2, 4) \Delta A + f(4, 2) \Delta A + f(4, 4) \Delta A + f(6, 2) \Delta A + f(6, 4) \Delta A$$

$$= 4(4) + 8(4) + 8(4) + 16(4) + 12(4) + 24(4) = 288$$

(b) $V \approx \sum_{i=1}^{3} \sum_{j=1}^{2} f(\overline{x}_i, \overline{y}_j) \Delta A = f(1, 1) \Delta A + f(1, 3) \Delta A + f(3, 1) \Delta A + f(3, 3) \Delta A + f(5, 1) \Delta A + f(5, 3) \Delta A$

$$= 1(4) + 3(4) + 3(4) + 9(4) + 5(4) + 15(4) = 144$$

3. (a) The subrectangles are shown in the figure. Since $\Delta A = \pi^2/4$, we estimate

$$\iint_R \sin(x + y) \, dA \approx \sum_{i=1}^{2} \sum_{j=1}^{2} f(x_{ij}^*, y_{ij}^*) \Delta A$$

$$= f(0, 0) \Delta A + f\left(0, \tfrac{\pi}{2}\right) \Delta A + f\left(\tfrac{\pi}{2}, 0\right) \Delta A + f\left(\tfrac{\pi}{2}, \tfrac{\pi}{2}\right) \Delta A$$

$$= 0\left(\tfrac{\pi^2}{4}\right) + 1\left(\tfrac{\pi^2}{4}\right) + 1\left(\tfrac{\pi^2}{4}\right) + 0\left(\tfrac{\pi^2}{4}\right) = \tfrac{\pi^2}{2} \approx 4.935$$

(b) $\iint_R \sin(x + y) \, dA \approx \sum_{i=1}^{2} \sum_{j=1}^{2} f(\overline{x}_i, \overline{y}_j) \Delta A$

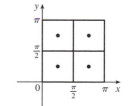

$$= f\left(\tfrac{\pi}{4}, \tfrac{\pi}{4}\right) \Delta A + f\left(\tfrac{\pi}{4}, \tfrac{3\pi}{4}\right) \Delta A + f\left(\tfrac{3\pi}{4}, \tfrac{\pi}{4}\right) \Delta A + f\left(\tfrac{3\pi}{4}, \tfrac{3\pi}{4}\right) \Delta A$$

$$= 1\left(\tfrac{\pi^2}{4}\right) + 0\left(\tfrac{\pi^2}{4}\right) + 0\left(\tfrac{\pi^2}{4}\right) + (-1)\left(\tfrac{\pi^2}{4}\right) = 0$$

5. (a) Each subrectangle and its midpoint are shown in the figure. The area of each

subrectangle is $\Delta A = 2$, so we evaluate f at each midpoint and estimate

$$\iint_R f(x, y) \, dA \approx \sum_{i=1}^{2} \sum_{j=1}^{2} f(\overline{x}_i, \overline{y}_j) \Delta A$$

$$= f(1.5, 1) \Delta A + f(1.5, 3) \Delta A$$

$$\qquad + \ f(2.5, 1) \Delta A + f(2.5, 3) \Delta A$$

$$= 1(2) + (-8)(2) + 5(2) + (-1)(2) = -6$$

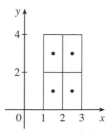

(b) The subrectangles are shown in the figure. In each subrectangle, the sample point farthest from the origin is the upper right corner, and the area of each subrectangle is $\Delta A = \frac{1}{2}$. Thus we estimate

$$\iint_R f(x, y)\, dA \approx \sum_{i=1}^{4} \sum_{j=1}^{4} f(x_i, y_j)\, \Delta A$$

$$= f(1.5, 1)\, \Delta A + f(1.5, 2)\, \Delta A + f(1.5, 3)\, \Delta A + f(1.5, 4)\, \Delta A$$

$$+ f(2, 1)\, \Delta A + f(2, 2)\, \Delta A + f(2, 3)\, \Delta A + f(2, 4)\, \Delta A$$

$$+ f(2.5, 1)\, \Delta A + f(2.5, 2)\, \Delta A + f(2.5, 3)\, \Delta A + f(2.5, 4)\, \Delta A$$

$$+ f(3, 1)\, \Delta A + f(3, 2)\, \Delta A + f(3, 3)\, \Delta A + f(3, 4)\, \Delta A$$

$$= 1\left(\tfrac{1}{2}\right) + (-4)\left(\tfrac{1}{2}\right) + (-8)\left(\tfrac{1}{2}\right) + (-6)\left(\tfrac{1}{2}\right) + 3\left(\tfrac{1}{2}\right) + 0\left(\tfrac{1}{2}\right) + (-5)\left(\tfrac{1}{2}\right) + (-8)\left(\tfrac{1}{2}\right)$$

$$+ 5\left(\tfrac{1}{2}\right) + 3\left(\tfrac{1}{2}\right) + (-1)\left(\tfrac{1}{2}\right) + (-4)\left(\tfrac{1}{2}\right) + 8\left(\tfrac{1}{2}\right) + 6\left(\tfrac{1}{2}\right) + 3\left(\tfrac{1}{2}\right) + 0\left(\tfrac{1}{2}\right)$$

$$= -3.5$$

7. The values of $f(x, y) = \sqrt{52 - x^2 - y^2}$ get smaller as we move farther from the origin, so on any of the subrectangles in the problem, the function will have its largest value at the lower left corner of the subrectangle and its smallest value at the upper right corner, and any other value will lie between these two. So using these subrectangles we have $U < V < L$. (Note that this is true no matter how R is divided into subrectangles.)

9. (a) With $m = n = 2$, we have $\Delta A = 4$. Using the contour map to estimate the value of f at the center of each subrectangle, we have

$$\iint_R f(x, y)\, dA \approx \sum_{i=1}^{2} \sum_{j=1}^{2} f\left(\overline{x}_i, \overline{y}_j\right) \Delta A = \Delta A[f(1, 1) + f(1, 3) + f(3, 1) + f(3, 3)]$$

$$\approx 4(27 + 4 + 14 + 17) = 248$$

(b) $f_{\text{ave}} = \frac{1}{A(R)} \iint_R f(x, y)\, dA \approx \frac{1}{16}(248) = 15.5$

11. $z = 3 > 0$, so we can interpret the integral as the volume of the solid S that lies below the plane $z = 3$ and above the rectangle $[-2, 2] \times [1, 6]$. S is a rectangular solid, thus $\iint_R 3\, dA = 4 \cdot 5 \cdot 3 = 60$.

13. $z = f(x, y) = 4 - 2y \geq 0$ for $0 \leq y \leq 1$. Thus the integral represents the volume of that part of the rectangular solid $[0, 1] \times [0, 1] \times [0, 4]$ which lies below the plane $z = 4 - 2y$. So

$$\iint_R (4 - 2y)\, dA = (1)(1)(2) + \tfrac{1}{2}(1)(1)(2) = 3$$

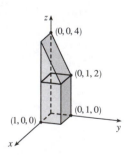

15. To calculate the estimates using a programmable calculator, we can use an algorithm similar to that of Exercise 5.1.7. In Maple, we can define the function $f(x, y) = \sqrt{1 + xe^{-y}}$ (calling it f), load the student package, and then use the command

$$\texttt{middlesum(middlesum(f,x=0..1,m),}$$
$$\texttt{y=0..1,m);}$$

to get the estimate with $n = m^2$ squares of equal size. Mathematica has no special Riemann sum command, but we can define f and then use nested Sum commands to calculate the estimates.

n	estimate
1	1.141606
4	1.143191
16	1.143535
64	1.143617
256	1.143637
1024	1.143642

17. If we divide R into mn subrectangles, $\iint_R k \, dA \approx \sum_{i=1}^{m} \sum_{j=1}^{n} f(x_{ij}^*, y_{ij}^*) \, \Delta A$ for any choice of sample points (x_{ij}^*, y_{ij}^*). But $f(x_{ij}^*, y_{ij}^*) = k$ always and $\sum_{i=1}^{m} \sum_{j=1}^{n} \Delta A = \text{area of } R = (b - a)(d - c)$. Thus, no matter how we choose the sample points,

$$\sum_{i=1}^{m} \sum_{j=1}^{n} f(x_{ij}^*, y_{ij}^*) \, \Delta A = k \sum_{i=1}^{m} \sum_{j=1}^{n} \Delta A = k(b - a)(d - c) \text{ and so}$$

$$\iint_R k \, dA = \lim_{m,n \to \infty} \sum_{i=1}^{m} \sum_{j=1}^{n} f(x_{ij}^*, y_{ij}^*) \, \Delta A = \lim_{m,n \to \infty} k \sum_{i=1}^{m} \sum_{j=1}^{n} \Delta A$$

$$= \lim_{m,n \to \infty} k(b - a)(d - c) = k(b - a)(d - c)$$

12.2   Iterated Integrals

1. $\int_0^3 (2x + 3x^2 y) \, dx = \left[x^2 + x^3 y \right]_{x=0}^{x=3} = (9 + 27y) - (0 + 0) = 9 + 27y$,

$\int_0^4 (2x + 3x^2 y) \, dy = \left[2xy + 3x^2 \dfrac{y^2}{2} \right]_{y=0}^{y=4} = \left(8x + 3x^2 \cdot \dfrac{16}{2} \right) - (0 + 0) = 8x + 24x^2$

3. $\int_1^3 \int_0^1 (1 + 4xy) \, dx \, dy = \int_1^3 \left[x + 2x^2 y \right]_{x=0}^{x=1} \, dy = \int_1^3 (1 + 2y) \, dy = \left[y + y^2 \right]_1^3 = (3 + 9) - (1 + 1) = 10$

5. $\int_0^2 \int_0^{\pi/2} x \sin y \, dy \, dx = \int_0^2 x \, dx \int_0^{\pi/2} \sin y \, dy$ [as in Example 5] $= \left[\dfrac{x^2}{2} \right]_0^2 \left[-\cos y \right]_0^{\pi/2} = (2 - 0)(0 + 1) = 2.$

7. $\int_0^2 \int_0^1 (2x + y)^8 \, dx \, dy = \int_0^2 \left[\dfrac{1}{2} \dfrac{(2x + y)^9}{9} \right]_{x=0}^{x=1} \, dy$ [substitute $u = 2x + y \Rightarrow dx = \tfrac{1}{2} du$]

$$= \dfrac{1}{18} \int_0^2 [(2 + y)^9 - (0 + y)^9] \, dy = \dfrac{1}{18} \left[\dfrac{(2 + y)^{10}}{10} - \dfrac{y^{10}}{10} \right]_0^2$$

$$= \tfrac{1}{180}[(4^{10} - 2^{10}) - (2^{10} - 0^{10})] = \dfrac{1,046,528}{180} = \dfrac{261,632}{45}$$

9. $\int_1^4 \int_1^2 \left(\dfrac{x}{y} + \dfrac{y}{x} \right) dy \, dx = \int_1^4 \left[x \ln|y| + \dfrac{1}{x} \cdot \dfrac{1}{2} y^2 \right]_{y=1}^{y=2} dx = \int_1^4 \left(x \ln 2 + \dfrac{3}{2x} \right) dx = \left[\dfrac{1}{2} x^2 \ln 2 + \dfrac{3}{2} \ln|x| \right]_1^4$

$$= 8 \ln 2 + \tfrac{3}{2} \ln 4 - \tfrac{1}{2} \ln 2 = \tfrac{15}{2} \ln 2 + 3 \ln 4^{1/2} = \tfrac{21}{2} \ln 2$$

11. $\int_0^{\ln 2} \int_0^{\ln 5} e^{2x - y} \, dx \, dy = \left(\int_0^{\ln 5} e^{2x} \, dx \right) \left(\int_0^{\ln 2} e^{-y} \, dy \right) = \left[\tfrac{1}{2} e^{2x} \right]_0^{\ln 5} \left[-e^{-y} \right]_0^{\ln 2} = \left(\tfrac{25}{2} - \tfrac{1}{2} \right) \left(-\tfrac{1}{2} + 1 \right) = 6$

13. $\iint_R \dfrac{xy^2}{x^2+1}\,dA = \int_0^1 \int_{-3}^3 \dfrac{xy^2}{x^2+1}\,dy\,dx = \int_0^1 \dfrac{x}{x^2+1}\,dx \int_{-3}^3 y^2\,dy$

$$= \left[\tfrac{1}{2}\ln(x^2+1)\right]_0^1 \left[\tfrac{1}{3}y^3\right]_{-3}^3 = \tfrac{1}{2}(\ln 2 - \ln 1)\cdot\tfrac{1}{3}(27+27) = 9\ln 2$$

15. $\int_0^{\pi/6}\int_0^{\pi/3} x\sin(x+y)\,dy\,dx$

$$= \int_0^{\pi/6}\left[-x\cos(x+y)\right]_{y=0}^{y=\pi/3}\,dx = \int_0^{\pi/6}\left[x\cos x - x\cos\left(x+\tfrac{\pi}{3}\right)\right]dx$$

$$= x\left[\sin x - \sin\left(x+\tfrac{\pi}{3}\right)\right]_0^{\pi/6} - \int_0^{\pi/6}\left[\sin x - \sin\left(x+\tfrac{\pi}{3}\right)\right]dx \quad \text{[by integrating by parts separately for each term]}$$

$$= \tfrac{\pi}{6}\left[\tfrac{1}{2}-1\right] - \left[-\cos x + \cos\left(x+\tfrac{\pi}{3}\right)\right]_0^{\pi/6} = -\tfrac{\pi}{12} - \left[-\tfrac{\sqrt{3}}{2}+0-\left(-1+\tfrac{1}{2}\right)\right] = \tfrac{\sqrt{3}-1}{2} - \tfrac{\pi}{12}$$

17. $\iint_R xye^{x^2y}\,dA = \int_0^2\int_0^1 xye^{x^2y}\,dx\,dy = \int_0^2\left[\tfrac{1}{2}e^{x^2y}\right]_{x=0}^{x=1}\,dy = \tfrac{1}{2}\int_0^2(e^y-1)\,dy$

$$= \tfrac{1}{2}\left[e^y - y\right]_0^2 = \tfrac{1}{2}[(e^2-2)-(1-0)] = \tfrac{1}{2}(e^2-3)$$

19. $z = f(x,y) = 4 - x - 2y \ge 0$ for $0 \le x \le 1$ and $0 \le y \le 1$. So the solid is the region in the first octant which lies below the plane $z = 4 - x - 2y$ and above $[0,1]\times[0,1]$.

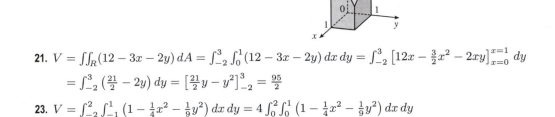

21. $V = \iint_R(12-3x-2y)\,dA = \int_{-2}^3\int_0^1(12-3x-2y)\,dx\,dy = \int_{-2}^3\left[12x-\tfrac{3}{2}x^2-2xy\right]_{x=0}^{x=1}\,dy$

$$= \int_{-2}^3\left(\tfrac{21}{2}-2y\right)dy = \left[\tfrac{21}{2}y - y^2\right]_{-2}^3 = \tfrac{95}{2}$$

23. $V = \int_{-2}^2\int_{-1}^1\left(1-\tfrac{1}{4}x^2-\tfrac{1}{9}y^2\right)dx\,dy = 4\int_0^2\int_0^1\left(1-\tfrac{1}{4}x^2-\tfrac{1}{9}y^2\right)dx\,dy$

$$= 4\int_0^2\left[x-\tfrac{1}{12}x^3-\tfrac{1}{9}y^2x\right]_{x=0}^{x=1}\,dy = 4\int_0^2\left(\tfrac{11}{12}-\tfrac{1}{9}y^2\right)dy = 4\left[\tfrac{11}{12}y-\tfrac{1}{27}y^3\right]_0^2 = 4\cdot\tfrac{83}{54} = \tfrac{166}{27}$$

25. Here we need the volume of the solid lying under the surface $z = x\sqrt{x^2+y}$ and above the square $R = [0,1]\times[0,1]$ in the xy-plane.

$$V\int_0^1\int_0^1 x\sqrt{x^2+y}\,dx\,dy = \int_0^1\tfrac{1}{3}\left[(x^2+y)^{3/2}\right]_{x=0}^{x=1}\,dy = \tfrac{1}{3}\int_0^1\left[(1+y)^{3/2}-y^{3/2}\right]dy$$

$$= \tfrac{1}{3}\cdot\tfrac{2}{5}\left[(1+y)^{5/2}-y^{5/2}\right]_0^1 = \tfrac{4}{15}(2\sqrt{2}-1)$$

27. In the first octant, $z \ge 0 \implies y \le 3$, so

$$V = \int_0^3\int_0^2(9-y^2)\,dx\,dy = \int_0^3\left[9x-y^2x\right]_{x=0}^{x=2}\,dy = \int_0^3(18-2y^2)\,dy = \left[18y-\tfrac{2}{3}y^3\right]_0^3 = 36$$

29. In Maple, we can calculate the integral by defining the integrand as `f` and then using the command `int(int(f,x=0..1),y=0..1);`. In Mathematica, we can use the command `Integrate[Integrate[f,{x,0,1}],{y,0,1}]`. We find that $\iint_R x^5y^3e^{xy}\,dA = 21e - 57 \approx 0.0839$. We can use `plot3d` (in Maple) or `Plot3d` (in Mathematica) to graph the function.

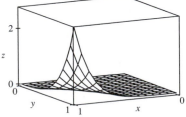

31. R is the rectangle $[-1, 1] \times [0, 5]$. Thus, $A(R) = 2 \cdot 5 = 10$ and

$$f_{\text{ave}} = \frac{1}{A(R)} \iint_R f(x, y)\, dA = \frac{1}{10} \int_0^5 \int_{-1}^1 x^2 y\, dx\, dy = \frac{1}{10} \int_0^5 \left[\frac{1}{3} x^3 y\right]_{x=-1}^{x=1} dy = \frac{1}{10} \int_0^5 \frac{2}{3} y\, dy = \frac{1}{10} \left[\frac{1}{3} y^2\right]_0^5 = \frac{5}{6}.$$

33. Let $f(x, y) = \dfrac{x - y}{(x + y)^3}$. Then a CAS gives $\int_0^1 \int_0^1 f(x, y)\, dy\, dx = \frac{1}{2}$ and $\int_0^1 \int_0^1 f(x, y)\, dx\, dy = -\frac{1}{2}$.

To explain the seeming violation of Fubini's Theorem, note that f has an infinite discontinuity at $(0, 0)$ and thus does not

satisfy the conditions of Fubini's Theorem. In fact, both iterated integrals involve improper integrals which diverge at their

lower limits of integration.

12.3   Double Integrals over General Regions

1. $\int_0^1 \int_0^{x^2} (x + 2y)\, dy\, dx = \int_0^1 \left[xy + y^2\right]_{y=0}^{y=x^2} dx = \int_0^1 \left[x(x^2) + (x^2)^2 - 0 - 0\right] dx$

$$= \int_0^1 (x^3 + x^4)\, dx = \left[\frac{1}{4} x^4 + \frac{1}{5} x^5\right]_0^1 = \frac{9}{20}$$

3. $\int_0^1 \int_y^{e^y} \sqrt{x}\, dx\, dy = \int_0^1 \left[\frac{2}{3} x^{3/2}\right]_{x=y}^{x=e^y} dy = \frac{2}{3} \int_0^1 (e^{3y/2} - y^{3/2})\, dy = \frac{2}{3} \left[\frac{2}{3} e^{3y/2} - \frac{2}{5} y^{5/2}\right]_0^1$

$$= \frac{2}{3} \left(\frac{2}{3} e^{3/2} - \frac{2}{5} - \frac{2}{3} e^0 + 0\right) = \frac{4}{9} e^{3/2} - \frac{32}{45}$$

5. $\int_0^{\pi/2} \int_0^{\cos\theta} e^{\sin\theta}\, dr\, d\theta = \int_0^{\pi/2} \left[r e^{\sin\theta}\right]_{r=0}^{r=\cos\theta} d\theta = \int_0^{\pi/2} (\cos\theta)\, e^{\sin\theta}\, d\theta = e^{\sin\theta}\Big]_0^{\pi/2} = e^{\sin(\pi/2)} - e^0 = e - 1$

7. $\iint_D x^3 y^2\, dA = \int_0^2 \int_{-x}^x x^3 y^2\, dy\, dx = \int_0^2 \left[\frac{1}{3} x^3 y^3\right]_{y=-x}^{y=x} dx = \frac{1}{3} \int_0^2 2x^6\, dx = \frac{2}{3} \left[\frac{1}{7} x^7\right]_0^2 = \frac{2}{21} \left[2^7 - 0\right] = \frac{256}{21}$

9. $\int_0^1 \int_0^{\sqrt{x}} \frac{2y}{x^2 + 1}\, dy\, dx = \int_0^1 \left[\frac{y^2}{x^2 + 1}\right]_{y=0}^{y=\sqrt{x}} dx = \int_0^1 \frac{x}{x^2 + 1}\, dx = \frac{1}{2} \ln\left|x^2 + 1\right|\Big|_0^1 = \frac{1}{2}(\ln 2 - \ln 1) = \frac{1}{2} \ln 2$

11. $\int_0^1 \int_0^{x^2} x \cos y\, dy\, dx = \int_0^1 \left[x \sin y\right]_{y=0}^{y=x^2} dx = \int_0^1 x \sin x^2\, dx = -\frac{1}{2} \cos x^2\Big]_0^1 = \frac{1}{2}(1 - \cos 1)$

13.

(0, 2) D (3, 2)

$x = 2 - y$ $x = 2y - 1$

(1, 1)

$$\int_1^2 \int_{2-y}^{2y-1} y^3\, dx\, dy = \int_1^2 \left[xy^3\right]_{x=2-y}^{x=2y-1} dy$$

$$= \int_1^2 \left[(2y - 1) - (2 - y)\right] y^3\, dy$$

$$= \int_1^2 (3y^4 - 3y^3)\, dy = \left[\frac{3}{5} y^5 - \frac{3}{4} y^4\right]_1^2$$

$$= \frac{96}{5} - 12 - \frac{3}{5} + \frac{3}{4} = \frac{147}{20}$$

15.

$y = \sqrt{4 - x^2}$

D

$y = -\sqrt{4 - x^2}$

$$\int_{-2}^2 \int_{-\sqrt{4-x^2}}^{\sqrt{4-x^2}} (2x - y)\, dy\, dx$$

$$= \int_{-2}^2 \left[2xy - \frac{1}{2} y^2\right]_{y=-\sqrt{4-x^2}}^{y=\sqrt{4-x^2}} dx$$

$$= \int_{-2}^2 \left[2x \sqrt{4 - x^2} - \frac{1}{2}(4 - x^2) + 2x \sqrt{4 - x^2} + \frac{1}{2}(4 - x^2)\right] dx$$

$$= \int_{-2}^2 4x \sqrt{4 - x^2}\, dx = -\frac{4}{3}(4 - x^2)^{3/2}\Big]_{-2}^2 = 0$$

(Or, note that $4x \sqrt{4 - x^2}$ is an odd function, so $\int_{-2}^2 4x \sqrt{4 - x^2}\, dx = 0$.)

17.

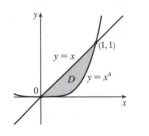

$V = \int_0^1 \int_{x^4}^x (x + 2y)\, dy\, dx$

$= \int_0^1 \left[xy + y^2 \right]_{y=x^4}^{y=x} dx = \int_0^1 (2x^2 - x^5 - x^8)\, dx$

$= \left[\frac{2}{3}x^3 - \frac{1}{6}x^6 - \frac{1}{9}x^9 \right]_0^1 = \frac{2}{3} - \frac{1}{6} - \frac{1}{9} = \frac{7}{18}$

19.

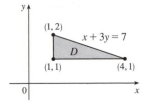

$V = \int_1^2 \int_1^{7-3y} xy\, dx\, dy = \int_1^2 \left[\frac{1}{2}x^2 y \right]_{x=1}^{x=7-3y} dy$

$= \frac{1}{2} \int_1^2 (48y - 42y^2 + 9y^3)\, dy$

$= \frac{1}{2} \left[24y^2 - 14y^3 + \frac{9}{4}y^4 \right]_1^2 = \frac{31}{8}$

21.

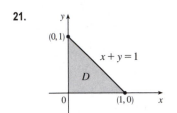

$V = \int_0^1 \int_0^{1-x} (1 - x - y)\, dy\, dx = \int_0^1 \left[y - xy - \frac{y^2}{2} \right]_{y=0}^{y=1-x} dx$

$= \int_0^1 \left[(1-x)^2 - \frac{1}{2}(1-x)^2 \right] dx$

$= \int_0^1 \frac{1}{2}(1-x)^2\, dx = \left[-\frac{1}{6}(1-x)^3 \right]_0^1 = \frac{1}{6}$

23.

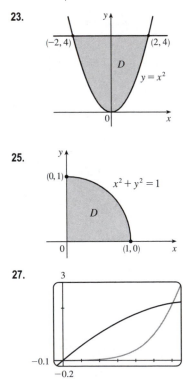

$V = \int_{-2}^2 \int_{x^2}^4 x^2\, dy\, dx$

$= \int_{-2}^2 x^2 \left[y \right]_{y=x^2}^{y=4} dx = \int_{-2}^2 (4x^2 - x^4)\, dx$

$= \left[\frac{4}{3}x^3 - \frac{1}{5}x^5 \right]_{-2}^2 = \frac{32}{3} - \frac{32}{5} + \frac{32}{3} - \frac{32}{5} = \frac{128}{15}$

25.

$V = \int_0^1 \int_0^{\sqrt{1-x^2}} y\, dy\, dx = \int_0^1 \left[\frac{y^2}{2} \right]_{y=0}^{y=\sqrt{1-x^2}} dx$

$= \int_0^1 \frac{1-x^2}{2}\, dx = \frac{1}{2} \left[x - \frac{1}{3}x^3 \right]_0^1 = \frac{1}{3}$

27.

From the graph, it appears that the two curves intersect at $x = 0$ and at $x \approx 1.213$. Thus the desired integral is

$\iint_D x\, dA \approx \int_0^{1.213} \int_{x^4}^{3x-x^2} x\, dy\, dx = \int_0^{1.213} \left[xy \right]_{y=x^4}^{y=3x-x^2} dx$

$= \int_0^{1.213} (3x^2 - x^3 - x^5)\, dx = \left[x^3 - \frac{1}{4}x^4 - \frac{1}{6}x^{6\approx 0.713} \right]_0^{1.213}$

≈ 0.713

29. The two bounding curves $y = 1 - x^2$ and $y = x^2 - 1$ intersect at $(\pm 1, 0)$ with $1 - x^2 \geq x^2 - 1$ on $[-1, 1]$. Within this region, the plane $z = 2x + 2y + 10$ is above the plane $z = 2 - x - y$, so

$$V = \int_{-1}^{1} \int_{x^2-1}^{1-x^2} (2x + 2y + 10) \, dy \, dx - \int_{-1}^{1} \int_{x^2-1}^{1-x^2} (2 - x - y) \, dy \, dx$$

$$= \int_{-1}^{1} \int_{x^2-1}^{1-x^2} (2x + 2y + 10 - (2 - x - y)) \, dy \, dx = \int_{-1}^{1} \int_{x^2-1}^{1-x^2} (3x + 3y + 8) \, dy \, dx$$

$$= \int_{-1}^{1} \left[3xy + \tfrac{3}{2}y^2 + 8y \right]_{y=x^2-1}^{y=1-x^2} dx$$

$$= \int_{-1}^{1} \left[3x(1 - x^2) + \tfrac{3}{2}(1 - x^2)^2 + 8(1 - x^2) - 3x(x^2 - 1) - \tfrac{3}{2}(x^2 - 1)^2 - 8(x^2 - 1) \right] dx$$

$$= \int_{-1}^{1} (-6x^3 - 16x^2 + 6x + 16) \, dx = \left[-\tfrac{3}{2}x^4 - \tfrac{16}{3}x^3 + 3x^2 + 16x \right]_{-1}^{1}$$

$$= -\tfrac{3}{2} - \tfrac{16}{3} + 3 + 16 + \tfrac{3}{2} - \tfrac{16}{3} - 3 + 16 = \tfrac{64}{3}$$

31. The two surfaces intersect in the circle $x^2 + y^2 = 1$, $z = 0$ and the region of integration is the disk $D: x^2 + y^2 \leq 1$. Using a CAS, the volume is $\iint_D (1 - x^2 - y^2) \, dA = \int_{-1}^{1} \int_{-\sqrt{1-x^2}}^{\sqrt{1-x^2}} (1 - x^2 - y^2) \, dy \, dx = \dfrac{\pi}{2}$.

33.

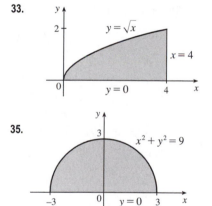

Because the region of integration is

$$D = \{(x, y) \mid 0 \leq y \leq \sqrt{x}, 0 \leq x \leq 4\} = \{(x, y) \mid y^2 \leq x \leq 4, 0 \leq y \leq 2\}$$

we have $\int_0^4 \int_0^{\sqrt{x}} f(x, y) \, dy \, dx = \iint_D f(x, y) \, dA = \int_0^2 \int_{y^2}^{4} f(x, y) \, dx \, dy$.

35.

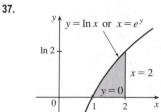

Because the region of integration is

$$D = \left\{ (x, y) \mid -\sqrt{9 - y^2} \leq x \leq \sqrt{9 - y^2}, 0 \leq y \leq 3 \right\}$$
$$= \left\{ (x, y) \mid 0 \leq y \leq \sqrt{9 - x^2}, -3 \leq x \leq 3 \right\}$$

we have

$$\int_0^3 \int_{-\sqrt{9-y^2}}^{\sqrt{9-y^2}} f(x, y) \, dx \, dy = \iint_D f(x, y) \, dA$$

$$= \int_{-3}^{3} \int_0^{\sqrt{9-x^2}} f(x, y) \, dy \, dx.$$

37.

Because the region of integration is

$$D = \{(x, y) \mid 0 \leq y \leq \ln x, 1 \leq x \leq 2\} = \{(x, y) \mid e^y \leq x \leq 2, 0 \leq y \leq \ln 2\}$$

we have

$$\int_1^2 \int_0^{\ln x} f(x, y) \, dy \, dx = \iint_D f(x, y) \, dA = \int_0^{\ln 2} \int_{e^y}^{2} f(x, y) \, dx \, dy.$$

39.

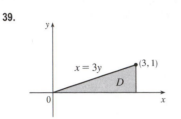

$$\int_0^1 \int_{3y}^3 e^{x^2}\, dx\, dy = \int_0^3 \int_0^{x/3} e^{x^2}\, dy\, dx$$

$$= \int_0^3 \left[e^{x^2} y\right]_{y=0}^{y=x/3} dx = \int_0^3 \left(\frac{x}{3}\right) e^{x^2}\, dx$$

$$= \tfrac{1}{6}\, e^{x^2}\Big]_0^3 = \frac{e^9 - 1}{6}$$

41.

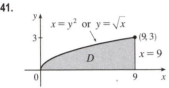

$$\int_0^3 \int_{y^2}^9 y\cos x^2\, dx\, dy = \int_0^9 \int_0^{\sqrt{x}} y\cos x^2\, dy\, dx$$

$$= \int_0^9 \cos x^2 \left[\frac{y^2}{2}\right]_{y=0}^{y=\sqrt{x}} dx = \int_0^9 \tfrac{1}{2} x\cos x^2\, dx$$

$$= \tfrac{1}{4}\sin x^2\Big]_0^9 = \tfrac{1}{4}\sin 81$$

43.

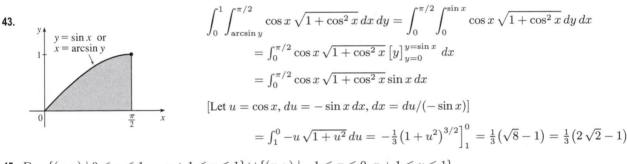

$$\int_0^1 \int_{\arcsin y}^{\pi/2} \cos x\, \sqrt{1+\cos^2 x}\, dx\, dy = \int_0^{\pi/2}\int_0^{\sin x} \cos x\, \sqrt{1+\cos^2 x}\, dy\, dx$$

$$= \int_0^{\pi/2} \cos x\, \sqrt{1+\cos^2 x}\, \left[y\right]_{y=0}^{y=\sin x} dx$$

$$= \int_0^{\pi/2} \cos x\, \sqrt{1+\cos^2 x}\, \sin x\, dx$$

[Let $u = \cos x$, $du = -\sin x\, dx$, $dx = du/(-\sin x)$]

$$= \int_1^0 -u\sqrt{1+u^2}\, du = -\tfrac{1}{3}\left(1+u^2\right)^{3/2}\Big]_1^0 = \tfrac{1}{3}\left(\sqrt{8}-1\right) = \tfrac{1}{3}\left(2\sqrt{2}-1\right)$$

45. $D = \{(x,y)\mid 0\le x\le 1,\ -x+1\le y\le 1\} \cup \{(x,y)\mid -1\le x\le 0,\ x+1\le y\le 1\}$

$$\cup\{(x,y)\mid 0\le x\le 1,\ -1\le y\le x-1\} \cup \{(x,y)\mid -1\le x\le 0,\ -1\le y\le -x-1\},$$

all type I.

$$\iint_D x^2\, dA = \int_0^1\int_{1-x}^1 x^2\, dy\, dx + \int_{-1}^0\int_{x+1}^1 x^2\, dy\, dx + \int_0^1\int_{-1}^{x-1} x^2\, dy\, dx + \int_{-1}^0\int_{-1}^{-x-1} x^2\, dy\, dx$$

$$= 4\int_0^1\int_{1-x}^1 x^2\, dy\, dx \qquad \text{[by symmetry of the regions and because } f(x,y) = x^2 \ge 0]$$

$$= 4\int_0^1 x^3\, dx = 4\left[\tfrac{1}{4}x^4\right]_0^1 = 1$$

47. For $D = [0,1]\times[0,1]$, $0\le \sqrt{x^3+y^3}\le \sqrt{2}$ and $A(D) = 1$, so $0\le \iint_D \sqrt{x^3+y^3}\, dA \le \sqrt{2}$.

49. Since $m\le f(x,y)\le M$, $\iint_D m\, dA \le \iint_D f(x,y)\, dA \le \iint_D M\, dA$ by (8) $\Rightarrow$

$m\iint_D 1\, dA \le \iint_D f(x,y)\, dA \le M\iint_D 1\, dA$ by (7) $\Rightarrow$ $mA(D)\le \iint_D f(x,y)\, dA \le MA(D)$ by (10).

51. $\iint_D (x^2\tan x + y^3 + 4)\, dA = \iint_D x^2\tan x\, dA + \iint_D y^3\, dA + \iint_D 4\, dA$. But $x^2\tan x$ is an odd function of x and D is

symmetric with respect to the y-axis, so $\iint_D x^2\tan x\, dA = 0$. Similarly, y^3 is an odd function of y and D is symmetric with

respect to the x-axis, so $\iint_D y^3\, dA = 0$. Thus $\iint_D (x^2\tan x + y^3 + 4)\, dA = 4\iint_D dA = 4(\text{area of } D) = 4\cdot\pi\left(\sqrt{2}\right)^2 = 8\pi$.

53. Since $\sqrt{1-x^2-y^2}\ge 0$, we can interpret $\iint_D \sqrt{1-x^2-y^2}\, dA$ as the volume of the solid that lies below the graph of

$z = \sqrt{1-x^2-y^2}$ and above the region D in the xy-plane. $z = \sqrt{1-x^2-y^2}$ is equivalent to $x^2+y^2+z^2 = 1$, $z\ge 0$

which meets the xy-plane in the circle $x^2+y^2 = 1$, the boundary of D. Thus, the solid is an upper hemisphere of radius 1

which has volume $\tfrac{1}{2}\left[\tfrac{4}{3}\pi(1)^3\right] = \tfrac{2}{3}\pi$.

12.4 Double Integrals in Polar Coordinates

1. The region R is more easily described by polar coordinates: $R = \{(r, \theta) \mid 0 \le r \le 2, 0 \le \theta \le 2\pi\}$.

Thus $\iint_R f(x, y)\, dA = \int_0^{2\pi} \int_0^2 f(r\cos\theta, r\sin\theta)\, r\, dr\, d\theta$.

3. The region R is more easily described by rectangular coordinates: $R = \{(x, y) \mid -2 \le x \le 2, x \le y \le 2\}$.

Thus $\iint_R f(x, y)\, dA = \int_{-2}^2 \int_x^2 f(x, y)\, dy\, dx$.

5. The region R is more easily described by polar coordinates: $R = \{(r, \theta) \mid 2 \le r \le 5, 0 \le \theta \le 2\pi\}$.

Thus $\iint_R f(x, y)\, dA = \int_0^{2\pi} \int_2^5 f(r\cos\theta, r\sin\theta)\, r\, dr\, d\theta$.

7. The integral $\int_\pi^{2\pi} \int_4^7 r\, dr\, d\theta$ represents the area of the region

$R = \{(r, \theta) \mid 4 \le r \le 7, \pi \le \theta \le 2\pi\}$, the lower half of a ring.

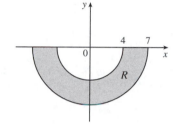

$$\int_\pi^{2\pi} \int_4^7 r\, dr\, d\theta = \left(\int_\pi^{2\pi} d\theta\right)\left(\int_4^7 r\, dr\right)$$

$$= \left[\theta\right]_\pi^{2\pi} \left[\tfrac{1}{2}r^2\right]_4^7 = \pi \cdot \tfrac{1}{2}(49 - 16) = \frac{33\pi}{2}$$

9. The disk D can be described in polar coordinates as $D = \{(r, \theta) \mid 0 \le r \le 3, 0 \le \theta \le 2\pi\}$. Then

$$\iint_D xy\, dA = \int_0^{2\pi} \int_0^3 (r\cos\theta)(r\sin\theta)\, r\, dr\, d\theta = \left(\int_0^{2\pi} \sin\theta\cos\theta\, d\theta\right)\left(\int_0^3 r^3\, dr\right) = \left[\tfrac{1}{2}\sin^2\theta\right]_0^{2\pi} \left[\tfrac{1}{4}r^4\right]_0^3 = 0.$$

11. $\iint_R \cos(x^2 + y^2)\, dA = \int_0^\pi \int_0^3 \cos(r^2)\, r\, dr\, d\theta = \left(\int_0^\pi d\theta\right)\left(\int_0^3 r\cos(r^2)\, dr\right)$

$$= \left[\theta\right]_0^\pi \left[\tfrac{1}{2}\sin(r^2)\right]_0^3 = \pi \cdot \tfrac{1}{2}(\sin 9 - \sin 0) = \tfrac{\pi}{2}\sin 9$$

13. R is the region shown in the figure, and can be described

by $R = \{(r, \theta) \mid 0 \le \theta \le \pi/4, 1 \le r \le 2\}$. Thus

$\iint_R \arctan(y/x)\, dA = \int_0^{\pi/4} \int_1^2 \arctan(\tan\theta)\, r\, dr\, d\theta$ since $y/x = \tan\theta$.

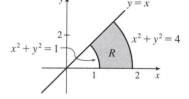

Also, $\arctan(\tan\theta) = \theta$ for $0 \le \theta \le \pi/4$, so the integral becomes

$\int_0^{\pi/4} \int_1^2 \theta\, r\, dr\, d\theta = \int_0^{\pi/4} \theta\, d\theta \int_1^2 r\, dr = \left[\tfrac{1}{2}\theta^2\right]_0^{\pi/4} \left[\tfrac{1}{2}r^2\right]_1^2 = \tfrac{\pi^2}{32} \cdot \tfrac{3}{2} = \tfrac{3}{64}\pi^2$.

15. $V = \iint_{x^2+y^2 \le 4} \sqrt{x^2 + y^2}\, dA = \int_0^{2\pi} \int_0^2 \sqrt{r^2}\, r\, dr\, d\theta = \int_0^{2\pi} d\theta \int_0^2 r^2\, dr = \left[\theta\right]_0^{2\pi} \left[\tfrac{1}{3}r^3\right]_0^2 = 2\pi\left(\tfrac{8}{3}\right) = \tfrac{16}{3}\pi$

17. By symmetry,

$$V = 2 \iint_{x^2+y^2 \le a^2} \sqrt{a^2 - x^2 - y^2}\, dA = 2\int_0^{2\pi} \int_0^a \sqrt{a^2 - r^2}\, r\, dr\, d\theta = 2\int_0^{2\pi} d\theta \int_0^a r\sqrt{a^2 - r^2}\, dr$$

$$= 2\left[\theta\right]_0^{2\pi} \left[-\tfrac{1}{3}(a^2 - r^2)^{3/2}\right]_0^a = 2(2\pi)\left(0 + \tfrac{1}{3}a^3\right) = \tfrac{4\pi}{3}a^3$$

19. The cone $z = \sqrt{x^2 + y^2}$ intersects the sphere $x^2 + y^2 + z^2 = 1$ when $x^2 + y^2 + \left(\sqrt{x^2 + y^2}\right)^2 = 1$ or $x^2 + y^2 = \tfrac{1}{2}$. So

$$V = \iint_{x^2+y^2 \le 1/2} \left(\sqrt{1 - x^2 - y^2} - \sqrt{x^2 + y^2}\right) dA = \int_0^{2\pi} \int_0^{1/\sqrt{2}} \left(\sqrt{1 - r^2} - r\right) r\, dr\, d\theta$$

$$= \int_0^{2\pi} d\theta \int_0^{1/\sqrt{2}} \left(r\sqrt{1 - r^2} - r^2\right) dr = \left[\theta\right]_0^{2\pi} \left[-\tfrac{1}{3}(1 - r^2)^{3/2} - \tfrac{1}{3}r^3\right]_0^{1/\sqrt{2}}$$

$$= 2\pi\left(-\tfrac{1}{3}\right)\left(\tfrac{1}{\sqrt{2}} - 1\right) = \tfrac{\pi}{3}\left(2 - \sqrt{2}\right)$$

21. The given solid is the region inside the cylinder $x^2 + y^2 = 4$ between the surfaces $z = \sqrt{64 - 4x^2 - 4y^2}$

and $z = -\sqrt{64 - 4x^2 - 4y^2}$. So

$$V = \iint\limits_{x^2 + y^2 \le 4} \left[\sqrt{64 - 4x^2 - 4y^2} - \left(-\sqrt{64 - 4x^2 - 4y^2} \right) \right] dA = \iint\limits_{x^2 + y^2 \le 4} 2\sqrt{64 - 4x^2 - 4y^2} \, dA$$

$$= 4 \int_0^{2\pi} \int_0^2 \sqrt{16 - r^2} \, r \, dr \, d\theta = 4 \int_0^{2\pi} d\theta \int_0^2 r\sqrt{16 - r^2} \, dr = 4 \left[\theta \right]_0^{2\pi} \left[-\tfrac{1}{3}(16 - r^2)^{3/2} \right]_0^2$$

$$= 8\pi \left(-\tfrac{1}{3} \right)(12^{3/2} - 16^{2/3}) = \tfrac{8\pi}{3} \left(64 - 24\sqrt{3} \right)$$

23. One loop is given by the region

$D = \{(r, \theta) \,|\, {-\pi/6} \le \theta \le \pi/6, 0 \le r \le \cos 3\theta \}$, so the area is

$$\iint\limits_D dA = \int_{-\pi/6}^{\pi/6} \int_0^{\cos 3\theta} r \, dr \, d\theta = \int_{-\pi/6}^{\pi/6} \left[\tfrac{1}{2} r^2 \right]_{r=0}^{r=\cos 3\theta} d\theta$$

$$= \int_{-\pi/6}^{\pi/6} \tfrac{1}{2} \cos^2 3\theta \, d\theta = 2 \int_0^{\pi/6} \tfrac{1}{2} \left(\frac{1 + \cos 6\theta}{2} \right) d\theta$$

$$= \tfrac{1}{2} \left[\theta + \tfrac{1}{6} \sin 6\theta \right]_0^{\pi/6} = \frac{\pi}{12}$$

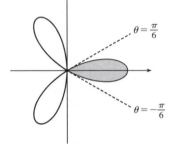

25.

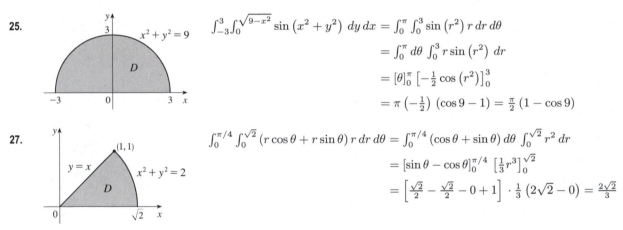

$x^2 + y^2 = 9$

D

$$\int_{-3}^3 \int_0^{\sqrt{9-x^2}} \sin\left(x^2 + y^2 \right) dy \, dx = \int_0^\pi \int_0^3 \sin\left(r^2 \right) r \, dr \, d\theta$$

$$= \int_0^\pi d\theta \int_0^3 r \sin\left(r^2 \right) dr$$

$$= \left[\theta \right]_0^\pi \left[-\tfrac{1}{2} \cos\left(r^2 \right) \right]_0^3$$

$$= \pi \left(-\tfrac{1}{2} \right)(\cos 9 - 1) = \tfrac{\pi}{2} \left(1 - \cos 9 \right)$$

27.

$y = x$

$(1, 1)$

$x^2 + y^2 = 2$

D

$\sqrt{2}$ x

$$\int_0^{\pi/4} \int_0^{\sqrt{2}} (r \cos \theta + r \sin \theta) \, r \, dr \, d\theta = \int_0^{\pi/4} (\cos \theta + \sin \theta) \, d\theta \int_0^{\sqrt{2}} r^2 \, dr$$

$$= \left[\sin \theta - \cos \theta \right]_0^{\pi/4} \left[\tfrac{1}{3} r^3 \right]_0^{\sqrt{2}}$$

$$= \left[\tfrac{\sqrt{2}}{2} - \tfrac{\sqrt{2}}{2} - 0 + 1 \right] \cdot \tfrac{1}{3} \left(2\sqrt{2} - 0 \right) = \tfrac{2\sqrt{2}}{3}$$

29. The surface of the water in the pool is a circular disk D with radius 20 ft. If we place D on coordinate axes with the origin at the center of D and define $f(x, y)$ to be the depth of the water at (x, y), then the volume of water in the pool is the volume of the solid that lies above $D = \{(x, y) \,|\, x^2 + y^2 \le 400\}$ and below the graph of $f(x, y)$. We can associate north with the positive y-direction, so we are given that the depth is constant in the x-direction and the depth increases linearly in the y-direction from $f(0, -20) = 2$ to $f(0, 20) = 7$. The trace in the yz-plane is a line segment from $(0, -20, 2)$ to $(0, 20, 7)$. The slope of this line is $\frac{7-2}{20-(-20)} = \tfrac{1}{8}$, so an equation of the line is $z - 7 = \tfrac{1}{8}(y - 20) \;\Rightarrow\; z = \tfrac{1}{8} y + \tfrac{9}{2}$. Since $f(x, y)$ is independent of x, $f(x, y) = \tfrac{1}{8} y + \tfrac{9}{2}$. Thus the volume is given by $\iint_D f(x, y) \, dA$, which is most conveniently evaluated using polar coordinates. Then $D = \{(r, \theta) \,|\, 0 \le r \le 20, 0 \le \theta \le 2\pi\}$ and substituting $x = r \cos \theta$, $y = r \sin \theta$ the integral becomes

$$\int_0^{2\pi} \int_0^{20} \left(\tfrac{1}{8} r \sin \theta + \tfrac{9}{2} \right) r \, dr \, d\theta = \int_0^{2\pi} \left[\tfrac{1}{24} r^3 \sin \theta + \tfrac{9}{4} r^2 \right]_{r=0}^{r=20} d\theta = \int_0^{2\pi} \left(\tfrac{1000}{3} \sin \theta + 900 \right) d\theta$$

$$= \left[-\tfrac{1000}{3} \cos \theta + 900\theta \right]_0^{2\pi} = 1800\pi$$

Thus the pool contains $1800\pi \approx 5655$ ft^3 of water.

31. $\int_{1/\sqrt{2}}^{1}\int_{\sqrt{1-x^2}}^{x} xy\,dy\,dx + \int_{1}^{\sqrt{2}}\int_{0}^{x} xy\,dy\,dx + \int_{\sqrt{2}}^{2}\int_{0}^{\sqrt{4-x^2}} xy\,dy\,dx$

$$= \int_{0}^{\pi/4}\int_{1}^{2} r^3\cos\theta\sin\theta\,dr\,d\theta = \int_{0}^{\pi/4}\left[\frac{r^4}{4}\cos\theta\sin\theta\right]_{r=1}^{r=2} d\theta$$

$$= \frac{15}{4}\int_{0}^{\pi/4}\sin\theta\cos\theta\,d\theta = \frac{15}{4}\left[\frac{\sin^2\theta}{2}\right]_{0}^{\pi/4} = \frac{15}{16}$$

33. (a) We integrate by parts with $u = x$ and $dv = xe^{-x^2}\,dx$. Then $du = dx$ and $v = -\frac{1}{2}e^{-x^2}$, so

$$\int_{0}^{\infty} x^2 e^{-x^2}\,dx = \lim_{t\to\infty}\int_{0}^{t} x^2 e^{-x^2}\,dx = \lim_{t\to\infty}\left(-\frac{1}{2}xe^{-x^2}\Big|_{0}^{t} + \int_{0}^{t}\frac{1}{2}e^{-x^2}\,dx\right)$$

$$= \lim_{t\to\infty}\left(-\frac{1}{2}te^{-t^2}\right) + \frac{1}{2}\int_{0}^{\infty} e^{-x^2}\,dx = 0 + \frac{1}{2}\int_{0}^{\infty} e^{-x^2}\,dx \qquad \text{[by l'Hospital's Rule]}$$

$$= \frac{1}{4}\int_{-\infty}^{\infty} e^{-x^2}\,dx \qquad \text{[since } e^{-x^2}\text{ is an even function]}$$

$$= \frac{1}{4}\sqrt{\pi} \qquad \text{[by Exercise 32(c)]}$$

(b) Let $u = \sqrt{x}$. Then $u^2 = x \Rightarrow dx = 2u\,du \Rightarrow$

$$\int_{0}^{\infty}\sqrt{x}\,e^{-x}\,dx = \lim_{t\to\infty}\int_{0}^{t}\sqrt{x}\,e^{-x}\,dx = \lim_{t\to\infty}\int_{0}^{\sqrt{t}} ue^{-u^2}2u\,du = 2\int_{0}^{\infty} u^2 e^{-u^2}\,du = 2\left(\frac{1}{4}\sqrt{\pi}\right) \text{ [by part(a)]}$$

$$= \frac{1}{2}\sqrt{\pi}$$

12.5   Applications of Double Integrals

1. $Q = \iint_D \sigma(x,y)\,dA = \int_1^3\int_0^2 (2xy + y^2)\,dy\,dx = \int_1^3\left[xy^2 + \frac{1}{3}y^3\right]_{y=0}^{y=2} dx$

$= \int_1^3\left(4x + \frac{8}{3}\right) dx = \left[2x^2 + \frac{8}{3}x\right]_1^3 = 16 + \frac{16}{3} = \frac{64}{3}$ C

3. $m = \iint_D \rho(x,y)\,dA = \int_0^2\int_{-1}^1 xy^2\,dy\,dx = \int_0^2 x\,dx\int_{-1}^1 y^2\,dy = \left[\frac{1}{2}x^2\right]_0^2\left[\frac{1}{3}y^3\right]_{-1}^1 = 2\cdot\frac{2}{3} = \frac{4}{3}$,

$\bar{x} = \frac{1}{m}\iint_D x\rho(x,y)\,dA = \frac{3}{4}\int_0^2\int_{-1}^1 x^2y^2\,dy\,dx = \frac{3}{4}\int_0^2 x^2\,dx\int_{-1}^1 y^2\,dy = \frac{3}{4}\left[\frac{1}{3}x^3\right]_0^2\left[\frac{1}{3}y^3\right]_{-1}^1 = \frac{3}{4}\cdot\frac{8}{3}\cdot\frac{2}{3} = \frac{4}{3}$,

$\bar{y} = \frac{1}{m}\iint_D y\rho(x,y)\,dA = \frac{3}{4}\int_0^2\int_{-1}^1 xy^3\,dy\,dx = \frac{3}{4}\int_0^2 x\,dx\int_{-1}^1 y^3\,dy = \frac{3}{4}\left[\frac{1}{2}x^2\right]_0^2\left[\frac{1}{4}y^4\right]_{-1}^1 = \frac{3}{4}\cdot 2\cdot 0 = 0$.

Hence, $(\bar{x},\bar{y}) = \left(\frac{4}{3}, 0\right)$.

5. $m = \int_0^2\int_{x/2}^{3-x}(x+y)\,dy\,dx = \int_0^2\left[xy + \frac{1}{2}y^2\right]_{y=x/2}^{y=3-x} dx = \int_0^2\left[x(3-\frac{3}{2}x) + \frac{1}{2}(3-x)^2 - \frac{1}{8}x^2\right] dx$

$= \int_0^2\left(-\frac{9}{8}x^2 + \frac{9}{2}\right) dx = \left[-\frac{9}{8}\left(\frac{1}{3}x^3\right) + \frac{9}{2}x\right]_0^2 = 6$,

$M_y = \int_0^2\int_{x/2}^{3-x}(x^2 + xy)\,dy\,dx = \int_0^2\left[x^2y + \frac{1}{2}xy^2\right]_{y=x/2}^{y=3-x} dx = \int_0^2\left(\frac{9}{2}x - \frac{9}{8}x^3\right) dx = \frac{9}{2}$, and

$M_x = \int_0^2\int_{x/2}^{3-x}(xy + y^2)\,dy\,dx = \int_0^2\left[\frac{1}{2}xy^2 + \frac{1}{3}y^3\right]_{y=x/2}^{y=3-x} dx = \int_0^2\left(9 - \frac{9}{2}x\right) dx = 9$.

Hence $m = 6$, $(\bar{x},\bar{y}) = \left(\frac{M_y}{m}, \frac{M_x}{m}\right) = \left(\frac{3}{4}, \frac{3}{2}\right)$.

7. $m = \int_0^1\int_0^{e^x} y\,dy\,dx = \int_0^1\left[\frac{1}{2}y^2\right]_{y=0}^{y=e^x} dx = \frac{1}{2}\int_0^1 e^{2x}\,dx = \frac{1}{4}e^{2x}\Big|_0^1 = \frac{1}{4}(e^2 - 1)$,

$M_y = \int_0^1\int_0^{e^x} xy\,dy\,dx = \frac{1}{2}\int_0^1 xe^{2x}\,dx = \frac{1}{2}\left[\frac{1}{2}xe^{2x} - \frac{1}{4}e^{2x}\right]_0^1 = \frac{1}{8}(e^2 + 1)$, and

$M_x = \int_0^1\int_0^{e^x} y^2\,dy\,dx = \int_0^1\left[\frac{1}{3}y^3\right]_{y=0}^{y=e^x} dx = \frac{1}{3}\int_0^1 e^{3x}\,dx = \frac{1}{3}\left[\frac{1}{3}e^{3x}\right]_0^1 = \frac{1}{9}(e^3 - 1)$.

Hence $m = \frac{1}{4}(e^2-1)$, $(\bar{x},\bar{y}) = \left(\frac{\frac{1}{8}(e^2+1)}{\frac{1}{4}(e^2-1)}, \frac{\frac{1}{9}(e^3-1)}{\frac{1}{4}(e^2-1)}\right) = \left(\frac{e^2+1}{2(e^2-1)}, \frac{4(e^3-1)}{9(e^2-1)}\right)$.

9.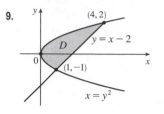

$$m = \int_{-1}^{2} \int_{y^2}^{y+2} 3 \, dx \, dy = \int_{-1}^{2} (3y + 6 - 3y^2) \, dy = \frac{27}{2},$$

$$M_y = \int_{-1}^{2} \int_{y^2}^{y+2} 3x \, dx \, dy = \int_{-1}^{2} \frac{3}{2} \left[(y+2)^2 - y^4 \right] dy$$

$$= \left[\frac{1}{2}(y+2)^3 - \frac{3}{10} y^5 \right]_{-1}^{2} = \frac{108}{5}$$

and

$$M_x = \int_{-1}^{2} \int_{y^2}^{y+2} 3y \, dx \, dy = \int_{-1}^{2} (3y^2 + 6y - 3y^3) \, dy$$

$$= \left[y^3 + 3y^2 - \frac{3}{4} y^4 \right]_{-1}^{2} = \frac{27}{4}$$

Hence $m = \frac{27}{2}$, $(\overline{x}, \overline{y}) = \left(\frac{8}{5}, \frac{1}{2} \right)$.

11. $\rho(x, y) = ky = kr \sin \theta$, $m = \int_{0}^{\pi/2} \int_{0}^{1} kr^2 \sin \theta \, dr \, d\theta = \frac{1}{3}k \int_{0}^{\pi/2} \sin \theta \, d\theta = \frac{1}{3}k \left[-\cos \theta \right]_{0}^{\pi/2} = \frac{1}{3}k$,

$M_y = \int_{0}^{\pi/2} \int_{0}^{1} kr^3 \sin \theta \cos \theta \, dr \, d\theta = \frac{1}{4}k \int_{0}^{\pi/2} \sin \theta \cos \theta \, d\theta = \frac{1}{8}k \left[-\cos 2\theta \right]_{0}^{\pi/2} = \frac{1}{8}k$,

$M_x = \int_{0}^{\pi/2} \int_{0}^{1} kr^3 \sin^2 \theta \, dr \, d\theta = \frac{1}{4}k \int_{0}^{\pi/2} \sin^2 \theta \, d\theta = \frac{1}{8}k \left[\theta + \sin 2\theta \right]_{0}^{\pi/2} = \frac{\pi}{16}k$.

Hence $(\overline{x}, \overline{y}) = \left(\frac{3}{8}, \frac{3\pi}{16} \right)$.

13. Placing the vertex opposite the hypotenuse at $(0, 0)$, $\rho(x, y) = k(x^2 + y^2)$. Then

$m = \int_{0}^{a} \int_{0}^{a-x} k(x^2 + y^2) \, dy \, dx = k \int_{0}^{a} \left[ax^2 - x^3 + \frac{1}{3}(a-x)^3 \right] dx = k \left[\frac{1}{3}ax^3 - \frac{1}{4}x^4 - \frac{1}{12}(a-x)^4 \right]_{0}^{a} = \frac{1}{6}ka^4$.

By symmetry,

$$M_y = M_x = \int_{0}^{a} \int_{0}^{a-x} ky(x^2 + y^2) \, dy \, dx = k \int_{0}^{a} \left[\frac{1}{2}(a-x)^2 x^2 + \frac{1}{4}(a-x)^4 \right] dx$$

$$= k \left[\frac{1}{6}a^2 x^3 - \frac{1}{4}ax^4 + \frac{1}{10}x^5 - \frac{1}{20}(a-x)^5 \right]_{0}^{a} = \frac{1}{15}ka^5$$

Hence $(\overline{x}, \overline{y}) = \left(\frac{2}{5}a, \frac{2}{5}a \right)$.

15. $I_x = \iint_{D} y^2 \rho(x, y) \, dA = \int_{0}^{1} \int_{0}^{e^x} y^2 \cdot y \, dy \, dx = \int_{0}^{1} \left[\frac{1}{4} y^4 \right]_{y=0}^{y=e^x} dx = \frac{1}{4} \int_{0}^{1} e^{4x} \, dx = \frac{1}{4} \left[\frac{1}{4} e^{4x} \right]_{0}^{1} = \frac{1}{16}(e^4 - 1)$,

$I_y = \iint_{D} x^2 \rho(x, y) \, dA = \int_{0}^{1} \int_{0}^{e^x} x^2 y \, dy \, dx = \int_{0}^{1} x^2 \left[\frac{1}{2} y^2 \right]_{y=0}^{y=e^x} dx = \frac{1}{2} \int_{0}^{1} x^2 e^{2x} \, dx$

$= \frac{1}{2} \left[\left(\frac{1}{2}x^2 - \frac{1}{2}x + \frac{1}{4} \right) e^{2x} \right]_{0}^{1}$ [integrate by parts twice]

$= \frac{1}{8}(e^2 - 1)$,

and $I_0 = I_x + I_y = \frac{1}{16}(e^4 - 1) + \frac{1}{8}(e^2 - 1) = \frac{1}{16}(e^4 + 2e^2 - 3)$.

17. As in Exercise 13, we place the vertex opposite the hypotenuse at $(0, 0)$ and the equal sides along the positive axes.

$$I_x = \int_{0}^{a} \int_{0}^{a-x} y^2 k(x^2 + y^2) \, dy \, dx = k \int_{0}^{a} \int_{0}^{a-x} (x^2 y^2 + y^4) \, dy \, dx$$

$$= k \int_{0}^{a} \left[\frac{1}{3}x^2 y^3 + \frac{1}{5} y^5 \right]_{y=0}^{y=a-x} dx = k \int_{0}^{a} \left[\frac{1}{3}x^2 (a-x)^3 + \frac{1}{5}(a-x)^5 \right] dx$$

$$= k \left[\frac{1}{3} \left(\frac{1}{3}a^3 x^3 - \frac{3}{4}a^2 x^4 + \frac{3}{5}ax^5 - \frac{1}{6}x^6 \right) - \frac{1}{30}(a-x)^6 \right]_{0}^{a} = \frac{7}{180}ka^6,$$

$$I_y = \int_{0}^{a} \int_{0}^{a-x} x^2 k(x^2 + y^2) \, dy \, dx = k \int_{0}^{a} \int_{0}^{a-x} (x^4 + x^2 y^2) \, dy \, dx$$

$$= k \int_{0}^{a} \left[x^4 y + \frac{1}{3}x^2 y^3 \right]_{y=0}^{y=a-x} dx = k \int_{0}^{a} \left[x^4 (a-x) + \frac{1}{3}x^2 (a-x)^3 \right] dx$$

$$= k \left[\frac{1}{5}ax^5 - \frac{1}{6}x^6 + \frac{1}{3} \left(\frac{1}{3}a^3 x^3 - \frac{3}{4}a^2 x^4 + \frac{3}{5}ax^5 - \frac{1}{6}x^6 \right) \right]_{0}^{a} = \frac{7}{180}ka^6,$$

and $I_0 = I_x + I_y = \frac{7}{90}ka^6$.

19. Using a CAS, we find $m = \iint_D \rho(x, y) \, dA = \int_0^\pi \int_0^{\sin x} xy \, dy \, dx = \dfrac{\pi^2}{8}$. Then

$$\bar{x} = \frac{1}{m} \iint_D x\rho(x, y) \, dA = \frac{8}{\pi^2} \int_0^\pi \int_0^{\sin x} x^2 y \, dy \, dx = \frac{2\pi}{3} - \frac{1}{\pi} \text{ and}$$

$$\bar{y} = \frac{1}{m} \iint_D y\rho(x, y) \, dA = \frac{8}{\pi^2} \int_0^\pi \int_0^{\sin x} xy^2 \, dy \, dx = \frac{16}{9\pi}, \text{ so } (\bar{x}, \bar{y}) = \left(\frac{2\pi}{3} - \frac{1}{\pi}, \frac{16}{9\pi} \right).$$

The moments of inertia are $I_x = \iint_D y^2 \rho(x, y) \, dA = \int_0^\pi \int_0^{\sin x} xy^3 \, dy \, dx = \dfrac{3\pi^2}{64}$,

$I_y = \iint_D x^2 \rho(x, y) \, dA = \int_0^\pi \int_0^{\sin x} x^3 y \, dy \, dx = \dfrac{\pi^2}{16}(\pi^2 - 3)$, and $I_0 = I_x + I_y = \dfrac{\pi^2}{64}(4\pi^2 - 9)$.

21. (a) $f(x, y)$ is a joint density function, so we know $\iint_{\mathbb{R}^2} f(x, y) \, dA = 1$. Since $f(x, y) = 0$ outside the rectangle

$[0, 1] \times [0, 2]$, we can say

$$\iint_{\mathbb{R}^2} f(x, y) \, dA = \int_{-\infty}^\infty \int_{-\infty}^\infty f(x, y) \, dy \, dx = \int_0^1 \int_0^2 Cx(1 + y) \, dy \, dx$$

$$= C \int_0^1 x \left[y + \tfrac{1}{2}y^2 \right]_{y=0}^{y=2} dx = C \int_0^1 4x \, dx = C \left[2x^2 \right]_0^1 = 2C$$

Then $2C = 1 \implies C = \frac{1}{2}$.

(b) $P(X \leq 1, Y \leq 1) = \int_{-\infty}^1 \int_{-\infty}^1 f(x, y) \, dy \, dx = \int_0^1 \int_0^1 \tfrac{1}{2}x(1 + y) \, dy \, dx$

$$= \int_0^1 \tfrac{1}{2}x \left[y + \tfrac{1}{2}y^2 \right]_{y=0}^{y=1} dx = \int_0^1 \tfrac{1}{2}x \left(\tfrac{3}{2} \right) dx = \tfrac{3}{4} \left[\tfrac{1}{2}x^2 \right]_0^1 = \tfrac{3}{8} \text{ or } 0.375$$

(c) $P(X + Y \leq 1) = P((X, Y) \in D)$ where D is the triangular region

shown in the figure. Thus

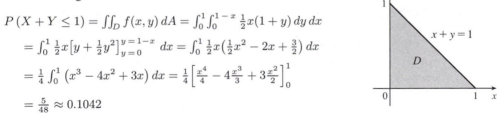

$$P(X + Y \leq 1) = \iint_D f(x, y) \, dA = \int_0^1 \int_0^{1-x} \tfrac{1}{2}x(1 + y) \, dy \, dx$$

$$= \int_0^1 \tfrac{1}{2}x \left[y + \tfrac{1}{2}y^2 \right]_{y=0}^{y=1-x} dx = \int_0^1 \tfrac{1}{2}x \left(\tfrac{1}{2}x^2 - 2x + \tfrac{3}{2} \right) dx$$

$$= \tfrac{1}{4} \int_0^1 (x^3 - 4x^2 + 3x) \, dx = \tfrac{1}{4} \left[\tfrac{x^4}{4} - 4\tfrac{x^3}{3} + 3\tfrac{x^2}{2} \right]_0^1$$

$$= \tfrac{5}{48} \approx 0.1042$$

23. (a) $f(x, y) \geq 0$, so f is a joint density function if $\iint_{\mathbb{R}^2} f(x, y) \, dA = 1$. Here, $f(x, y) = 0$ outside the first quadrant, so

$$\iint_{\mathbb{R}^2} f(x, y) \, dA = \int_0^\infty \int_0^\infty 0.1e^{-(0.5x + 0.2y)} \, dy \, dx = 0.1 \int_0^\infty \int_0^\infty e^{-0.5x} e^{-0.2y} \, dy \, dx$$

$$= 0.1 \int_0^\infty e^{-0.5x} \, dx \int_0^\infty e^{-0.2y} \, dy = 0.1 \lim_{t \to \infty} \int_0^t e^{-0.5x} \, dx \lim_{t \to \infty} \int_0^t e^{-0.2y} \, dy$$

$$= 0.1 \lim_{t \to \infty} \left[-2e^{-0.5x} \right]_0^t \lim_{t \to \infty} \left[-5e^{-0.2y} \right]_0^t = 0.1 \lim_{t \to \infty} \left[-2(e^{-0.5t} - 1) \right] \lim_{t \to \infty} \left[-5(e^{-0.2t} - 1) \right]$$

$$= (0.1) \cdot (-2)(0 - 1) \cdot (-5)(0 - 1) = 1$$

Thus $f(x, y)$ is a joint density function.

(b) (i) No restriction is placed on X, so

$$P(Y \geq 1) = \int_{-\infty}^{\infty} \int_{1}^{\infty} f(x,y) \, dy \, dx = \int_{0}^{\infty} \int_{1}^{\infty} 0.1 e^{-(0.5x + 0.2y)} \, dy \, dx$$

$$= 0.1 \int_{0}^{\infty} e^{-0.5x} \, dx \int_{1}^{\infty} e^{-0.2y} \, dy = 0.1 \lim_{t \to \infty} \int_{0}^{t} e^{-0.5x} \, dx \lim_{t \to \infty} \int_{1}^{t} e^{-0.2y} \, dy$$

$$= 0.1 \lim_{t \to \infty} \left[-2e^{-0.5x} \right]_{0}^{t} \lim_{t \to \infty} \left[-5e^{-0.2y} \right]_{1}^{t} = 0.1 \lim_{t \to \infty} \left[-2(e^{-0.5t} - 1) \right] \lim_{t \to \infty} \left[-5(e^{-0.2t} - e^{-0.2}) \right]$$

$$= (0.1) \cdot (-2)(0 - 1) \cdot (-5)(0 - e^{-0.2}) = e^{-0.2} \approx 0.8187$$

(ii) $P(X \leq 2, Y \leq 4) = \int_{-\infty}^{2} \int_{-\infty}^{4} f(x,y) \, dy \, dx = \int_{0}^{2} \int_{0}^{4} 0.1 e^{-(0.5x + 0.2y)} \, dy \, dx$

$$= 0.1 \int_{0}^{2} e^{-0.5x} \, dx \int_{0}^{4} e^{-0.2y} \, dy = 0.1 \left[-2e^{-0.5x} \right]_{0}^{2} \left[-5e^{-0.2y} \right]_{0}^{4}$$

$$= (0.1) \cdot (-2)(e^{-1} - 1) \cdot (-5)(e^{-0.8} - 1)$$

$$= (e^{-1} - 1)(e^{-0.8} - 1) = 1 + e^{-1.8} - e^{-0.8} - e^{-1} \approx 0.3481$$

(c) The expected value of X is given by

$$\mu_1 = \iint_{\mathbb{R}^2} x \, f(x,y) \, dA = \int_{0}^{\infty} \int_{0}^{\infty} x \left[0.1 e^{-(0.5x + 0.2y)} \right] \, dy \, dx$$

$$= 0.1 \int_{0}^{\infty} x e^{-0.5x} \, dx \int_{0}^{\infty} e^{-0.2y} \, dy = 0.1 \lim_{t \to \infty} \int_{0}^{t} x e^{-0.5x} \, dx \lim_{t \to \infty} \int_{0}^{t} e^{-0.2y} \, dy$$

To evaluate the first integral, we integrate by parts with $u = x$ and $dv = e^{-0.5x} \, dx$ (or we can use Formula 96 in the Table of Integrals): $\int x e^{-0.5x} \, dx = -2xe^{-0.5x} - \int -2e^{-0.5x} \, dx = -2xe^{-0.5x} - 4e^{-0.5x} = -2(x+2)e^{-0.5x}$. Thus

$$\mu_1 = 0.1 \lim_{t \to \infty} \left[-2(x+2)e^{-0.5x} \right]_{0}^{t} \lim_{t \to \infty} \left[-5e^{-0.2y} \right]_{0}^{t}$$

$$= 0.1 \lim_{t \to \infty} (-2) \left[(t+2)e^{-0.5t} - 2 \right] \lim_{t \to \infty} (-5) \left[e^{-0.2t} - 1 \right]$$

$$= 0.1(-2) \left(\lim_{t \to \infty} \frac{t+2}{e^{0.5t}} - 2 \right)(-5)(-1) = 2 \qquad \text{[by l'Hospital's Rule]}$$

The expected value of Y is given by

$$\mu_2 = \iint_{\mathbb{R}^2} y \, f(x,y) \, dA = \int_{0}^{\infty} \int_{0}^{\infty} y \left[0.1 e^{-(0.5 + 0.2y)} \right] \, dy \, dx$$

$$= 0.1 \int_{0}^{\infty} e^{-0.5x} \, dx \int_{0}^{\infty} y e^{-0.2y} \, dy = 0.1 \lim_{t \to \infty} \int_{0}^{t} e^{-0.5x} \, dx \lim_{t \to \infty} \int_{0}^{t} y e^{-0.2y} \, dy$$

To evaluate the second integral, we integrate by parts with $u = y$ and $dv = e^{-0.2y} \, dy$ (or again we can use Formula 96 in the Table of Integrals) which gives $\int y e^{-0.2y} \, dy = -5y e^{-0.2y} + \int 5e^{-0.2y} \, dy = -5(y+5)e^{-0.2y}$. Then

$$\mu_2 = 0.1 \lim_{t \to \infty} \left[-2e^{-0.5x} \right]_{0}^{t} \lim_{t \to \infty} \left[-5(y+5)e^{-0.2y} \right]_{0}^{t}$$

$$= 0.1 \lim_{t \to \infty} \left[-2(e^{-0.5t} - 1) \right] \lim_{t \to \infty} \left(-5 \left[(t+5)e^{-0.2t} - 5 \right] \right)$$

$$= 0.1(-2)(-1) \cdot (-5) \left(\lim_{t \to \infty} \frac{t+5}{e^{0.2t}} - 5 \right) = 5 \qquad \text{[by l'Hospital's Rule]}$$

25. (a) The random variables X and Y are normally distributed with $\mu_1 = 45$, $\mu_2 = 20$, $\sigma_1 = 0.5$, and $\sigma_2 = 0.1$. The individual density functions for X and Y, then, are $f_1(x) = \dfrac{1}{0.5\sqrt{2\pi}} e^{-(x-45)^2/0.5}$ and

$f_2(y) = \dfrac{1}{0.1\sqrt{2\pi}} e^{-(y-20)^2/0.02}$. Since X and Y are independent, joint density function is the product

$$f(x,y) = f_1(x)f_2(y) = \frac{1}{0.5\sqrt{2\pi}} e^{-(x-45)^2/0.5} \frac{1}{0.1\sqrt{2\pi}} e^{-(y-20)^2/0.02} = \frac{10}{\pi} e^{-2(x-45)^2 - 50(y-20)^2}.$$

Then $P(40 \leq X \leq 50, 20 \leq Y \leq 25) = \int_{40}^{50} \int_{20}^{25} f(x,y) \, dy \, dx = \dfrac{10}{\pi} \int_{40}^{50} \int_{20}^{25} e^{-2(x-45)^2 - 50(y-20)^2} \, dy \, dx$.

Using a CAS or calculator to evaluate the integral, we get $P(40 \leq X \leq 50, 20 \leq Y \leq 25) \approx 0.500$.

(b) $P(4(X-45)^2 + 100(Y-20)^2 \le 2) = \iint_D \frac{10}{\pi} e^{-2(x-45)^2 - 50(y-20)^2} \, dA$, where D is the region enclosed by the ellipse

$4(x-45)^2 + 100(y-20)^2 = 2$. Solving for y gives $y = 20 \pm \frac{1}{10}\sqrt{2 - 4(x-45)^2}$, the upper and lower halves of the

ellipse, and these two halves meet where $y = 20$ [since the ellipse is centered at $(45, 20)$] $\Rightarrow$ $4(x-45)^2 = 2$ $\Rightarrow$

$x = 45 \pm \frac{1}{\sqrt{2}}$. Thus

$$\iint_D \frac{10}{\pi} e^{-2(x-45)^2 - 50(y-20)^2} \, dA = \frac{10}{\pi} \int_{45-1/\sqrt{2}}^{45+1/\sqrt{2}} \int_{20-\frac{1}{10}\sqrt{2-4(x-45)^2}}^{20+\frac{1}{10}\sqrt{2-4(x-45)^2}} e^{-2(x-45)^2 - 50(y-20)^2} \, dy \, dx.$$

Using a CAS or calculator to evaluate the integral, we get $P(4(X-45)^2 + 100(Y-20)^2 \le 2) \approx 0.632$.

27. (a) If $f(P, A)$ is the probability that an individual at A will be infected by an individual at P, and $k\,dA$ is the number of

infected individuals in an element of area dA, then $f(P, A)k\,dA$ is the number of infections that should result from

exposure of the individual at A to infected people in the element of area dA. Integration over D gives the number of

infections of the person at A due to all the infected people in D. In rectangular coordinates (with the origin at the city's

center), the exposure of a person at A is

$$E = \iint_D k f(P, A) \, dA = k \iint_D \frac{20 - d(P, A)}{20} \, dA = k \iint_D \left[1 - \frac{\sqrt{(x - x_0)^2 + (y - y_0)^2}}{20} \right] dx \, dy$$

(b) If $A = (0, 0)$, then

$$E = k \iint_D \left[1 - \frac{1}{20}\sqrt{x^2 + y^2} \right] dx \, dy$$

$r = 20 \cos \theta$

$$= k \int_0^{2\pi} \int_0^{10} \left(1 - \frac{r}{20} \right) r \, dr \, d\theta = 2\pi k \left[\frac{r^2}{2} - \frac{r^3}{60} \right]_0^{10}$$

$$= 2\pi k \left(50 - \frac{50}{3} \right) = \frac{200}{3}\pi k \approx 209k$$

For A at the edge of the city, it is convenient to use a polar coordinate system centered at A. Then the polar equation for

the circular boundary of the city becomes $r = 20 \cos \theta$ instead of $r = 10$, and the distance from A to a point P in the city

is again r (see the figure). So

$$E = k \int_{-\pi/2}^{\pi/2} \int_0^{20 \cos \theta} \left(1 - \frac{r}{20} \right) r \, dr \, d\theta = k \int_{-\pi/2}^{\pi/2} \left[\frac{r^2}{2} - \frac{r^3}{60} \right]_{r=0}^{r=20 \cos \theta} d\theta$$

$$= k \int_{-\pi/2}^{\pi/2} \left(200 \cos^2 \theta - \frac{400}{3} \cos^3 \theta \right) d\theta = 200k \int_{-\pi/2}^{\pi/2} \left[\frac{1}{2} + \frac{1}{2} \cos 2\theta - \frac{2}{3}(1 - \sin^2 \theta) \cos \theta \right] d\theta$$

$$= 200k \left[\frac{1}{2}\theta + \frac{1}{4} \sin 2\theta - \frac{2}{3} \sin \theta + \frac{2}{3} \cdot \frac{1}{3} \sin^3 \theta \right]_{-\pi/2}^{\pi/2} = 200k \left[\frac{\pi}{4} + 0 - \frac{2}{3} + \frac{2}{9} + \frac{\pi}{4} + 0 - \frac{2}{3} + \frac{2}{9} \right]$$

$$= 200k \left(\frac{\pi}{2} - \frac{8}{9} \right) \approx 136k$$

Therefore the risk of infection is much lower at the edge of the city than in the middle, so it is better to live at the edge.

12.6 Surface Area

1. Here $z = f(x, y) = 2 + 3x + 4y$ and D is the rectangle $[0, 5] \times [1, 4]$, so by Formula 6 the area of the surface is

$$A(S) = \iint_D \sqrt{1 + \left(\frac{\partial z}{\partial x}\right)^2 + \left(\frac{\partial z}{\partial y}\right)^2}\, dA = \iint_D \sqrt{1 + 3^2 + 4^2}\, dA = \sqrt{26} \iint_D dA = \sqrt{26}\, A(D)$$

$$= \sqrt{26}\, (5)(3) = 15\sqrt{26}$$

3. $z = f(x, y) = 6 - 3x - 2y$ which intersects the xy-plane in the line $3x + 2y = 6$, so D is the triangular region given by $\left\{ (x, y) \mid 0 \le x \le 2, 0 \le y \le 3 - \frac{3}{2}x \right\}$. Thus

$A(S) = \iint_D \sqrt{1 + (-3)^2 + (-2)^2}\, dA = \sqrt{14} \iint_D dA = \sqrt{14}\, A(D) = \sqrt{14} \left(\frac{1}{2} \cdot 2 \cdot 3 \right) = 3\sqrt{14}.$

5. $z = f(x, y) = y^2 - x^2$ with $1 \le x^2 + y^2 \le 4$. Then

$$A(S) = \iint_D \sqrt{1 + 4x^2 + 4y^2}\, dA = \int_0^{2\pi} \int_1^2 \sqrt{1 + 4r^2}\, r\, dr\, d\theta = \int_0^{2\pi} d\theta \int_1^2 r\sqrt{1 + 4r^2}\, dr$$

$$= \left[\theta \right]_0^{2\pi} \left[\frac{1}{12}(1 + 4r^2)^{3/2} \right]_1^2 = \frac{\pi}{6}\left(17\sqrt{17} - 5\sqrt{5} \right)$$

7. $\mathbf{r}_u = \langle 2u, v, 0 \rangle$, $\mathbf{r}_v = \langle 0, u, v \rangle$, and $\mathbf{r}_u \times \mathbf{r}_v = \langle v^2, -2uv, 2u^2 \rangle$. Then

$$A(S) = \iint_D |\mathbf{r}_u \times \mathbf{r}_v|\, dA = \int_0^1 \int_0^2 \sqrt{v^4 + 4u^2 v^2 + 4u^4}\, dv\, du$$

$$= \int_0^1 \int_0^2 \sqrt{(v^2 + 2u^2)^2}\, dv\, du = \int_0^1 \int_0^2 (v^2 + 2u^2)\, dv\, du$$

$$= \int_0^1 \left[\frac{1}{3}v^3 + 2u^2 v \right]_{v=0}^{v=2} du = \int_0^1 \left(\frac{8}{3} + 4u^2 \right) du = \left[\frac{8}{3}u + \frac{4}{3}u^3 \right]_0^1 = 4$$

9. A parametric representation of the surface is $x = x$, $y = 4x + z^2$, $z = z$ with $0 \le x \le 1, 0 \le z \le 1$.
Hence $\mathbf{r}_x \times \mathbf{r}_z = (\mathbf{i} + 4\mathbf{j}) \times (2z\,\mathbf{j} + \mathbf{k}) = 4\,\mathbf{i} - \mathbf{j} + 2z\,\mathbf{k}$.

Note: In general, if $y = f(x, z)$ then $\mathbf{r}_x \times \mathbf{r}_z = \dfrac{\partial f}{\partial x}\mathbf{i} - \mathbf{j} + \dfrac{\partial f}{\partial z}\mathbf{k}$ and $A(S) = \iint_D \sqrt{1 + \left(\dfrac{\partial f}{\partial x}\right)^2 + \left(\dfrac{\partial f}{\partial z}\right)^2}\, dA$. Then

$$A(S) = \int_0^1 \int_0^1 \sqrt{17 + 4z^2}\, dx\, dz = \int_0^1 \sqrt{17 + 4z^2}\, dz$$

$$= \frac{1}{2}\left(z\sqrt{17 + 4z^2} + \frac{17}{2}\ln\left|2z + \sqrt{4z^2 + 17}\right| \right)\Big]_0^1 = \frac{\sqrt{21}}{2} + \frac{17}{4}\left[\ln\left(2 + \sqrt{21}\right) - \ln\sqrt{17}\right]$$

11. $z = f(x, y) = xy$ with $0 \le x^2 + y^2 \le 1$, so $f_x = y$, $f_y = x$ $\Rightarrow$

$$A(S) = \iint_D \sqrt{1 + y^2 + x^2}\, dA = \int_0^{2\pi} \int_0^1 \sqrt{r^2 + 1}\, r\, dr\, d\theta = \int_0^{2\pi} \left[\frac{1}{3}(r^2 + 1)^{3/2} \right]_{r=0}^{r=1} d\theta$$

$$= \int_0^{2\pi} \frac{1}{3}(2\sqrt{2} - 1)\, d\theta = \frac{2\pi}{3}(2\sqrt{2} - 1)$$

13. $z = f(x, y) = e^{-x^2 - y^2}$, $f_x = -2xe^{-x^2 - y^2}$, $f_y = -2ye^{-x^2 - y^2}$. Then

$A(S) = \displaystyle\iint_{x^2 + y^2 \le 4} \sqrt{1 + (-2xe^{-x^2 - y^2})^2 + (-2ye^{-x^2 - y^2})^2}\, dA = \iint_{x^2 + y^2 \le 4} \sqrt{1 + 4(x^2 + y^2)e^{-2(x^2 + y^2)}}\, dA.$

Converting to polar coordinates we have

$$A(S) = \int_0^{2\pi} \int_0^2 \sqrt{1 + 4r^2 e^{-2r^2}}\, r\, dr\, d\theta = \int_0^{2\pi} d\theta \int_0^2 r\sqrt{1 + 4r^2 e^{-2r^2}}\, dr$$

$$= 2\pi \int_0^2 r\sqrt{1 + 4r^2 e^{-2r^2}}\, dr \approx 13.9783 \text{ using a calculator.}$$

15. (a) $A(S) = \iint_D \sqrt{1 + \left(\dfrac{\partial z}{\partial x}\right)^2 + \left(\dfrac{\partial z}{\partial y}\right)^2}\, dA = \int_0^6 \int_0^4 \sqrt{1 + \dfrac{4x^2 + 4y^2}{(1 + x^2 + y^2)^4}}\, dy\, dx.$ Using the Midpoint Rule with

$f(x, y) = \sqrt{1 + \dfrac{4x^2 + 4y^2}{(1 + x^2 + y^2)^4}}$, $m = 3, n = 2$ we have

$$A(S) \approx \sum_{i=1}^3 \sum_{j=1}^2 f\left(\overline{x}_i, \overline{y}_j\right) \Delta A = 4\left[f\left(1, 1\right) + f\left(1, 3\right) + f\left(3, 1\right) + f\left(3, 3\right) + f\left(5, 1\right) + f\left(5, 3\right)\right] \approx 24.2055$$

(b) Using a CAS we have $A(S) = \int_0^6 \int_0^4 \sqrt{1 + \dfrac{4x^2 + 4y^2}{(1 + x^2 + y^2)^4}}\, dy\, dx \approx 24.2476.$ This agrees with the estimate in part (a)

to the first decimal place.

17. $\mathbf{r}(u, v) = \left\langle \cos^3 u \cos^3 v, \sin^3 u \cos^3 v, \sin^3 v \right\rangle$, so $\mathbf{r}_u = \left\langle -3\cos^2 u \sin u \cos^3 v, 3\sin^2 u \cos u \cos^3 v, 0 \right\rangle$,

$\mathbf{r}_v = \left\langle -3\cos^3 u \cos^2 v \sin v, -3\sin^3 u \cos^2 v \sin v, 3\sin^2 v \cos v \right\rangle$, and

$\mathbf{r}_u \times \mathbf{r}_v = \left\langle 9\cos u \sin^2 u \cos^4 v \sin^2 v, 9\cos^2 u \sin u \cos^4 v \sin^2 v, 9\cos^2 u \sin^2 u \cos^5 v \sin v \right\rangle$. Then

$$|\mathbf{r}_u \times \mathbf{r}_v| = 9\sqrt{\cos^2 u \sin^4 u \cos^8 v \sin^4 v + \cos^4 u \sin^2 u \cos^8 v \sin^4 v + \cos^4 u \sin^4 u \cos^{10} v \sin^2 v}$$

$$= 9\sqrt{\cos^2 u \sin^2 u \cos^8 v \sin^2 v \left(\sin^2 v + \cos^2 u \sin^2 u \cos^2 v\right)}$$

$$= 9\cos^4 v\, |\cos u \sin u \sin v|\, \sqrt{\sin^2 v + \cos^2 u \sin^2 u \cos^2 v}$$

Using a CAS, we have $A(S) = \int_0^\pi \int_0^{2\pi} 9\cos^4 v\, |\cos u \sin u \sin v|\, \sqrt{\sin^2 v + \cos^2 u \sin^2 u \cos^2 v}\, dv\, du \approx 4.4506.$

19. $z = 1 + 2x + 3y + 4y^2$, so

$$A(S) = \iint_D \sqrt{1 + \left(\dfrac{\partial z}{\partial x}\right)^2 + \left(\dfrac{\partial z}{\partial y}\right)^2}\, dA = \int_1^4 \int_0^1 \sqrt{1 + 4 + (3 + 8y)^2}\, dy\, dx = \int_1^4 \int_0^1 \sqrt{14 + 48y + 64y^2}\, dy\, dx.$$

Using a CAS, we have

$\int_1^4 \int_0^1 \sqrt{14 + 48y + 64y^2}\, dy\, dx = \dfrac{45}{8}\sqrt{14} + \dfrac{15}{16}\ln\left(11\sqrt{5} + 3\sqrt{14}\sqrt{5}\right) - \dfrac{15}{16}\ln\left(3\sqrt{5} + \sqrt{14}\sqrt{5}\right)$

or $\dfrac{45}{8}\sqrt{14} + \dfrac{15}{16}\ln\dfrac{11\sqrt{5} + 3\sqrt{70}}{3\sqrt{5} + \sqrt{70}}.$

21. (a) $x = a\sin u \cos v$, $y = b\sin u \sin v$, $z = c\cos u$ $\Rightarrow$

$$\dfrac{x^2}{a^2} + \dfrac{y^2}{b^2} + \dfrac{z^2}{c^2} = (\sin u \cos v)^2 + (\sin u \sin v)^2 + (\cos u)^2$$

$$= \sin^2 u + \cos^2 u = 1$$

and since the ranges of u and v are sufficient to generate the entire graph,

the parametric equations represent an ellipsoid.

(c) From the parametric equations (with $a = 1$, $b = 2$, and $c = 3$),

we calculate $\mathbf{r}_u = \cos u \cos v\, \mathbf{i} + 2\cos u \sin v\, \mathbf{j} - 3\sin u\, \mathbf{k}$ and

$\mathbf{r}_v = -\sin u \sin v\, \mathbf{i} + 2\sin u \cos v\, \mathbf{j}$. So $\mathbf{r}_u \times \mathbf{r}_v = 6\sin^2 u \cos v\, \mathbf{i} + 3\sin^2 u \sin v\, \mathbf{j} + 2\sin u \cos u\, \mathbf{k}$, and the surface

area is given by $A(S) = \int_0^{2\pi} \int_0^\pi |\mathbf{r}_u \times \mathbf{r}_v|\, du\, dv = \int_0^{2\pi} \int_0^\pi \sqrt{36\sin^4 u \cos^2 v + 9\sin^4 u \sin^2 v + 4\cos^2 u \sin^2 u}\, du\, dv.$

(b)

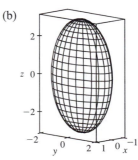

23. To find the region D: $z = x^2 + y^2$ implies $z + z^2 = 4z$ or $z^2 - 3z = 0$. Thus $z = 0$ or $z = 3$ are the planes where the surfaces intersect. But $x^2 + y^2 + z^2 = 4z$ implies $x^2 + y^2 + (z-2)^2 = 4$, so $z = 3$ intersects the upper hemisphere. Thus $(z-2)^2 = 4 - x^2 - y^2$ or $z = 2 + \sqrt{4 - x^2 - y^2}$. Therefore D is the region inside the circle $x^2 + y^2 + (3-2)^2 = 4$, that is, $D = \{(x, y) \mid x^2 + y^2 \le 3\}$.

$$A(S) = \iint_D \sqrt{1 + [(-x)(4 - x^2 - y^2)^{-1/2}]^2 + [(-y)(4 - x^2 - y^2)^{-1/2}]^2} \, dA$$

$$= \int_0^{2\pi} \int_0^{\sqrt{3}} \sqrt{1 + \frac{r^2}{4 - r^2}} \, r \, dr \, d\theta = \int_0^{2\pi} \int_0^{\sqrt{3}} \frac{2r \, dr}{\sqrt{4 - r^2}} \, d\theta = \int_0^{2\pi} \left[-2(4 - r^2)^{1/2} \right]_{r=0}^{r=\sqrt{3}} d\theta$$

$$= \int_0^{2\pi} (-2 + 4) \, d\theta = 2\theta \Big]_0^{2\pi} = 4\pi$$

25. If we revolve the curve $y = f(x)$, $a \le x \le b$ about the x-axis, where $f(x) \ge 0$, then from Equations 10.5.3 we know we can parametrize the surface using $x = x$, $y = f(x) \cos\theta$, and $z = f(x) \sin\theta$, where $a \le x \le b$ and $0 \le \theta \le 2\pi$. Thus we can say the surface is represented by $\mathbf{r}(x, \theta) = x\,\mathbf{i} + f(x) \cos\theta\,\mathbf{j} + f(x) \sin\theta\,\mathbf{k}$, with $a \le x \le b$ and $0 \le \theta \le 2\pi$. Then by (4), the surface area is given by $A(S) = \iint_D |\mathbf{r}_x \times \mathbf{r}_\theta| \, dA$ where D is the rectangular parameter region $[a, b] \times [0, 2\pi]$. Here, $\mathbf{r}_x(x, \theta) = \mathbf{i} + f'(x) \cos\theta\,\mathbf{j} + f'(x) \sin\theta\,\mathbf{k}$ and $\mathbf{r}_\theta(x) = -f(x) \sin\theta\,\mathbf{j} + f(x) \cos\theta\,\mathbf{k}$. So

$$\mathbf{r}_x \times \mathbf{r}_\theta = \begin{vmatrix} \mathbf{i} & \mathbf{j} & \mathbf{k} \\ 1 & f'(x) \cos\theta & f'(x) \sin\theta \\ 0 & -f(x) \sin\theta & f(x) \cos\theta \end{vmatrix} = \left[f(x)f'(x) \cos^2\theta + f(x)f'(x) \sin^2\theta \right] \mathbf{i} - f(x) \cos\theta\,\mathbf{j} - f(x) \sin\theta\,\mathbf{k}$$

$$= f(x)f'(x)\mathbf{i} - f(x) \cos\theta\,\mathbf{j} - f(x) \sin\theta\,\mathbf{k} \text{ and}$$

$$|\mathbf{r}_x \times \mathbf{r}_\theta| = \sqrt{[f(x)f'(x)]^2 + [f(x)]^2 \cos^2\theta + [f(x)]^2 \sin^2\theta}$$

$$= \sqrt{[f(x)]^2 ([f'(x)]^2 + 1)} = f(x)\sqrt{1 + [f'(x)]^2} \text{ [since } f(x) \ge 0]. \text{ Thus}$$

$$A(S) = \iint_D |\mathbf{r}_x \times \mathbf{r}_\theta| \, dA = \int_a^b \int_0^{2\pi} f(x)\sqrt{1 + [f'(x)]^2} \, d\theta \, dx$$

$$= \int_a^b f(x)\sqrt{1 + [f'(x)]^2} \, [\theta]_0^{2\pi} \, dx = 2\pi \int_a^b f(x)\sqrt{1 + [f'(x)]^2} \, dx$$

27. $y = \sqrt{x} \;\; \Rightarrow \;\; 1 + \left(\dfrac{dy}{dx}\right)^2 = 1 + \left(\dfrac{1}{2\sqrt{x}}\right)^2 = 1 + \dfrac{1}{4x}$. So

$$S = \int_4^9 2\pi y \sqrt{1 + \left(\frac{dy}{dx}\right)^2} \, dx = \int_4^9 2\pi \sqrt{x} \sqrt{1 + \frac{1}{4x}} \, dx = 2\pi \int_4^9 \left(x + \tfrac{1}{4}\right) dx$$

$$= 2\pi \left[\tfrac{2}{3} \left(x + \tfrac{1}{4}\right)^{3/2} \right]_4^9 = \tfrac{4\pi}{3} \left[\tfrac{1}{8}(4x + 1)^{3/2} \right]_4^9 = \tfrac{\pi}{6} \left(37\sqrt{37} - 17\sqrt{17} \right)$$

12.7   Triple Integrals

1. $\iiint_B xyz^2 \, dV = \int_0^1 \int_{-1}^2 \int_0^3 xyz^2 \, dz \, dx \, dy = \int_0^1 \int_{-1}^2 xy \left[\frac{1}{3}z^3\right]_{z=0}^{z=3} dx \, dy = \int_0^1 \int_{-1}^2 9xy \, dx \, dy$

$\qquad = \int_0^1 \left[\frac{9}{2}x^2 y\right]_{x=-1}^{x=2} dy = \int_0^1 \frac{27}{2} y \, dy = \frac{27}{4} y^2 \Big]_0^1 = \frac{27}{4}$

3. $\int_0^1 \int_0^z \int_0^{x+z} 6xz \, dy \, dx \, dz = \int_0^1 \int_0^z \left[6xyz\right]_{y=0}^{y=x+z} dx \, dz = \int_0^1 \int_0^z 6xz(x+z) \, dx \, dz$

$\qquad = \int_0^1 \left[2x^3 z + 3x^2 z^2\right]_{x=0}^{x=z} dz = \int_0^1 (2z^4 + 3z^4) \, dz = \int_0^1 5z^4 \, dz = z^5 \Big]_0^1 = 1$

5. $\int_0^3 \int_0^1 \int_0^{\sqrt{1-z^2}} ze^y \, dx \, dz \, dy = \int_0^3 \int_0^1 \left[xze^y\right]_{x=0}^{x=\sqrt{1-z^2}} dz \, dy = \int_0^3 \int_0^1 ze^y \sqrt{1-z^2} \, dz \, dy$

$\qquad = \int_0^3 \left[-\frac{1}{3}(1-z^2)^{3/2} e^y\right]_{z=0}^{z=1} dy = \int_0^3 \frac{1}{3} e^y \, dy = \frac{1}{3} e^y \Big]_0^3 = \frac{1}{3}(e^3 - 1)$

7. $\iiint_E 2x \, dV = \int_0^2 \int_0^{\sqrt{4-y^2}} \int_0^y 2x \, dz \, dx \, dy = \int_0^2 \int_0^{\sqrt{4-y^2}} \left[2xz\right]_{z=0}^{z=y} dx \, dy = \int_0^2 \int_0^{\sqrt{4-y^2}} 2xy \, dx \, dy$

$\qquad = \int_0^2 \left[x^2 y\right]_{x=0}^{x=\sqrt{4-y^2}} dy = \int_0^2 (4-y^2)y \, dy = \left[2y^2 - \frac{1}{4}y^4\right]_0^2 = 4$

9. Here $E = \{(x, y, z) \mid 0 \le x \le 1, 0 \le y \le \sqrt{x}, 0 \le z \le 1 + x + y\}$, so

$\qquad \iiint_E 6xy \, dV = \int_0^1 \int_0^{\sqrt{x}} \int_0^{1+x+y} 6xy \, dz \, dy \, dx = \int_0^1 \int_0^{\sqrt{x}} \left[6xyz\right]_{z=0}^{z=1+x+y} dy \, dx$

$\qquad\qquad = \int_0^1 \int_0^{\sqrt{x}} 6xy(1 + x + y) \, dy \, dx = \int_0^1 \left[3xy^2 + 3x^2 y^2 + 2xy^3\right]_{y=0}^{y=\sqrt{x}} dx$

$\qquad\qquad = \int_0^1 (3x^2 + 3x^3 + 2x^{5/2}) \, dx = \left[x^3 + \frac{3}{4}x^4 + \frac{4}{7}x^{7/2}\right]_0^1 = \frac{65}{28}$

11. Here E is the region that lies below the plane with x-, y-, and z-intercepts 1, 2, and 3 respectively, that is, below the plane $2z + 6x + 3y = 6$ and above the region in the xy-plane bounded by the lines $x = 0$, $y = 0$ and $6x + 3y = 6$. So

$\qquad \iiint_E xy \, dV = \int_0^1 \int_0^{2-2x} \int_0^{3-3x-3y/2} xy \, dz \, dy \, dx = \int_0^1 \int_0^{2-2x} \left(3xy - 3x^2 y - \frac{3}{2}xy^2\right) dy \, dx$

$\qquad\qquad = \int_0^1 \left[\frac{3}{2}xy^2 - \frac{3}{2}x^2 y^2 - \frac{1}{2}xy^3\right]_{y=0}^{y=2-2x} dx = \int_0^1 (2x - 6x^2 + 6x^3 - 2x^4) \, dx$

$\qquad\qquad = \left[x^2 - 2x^3 + \frac{3}{2}x^4 - \frac{2}{5}x^5\right]_0^1 = \frac{1}{10}$

13.

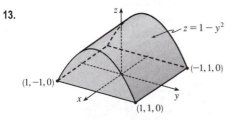

E is the region below the parabolic cylinder $z = 1 - y^2$ and above the square $[-1, 1] \times [-1, 1]$ in the xy-plane.

$\qquad \iiint_E x^2 e^y \, dV = \int_{-1}^1 \int_{-1}^1 \int_0^{1-y^2} x^2 e^y \, dz \, dy \, dx$

$\qquad\qquad = \int_{-1}^1 \int_{-1}^1 x^2 e^y (1 - y^2) \, dy \, dx$

$\qquad\qquad = \int_{-1}^1 x^2 \, dx \int_{-1}^1 (e^y - y^2 e^y) \, dy$

$\qquad\qquad = \left[\frac{1}{3}x^3\right]_{-1}^1 \left[e^y - (y^2 - 2y + 2)e^y\right]_{-1}^1$

$\qquad\qquad\qquad \text{[integrate by parts twice]}$

$\qquad\qquad = \frac{1}{3}(2)[e - e - e^{-1} + 5e^{-1}] = \frac{8}{3e}$

15.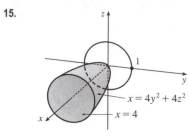

The projection E on the yz-plane is the disk $y^2 + z^2 \le 1$.

Using polar coordinates $y = r \cos \theta$ and $z = r \sin \theta$, we get

$$\iiint_E x \, dV = \iint_D \left[\int_{4y^2 + 4z^2}^4 x \, dx \right] dA$$

$$= \tfrac{1}{2} \iint_D \left[4^2 - (4y^2 + 4z^2)^2 \right] dA = 8 \int_0^{2\pi} \int_0^1 (1 - r^4) \, r \, dr \, d\theta$$

$$= 8 \int_0^{2\pi} d\theta \int_0^1 (r - r^5) \, dr = 8(2\pi) \left[\tfrac{1}{2} r^2 - \tfrac{1}{6} r^6 \right]_0^1 = \tfrac{16\pi}{3}$$

17. The plane $2x + y + z = 4$ intersects the xy-plane when

$2x + y + 0 = 4 \implies y = 4 - 2x$, so

$E = \{(x, y, z) \mid 0 \le x \le 2, 0 \le y \le 4 - 2x, 0 \le z \le 4 - 2x - y\}$ and

$$V = \int_0^2 \int_0^{4-2x} \int_0^{4-2x-y} dz \, dy \, dx = \int_0^2 \int_0^{4-2x} (4 - 2x - y) \, dy \, dx$$

$$= \int_0^2 \left[4y - 2xy - \tfrac{1}{2} y^2 \right]_{y=0}^{y=4-2x} dx$$

$$= \int_0^2 \left[4(4 - 2x) - 2x(4 - 2x) - \tfrac{1}{2}(4 - 2x)^2 \right] dx$$

$$= \int_0^2 (2x^2 - 8x + 8) \, dx = \left[\tfrac{2}{3} x^3 - 4x^2 + 8x \right]_0^2 = \tfrac{16}{3}$$

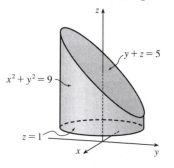

19. $V = \displaystyle\int_{-3}^3 \int_{-\sqrt{9-x^2}}^{\sqrt{9-x^2}} \int_1^{5-y} dz \, dy \, dx = \int_{-3}^3 \int_{-\sqrt{9-x^2}}^{\sqrt{9-x^2}} (5 - y - 1) \, dy \, dx = \int_{-3}^3 \left[4y - \tfrac{1}{2} y^2 \right]_{y=-\sqrt{9-x^2}}^{y=\sqrt{9-x^2}} dx$

$= \int_{-3}^3 8\sqrt{9 - x^2} \, dx = 8 \left[\tfrac{x}{2} \sqrt{9 - x^2} + \tfrac{9}{2} \sin^{-1} \left(\tfrac{x}{3} \right) \right]_{-3}^3 \quad \begin{bmatrix} \text{using trigonometric substitution} \\ \text{or Formula 30 in the Table of Integrals} \end{bmatrix}$

$= 8 \left[\tfrac{9}{2} \sin^{-1}(1) - \tfrac{9}{2} \sin^{-1}(-1) \right] = 36 \left(\tfrac{\pi}{2} - \left(-\tfrac{\pi}{2} \right) \right) = 36\pi$

Alternatively, use polar coordinates to evaluate the double integral:

$$\int_{-3}^3 \int_{-\sqrt{9-x^2}}^{\sqrt{9-x^2}} (4 - y) \, dy \, dx = \int_0^{2\pi} \int_0^3 (4 - r \sin \theta) \, r \, dr \, d\theta$$

$$= \int_0^{2\pi} \left[2r^2 - \tfrac{1}{3} r^3 \sin \theta \right]_{r=0}^{r=3} d\theta$$

$$= \int_0^{2\pi} (18 - 9 \sin \theta) \, d\theta$$

$$= 18\theta + 9 \cos \theta \Big]_0^{2\pi} = 36\pi$$

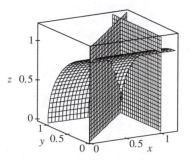

21. (a) The wedge can be described as the region

$$D = \left\{ (x, y, z) \mid y^2 + z^2 \le 1, 0 \le x \le 1, 0 \le y \le x \right\}$$

$$= \left\{ (x, y, z) \mid 0 \le x \le 1, 0 \le y \le x, 0 \le z \le \sqrt{1 - y^2} \right\}$$

So the integral expressing the volume of the wedge is

$$\iiint_D dV = \int_0^1 \int_0^x \int_0^{\sqrt{1-y^2}} dz \, dy \, dx.$$

(b) A CAS gives $\int_0^1 \int_0^x \int_0^{\sqrt{1-y^2}} dz \, dy \, dx = \tfrac{\pi}{4} - \tfrac{1}{3}$.

(Or use Formulas 30 and 87 from the Table of Integrals.)

23. Here $f(x, y, z) = \dfrac{1}{\ln(1 + x + y + z)}$ and $\Delta V = 2 \cdot 4 \cdot 2 = 16$, so the Midpoint Rule gives

$$\iiint_B f(x, y, z)\, dV \approx \sum_{i=1}^{l} \sum_{j=1}^{m} \sum_{k=1}^{n} f\left(\overline{x}_i, \overline{y}_j, \overline{z}_k\right) \Delta V$$

$$= 16\left[f(1, 2, 1) + f(1, 2, 3) + f(1, 6, 1) + f(1, 6, 3)\right.$$

$$\left. + f(3, 2, 1) + f(3, 2, 3) + f(3, 6, 1) + f(3, 6, 3)\right]$$

$$= 16\left[\frac{1}{\ln 5} + \frac{1}{\ln 7} + \frac{1}{\ln 9} + \frac{1}{\ln 11} + \frac{1}{\ln 7} + \frac{1}{\ln 9} + \frac{1}{\ln 11} + \frac{1}{\ln 13}\right] \approx 60.533$$

25. $E = \{(x, y, z) \mid 0 \le x \le 1, 0 \le z \le 1 - x, 0 \le y \le 2 - 2z\}$,

the solid bounded by the three coordinate planes and the planes

$z = 1 - x$, $y = 2 - 2z$.

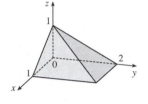

27.

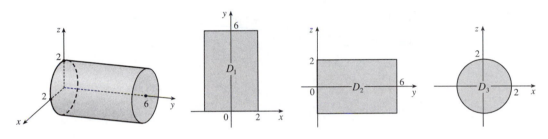

If D_1, D_2, D_3 are the projections of E on the xy-, yz-, and xz-planes, then

$$D_1 = \{(x, y) \mid -2 \le x \le 2, 0 \le y \le 6\}$$

$$D_2 = \{(y, z) \mid -2 \le z \le 2, 0 \le y \le 6\}$$

$$D_3 = \{(x, z) \mid x^2 + z^2 \le 4\}$$

Therefore

$$E = \left\{(x, y, z) \mid -\sqrt{4 - x^2} \le z \le \sqrt{4 - x^2},\ -2 \le x \le 2, 0 \le y \le 6\right\}$$

$$= \left\{(x, y, z) \mid -\sqrt{4 - z^2} \le x \le \sqrt{4 - z^2},\ -2 \le z \le 2, 0 \le y \le 6\right\}$$

$$\iiint_E f(x, y, z)\, dV = \int_{-2}^{2} \int_{0}^{6} \int_{-\sqrt{4-x^2}}^{\sqrt{4-x^2}} f(x, y, z)\, dz\, dy\, dx = \int_{0}^{6} \int_{-2}^{2} \int_{-\sqrt{4-x^2}}^{\sqrt{4-x^2}} f(x, y, z)\, dz\, dx\, dy$$

$$= \int_{0}^{6} \int_{-2}^{2} \int_{-\sqrt{4-z^2}}^{\sqrt{4-z^2}} f(x, y, z)\, dx\, dz\, dy = \int_{-2}^{2} \int_{0}^{6} \int_{-\sqrt{4-z^2}}^{\sqrt{4-z^2}} f(x, y, z)\, dx\, dy\, dz$$

$$= \int_{-2}^{2} \int_{-\sqrt{4-x^2}}^{\sqrt{4-x^2}} \int_{0}^{6} f(x, y, z)\, dy\, dz\, dx = \int_{-2}^{2} \int_{-\sqrt{4-z^2}}^{\sqrt{4-z^2}} \int_{0}^{6} f(x, y, z)\, dy\, dx\, dz$$

29.

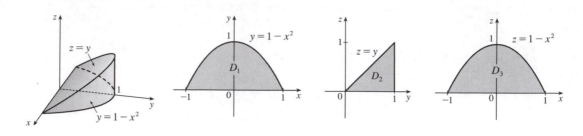

If D_1, D_2, and D_3 are the projections of E on the xy-, yz-, and xz-planes, then

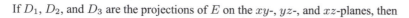

$$D_1 = \left\{(x,y) \mid -1 \le x \le 1, 0 \le y \le 1-x^2\right\} = \left\{(x,y) \mid 0 \le y \le 1, -\sqrt{1-y} \le x \le \sqrt{1-y}\right\},$$

$$D_2 = \left\{(y,z) \mid 0 \le y \le 1, 0 \le z \le y\right\} = \left\{(y,z) \mid 0 \le z \le 1, z \le y \le 1\right\}, \text{ and}$$

$$D_3 = \left\{(x,z) \mid -1 \le x \le 1, 0 \le z \le 1-x^2\right\} = \left\{(x,z) \mid 0 \le z \le 1, -\sqrt{1-z} \le x \le \sqrt{1-z}\right\}$$

Therefore

$$
\begin{aligned}
E &= \left\{(x,y,z) \mid -1 \le x \le 1, 0 \le y \le 1-x^2, 0 \le z \le y\right\} \\
&= \left\{(x,y,z) \mid 0 \le y \le 1, -\sqrt{1-y} \le x \le \sqrt{1-y}, 0 \le z \le y\right\} \\
&= \left\{(x,y,z) \mid 0 \le y \le 1, 0 \le z \le y, -\sqrt{1-y} \le x \le \sqrt{1-y}\right\} \\
&= \left\{(x,y,z) \mid 0 \le z \le 1, z \le y \le 1, -\sqrt{1-y} \le x \le \sqrt{1-y}\right\} \\
&= \left\{(x,y,z) \mid -1 \le x \le 1, 0 \le z \le 1-x^2, z \le y \le 1-x^2\right\} \\
&= \left\{(x,y,z) \mid 0 \le z \le 1, -\sqrt{1-z} \le x \le \sqrt{1-z}, z \le y \le 1-x^2\right\}
\end{aligned}
$$

Then

$$
\begin{aligned}
\iiint_E f(x,y,z)\,dV &= \int_{-1}^{1}\int_{0}^{1-x^2}\int_{0}^{y} f(x,y,z)\,dz\,dy\,dx = \int_{0}^{1}\int_{-\sqrt{1-y}}^{\sqrt{1-y}}\int_{0}^{y} f(x,y,z)\,dz\,dx\,dy \\
&= \int_{0}^{1}\int_{0}^{y}\int_{-\sqrt{1-y}}^{\sqrt{1-y}} f(x,y,z)\,dx\,dz\,dy = \int_{0}^{1}\int_{z}^{1}\int_{-\sqrt{1-y}}^{\sqrt{1-y}} f(x,y,z)\,dx\,dy\,dz \\
&= \int_{-1}^{1}\int_{0}^{1-x^2}\int_{z}^{1-x^2} f(x,y,z)\,dy\,dz\,dx = \int_{0}^{1}\int_{-\sqrt{1-z}}^{\sqrt{1-z}}\int_{z}^{1-x^2} f(x,y,z)\,dy\,dx\,dz
\end{aligned}
$$

31.

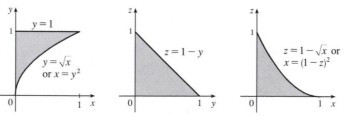

The diagrams show the projections of E on the xy-, yz-, and xz-planes. Therefore

$$\int_{0}^{1}\int_{\sqrt{x}}^{1}\int_{0}^{1-y} f(x,y,z)\,dz\,dy\,dx = \int_{0}^{1}\int_{0}^{y^2}\int_{0}^{1-y} f(x,y,z)\,dz\,dx\,dy = \int_{0}^{1}\int_{0}^{1-z}\int_{0}^{y^2} f(x,y,z)\,dx\,dy\,dz$$

$$= \int_{0}^{1}\int_{0}^{1-y}\int_{0}^{y^2} f(x,y,z)\,dx\,dz\,dy = \int_{0}^{1}\int_{0}^{1-\sqrt{x}}\int_{\sqrt{x}}^{1-z} f(x,y,z)\,dy\,dz\,dx$$

$$= \int_{0}^{1}\int_{0}^{(1-z)^2}\int_{\sqrt{x}}^{1-z} f(x,y,z)\,dy\,dx\,dz$$

33.

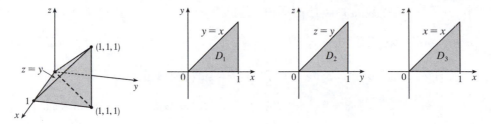

$\int_0^1 \int_y^1 \int_0^y f(x, y, z)\, dz\, dx\, dy = \iiint_E f(x, y, z)\, dV$ where $E = \{(x, y, z) \mid 0 \le z \le y, y \le x \le 1, 0 \le y \le 1\}$.

If D_1, D_2, and D_3 are the projections of E on the xy-, yz- and xz-planes then

$$D_1 = \{(x, y) \mid 0 \le y \le 1, y \le x \le 1\} = \{(x, y) \mid 0 \le x \le 1, 0 \le y \le x\},$$

$$D_2 = \{(y, z) \mid 0 \le y \le 1, 0 \le z \le y\} = \{(y, z) \mid 0 \le z \le 1, z \le y \le 1\}, \text{ and}$$

$$D_3 = \{(x, z) \mid 0 \le x \le 1, 0 \le z \le x\} = \{(x, z) \mid 0 \le z \le 1, z \le x \le 1\}.$$

Thus we also have

$$E = \{(x, y, z) \mid 0 \le x \le 1, 0 \le y \le x, 0 \le z \le y\} = \{(x, y, z) \mid 0 \le y \le 1, 0 \le z \le y, y \le x \le 1\}$$

$$= \{(x, y, z) \mid 0 \le z \le 1, z \le y \le 1, y \le x \le 1\} = \{(x, y, z) \mid 0 \le x \le 1, 0 \le z \le x, z \le y \le x\}$$

$$= \{(x, y, z) \mid 0 \le z \le 1, z \le x \le 1, z \le y \le x\}.$$

Then

$$\int_0^1 \int_y^1 \int_0^y f(x, y, z)\, dz\, dx\, dy = \int_0^1 \int_0^x \int_0^y f(x, y, z)\, dz\, dy\, dx = \int_0^1 \int_0^y \int_y^1 f(x, y, z)\, dx\, dz\, dy$$

$$= \int_0^1 \int_z^1 \int_y^1 f(x, y, z)\, dx\, dy\, dz = \int_0^1 \int_0^x \int_z^x f(x, y, z)\, dy\, dz\, dx$$

$$= \int_0^1 \int_z^1 \int_z^x f(x, y, z)\, dy\, dx\, dz$$

35. $m = \iiint_E \rho(x, y, z)\, dV = \int_0^1 \int_0^{\sqrt{x}} \int_0^{1+x+y} 2\, dz\, dy\, dx = \int_0^1 \int_0^{\sqrt{x}} 2(1 + x + y)\, dy\, dx$

$= \int_0^1 \left[2y + 2xy + y^2\right]_{y=0}^{y=\sqrt{x}} dx = \int_0^1 \left(2\sqrt{x} + 2x^{3/2} + x\right) dx = \left[\frac{4}{3}x^{3/2} + \frac{4}{5}x^{5/2} + \frac{1}{2}x^2\right]_0^1 = \frac{79}{30}$

$M_{yz} = \iiint_E x\rho(x, y, z)\, dV = \int_0^1 \int_0^{\sqrt{x}} \int_0^{1+x+y} 2x\, dz\, dy\, dx = \int_0^1 \int_0^{\sqrt{x}} 2x(1 + x + y)\, dy\, dx$

$= \int_0^1 \left[2xy + 2x^2y + xy^2\right]_{y=0}^{y=\sqrt{x}} dx = \int_0^1 (2x^{3/2} + 2x^{5/2} + x^2)\, dx = \left[\frac{4}{5}x^{5/2} + \frac{4}{7}x^{7/2} + \frac{1}{3}x^3\right]_0^1 = \frac{179}{105}$

$M_{xz} = \iiint_E y\rho(x, y, z)\, dV = \int_0^1 \int_0^{\sqrt{x}} \int_0^{1+x+y} 2y\, dz\, dy\, dx = \int_0^1 \int_0^{\sqrt{x}} 2y(1 + x + y)\, dy\, dx$

$= \int_0^1 \left[y^2 + xy^2 + \frac{2}{3}y^3\right]_{y=0}^{y=\sqrt{x}} dx = \int_0^1 \left(x + x^2 + \frac{2}{3}x^{3/2}\right) dx = \left[\frac{1}{2}x^2 + \frac{1}{3}x^3 + \frac{4}{15}x^{5/2}\right]_0^1 = \frac{11}{10}$

$M_{xy} = \iiint_E z\rho(x, y, z)\, dV = \int_0^1 \int_0^{\sqrt{x}} \int_0^{1+x+y} 2z\, dz\, dy\, dx = \int_0^1 \int_0^{\sqrt{x}} \left[z^2\right]_{z=0}^{z=1+x+y} dy\, dx = \int_0^1 \int_0^{\sqrt{x}} (1 + x + y)^2\, dy\, dx$

$= \int_0^1 \int_0^{\sqrt{x}} (1 + 2x + 2y + 2xy + x^2 + y^2)\, dy\, dx = \int_0^1 \left[y + 2xy + y^2 + xy^2 + x^2y + \frac{1}{3}y^3\right]_{y=0}^{y=\sqrt{x}} dx$

$= \int_0^1 \left(\sqrt{x} + \frac{7}{3}x^{3/2} + x + x^2 + x^{5/2}\right) dx = \left[\frac{2}{3}x^{3/2} + \frac{14}{15}x^{5/2} + \frac{1}{2}x^2 + \frac{1}{3}x^3 + \frac{2}{7}x^{7/2}\right]_0^1 = \frac{571}{210}$

Thus the mass is $\frac{79}{30}$ and the center of mass is $(\overline{x}, \overline{y}, \overline{z}) = \left(\dfrac{M_{yz}}{m}, \dfrac{M_{xz}}{m}, \dfrac{M_{xy}}{m}\right) = \left(\dfrac{358}{553}, \dfrac{33}{79}, \dfrac{571}{553}\right)$.

37. $m = \int_0^a \int_0^a \int_0^a (x^2 + y^2 + z^2)\, dx\, dy\, dz = \int_0^a \int_0^a \left[\frac{1}{3}x^3 + xy^2 + xz^2\right]_{x=0}^{x=a} dy\, dz = \int_0^a \int_0^a \left(\frac{1}{3}a^3 + ay^2 + az^2\right) dy\, dz$

$$= \int_0^a \left[\frac{1}{3}a^3 y + \frac{1}{3}ay^3 + ayz^2\right]_{y=0}^{y=a} dz = \int_0^a \left(\frac{2}{3}a^4 + a^2 z^2\right) dz = \left[\frac{2}{3}a^4 z + \frac{1}{3}a^2 z^3\right]_0^a = \frac{2}{3}a^5 + \frac{1}{3}a^5 = a^5$$

$$M_{yz} = \int_0^a \int_0^a \int_0^a \left[x^3 + x(y^2 + z^2)\right] dx\, dy\, dz = \int_0^a \int_0^a \left[\frac{1}{4}a^4 + \frac{1}{2}a^2(y^2 + z^2)\right] dy\, dz$$

$$= \int_0^a \left(\frac{1}{4}a^5 + \frac{1}{6}a^5 + \frac{1}{2}a^3 z^2\right) dz = \frac{1}{4}a^6 + \frac{1}{3}a^6 = \frac{7}{12}a^6 = M_{xz} = M_{xy} \text{ by symmetry of } E \text{ and } \rho(x,y,z)$$

Hence $(\overline{x}, \overline{y}, \overline{z}) = \left(\frac{7}{12}a, \frac{7}{12}a, \frac{7}{12}a\right)$.

39. (a) $m = \int_{-3}^3 \int_{-\sqrt{9-x^2}}^{\sqrt{9-x^2}} \int_1^{5-y} \sqrt{x^2 + y^2}\, dz\, dy\, dx$

(b) $(\overline{x}, \overline{y}, \overline{z}) = \left(\dfrac{M_{yz}}{m}, \dfrac{M_{xz}}{m}, \dfrac{M_{xy}}{m}\right)$ where

$$M_{yz} = \int_{-3}^3 \int_{-\sqrt{9-x^2}}^{\sqrt{9-x^2}} \int_1^{5-y} x\sqrt{x^2+y^2}\, dz\, dy\, dx, \; M_{xz} = \int_{-3}^3 \int_{-\sqrt{9-x^2}}^{\sqrt{9-x^2}} \int_1^{5-y} y\sqrt{x^2+y^2}\, dz\, dy\, dx, \text{ and}$$

$$M_{xy} = \int_{-3}^3 \int_{-\sqrt{9-x^2}}^{\sqrt{9-x^2}} \int_1^{5-y} z\sqrt{x^2+y^2}\, dz\, dy\, dx.$$

(c) $I_z = \int_{-3}^3 \int_{-\sqrt{9-x^2}}^{\sqrt{9-x^2}} \int_1^{5-y} (x^2+y^2)\sqrt{x^2+y^2}\, dz\, dy\, dx = \int_{-3}^3 \int_{-\sqrt{9-x^2}}^{\sqrt{9-x^2}} \int_1^{5-y} (x^2+y^2)^{3/2}\, dz\, dy\, dx$

41. (a) $m = \int_0^1 \int_0^{\sqrt{1-x^2}} \int_0^y (1 + x + y + z)\, dz\, dy\, dx = \frac{3\pi}{32} + \frac{11}{24}$

(b) $(\overline{x}, \overline{y}, \overline{z}) = \Big(m^{-1} \int_0^1 \int_0^{\sqrt{1-x^2}} \int_0^y x(1 + x + y + z)\, dz\, dy\, dx,$

$$m^{-1} \int_0^1 \int_0^{\sqrt{1-x^2}} \int_0^y y(1 + x + y + z)\, dz\, dy\, dx,$$

$$m^{-1} \int_0^1 \int_0^{\sqrt{1-x^2}} \int_0^y z(1 + x + y + z)\, dz\, dy\, dx\Big)$$

$$= \left(\frac{28}{9\pi + 44}, \frac{30\pi + 128}{45\pi + 220}, \frac{45\pi + 208}{135\pi + 660}\right)$$

(c) $I_z = \int_0^1 \int_0^{\sqrt{1-x^2}} \int_0^y (x^2 + y^2)(1 + x + y + z)\, dz\, dy\, dx = \dfrac{68 + 15\pi}{240}$

43. $I_x = \int_0^L \int_0^L \int_0^L k(y^2 + z^2)\, dz\, dy\, dx = k \int_0^L \int_0^L \left(Ly^2 + \frac{1}{3}L^3\right) dy\, dx = k \int_0^L \frac{2}{3}L^4\, dx = \frac{2}{3}kL^5$.

By symmetry, $I_x = I_y = I_z = \frac{2}{3}kL^5$.

45. (a) $f(x,y,z)$ is a joint density function, so we know $\iiint_{\mathbb{R}^3} f(x,y,z)\, dV = 1$. Here we have

$$\iiint_{\mathbb{R}^3} f(x,y,z)\, dV = \int_{-\infty}^\infty \int_{-\infty}^\infty \int_{-\infty}^\infty f(x,y,z)\, dz\, dy\, dx = \int_0^2 \int_0^2 \int_0^2 Cxyz\, dz\, dy\, dx$$

$$= C \int_0^2 x\, dx \int_0^2 y\, dy \int_0^2 z\, dz = C \left[\frac{x^2}{2}\right]_0^2 \left[\frac{y^2}{2}\right]_0^2 \left[\frac{z^2}{2}\right]_0^2 = 8C$$

Then we must have $8C = 1 \; \Rightarrow \; C = \frac{1}{8}$.

(b) $P(X \le 1, Y \le 1, Z \le 1) = \int_{-\infty}^1 \int_{-\infty}^1 \int_{-\infty}^1 f(x,y,z)\, dz\, dy\, dx = \int_0^1 \int_0^1 \int_0^1 \frac{1}{8}xyz\, dz\, dy\, dx$

$$= \frac{1}{8} \int_0^1 x\, dx \int_0^1 y\, dy \int_0^1 z\, dz = \frac{1}{8} \left[\frac{x^2}{2}\right]_0^1 \left[\frac{y^2}{2}\right]_0^1 \left[\frac{z^2}{2}\right]_0^1 = \frac{1}{8}\left(\frac{1}{2}\right)^3 = \frac{1}{64}$$

(c) $P(X + Y + Z \le 1) = P((X, Y, Z) \in E)$ where E is the solid region in the first octant bounded by the coordinate planes and the plane $x + y + z = 1$. The plane $x + y + z = 1$ meets the xy-plane in the line $x + y = 1$, so we have

$$P(X + Y + Z \le 1) = \iiint_E f(x, y, z)\, dV = \int_0^1 \int_0^{1-x} \int_0^{1-x-y} \tfrac{1}{8} xyz\, dz\, dy\, dx$$

$$= \tfrac{1}{8} \int_0^1 \int_0^{1-x} xy\left[\tfrac{1}{2} z^2\right]_{z=0}^{z=1-x-y} dy\, dx = \tfrac{1}{16} \int_0^1 \int_0^{1-x} xy(1 - x - y)^2\, dy\, dx$$

$$= \tfrac{1}{16} \int_0^1 \int_0^{1-x} \left[(x^3 - 2x^2 + x)y + (2x^2 - 2x)y^2 + xy^3\right] dy\, dx$$

$$= \tfrac{1}{16} \int_0^1 \left[(x^3 - 2x^2 + x)\tfrac{1}{2}y^2 + (2x^2 - 2x)\tfrac{1}{3}y^3 + x\left(\tfrac{1}{4}y^4\right)\right]_{y=0}^{y=1-x} dx$$

$$= \tfrac{1}{192} \int_0^1 (x - 4x^2 + 6x^3 - 4x^4 + x^5)\, dx = \tfrac{1}{192}\left(\tfrac{1}{30}\right) = \tfrac{1}{5760}$$

47. $V(E) = L^3$,

$$f_{\text{ave}} = \frac{1}{L^3} \int_0^L \int_0^L \int_0^L xyz\, dx\, dy\, dz = \frac{1}{L^3} \int_0^L x\, dx \int_0^L y\, dy \int_0^L z\, dz$$

$$= \frac{1}{L^3} \left[\frac{x^2}{2}\right]_0^L \left[\frac{y^2}{2}\right]_0^L \left[\frac{z^2}{2}\right]_0^L = \frac{1}{L^3} \frac{L^2}{2} \frac{L^2}{2} \frac{L^2}{2} = \frac{L^3}{8}$$

49. The triple integral will attain its maximum when the integrand $1 - x^2 - 2y^2 - 3z^2$ is positive in the region E and negative everywhere else. For if E contains some region F where the integrand is negative, the integral could be increased by excluding F from E, and if E fails to contain some part G of the region where the integrand is positive, the integral could be increased by including G in E. So we require that $x^2 + 2y^2 + 3z^2 \le 1$. This describes the region bounded by the ellipsoid $x^2 + 2y^2 + 3z^2 = 1$.

12.8   Triple Integrals in Cylindrical and Spherical Coordinates

1.

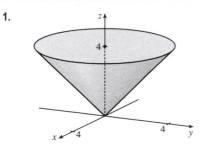

The region of integration is given in cylindrical coordinates by

$E = \{(r, \theta, z) \mid 0 \le \theta \le 2\pi, 0 \le r \le 4, r \le z \le 4\}$. This represents the solid region bounded below by the cone $z = r$ and above by the horizontal plane $z = 4$.

$$\int_0^4 \int_0^{2\pi} \int_r^4 r\, dz\, d\theta\, dr = \int_0^4 \int_0^{2\pi} [rz]_{z=r}^{z=4}\, d\theta\, dr = \int_0^4 \int_0^{2\pi} r(4 - r)\, d\theta\, dr$$

$$= \int_0^4 (4r - r^2)\, dr \int_0^{2\pi} d\theta = \left[2r^2 - \tfrac{1}{3}r^3\right]_0^4 [\theta]_0^{2\pi}$$

$$= \left(32 - \tfrac{64}{3}\right)(2\pi) = \tfrac{64\pi}{3}$$

3.

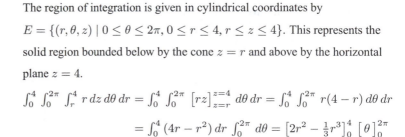

The region of integration is given in spherical coordinates by

$E = \{(\rho, \theta, \phi) \mid 0 \le \rho \le 3, 0 \le \theta \le \pi/2, 0 \le \phi \le \pi/6\}$. This represents the solid region in the first octant bounded above by the sphere $\rho = 3$ and below by the cone $\phi = \pi/6$.

$$\int_0^{\pi/6} \int_0^{\pi/2} \int_0^3 \rho^2 \sin\phi\, d\rho\, d\theta\, d\phi = \int_0^{\pi/6} \sin\phi\, d\phi \int_0^{\pi/2} d\theta \int_0^3 \rho^2\, d\rho$$

$$= [-\cos\phi]_0^{\pi/6} [\theta]_0^{\pi/2} \left[\tfrac{1}{3}\rho^3\right]_0^3$$

$$= \left(1 - \frac{\sqrt{3}}{2}\right)\left(\frac{\pi}{2}\right)(9) = \frac{9\pi}{4}(2 - \sqrt{3})$$

5. The solid E is most conveniently described if we use cylindrical coordinates:

$E = \{(r, \theta, z) \mid 0 \le \theta \le \frac{\pi}{2}, 0 \le r \le 3, 0 \le z \le 2\}$. Then $\iiint_E f(x, y, z)\, dV = \int_0^{\pi/2} \int_0^3 \int_0^2 f(r \cos\theta, r \sin\theta, z)\, r\, dz\, dr\, d\theta$.

7. In cylindrical coordinates, E is given by $\{(r, \theta, z) \mid 0 \le \theta \le 2\pi, 0 \le r \le 4, -5 \le z \le 4\}$. So

$\iiint_E \sqrt{x^2 + y^2}\, dV = \int_0^{2\pi} \int_0^4 \int_{-5}^4 \sqrt{r^2}\, r\, dz\, dr\, d\theta = \int_0^{2\pi} d\theta \int_0^4 r^2\, dr \int_{-5}^4 dz$

$\qquad = \left[\theta\right]_0^{2\pi} \left[\frac{1}{3}r^3\right]_0^4 \left[z\right]_{-5}^4 = (2\pi)\left(\frac{64}{3}\right)(9) = 384\pi$

9. In cylindrical coordinates E is bounded by the paraboloid $z = 1 + r^2$, the cylinder $r^2 = 5$ or $r = \sqrt{5}$, and the xy-plane, so E is given by $\{(r, \theta, z) \mid 0 \le \theta \le 2\pi, 0 \le r \le \sqrt{5}, 0 \le z \le 1 + r^2\}$. Thus

$\iiint_E e^z\, dV = \int_0^{2\pi} \int_0^{\sqrt{5}} \int_0^{1+r^2} e^z\, r\, dz\, dr\, d\theta = \int_0^{2\pi} \int_0^{\sqrt{5}} r\left[e^z\right]_{z=0}^{z=1+r^2}\, dr\, d\theta = \int_0^{2\pi} \int_0^{\sqrt{5}} r(e^{1+r^2} - 1)\, dr\, d\theta$

$\qquad = \int_0^{2\pi} d\theta \int_0^{\sqrt{5}} \left(re^{1+r^2} - r\right) dr = 2\pi\left[\frac{1}{2}e^{1+r^2} - \frac{1}{2}r^2\right]_0^{\sqrt{5}} = \pi(e^6 - e - 5)$

11. In cylindrical coordinates, E is bounded by the cylinder $r = 1$, the plane $z = 0$, and the cone $z = 2r$. So $E = \{(r, \theta, z) \mid 0 \le \theta \le 2\pi, 0 \le r \le 1, 0 \le z \le 2r\}$ and

$\iiint_E x^2\, dV = \int_0^{2\pi} \int_0^1 \int_0^{2r} r^2 \cos^2\theta\, r\, dz\, dr\, d\theta = \int_0^{2\pi} \int_0^1 \left[r^3 \cos^2\theta\, z\right]_{z=0}^{z=2r}\, dr\, d\theta$

$\qquad = \int_0^{2\pi} \int_0^1 2r^4 \cos^2\theta\, dr\, d\theta = \int_0^{2\pi} \left[\frac{2}{5}r^5 \cos^2\theta\right]_{r=0}^{r=1}\, d\theta = \frac{2}{5}\int_0^{2\pi} \cos^2\theta\, d\theta$

$\qquad = \frac{2}{5} \int_0^{2\pi} \frac{1 + \cos 2\theta}{2}\, d\theta = \frac{1}{5}\left[\theta + \frac{1}{2}\sin 2\theta\right]_0^{2\pi} = \frac{2\pi}{5}$

13. (a) The paraboloids intersect when $x^2 + y^2 = 36 - 3x^2 - 3y^2 \;\Rightarrow\; x^2 + y^2 = 9$, so the region of integration is $D = \{(x, y) \mid x^2 + y^2 \le 9\}$. Then, in cylindrical coordinates, $E = \{(r, \theta, z) \mid r^2 \le z \le 36 - 3r^2, 0 \le r \le 3, 0 \le \theta \le 2\pi\}$ and

$V = \int_0^{2\pi} \int_0^3 \int_{r^2}^{36-3r^2} r\, dz\, dr\, d\theta = \int_0^{2\pi} \int_0^3 \left(36r - 4r^3\right) dr\, d\theta$

$\qquad = \int_0^{2\pi} \left[18r^2 - r^4\right]_{r=0}^{r=3}\, d\theta = \int_0^{2\pi} 81\, d\theta = 162\pi$

(b) For constant density K, $m = KV = 162\pi K$ from part (a). Since the region is homogeneous and symmetric, $M_{yz} = M_{xz} = 0$ and

$M_{xy} = \int_0^{2\pi} \int_0^3 \int_{r^2}^{36-3r^2} (zK)\, r\, dz\, dr\, d\theta = K \int_0^{2\pi} \int_0^3 r\left[\frac{1}{2}z^2\right]_{z=r^2}^{z=36-3r^2}\, dr\, d\theta$

$\qquad = \frac{K}{2} \int_0^{2\pi} \int_0^3 r((36 - 3r^2)^2 - r^4)\, dr\, d\theta = \frac{K}{2} \int_0^{2\pi} d\theta \int_0^3 (8r^5 - 216r^3 + 1296r)\, dr$

$\qquad = \frac{K}{2}(2\pi)\left[\frac{8}{6}r^6 - \frac{216}{4}r^4 + \frac{1296}{2}r^2\right]_0^3 = \pi K(2430) = 2430\pi K$

Thus $(\overline{x}, \overline{y}, \overline{z}) = \left(\dfrac{M_{yz}}{m}, \dfrac{M_{xz}}{m}, \dfrac{M_{xy}}{m}\right) = \left(0, 0, \dfrac{2430\pi K}{162\pi K}\right) = (0, 0, 15)$.

15. The paraboloid $z = 4x^2 + 4y^2$ intersects the plane $z = a$ when $a = 4x^2 + 4y^2$ or $x^2 + y^2 = \frac{1}{4}a$. So, in cylindrical coordinates, $E = \{(r, \theta, z) \mid 0 \le r \le \frac{1}{2}\sqrt{a}, 0 \le \theta \le 2\pi, 4r^2 \le z \le a\}$. Thus

$m = \int_0^{2\pi} \int_0^{\sqrt{a}/2} \int_{4r^2}^a Kr\, dz\, dr\, d\theta = K \int_0^{2\pi} \int_0^{\sqrt{a}/2} (ar - 4r^3)\, dr\, d\theta$

$\qquad = K \int_0^{2\pi} \left[\frac{1}{2}ar^2 - r^4\right]_{r=0}^{r=\sqrt{a}/2}\, d\theta = K \int_0^{2\pi} \frac{1}{16}a^2\, d\theta = \frac{1}{8}a^2 \pi K$

Since the region is homogeneous and symmetric, $M_{yz} = M_{xz} = 0$ and

$M_{xy} = \int_0^{2\pi} \int_0^{\sqrt{a}/2} \int_{4r^2}^a Krz\, dz\, dr\, d\theta = K \int_0^{2\pi} \int_0^{\sqrt{a}/2} \left(\frac{1}{2}a^2 r - 8r^5\right) dr\, d\theta$

$\qquad = K \int_0^{2\pi} \left[\frac{1}{4}a^2 r^2 - \frac{4}{3}r^6\right]_{r=0}^{r=\sqrt{a}/2}\, d\theta = K \int_0^{2\pi} \frac{1}{24}a^3\, d\theta = \frac{1}{12}a^3 \pi K$

Hence $(\overline{x}, \overline{y}, \overline{z}) = \left(0, 0, \frac{2}{3}a\right)$.

17. In spherical coordinates, B is represented by $\{(\rho, \theta, \phi) \mid 0 \le \rho \le 1, 0 \le \theta \le 2\pi, 0 \le \phi \le \pi\}$. Thus

$$\iiint_B (x^2 + y^2 + z^2)\, dV = \int_0^\pi \int_0^{2\pi} \int_0^1 (\rho^2)\rho^2 \sin\phi \, d\rho\, d\theta\, d\phi = \int_0^\pi \sin\phi\, d\phi \int_0^{2\pi} d\theta \int_0^1 \rho^4\, d\rho$$

$$= \left[-\cos\phi\right]_0^\pi \left[\theta\right]_0^{2\pi} \left[\tfrac{1}{5}\rho^5\right]_0^1 = (2)(2\pi)\left(\tfrac{1}{5}\right) = \tfrac{4\pi}{5}$$

19. In spherical coordinates, E is represented by $\{(\rho, \theta, \phi) \mid 1 \le \rho \le 2, 0 \le \theta \le \frac{\pi}{2}, 0 \le \phi \le \frac{\pi}{2}\}$. Thus

$$\iiint_E z\, dV = \int_0^{\pi/2} \int_0^{\pi/2} \int_1^2 (\rho \cos\phi)\, \rho^2 \sin\phi\, d\rho\, d\theta\, d\phi$$

$$= \int_0^{\pi/2} \cos\phi \sin\phi\, d\phi \int_0^{\pi/2} d\theta \int_1^2 \rho^3\, d\rho = \left[\tfrac{1}{2}\sin^2\phi\right]_0^{\pi/2} \left[\theta\right]_0^{\pi/2} \left[\tfrac{1}{4}\rho^4\right]_1^2$$

$$= \left(\tfrac{1}{2}\right)\left(\tfrac{\pi}{2}\right)\left(\tfrac{15}{4}\right) = \tfrac{15\pi}{16}$$

21. $\iiint_E x^2\, dV = \int_0^\pi \int_0^\pi \int_3^4 (\rho \sin\phi \cos\theta)^2\, \rho^2 \sin\phi\, d\rho\, d\phi\, d\theta = \int_0^\pi \cos^2\theta\, d\theta \int_0^\pi \sin^3\phi\, d\phi \int_3^4 \rho^4\, d\rho$

$$= \left[\tfrac{1}{2}\theta + \tfrac{1}{4}\sin 2\theta\right]_0^\pi \left[-\tfrac{1}{3}(2 + \sin^2\phi)\cos\phi\right]_0^\pi \left[\tfrac{1}{5}\rho^5\right]_3^4 = \left(\tfrac{\pi}{2}\right)\left(\tfrac{2}{3} + \tfrac{2}{3}\right)\tfrac{1}{5}(4^5 - 3^5) = \tfrac{1562}{15}\pi$$

23. (a) Since $\rho = 4\cos\phi$ implies $\rho^2 = 4\rho\cos\phi$, the equation is that of a sphere of radius 2 with center at $(0, 0, 2)$. Thus

$$V = \int_0^{2\pi} \int_0^{\pi/3} \int_0^{4\cos\phi} \rho^2 \sin\phi\, d\rho\, d\phi\, d\theta = \int_0^{2\pi} \int_0^{\pi/3} \left[\tfrac{1}{3}\rho^3\right]_{\rho=0}^{\rho=4\cos\phi} \sin\phi\, d\phi\, d\theta = \int_0^{2\pi} \int_0^{\pi/3} \left(\tfrac{64}{3}\cos^3\phi\right)\sin\phi\, d\phi\, d\theta$$

$$= \int_0^{2\pi} \left[-\tfrac{16}{3}\cos^4\phi\right]_{\phi=0}^{\phi=\pi/3} d\theta = \int_0^{2\pi} -\tfrac{16}{3}\left(\tfrac{1}{16} - 1\right) d\theta = 5\theta\Big]_0^{2\pi} = 10\pi$$

(b) By the symmetry of the problem $M_{yz} = M_{xz} = 0$. Then

$$M_{xy} = \int_0^{2\pi} \int_0^{\pi/3} \int_0^{4\cos\phi} \rho^3 \cos\phi \sin\phi\, d\rho\, d\phi\, d\theta = \int_0^{2\pi} \int_0^{\pi/3} \cos\phi \sin\phi \left(64\cos^4\phi\right) d\phi\, d\theta$$

$$= \int_0^{2\pi} 64\left[-\tfrac{1}{6}\cos^6\phi\right]_{\phi=0}^{\phi=\pi/3} d\theta = \int_0^{2\pi} \tfrac{21}{2}\, d\theta = 21\pi$$

Hence $(\overline{x}, \overline{y}, \overline{z}) = (0, 0, 2.1)$.

25. (a) The density function is $\rho(x, y, z) = K$, a constant, and by the symmetry of the problem $M_{xz} = M_{yz} = 0$. Then

$$M_{xy} = \int_0^{2\pi} \int_0^{\pi/2} \int_0^a K\rho^3 \sin\phi \cos\phi\, d\rho\, d\phi\, d\theta = \tfrac{1}{2}\pi K a^4 \int_0^{\pi/2} \sin\phi\cos\phi\, d\phi = \tfrac{1}{8}\pi K a^4. \text{ But the mass is}$$

$K(\text{volume of the hemisphere}) = \tfrac{2}{3}\pi K a^3$, so the centroid is $\left(0, 0, \tfrac{3}{8}a\right)$.

(b) Place the center of the base at $(0, 0, 0)$; the density function is $\rho(x, y, z) = K$. By symmetry, the moments of inertia about any two such diameters will be equal, so we just need to find I_x:

$$I_x = \int_0^{2\pi} \int_0^{\pi/2} \int_0^a \left(K\rho^2 \sin\phi\right)\rho^2 \left(\sin^2\phi \sin^2\theta + \cos^2\phi\right) d\rho\, d\phi\, d\theta$$

$$= K \int_0^{2\pi} \int_0^{\pi/2} \left(\sin^3\phi \sin^2\theta + \sin\phi \cos^2\phi\right)\left(\tfrac{1}{5}a^5\right) d\phi\, d\theta$$

$$= \tfrac{1}{5}Ka^5 \int_0^{2\pi} \left[\sin^2\theta\left(-\cos\phi + \tfrac{1}{3}\cos^3\phi\right) + \left(-\tfrac{1}{3}\cos^3\phi\right)\right]_{\phi=0}^{\phi=\pi/2} d\theta = \tfrac{1}{5}Ka^5 \int_0^{2\pi} \left[\tfrac{2}{3}\sin^2\theta + \tfrac{1}{3}\right] d\theta$$

$$= \tfrac{1}{5}Ka^5 \left[\tfrac{2}{3}\left(\tfrac{1}{2}\theta - \tfrac{1}{4}\sin 2\theta\right) + \tfrac{1}{3}\theta\right]_0^{2\pi} = \tfrac{1}{5}Ka^5 \left[\tfrac{2}{3}(\pi - 0) + \tfrac{1}{3}(2\pi - 0)\right] = \tfrac{4}{15}Ka^5\pi$$

27. In spherical coordinates $z = \sqrt{x^2 + y^2}$ becomes $\cos\phi = \sin\phi$ or $\phi = \frac{\pi}{4}$. Then

$$V = \int_0^{2\pi} \int_0^{\pi/4} \int_0^1 \rho^2 \sin\phi\, d\rho\, d\phi\, d\theta = \int_0^{2\pi} d\theta \int_0^{\pi/4} \sin\phi\, d\phi \int_0^1 \rho^2\, d\rho = \tfrac{1}{3}\pi\left(2 - \sqrt{2}\right),$$

$$M_{xy} = \int_0^{2\pi} \int_0^{\pi/4} \int_0^1 \rho^3 \sin\phi \cos\phi\, d\rho\, d\phi\, d\theta = 2\pi\left[-\tfrac{1}{4}\cos 2\phi\right]_0^{\pi/4}\left(\tfrac{1}{4}\right) = \tfrac{\pi}{8} \text{ and by symmetry } M_{yz} = M_{xz} = 0.$$

Hence $(\overline{x}, \overline{y}, \overline{z}) = \left(0, 0, \dfrac{3}{8(2 - \sqrt{2})}\right)$.

29. In cylindrical coordinates the paraboloid is given by $z = r^2$ and the plane by $z = 2r\sin\theta$ and they intersect in the circle

$r = 2\sin\theta$. Then $\iiint_E z\,dV = \int_0^\pi \int_0^{2\sin\theta} \int_{r^2}^{2r\sin\theta} rz\,dz\,dr\,d\theta = \frac{5\pi}{6}$ [using a CAS].

31. The region of integration is the region above the cone $z = \sqrt{x^2+y^2}$, or $z = r$, and below the plane $z = 2$. Also, we have

$-2 \le y \le 2$ with $-\sqrt{4-y^2} \le x \le \sqrt{4-y^2}$ which describes a circle of radius 2 in the xy-plane centered at $(0,0)$. Thus,

$$\int_{-2}^{2}\int_{-\sqrt{4-y^2}}^{\sqrt{4-y^2}}\int_{\sqrt{x^2+y^2}}^{2} xz\,dz\,dx\,dy = \int_0^{2\pi}\int_0^2\int_r^2 (r\cos\theta)\,zr\,dz\,dr\,d\theta = \int_0^{2\pi}\int_0^2\int_r^2 r^2(\cos\theta)\,z\,dz\,dr\,d\theta$$

$$= \int_0^{2\pi}\int_0^2 r^2(\cos\theta)\left[\tfrac{1}{2}z^2\right]_{z=r}^{z=2}\,dr\,d\theta = \tfrac{1}{2}\int_0^{2\pi}\int_0^2 r^2(\cos\theta)\left(4-r^2\right)\,dr\,d\theta$$

$$= \tfrac{1}{2}\int_0^{2\pi}\cos\theta\,d\theta\int_0^2\left(4r^2-r^4\right)\,dr = \tfrac{1}{2}\left[\sin\theta\right]_0^{2\pi}\left[\tfrac{4}{3}r^3-\tfrac{1}{5}r^5\right]_0^2 = 0$$

33. The region E of integration is the region above the cone $z = \sqrt{x^2+y^2}$ and below the sphere $x^2+y^2+z^2 = 2$ in the first

octant. Because E is in the first octant we have $0 \le \theta \le \frac{\pi}{2}$. The cone has equation $\phi = \frac{\pi}{4}$ (as in Example 4), so $0 \le \phi \le \frac{\pi}{4}$,

and $0 \le \rho \le \sqrt{2}$. So the integral becomes

$\int_0^{\pi/4}\int_0^{\pi/2}\int_0^{\sqrt{2}} (\rho\sin\phi\cos\theta)(\rho\sin\phi\sin\theta)\,\rho^2\sin\phi\,d\rho\,d\theta\,d\phi$

$$= \int_0^{\pi/4}\sin^3\phi\,d\phi\int_0^{\pi/2}\sin\theta\cos\theta\,d\theta\int_0^{\sqrt{2}}\rho^4\,d\rho = \left(\int_0^{\pi/4}(1-\cos^2\phi)\sin\phi\,d\phi\right)\left[\tfrac{1}{2}\sin^2\theta\right]_0^{\pi/2}\left[\tfrac{1}{5}\rho^5\right]_0^{\sqrt{2}}$$

$$= \left[\tfrac{1}{3}\cos^3\phi - \cos\phi\right]_0^{\pi/4}\cdot\tfrac{1}{2}\cdot\tfrac{1}{5}\left(\sqrt{2}\right)^5 = \left[\tfrac{\sqrt{2}}{12}-\tfrac{\sqrt{2}}{2}-\left(\tfrac{1}{3}-1\right)\right]\cdot\tfrac{2\sqrt{2}}{5} = \tfrac{4\sqrt{2}-5}{15}$$

35. If E is the solid enclosed by the surface $\rho = 1+\frac{1}{5}\sin 6\theta\sin 5\phi$, it can be described in spherical coordinates as

$E = \left\{(\rho,\theta,\phi)\mid 0\le\rho\le 1+\tfrac{1}{5}\sin 6\theta\sin 5\phi, 0\le\theta\le 2\pi, 0\le\phi\le\pi\right\}$. Its volume is given by

$V(E) = \iiint_E dV = \int_0^\pi\int_0^{2\pi}\int_0^{1+(\sin 6\theta\sin 5\phi)/5}\rho^2\sin\phi\,d\rho\,d\theta\,d\phi = \frac{136\pi}{99}$ [using a CAS].

37. (a) The mountain comprises a solid conical region C. The work done in lifting a small volume of material ΔV with density

$g(P)$ to a height $h(P)$ above sea level is $h(P)g(P)\,\Delta V$. Summing over the whole mountain we get

$W = \iiint_C h(P)g(P)\,\Delta V$.

(b) Here C is a solid right circular cone with radius $R = 62{,}000$ ft, height $H = 12{,}400$ ft, and density $g(P) = 200$ lb/ft^3 at

all points P in C. We use cylindrical coordinates:

$$W = \int_0^{2\pi}\int_0^H\int_0^{R(1-z/H)} z\cdot 200r\,dr\,dz\,d\theta = 2\pi\int_0^H 200z\left[\tfrac{1}{2}r^2\right]_{r=0}^{r=R(1-z/H)}dz$$

$$= 400\pi\int_0^H z\,\frac{R^2}{2}\left(1-\frac{z}{H}\right)^2 dz = 200\pi R^2\int_0^H\left(z-\frac{2z^2}{H}+\frac{z^3}{H^2}\right)dz$$

$$= 200\pi R^2\left[\frac{z^2}{2}-\frac{2z^3}{3H}+\frac{z^4}{4H^2}\right]_0^H = 200\pi R^2\left(\frac{H^2}{2}-\frac{2H^2}{3}+\frac{H^2}{4}\right)$$

$$= \tfrac{50}{3}\pi R^2 H^2 = \tfrac{50}{3}\pi(62{,}000)^2(12{,}400)^2 \approx 3.1\times 10^{19}\text{ ft-lb}$$

$$\frac{r}{R} = \frac{H-z}{H} = 1-\frac{z}{H}$$

12.9   Change of Variables in Multiple Integrals

1. $x = u + 4v$, $y = 3u - 2v$.   The Jacobian is $\dfrac{\partial(x, y)}{\partial(u, v)} = \begin{vmatrix} \partial x/\partial u & \partial x/\partial v \\ \partial y/\partial u & \partial y/\partial v \end{vmatrix} = \begin{vmatrix} 1 & 4 \\ 3 & -2 \end{vmatrix} = 1(-2) - 4(3) = -14.$

3. $\dfrac{\partial(x, y)}{\partial(u, v)} = \begin{vmatrix} \dfrac{\partial x}{\partial u} & \dfrac{\partial x}{\partial v} \\[2mm] \dfrac{\partial y}{\partial u} & \dfrac{\partial y}{\partial v} \end{vmatrix} = \begin{vmatrix} \dfrac{v}{(u+v)^2} & -\dfrac{u}{(u+v)^2} \\[2mm] -\dfrac{v}{(u-v)^2} & \dfrac{u}{(u-v)^2} \end{vmatrix} = \dfrac{uv}{(u+v)^2(u-v)^2} - \dfrac{uv}{(u+v)^2(u-v)^2} = 0$

5. $\dfrac{\partial(x, y, z)}{\partial(u, v, w)} = \begin{vmatrix} \partial x/\partial u & \partial x/\partial v & \partial x/\partial w \\ \partial y/\partial u & \partial y/\partial v & \partial y/\partial w \\ \partial z/\partial u & \partial z/\partial v & \partial z/\partial w \end{vmatrix} = \begin{vmatrix} v & u & 0 \\ 0 & w & v \\ w & 0 & u \end{vmatrix} = v\begin{vmatrix} w & v \\ 0 & u \end{vmatrix} - u\begin{vmatrix} 0 & v \\ w & u \end{vmatrix} + 0\begin{vmatrix} 0 & w \\ w & 0 \end{vmatrix}$

$= v(uw - 0) - u(0 - vw) = 2uvw$

7. The transformation maps the boundary of S to the boundary of the image R, so we first look at side S_1 in the uv-plane. S_1 is described by $v = 0$ $(0 \le u \le 3)$, so $x = 2u + 3v = 2u$ and $y = u - v = u$. Eliminating u, we have $x = 2y$, $0 \le x \le 6$. S_2 is the line segment $u = 3$, $0 \le v \le 2$, so $x = 6 + 3v$ and $y = 3 - v$. Then $v = 3 - y$ $\Rightarrow$ $x = 6 + 3(3 - y) = 15 - 3y$, $6 \le x \le 12$. S_3 is the line segment $v = 2$, $0 \le u \le 3$, so $x = 2u + 6$ and $y = u - 2$, giving $u = y + 2$ $\Rightarrow$ $x = 2y + 10$, $6 \le x \le 12$. Finally, S_4 is the segment $u = 0$, $0 \le v \le 2$, so $x = 3v$ and $y = -v$ $\Rightarrow$ $x = -3y$, $0 \le x \le 6$. The image of set S is the region R shown in the xy-plane, a parallelogram bounded by these four segments.

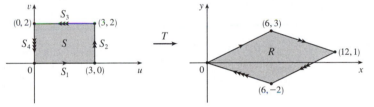

9. S_1 is the line segment $u = v$, $0 \le u \le 1$, so $y = v = u$ and $x = u^2 = y^2$. Since $0 \le u \le 1$, the image is the portion of the parabola $x = y^2$, $0 \le y \le 1$. S_2 is the segment $v = 1$, $0 \le u \le 1$, thus $y = v = 1$ and $x = u^2$, so $0 \le x \le 1$. The image is the line segment $y = 1$, $0 \le x \le 1$. S_3 is the segment $u = 0$, $0 \le v \le 1$, so $x = u^2 = 0$ and $y = v$ $\Rightarrow$ $0 \le y \le 1$. The image is the segment $x = 0$, $0 \le y \le 1$. Thus, the image of S is the region R in the first quadrant bounded by the parabola $x = y^2$, the y-axis, and the line $y = 1$.

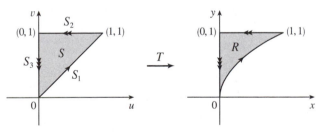

11. $\dfrac{\partial(x, y)}{\partial(u, v)} = \begin{vmatrix} 2 & 1 \\ 1 & 2 \end{vmatrix} = 3$ and $x - 3y = (2u + v) - 3(u + 2v) = -u - 5v$. To find the region S in the uv-plane that corresponds to R we first find the corresponding boundary under the given transformation. The line through $(0, 0)$ and $(2, 1)$ is $y = \frac{1}{2}x$ which is the image of $u + 2v = \frac{1}{2}(2u + v)$ $\Rightarrow$ $v = 0$; the line through $(2, 1)$ and $(1, 2)$ is $x + y = 3$ which is the image of $(2u + v) + (u + 2v) = 3$ $\Rightarrow$ $u + v = 1$; the line through $(0, 0)$ and $(1, 2)$ is $y = 2x$ which is the image of $u + 2v = 2(2u + v)$ $\Rightarrow$ $u = 0$. Thus S is the triangle $0 \le v \le 1 - u$, $0 \le u \le 1$ in the uv-plane and

$$\iint_R (x - 3y)\, dA = \int_0^1 \int_0^{1-u} (-u - 5v)\, |3|\, dv\, du = -3\int_0^1 \left[uv + \tfrac{5}{2}v^2 \right]_{v=0}^{v=1-u} du$$

$$= -3\int_0^1 \left(u - u^2 + \tfrac{5}{2}(1 - u)^2 \right) du = -3\left[\tfrac{1}{2}u^2 - \tfrac{1}{3}u^3 - \tfrac{5}{6}(1 - u)^3 \right]_0^1 = -3\left(\tfrac{1}{2} - \tfrac{1}{3} + \tfrac{5}{6} \right) = -3$$

13. $\dfrac{\partial(x, y)}{\partial(u, v)} = \begin{vmatrix} 2 & 0 \\ 0 & 3 \end{vmatrix} = 6$, $x^2 = 4u^2$ and the planar ellipse $9x^2 + 4y^2 \le 36$ is the image of the disk $u^2 + v^2 \le 1$. Thus

$$\iint_R x^2\, dA = \iint_{u^2+v^2\le 1} (4u^2)(6)\, du\, dv = \int_0^{2\pi}\int_0^1 (24r^2\cos^2\theta)\, r\, dr\, d\theta = 24\int_0^{2\pi}\cos^2\theta\, d\theta \int_0^1 r^3\, dr$$

$$= 24\left[\tfrac{1}{2}x + \tfrac{1}{4}\sin 2x\right]_0^{2\pi} \left[\tfrac{1}{4}r^4\right]_0^1 = 24(\pi)\left(\tfrac{1}{4}\right) = 6\pi$$

15. $\dfrac{\partial(x, y)}{\partial(u, v)} = \begin{vmatrix} 1/v & -u/v^2 \\ 0 & 1 \end{vmatrix} = \dfrac{1}{v}$, $xy = u$, $y = x$ is the image of the parabola $v^2 = u$, $y = 3x$ is the image of the parabola $v^2 = 3u$, and the hyperbolas $xy = 1$, $xy = 3$ are the images of the lines $u = 1$ and $u = 3$ respectively. Thus

$$\iint_R xy\, dA = \int_1^3 \int_{\sqrt{u}}^{\sqrt{3u}} u\left(\dfrac{1}{v}\right) dv\, du = \int_1^3 u\left(\ln\sqrt{3u} - \ln\sqrt{u}\right) du = \int_1^3 u\ln\sqrt{3}\, du = 4\ln\sqrt{3} = 2\ln 3.$$

17. (a) $\dfrac{\partial(x, y, z)}{\partial(u, v, w)} = \begin{vmatrix} a & 0 & 0 \\ 0 & b & 0 \\ 0 & 0 & c \end{vmatrix} = abc$ and since $u = \dfrac{x}{a}$, $v = \dfrac{y}{b}$, $w = \dfrac{z}{c}$ the solid enclosed by the ellipsoid is the image of the

ball $u^2 + v^2 + w^2 \le 1$. So

$$\iiint_E dV = \iiint_{u^2+v^2+w^2\le 1} abc\, du\, dv\, dw = (abc)(\text{volume of the ball}) = \tfrac{4}{3}\pi abc$$

(b) If we approximate the surface of the Earth by the ellipsoid $\dfrac{x^2}{6378^2} + \dfrac{y^2}{6378^2} + \dfrac{z^2}{6356^2} = 1$, then we can estimate the volume of the Earth by finding the volume of the solid E enclosed by the ellipsoid. From part (a), this is $\iiint_E dV = \tfrac{4}{3}\pi(6378)(6378)(6356) \approx 1.083 \times 10^{12} \text{ km}^3$.

19. Letting $u = x - 2y$ and $v = 3x - y$, we have $x = \tfrac{1}{5}(2v - u)$ and $y = \tfrac{1}{5}(v - 3u)$. Then $\dfrac{\partial(x, y)}{\partial(u, v)} = \begin{vmatrix} -1/5 & 2/5 \\ -3/5 & 1/5 \end{vmatrix} = \dfrac{1}{5}$

and R is the image of the rectangle enclosed by the lines $u = 0$, $u = 4$, $v = 1$, and $v = 8$. Thus

$$\iint_R \dfrac{x - 2y}{3x - y}\, dA = \int_0^4 \int_1^8 \dfrac{u}{v}\left|\dfrac{1}{5}\right| dv\, du = \dfrac{1}{5}\int_0^4 u\, du \int_1^8 \dfrac{1}{v}\, dv = \tfrac{1}{5}\left[\tfrac{1}{2}u^2\right]_0^4 \left[\ln|v|\right]_1^8 = \tfrac{8}{5}\ln 8.$$

21. Letting $u = y - x$, $v = y + x$, we have $y = \tfrac{1}{2}(u + v)$, $x = \tfrac{1}{2}(v - u)$. Then $\dfrac{\partial(x, y)}{\partial(u, v)} = \begin{vmatrix} -1/2 & 1/2 \\ 1/2 & 1/2 \end{vmatrix} = -\dfrac{1}{2}$ and R is the

image of the trapezoidal region with vertices $(-1, 1)$, $(-2, 2)$, $(2, 2)$, and $(1, 1)$. Thus

$$\iint_R \cos\dfrac{y - x}{y + x}\, dA = \int_1^2 \int_{-v}^{v} \cos\dfrac{u}{v}\left|-\dfrac{1}{2}\right| du\, dv = \dfrac{1}{2}\int_1^2 \left[v\sin\dfrac{u}{v}\right]_{u=-v}^{u=v} dv$$

$$= \tfrac{1}{2}\int_1^2 2v\sin(1)\, dv = \tfrac{3}{2}\sin 1$$

23. Let $u = x + y$ and $v = -x + y$. Then $u + v = 2y \Rightarrow y = \tfrac{1}{2}(u + v)$ and

$u - v = 2x \Rightarrow x = \tfrac{1}{2}(u - v)$. $\dfrac{\partial(x, y)}{\partial(u, v)} = \begin{vmatrix} 1/2 & -1/2 \\ 1/2 & 1/2 \end{vmatrix} = \dfrac{1}{2}$.

Now $|u| = |x + y| \le |x| + |y| \le 1 \Rightarrow -1 \le u \le 1$, and

$|v| = |-x + y| \le |x| + |y| \le 1 \Rightarrow -1 \le v \le 1$. R is the image of

the square region with vertices $(1, 1)$, $(1, -1)$, $(-1, -1)$, and $(-1, 1)$.

So $\iint_R e^{x+y}\, dA = \tfrac{1}{2}\int_{-1}^1 \int_{-1}^1 e^u\, du\, dv = \tfrac{1}{2}\left[e^u\right]_{-1}^1 \left[v\right]_{-1}^1 = e - e^{-1}$.

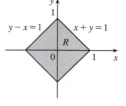

12   Review

CONCEPT CHECK

1. (a) A double Riemann sum of f is $\sum\limits_{i=1}^{m} \sum\limits_{j=1}^{n} f(x_{ij}^*, y_{ij}^*) \, \Delta A$, where ΔA is the area of each subrectangle and (x_{ij}^*, y_{ij}^*) is a sample point in each subrectangle. If $f(x,y) \geq 0$, this sum represents an approximation to the volume of the solid that lies above the rectangle R and below the graph of f.

(b) $\iint_R f(x,y) \, dA = \lim\limits_{m,n\to\infty} \sum\limits_{i=1}^{m} \sum\limits_{j=1}^{n} f(x_{ij}^*, y_{ij}^*) \, \Delta A$

(c) If $f(x,y) \geq 0$, $\iint_R f(x,y) \, dA$ represents the volume of the solid that lies above the rectangle R and below the surface $z = f(x,y)$. If f takes on both positive and negative values, $\iint_R f(x,y) \, dA$ is the difference of the volume above R but below the surface $z = f(x,y)$ and the volume below R but above the surface $z = f(x,y)$.

(d) We usually evaluate $\iint_R f(x,y) \, dA$ as an iterated integral according to Fubini's Theorem (see Theorem 12.2.4).

(e) The Midpoint Rule for Double Integrals says that we approximate the double integral $\iint_R f(x,y) \, dA$ by the double Riemann sum $\sum\limits_{i=1}^{m} \sum\limits_{j=1}^{n} f(\overline{x}_i, \overline{y}_j) \, \Delta A$ where the sample points $(\overline{x}_i, \overline{y}_j)$ are the centers of the subrectangles.

(f) $f_{\text{ave}} = \dfrac{1}{A(R)} \iint_R f(x,y) \, dA$ where $A(R)$ is the area of R.

2. (a) See (1) and (2) and the accompanying discussion in Section 12.3.

(b) See (3) and the accompanying discussion in Section 12.3.

(c) See (5) and the preceding discussion in Section 12.3.

(d) See (6)–(11) in Section 12.3.

3. We may want to change from rectangular to polar coordinates in a double integral if the region R of integration is more easily described in polar coordinates. To accomplish this, we use $\iint_R f(x,y) \, dA = \int_\alpha^\beta \int_a^b f(r\cos\theta, r\sin\theta) \, r \, dr \, d\theta$ where R is given by $0 \leq a \leq r \leq b$, $\alpha \leq \theta \leq \beta$.

4. (a) $m = \iint_D \rho(x,y) \, dA$

(b) $M_x = \iint_D y\rho(x,y) \, dA$, $M_y = \iint_D x\rho(x,y) \, dA$

(c) The center of mass is $(\overline{x}, \overline{y})$ where $\overline{x} = \dfrac{M_y}{m}$ and $\overline{y} = \dfrac{M_x}{m}$.

(d) $I_x = \iint_D y^2\rho(x,y) \, dA$, $I_y = \iint_D x^2\rho(x,y) \, dA$, $I_0 = \iint_D (x^2+y^2)\rho(x,y) \, dA$

5. (a) $P(a \leq X \leq b, c \leq Y \leq d) = \int_a^b \int_c^d f(x,y) \, dy \, dx$

(b) $f(x,y) \geq 0$ and $\iint_{\mathbb{R}^2} f(x,y) \, dA = 1$.

(c) The expected value of X is $\mu_1 = \iint_{\mathbb{R}^2} x f(x,y) \, dA$; the expected value of Y is $\mu_2 = \iint_{\mathbb{R}^2} y f(x,y) \, dA$.

6. (a) $A(S) = \iint_D |\mathbf{r}_u \times \mathbf{r}_v| \, dA$

(b) $A(S) = \iint_D \sqrt{1 + \left(\dfrac{\partial z}{\partial x}\right)^2 + \left(\dfrac{\partial z}{\partial y}\right)^2} \, dA$

(c) $A(S) = 2\pi \int_a^b f(x)\sqrt{1 + [f'(x)]^2} \, dx$

7. (a) $\iiint_B f(x,y,z) \, dV = \lim\limits_{l,m,n\to\infty} \sum\limits_{i=1}^{l} \sum\limits_{j=1}^{m} \sum\limits_{k=1}^{n} f(x_{ijk}^*, y_{ijk}^*, z_{ijk}^*) \, \Delta V$

(b) We usually evaluate $\iiint_B f(x,y,z) \, dV$ as an iterated integral according to Fubini's Theorem for Triple Integrals (see Theorem 12.7.4).

(c) See the paragraph following Example 12.7.1.

(d) See (5) and (6) and the accompanying discussion in Section 12.7.

(e) See (10) and the accompanying discussion in Section 12.7.

(f) See (11) and the preceding discussion in Section 12.7.

8. (a) $m = \iiint_E \rho(x, y, z)\, dV$

(b) $M_{yz} = \iiint_E x\rho(x, y, z)\, dV$, $M_{xz} = \iiint_E y\rho(x, y, z)\, dV$, $M_{xy} = \iiint_E z\rho(x, y, z)\, dV$.

(c) The center of mass is $(\overline{x}, \overline{y}, \overline{z})$ where $\overline{x} = \dfrac{M_{yz}}{m}$, $\overline{y} = \dfrac{M_{xz}}{m}$, and $\overline{z} = \dfrac{M_{xy}}{m}$.

(d) $I_x = \iiint_E (y^2 + z^2)\rho(x, y, z)\, dV$, $I_y = \iiint_E (x^2 + z^2)\rho(x, y, z)\, dV$, $I_z = \iiint_E (x^2 + y^2)\rho(x, y, z)\, dV$.

9. (a) See Formula 12.8.2 and the accompanying discussion.

(b) See Formula 12.8.4 and the accompanying discussion.

(c) We may want to change from rectangular to cylindrical or spherical coordinates in a triple integral if the region E of integration is more easily described in cylindrical or spherical coordinates or if the triple integral is easier to evaluate using cylindrical or spherical coordinates.

10. (a) $\dfrac{\partial (x, y)}{\partial (u, v)} = \begin{vmatrix} \partial x/\partial u & \partial x/\partial v \\ \partial y/\partial u & \partial y/\partial v \end{vmatrix} = \dfrac{\partial x}{\partial u}\dfrac{\partial y}{\partial v} - \dfrac{\partial x}{\partial v}\dfrac{\partial y}{\partial u}$

(b) See (9) and the accompanying discussion in Section 12.9.

(c) See (13) and the accompanying discussion in Section 12.9.

TRUE-FALSE QUIZ

1. This is true by Fubini's Theorem.

3. True. See the discussion following Example 4 on page 841.

5. True:

$$\iint_D \sqrt{4 - x^2 - y^2}\, dA = \text{the volume under the surface } x^2 + y^2 + z^2 = 4 \text{ and above the } xy\text{-plane}$$
$$= \tfrac{1}{2}\left(\text{the volume of the sphere } x^2 + y^2 + z^2 = 4\right) = \tfrac{1}{2}\cdot\tfrac{4}{3}\pi(2)^3 = \tfrac{16}{3}\pi$$

7. The volume enclosed by the cone $z = \sqrt{x^2 + y^2}$ and the plane $z = 2$ is, in cylindrical coordinates,

$V = \int_0^{2\pi}\int_0^2\int_r^2 r\, dz\, dr\, d\theta \neq \int_0^{2\pi}\int_0^2\int_r^2 dz\, dr\, d\theta$, so the assertion is false.

EXERCISES

1. As shown in the contour map, we divide R into 9 equally sized subsquares, each with area $\Delta A = 1$. Then we approximate $\iint_R f(x, y)\, dA$ by a Riemann sum with $m = n = 3$ and the sample points the upper right corners of each square, so

$$\iint_R f(x, y)\, dA \approx \sum_{i=1}^{3}\sum_{j=1}^{3} f(x_i, y_j)\, \Delta A$$

$$= \Delta A\left[f(1, 1) + f(1, 2) + f(1, 3) + f(2, 1) + f(2, 2) + f(2, 3) + f(3, 1) + f(3, 2) + f(3, 3)\right]$$

Using the contour lines to estimate the function values, we have

$$\iint_R f(x, y)\, dA \approx 1[2.7 + 4.7 + 8.0 + 4.7 + 6.7 + 10.0 + 6.7 + 8.6 + 11.9] \approx 64.0$$

3. $\int_1^2\int_0^2 (y + 2xe^y)\, dx\, dy = \int_1^2 \left[xy + x^2 e^y\right]_{x=0}^{x=2} dy = \int_1^2 (2y + 4e^y)\, dy = \left[y^2 + 4e^y\right]_1^2$
$$= 4 + 4e^2 - 1 - 4e = 4e^2 - 4e + 3$$

5. $\int_0^1 \int_0^x \cos(x^2) \, dy \, dx = \int_0^1 \left[\cos(x^2) y \right]_{y=0}^{y=x} dx = \int_0^1 x \cos(x^2) \, dx = \frac{1}{2} \sin(x^2) \Big]_0^1 = \frac{1}{2} \sin 1$

7. $\int_0^\pi \int_0^1 \int_0^{\sqrt{1-y^2}} y \sin x \, dz \, dy \, dx = \int_0^\pi \int_0^1 \left[(y \sin x) z \right]_{z=0}^{z=\sqrt{1-y^2}} dy \, dx = \int_0^\pi \int_0^1 y \sqrt{1-y^2} \sin x \, dy \, dx$

$= \int_0^\pi \left[-\frac{1}{3}(1-y^2)^{3/2} \sin x \right]_{y=0}^{y=1} dx = \int_0^\pi \frac{1}{3} \sin x \, dx = -\frac{1}{3} \cos x \Big]_0^\pi = \frac{2}{3}$

9. The region R is more easily described by polar coordinates: $R = \{(r, \theta) \mid 2 \leq r \leq 4, 0 \leq \theta \leq \pi\}$. Thus

$\iint_R f(x, y) \, dA = \int_0^\pi \int_2^4 f(r \cos \theta, r \sin \theta) \, r \, dr \, d\theta$.

11.

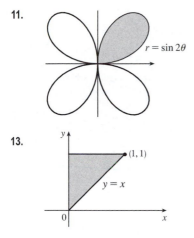

$r = \sin 2\theta$

The region whose area is given by $\int_0^{\pi/2} \int_0^{\sin 2\theta} r \, dr \, d\theta$ is

$\left\{ (r, \theta) \mid 0 \leq \theta \leq \frac{\pi}{2}, 0 \leq r \leq \sin 2\theta \right\}$, which is the region

contained in the loop in the first quadrant of the four-leaved rose

$r = \sin 2\theta$.

13.

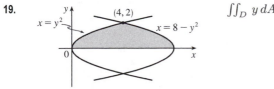

$y = x$

$(1, 1)$

$\int_0^1 \int_x^1 \cos(y^2) \, dy \, dx = \int_0^1 \int_0^y \cos(y^2) \, dx \, dy$

$= \int_0^1 \cos(y^2) \left[x \right]_{x=0}^{x=y} dy = \int_0^1 y \cos(y^2) \, dy$

$= \left[\frac{1}{2} \sin(y^2) \right]_0^1 = \frac{1}{2} \sin 1$

15. $\iint_R y e^{xy} \, dA = \int_0^3 \int_0^2 y e^{xy} \, dx \, dy = \int_0^3 \left[e^{xy} \right]_{x=0}^{x=2} dy = \int_0^3 (e^{2y} - 1) \, dy = \left[\frac{1}{2} e^{2y} - y \right]_0^3$

$= \frac{1}{2} e^6 - 3 - \frac{1}{2} = \frac{1}{2} e^6 - \frac{7}{2}$

17.

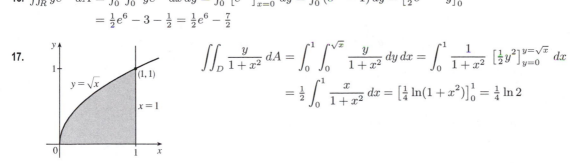

$y = \sqrt{x}$

$(1, 1)$

$x = 1$

$\iint_D \frac{y}{1+x^2} \, dA = \int_0^1 \int_0^{\sqrt{x}} \frac{y}{1+x^2} \, dy \, dx = \int_0^1 \frac{1}{1+x^2} \left[\frac{1}{2} y^2 \right]_{y=0}^{y=\sqrt{x}} dx$

$= \frac{1}{2} \int_0^1 \frac{x}{1+x^2} \, dx = \left[\frac{1}{4} \ln(1+x^2) \right]_0^1 = \frac{1}{4} \ln 2$

19.

$x = y^2$

$(4, 2)$

$x = 8 - y^2$

$\iint_D y \, dA = \int_0^2 \int_{y^2}^{8-y^2} y \, dx \, dy$

$= \int_0^2 y \left[x \right]_{x=y^2}^{x=8-y^2} dy = \int_0^2 y(8 - y^2 - y^2) \, dy$

$= \int_0^2 (8y - 2y^3) \, dy = \left[4y^2 - \frac{1}{2} y^4 \right]_0^2 = 8$

21.

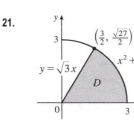

$\left(\frac{3}{2}, \frac{\sqrt{27}}{2} \right)$

$x^2 + y^2 = 9$

$y = \sqrt{3} x$

D

$\iint_D (x^2 + y^2)^{3/2} \, dA = \int_0^{\pi/3} \int_0^3 (r^2)^{3/2} r \, dr \, d\theta$

$= \int_0^{\pi/3} d\theta \int_0^3 r^4 \, dr = \left[\theta \right]_0^{\pi/3} \left[\frac{1}{5} r^5 \right]_0^3$

$= \frac{\pi}{3} \frac{3^5}{5} = \frac{81\pi}{5}$

23. $\iiint_E xy\, dV = \int_0^3 \int_0^x \int_0^{x+y} xy\, dz\, dy\, dx = \int_0^3 \int_0^x xy\, [\,z\,]_{z=0}^{z=x+y}\, dy\, dx = \int_0^3 \int_0^x xy(x+y)\, dy\, dx$

$\qquad = \int_0^3 \int_0^x (x^2 y + xy^2)\, dy\, dx = \int_0^3 \left[\frac{1}{2}x^2 y^2 + \frac{1}{3}xy^3\right]_{y=0}^{y=x} dx = \int_0^3 \left(\frac{1}{2}x^4 + \frac{1}{3}x^4\right) dx$

$\qquad = \frac{5}{6}\int_0^3 x^4\, dx = \left[\frac{1}{6}x^5\right]_0^3 = \frac{81}{2} = 40.5$

25. $\iiint_E y^2 z^2\, dV = \int_{-1}^1 \int_{-\sqrt{1-y^2}}^{\sqrt{1-y^2}} \int_0^{1-y^2-z^2} y^2 z^2\, dx\, dz\, dy = \int_{-1}^1 \int_{-\sqrt{1-y^2}}^{\sqrt{1-y^2}} y^2 z^2 (1-y^2-z^2)\, dz\, dy$

$\qquad = \int_0^{2\pi} \int_0^1 (r^2 \cos^2 \theta)(r^2 \sin^2 \theta)(1-r^2)\, r\, dr\, d\theta = \int_0^{2\pi} \int_0^1 \frac{1}{4}\sin^2 2\theta (r^5 - r^7)\, dr\, d\theta$

$\qquad = \int_0^{2\pi} \frac{1}{8}(1-\cos 4\theta)\left[\frac{1}{6}r^6 - \frac{1}{8}r^8\right]_{r=0}^{r=1} d\theta = \frac{1}{192}\left[\theta - \frac{1}{4}\sin 4\theta\right]_0^{2\pi} = \frac{2\pi}{192} = \frac{\pi}{96}$

27. $\iiint_E yz\, dV = \int_{-2}^2 \int_0^{\sqrt{4-x^2}} \int_0^y yz\, dz\, dy\, dx = \int_{-2}^2 \int_0^{\sqrt{4-x^2}} \frac{1}{2}y^3\, dy\, dx = \int_0^\pi \int_0^2 \frac{1}{2}r^3 (\sin^3 \theta)\, r\, dr\, d\theta$

$\qquad = \frac{16}{5}\int_0^\pi \sin^3 \theta\, d\theta = \frac{16}{5}\left[-\cos\theta + \frac{1}{3}\cos^3\theta\right]_0^\pi = \frac{64}{15}$

29. $V = \int_0^2 \int_1^4 (x^2 + 4y^2)\, dy\, dx = \int_0^2 \left[x^2 y + \frac{4}{3}y^3\right]_{y=1}^{y=4} dx = \int_0^2 (3x^2 + 84)\, dx = 176$

31.

$(0,0,1)$, $y + 2z = 2$, 0, $(0,2,0)$, $x = y$, $-y = 2$, y, x, $(2,2,0)$

$V = \int_0^2 \int_0^y \int_0^{(2-y)/2} dz\, dx\, dy$

$\quad = \int_0^2 \int_0^y \left(1 - \frac{1}{2}y\right) dx\, dy$

$\quad = \int_0^2 \left(y - \frac{1}{2}y^2\right) dy = \frac{2}{3}$

33. Using the wedge above the plane $z = 0$ and below the plane $z = mx$ and noting that we have the same volume for $m < 0$ as for $m > 0$ (so use $m > 0$), we have

$V = 2\int_0^{a/3} \int_0^{\sqrt{a^2 - 9y^2}} mx\, dx\, dy = 2\int_0^{a/3} \frac{1}{2}m(a^2 - 9y^2)\, dy = m\left[a^2 y - 3y^3\right]_0^{a/3} = m\left(\frac{1}{3}a^3 - \frac{1}{9}a^3\right) = \frac{2}{9}ma^3.$

35. (a) $m = \int_0^1 \int_0^{1-y^2} y\, dx\, dy = \int_0^1 (y - y^3)\, dy = \frac{1}{2} - \frac{1}{4} = \frac{1}{4}$

(b) $M_y = \int_0^1 \int_0^{1-y^2} xy\, dx\, dy = \int_0^1 \frac{1}{2}y(1-y^2)^2\, dy = -\frac{1}{12}(1-y^2)^3\Big]_0^1 = \frac{1}{12}$,

$\qquad M_x = \int_0^1 \int_0^{1-y^2} y^2\, dx\, dy = \int_0^1 (y^2 - y^4)\, dy = \frac{2}{15}$. Hence $(\overline{x}, \overline{y}) = \left(\frac{1}{3}, \frac{8}{15}\right)$.

(c) $I_x = \int_0^1 \int_0^{1-y^2} y^3\, dx\, dy = \int_0^1 (y^3 - y^5)\, dy = \frac{1}{12}$,

$\qquad I_y = \int_0^1 \int_0^{1-y^2} yx^2\, dx\, dy = \int_0^1 \frac{1}{3}y(1-y^2)^3\, dy = -\frac{1}{24}(1-y^2)^4\Big]_0^1 = \frac{1}{24}$,

$\qquad I_0 = I_x + I_y = \frac{1}{8}, \overline{\overline{y}}^2 = \frac{1/12}{1/4} = \frac{1}{3} \Rightarrow \overline{\overline{y}} = \frac{1}{\sqrt{3}}$, and $\overline{\overline{x}}^2 = \frac{1/24}{1/4} = \frac{1}{6} \Rightarrow \overline{\overline{x}} = \frac{1}{\sqrt{6}}$.

37. (a) The equation of the cone with the suggested orientation is $(h - z) = \frac{h}{a}\sqrt{x^2 + y^2}$, $0 \le z \le h$. Then $V = \frac{1}{3}\pi a^2 h$ is the volume of one frustum of a cone; by symmetry $M_{yz} = M_{xz} = 0$; and

$$M_{xy} = \iint_{x^2+y^2 \le a^2} \int_0^{h - (h/a)\sqrt{x^2+y^2}} z\, dz\, dA = \int_0^{2\pi} \int_0^a \int_0^{(h/a)(a-r)} rz\, dz\, dr\, d\theta = \pi \int_0^a r\frac{h^2}{a^2}(a-r)^2\, dr$$

$$= \frac{\pi h^2}{a^2}\int_0^a (a^2 r - 2ar^2 + r^3)\, dr = \frac{\pi h^2}{a^2}\left(\frac{a^4}{2} - \frac{2a^4}{3} + \frac{a^4}{4}\right) = \frac{\pi h^2 a^2}{12}$$

Hence the centroid is $(\overline{x}, \overline{y}, \overline{z}) = \left(0, 0, \frac{1}{4}h\right)$.

(b) $I_z = \int_0^{2\pi} \int_0^a \int_0^{(h/a)(a-r)} r^3\, dz\, dr\, d\theta = 2\pi \int_0^a \frac{h}{a}(ar^3 - r^4)\, dr = \frac{2\pi h}{a}\left(\frac{a^5}{4} - \frac{a^5}{5}\right) = \frac{\pi a^4 h}{10}$

39. Let D represent the given triangle; then D can be described as the area enclosed by the x- and y-axes and the line $y = 2 - 2x$, or equivalently $D = \{(x, y) \mid 0 \le x \le 1, 0 \le y \le 2 - 2x\}$. We want to find the surface area of the part of the graph of $z = x^2 + y$ that lies over D, so using Equation 12.6.6 we have

$$A(S) = \iint_D \sqrt{1 + \left(\frac{\partial z}{\partial x}\right)^2 + \left(\frac{\partial z}{\partial y}\right)^2}\, dA = \iint_D \sqrt{1 + (2x)^2 + (1)^2}\, dA = \int_0^1 \int_0^{2-2x} \sqrt{2 + 4x^2}\, dy\, dx$$

$$= \int_0^1 \sqrt{2 + 4x^2}\, [y]_{y=0}^{y=2-2x}\, dx = \int_0^1 (2 - 2x)\sqrt{2 + 4x^2}\, dx = \int_0^1 2\sqrt{2 + 4x^2}\, dx - \int_0^1 2x\sqrt{2 + 4x^2}\, dx$$

Using Formula 21 in the Table of Integrals with $a = \sqrt{2}$, $u = 2x$, and $du = 2\, dx$, we have

$\int 2\sqrt{2 + 4x^2}\, dx = x\sqrt{2 + 4x^2} + \ln\left(2x + \sqrt{2 + 4x^2}\right)$. If we substitute $u = 2 + 4x^2$ in the second integral, then

$du = 8x\, dx$ and $\int 2x\sqrt{2 + 4x^2}\, dx = \frac{1}{4}\int \sqrt{u}\, du = \frac{1}{4} \cdot \frac{2}{3}u^{3/2} = \frac{1}{6}(2 + 4x^2)^{3/2}$. Thus

$$A(S) = \left[x\sqrt{2 + 4x^2} + \ln\left(2x + \sqrt{2 + 4x^2}\right) - \frac{1}{6}(2 + 4x^2)^{3/2}\right]_0^1$$

$$= \sqrt{6} + \ln\left(2 + \sqrt{6}\right) - \frac{1}{6}(6)^{3/2} - \ln\sqrt{2} + \frac{\sqrt{2}}{3} = \ln\frac{2+\sqrt{6}}{\sqrt{2}} + \frac{\sqrt{2}}{3}$$

$$= \ln\left(\sqrt{2} + \sqrt{3}\right) + \frac{\sqrt{2}}{3} \approx 1.6176$$

41.

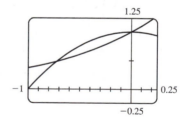

$$\int_0^3 \int_{-\sqrt{9-x^2}}^{\sqrt{9-x^2}} (x^3 + xy^2)\, dy\, dx = \int_0^3 \int_{-\sqrt{9-x^2}}^{\sqrt{9-x^2}} x(x^2 + y^2)\, dy\, dx$$

$$= \int_{-\pi/2}^{\pi/2} \int_0^3 (r\cos\theta)(r^2)\, r\, dr\, d\theta$$

$$= \int_{-\pi/2}^{\pi/2} \cos\theta\, d\theta \int_0^3 r^4\, dr$$

$$= \left[\sin\theta\right]_{-\pi/2}^{\pi/2} \left[\frac{1}{5}r^5\right]_0^3 = 2 \cdot \frac{1}{5}(243) = \frac{486}{5} = 97.2$$

43. From the graph, it appears that $1 - x^2 = e^x$ at $x \approx -0.71$ and at $x = 0$, with $1 - x^2 > e^x$ on $(-0.71, 0)$. So the desired integral is

$$\iint_D y^2\, dA \approx \int_{-0.71}^0 \int_{e^x}^{1-x^2} y^2\, dy\, dx$$

$$= \frac{1}{3}\int_{-0.71}^0 [(1 - x^2)^3 - e^{3x}]\, dx$$

$$= \frac{1}{3}\left[x - x^3 + \frac{3}{5}x^5 - \frac{1}{7}x^7 - \frac{1}{3}e^{3x}\right]_{-0.71}^0 \approx 0.0512$$

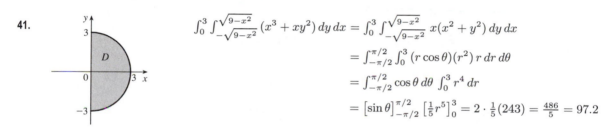

45. (a) $f(x, y)$ is a joint density function, so we know that $\iint_{\mathbb{R}^2} f(x, y)\, dA = 1$. Since $f(x, y) = 0$ outside the rectangle $[0, 3] \times [0, 2]$, we can say

$$\iint_{\mathbb{R}^2} f(x, y)\, dA = \int_{-\infty}^{\infty} \int_{-\infty}^{\infty} f(x, y)\, dy\, dx = \int_0^3 \int_0^2 C(x + y)\, dy\, dx$$

$$= C\int_0^3 \left[xy + \frac{1}{2}y^2\right]_{y=0}^{y=2}\, dx = C\int_0^3 (2x + 2)\, dx = C\left[x^2 + 2x\right]_0^3 = 15C$$

Then $15C = 1 \;\Rightarrow\; C = \frac{1}{15}$.

(b) $P(X \le 2, Y \ge 1) = \int_{-\infty}^2 \int_1^{\infty} f(x, y)\, dy\, dx = \int_0^2 \int_1^2 \frac{1}{15}(x, y)\, dy\, dx = \frac{1}{15}\int_0^2 \left[xy + \frac{1}{2}y^2\right]_{y=1}^{y=2}\, dx$

$= \frac{1}{15}\int_0^2 \left(x + \frac{3}{2}\right)\, dx = \frac{1}{15}\left[\frac{1}{2}x^2 + \frac{3}{2}x\right]_0^2 = \frac{1}{3}$

(c) $P(X + Y \le 1) = P((X, Y) \in D)$ where D is the triangular region shown in the figure. Thus

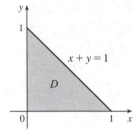

$$P(X + Y \le 1) = \iint_D f(x, y)\, dA = \int_0^1 \int_0^{1-x} \tfrac{1}{15}(x+y)\, dy\, dx$$

$$= \tfrac{1}{15} \int_0^1 \left[xy + \tfrac{1}{2}y^2\right]_{y=0}^{y=1-x} dx$$

$$= \tfrac{1}{15} \int_0^1 \left[x(1-x) + \tfrac{1}{2}(1-x)^2\right] dx$$

$$= \tfrac{1}{30} \int_0^1 (1 - x^2)\, dx = \tfrac{1}{30}\left[x - \tfrac{1}{3}x^3\right]_0^1 = \tfrac{1}{45}$$

47.

$$\int_{-1}^{1} \int_{x^2}^{1} \int_{0}^{1-y} f(x, y, z)\, dz\, dy\, dx = \int_0^1 \int_0^{1-z} \int_{-\sqrt{y}}^{\sqrt{y}} f(x, y, z)\, dx\, dy\, dz$$

49. Since $u = x - y$ and $v = x + y$, $x = \tfrac{1}{2}(u + v)$ and $y = \tfrac{1}{2}(v - u)$.

Thus $\dfrac{\partial(x, y)}{\partial(u, v)} = \begin{vmatrix} 1/2 & 1/2 \\ -1/2 & 1/2 \end{vmatrix} = \dfrac{1}{2}$ and $\displaystyle\iint_R \dfrac{x - y}{x + y}\, dA = \int_2^4 \int_{-2}^0 \dfrac{u}{v}\left(\dfrac{1}{2}\right) du\, dv = -\int_2^4 \dfrac{dv}{v} = -\ln 2$.

51. Let $u = y - x$ and $v = y + x$ so $x = y - u = (v - x) - u \Rightarrow x = \tfrac{1}{2}(v - u)$ and $y = v - \tfrac{1}{2}(v - u) = \tfrac{1}{2}(v + u)$.

$\left|\dfrac{\partial(x, y)}{\partial(u, v)}\right| = \left|\dfrac{\partial x}{\partial u}\dfrac{\partial y}{\partial v} - \dfrac{\partial x}{\partial v}\dfrac{\partial y}{\partial u}\right| = \left|-\tfrac{1}{2}\left(\tfrac{1}{2}\right) - \tfrac{1}{2}\left(\tfrac{1}{2}\right)\right| = \left|-\tfrac{1}{2}\right| = \tfrac{1}{2}$. R is the image under this transformation of the square

with vertices $(u, v) = (0, 0), (-2, 0), (0, 2)$, and $(-2, 2)$. So

$$\iint_R xy\, dA = \int_0^2 \int_{-2}^0 \dfrac{v^2 - u^2}{4}\left(\dfrac{1}{2}\right) du\, dv = \tfrac{1}{8}\int_0^2 \left[v^2 u - \tfrac{1}{3}u^3\right]_{u=-2}^{u=0} dv = \tfrac{1}{8}\int_0^2 \left(2v^2 - \tfrac{8}{3}\right) dv = \tfrac{1}{8}\left[\tfrac{2}{3}v^3 - \tfrac{8}{3}v\right]_0^2 = 0$$

This result could have been anticipated by symmetry, since the integrand is an odd function of y and R is symmetric about the x-axis.

FOCUS ON PROBLEM SOLVING

1.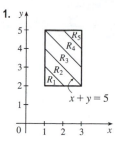

Let $R = \bigcup_{i=1}^{5} R_i$, where

$$R_i = \{(x, y) \mid x + y \geq i + 2, x + y < i + 3, 1 \leq x \leq 3, 2 \leq y \leq 5\}.$$

$\iint_R [\![x + y]\!]\, dA = \sum_{i=1}^{5} \iint_{R_i} [\![x + y]\!]\, dA = \sum_{i=1}^{5} [\![x + y]\!] \iint_{R_i} dA$, since

$[\![x + y]\!] = \text{constant} = i + 2$ for $(x, y) \in R_i$. Therefore

$$\iint_R [\![x + y]\!]\, dA = \sum_{i=1}^{5} (i + 2)\,[A(R_i)]$$

$$= 3A(R_1) + 4A(R_2) + 5A(R_3) + 6A(R_4) + 7A(R_5)$$

$$= 3\left(\tfrac{1}{2}\right) + 4\left(\tfrac{3}{2}\right) + 5(2) + 6\left(\tfrac{3}{2}\right) + 7\left(\tfrac{1}{2}\right) = 30$$

3. $f_{\text{ave}} = \dfrac{1}{b - a} \displaystyle\int_a^b f(x)\, dx = \dfrac{1}{1 - 0} \int_0^1 \left[\int_x^1 \cos(t^2)\, dt\right] dx$

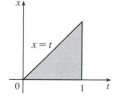

$= \int_0^1 \int_x^1 \cos(t^2)\, dt\, dx$

$= \int_0^1 \int_0^t \cos(t^2)\, dx\, dt$ [changing the order of integration]

$= \int_0^1 t \cos(t^2)\, dt = \tfrac{1}{2} \sin(t^2)\big]_0^1 = \tfrac{1}{2} \sin 1$

5. Since $|xy| < 1$, except at $(1, 1)$, the formula for the sum of a geometric series gives $\dfrac{1}{1 - xy} = \sum\limits_{n=0}^{\infty} (xy)^n$, so

$$\int_0^1 \int_0^1 \frac{1}{1 - xy}\, dx\, dy = \int_0^1 \int_0^1 \sum_{n=0}^{\infty} (xy)^n\, dx\, dy = \sum_{n=0}^{\infty} \int_0^1 \int_0^1 (xy)^n\, dx\, dy$$

$$= \sum_{n=0}^{\infty} \left[\int_0^1 x^n\, dx\right]\left[\int_0^1 y^n\, dy\right] = \sum_{n=0}^{\infty} \frac{1}{n+1} \cdot \frac{1}{n+1}$$

$$= \sum_{n=0}^{\infty} \frac{1}{(n+1)^2} = \frac{1}{1^2} + \frac{1}{2^2} + \frac{1}{3^2} + \cdots = \sum_{n=1}^{\infty} \frac{1}{n^2}$$

7. (a) Since $|xyz| < 1$ except at $(1, 1, 1)$, the formula for the sum of a geometric series gives $\dfrac{1}{1 - xyz} = \sum\limits_{n=0}^{\infty} (xyz)^n$,

so

$$\int_0^1 \int_0^1 \int_0^1 \frac{1}{1 - xyz}\, dx\, dy\, dz = \int_0^1 \int_0^1 \int_0^1 \sum_{n=0}^{\infty} (xyz)^n\, dx\, dy\, dz$$

$$= \sum_{n=0}^{\infty} \int_0^1 \int_0^1 \int_0^1 (xyz)^n\, dx\, dy\, dz$$

$$= \sum_{n=0}^{\infty} \left[\int_0^1 x^n\, dx\right]\left[\int_0^1 y^n\, dy\right]\left[\int_0^1 z^n\, dz\right]$$

$$= \sum_{n=0}^{\infty} \frac{1}{n+1} \cdot \frac{1}{n+1} \cdot \frac{1}{n+1}$$

$$= \sum_{n=0}^{\infty} \frac{1}{(n+1)^3} = \frac{1}{1^3} + \frac{1}{2^3} + \frac{1}{3^3} + \cdots = \sum_{n=1}^{\infty} \frac{1}{n^3}$$

(b) Since $|-xyz| < 1$, except at $(1, 1, 1)$, the formula for the sum of a geometric series gives $\dfrac{1}{1+xyz} = \displaystyle\sum_{n=0}^{\infty}(-xyz)^n$,

so

$$\int_0^1\int_0^1\int_0^1 \frac{1}{1+xyz}\,dx\,dy\,dz = \int_0^1\int_0^1\int_0^1 \sum_{n=0}^{\infty}(-xyz)^n\,dx\,dy\,dz$$

$$= \sum_{n=0}^{\infty}\int_0^1\int_0^1\int_0^1(-xyz)^n\,dx\,dy\,dz$$

$$= \sum_{n=0}^{\infty}(-1)^n\left[\int_0^1 x^n\,dx\right]\left[\int_0^1 y^n\,dy\right]\left[\int_0^1 z^n\,dz\right]$$

$$= \sum_{n=0}^{\infty}(-1)^n\frac{1}{n+1}\cdot\frac{1}{n+1}\cdot\frac{1}{n+1}$$

$$= \sum_{n=0}^{\infty}\frac{(-1)^n}{(n+1)^3} = \frac{1}{1^3} - \frac{1}{2^3} + \frac{1}{3^3} - \cdots = \sum_{n=1}^{\infty}\frac{(-1)^{n-1}}{n^3}$$

To evaluate this sum, we first write out a few terms: $s = 1 - \dfrac{1}{2^3} + \dfrac{1}{3^3} - \dfrac{1}{4^3} + \dfrac{1}{5^3} - \dfrac{1}{6^3} \approx 0.8998$. Notice that

$a_7 = \dfrac{1}{7^3} < 0.003$. By the Alternating Series Estimation Theorem from Section 8.4, we have $|s - s_6| \le a_7 < 0.003$. This

error of 0.003 will not affect the second decimal place, so we have $s \approx 0.90$.

9. $\int_0^x\int_0^y\int_0^z f(t)\,dt\,dz\,dy = \iiint_E f(t)\,dV$, where

$E = \{(t, z, y) \mid 0 \le t \le z, 0 \le z \le y, 0 \le y \le x\}$.

If we let D be the projection of E on the yt-plane then

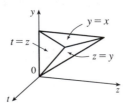

$D = \{(y, t) \mid 0 \le t \le x, t \le y \le x\}$. And we see from the diagram

that $E = \{(t, z, y) \mid t \le z \le y, t \le y \le x, 0 \le t \le x\}$. So

$$\int_0^x\int_0^y\int_0^z f(t)\,dt\,dz\,dy = \int_0^x\int_t^x\int_t^y f(t)\,dz\,dy\,dt = \int_0^x\left[\int_t^x(y-t)f(t)\,dy\right]dt$$

$$= \int_0^x\left[(\tfrac{1}{2}y^2 - ty)f(t)\right]_{y=t}^{y=x}dt = \int_0^x\left[\tfrac{1}{2}x^2 - tx - \tfrac{1}{2}t^2 + t^2\right]f(t)\,dt$$

$$= \int_0^x\left[\tfrac{1}{2}x^2 - tx + \tfrac{1}{2}t^2\right]f(t)\,dt = \int_0^x\left(\tfrac{1}{2}x^2 - 2tx + t^2\right)f(t)\,dt$$

$$= \tfrac{1}{2}\int_0^x(x-t)^2 f(t)\,dt$$

13 ☐   VECTOR CALCULUS

13.1   Vector Fields

1. $\mathbf{F}(x, y) = \frac{1}{2}(\mathbf{i} + \mathbf{j})$

All vectors in this field are identical, with length $\frac{1}{\sqrt{2}}$ and direction parallel to the line $y = x$.

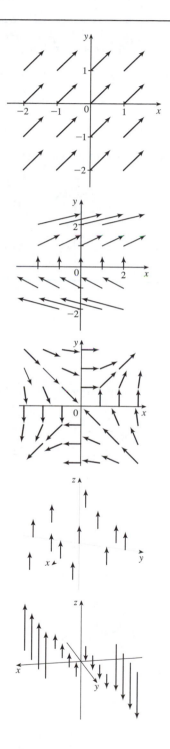

3. $\mathbf{F}(x, y) = y\,\mathbf{i} + \frac{1}{2}\,\mathbf{j}$

The length of the vector $y\,\mathbf{i} + \frac{1}{2}\,\mathbf{j}$ is $\sqrt{y^2 + \frac{1}{4}}$. Vectors are tangent to parabolas opening about the x-axis.

5. $\mathbf{F}(x, y) = \dfrac{y\,\mathbf{i} + x\,\mathbf{j}}{\sqrt{x^2 + y^2}}$

The length of the vector $\dfrac{y\,\mathbf{i} + x\,\mathbf{j}}{\sqrt{x^2 + y^2}}$ is 1.

7. $\mathbf{F}(x, y, z) = \mathbf{k}$

All vectors in this field are parallel to the z-axis and have length 1.

9. $\mathbf{F}(x, y, z) = x\,\mathbf{k}$

At each point (x, y, z), $\mathbf{F}(x, y, z)$ is a vector of length $|x|$. For $x > 0$, all point in the direction of the positive z-axis, while for $x < 0$, all are in the direction of the negative z-axis. In each plane $x = k$, all the vectors are identical.

11. $\mathbf{F}(x, y) = \langle y, x \rangle$ corresponds to graph II. In the first quadrant all the vectors have positive x- and y-components, in the second quadrant all vectors have positive x-components and negative y-components, in the third quadrant all vectors have negative x- and y-components, and in the fourth quadrant all vectors have negative x-components and positive y-components. In addition, the vectors get shorter as we approach the origin.

13. $\mathbf{F}(x, y) = \langle x - 2, x + 1 \rangle$ corresponds to graph I since the vectors are independent of y (vectors along vertical lines are identical) and, as we move to the right, both the x- and the y-components get larger.

15. $\mathbf{F}(x, y, z) = \mathbf{i} + 2\mathbf{j} + 3\mathbf{k}$ corresponds to graph IV, since all vectors have identical length and direction.

17. $\mathbf{F}(x, y, z) = x\mathbf{i} + y\mathbf{j} + 3\mathbf{k}$ corresponds to graph III; the projection of each vector onto the xy-plane is $x\mathbf{i} + y\mathbf{j}$, which points away from the origin, and the vectors point generally upward because their z-components are all 3.

19.

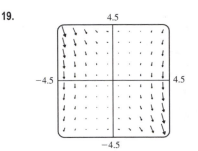

The vector field seems to have very short vectors near the line $y = 2x$.

For $\mathbf{F}(x, y) = \langle 0, 0 \rangle$ we must have $y^2 - 2xy = 0$ and $3xy - 6x^2 = 0$.

The first equation holds if $y = 0$ or $y = 2x$, and the second holds if $x = 0$ or $y = 2x$. So both equations hold [and thus $\mathbf{F}(x, y) = \mathbf{0}$] along the line $y = 2x$.

21. $\nabla f(x, y) = f_x(x, y)\mathbf{i} + f_y(x, y)\mathbf{j} = \dfrac{1}{x + 2y}\mathbf{i} + \dfrac{2}{x + 2y}\mathbf{j}$

23. $\nabla f(x, y, z) = f_x(x, y, z)\mathbf{i} + f_y(x, y, z)\mathbf{j} + f_z(x, y, z)\mathbf{k} = \dfrac{x}{\sqrt{x^2 + y^2 + z^2}}\mathbf{i} + \dfrac{y}{\sqrt{x^2 + y^2 + z^2}}\mathbf{j} + \dfrac{z}{\sqrt{x^2 + y^2 + z^2}}\mathbf{k}$

25. $f(x, y) = xy - 2x \Rightarrow \nabla f(x, y) = (y - 2)\mathbf{i} + x\mathbf{j}$.

The length of $\nabla f(x, y)$ is $\sqrt{(y - 2)^2 + x^2}$ and

$\nabla f(x, y)$ terminates on the line $y = x + 2$ at the point

$(x + y - 2, x + y)$.

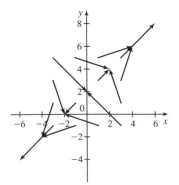

27. We graph ∇f along with a contour map of f.

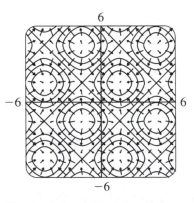

The graph shows that the gradient vectors are perpendicular to the level curves. Also, the gradient vectors point in the direction in which f is increasing and are longer where the level curves are closer together.

29. $f(x, y) = xy \Rightarrow \nabla f(x, y) = y\,\mathbf{i} + x\,\mathbf{j}$. In the first quadrant, both components of each vector are positive, while in the third quadrant both components are negative. However, in the second quadrant each vector's x-component is positive while its y-component is negative (and vice versa in the fourth quadrant). Thus, ∇f is graph IV.

31. $f(x, y) = x^2 + y^2 \Rightarrow \nabla f(x, y) = 2x\,\mathbf{i} + 2y\,\mathbf{j}$. Thus, each vector $\nabla f(x, y)$ has the same direction and twice the length of the position vector of the point (x, y), so the vectors all point directly away from the origin and their lengths increase as we move away from the origin. Hence, ∇f is graph II.

33. At $t = 3$ the particle is at $(2, 1)$ so its velocity is $\mathbf{V}(2, 1) = \langle 4, 3 \rangle$. After 0.01 units of time, the particle's change in location should be approximately $0.01\mathbf{V}(2, 1) = 0.01\langle 4, 3 \rangle = \langle 0.04, 0.03 \rangle$, so the particle should be approximately at the point $(2.04, 1.03)$.

35. (a) We sketch the vector field $\mathbf{F}(x, y) = x\,\mathbf{i} - y\,\mathbf{j}$ along with several approximate flow lines. The flow lines appear to be hyperbolas with shape similar to the graph of $y = \pm 1/x$, so we might guess that the flow lines have equations $y = C/x$.

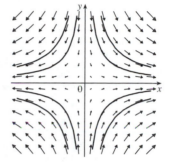

(b) If $x = x(t)$ and $y = y(t)$ are parametric equations of a flow line, then the velocity vector of the flow line at the point (x, y) is $x'(t)\,\mathbf{i} + y'(t)\,\mathbf{j}$. Since the velocity vectors coincide with the vectors in the vector field, we have
$x'(t)\,\mathbf{i} + y'(t)\,\mathbf{j} = x\,\mathbf{i} - y\,\mathbf{j} \Rightarrow dx/dt = x,\ dy/dt = -y$. To solve these differential equations, we know
$dx/dt = x \Rightarrow dx/x = dt \Rightarrow \ln|x| = t + C \Rightarrow x = \pm e^{t+C} = Ae^t$ for some constant A, and
$dy/dt = -y \Rightarrow dy/y = -dt \Rightarrow \ln|y| = -t + K \Rightarrow y = \pm e^{-t+K} = Be^{-t}$ for some constant B. Therefore
$xy = Ae^t Be^{-t} = AB = $ constant. If the flow line passes through $(1, 1)$ then $(1)(1) = $ constant $= 1 \Rightarrow xy = 1 \Rightarrow y = 1/x,\ x > 0$.

13.2   Line Integrals

1. $x = t^2$ and $y = t$, $0 \le t \le 2$, so by Formula 3

$$\int_C y\,ds = \int_0^2 t\sqrt{\left(\frac{dx}{dt}\right)^2 + \left(\frac{dy}{dt}\right)^2}\,dt = \int_0^2 t\sqrt{(2t)^2 + (1)^2}\,dt$$

$$= \int_0^2 t\sqrt{4t^2 + 1}\,dt = \tfrac{1}{12}\left(4t^2 + 1\right)^{3/2}\Big]_0^2 = \tfrac{1}{12}\left(17\sqrt{17} - 1\right)$$

3. Parametric equations for C are $x = 4\cos t$, $y = 4\sin t$, $-\frac{\pi}{2} \le t \le \frac{\pi}{2}$. Then

$$\int_C xy^4\,ds = \int_{-\pi/2}^{\pi/2}(4\cos t)(4\sin t)^4\sqrt{(-4\sin t)^2 + (4\cos t)^2}\,dt = \int_{-\pi/2}^{\pi/2} 4^5\cos t\sin^4 t\sqrt{16\left(\sin^2 t + \cos^2 t\right)}\,dt$$

$$= 4^5\int_{-\pi/2}^{\pi/2}\left(\sin^4 t\cos t\right)(4)\,dt = (4)^6\left[\tfrac{1}{5}\sin^5 t\right]_{-\pi/2}^{\pi/2} = \tfrac{2 \cdot 4^6}{5} = 1638.4$$

5.

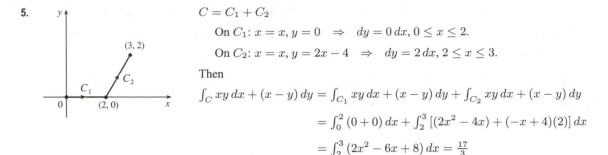

$$C = C_1 + C_2$$

On C_1: $x = x, y = 0 \implies dy = 0\,dx, 0 \le x \le 2$.

On C_2: $x = x, y = 2x - 4 \implies dy = 2\,dx, 2 \le x \le 3$.

Then

$$\int_C xy\,dx + (x - y)\,dy = \int_{C_1} xy\,dx + (x - y)\,dy + \int_{C_2} xy\,dx + (x - y)\,dy$$

$$= \int_0^2 (0 + 0)\,dx + \int_2^3 [(2x^2 - 4x) + (-x + 4)(2)]\,dx$$

$$= \int_2^3 (2x^2 - 6x + 8)\,dx = \tfrac{17}{3}$$

7. $x = 4\sin t, y = 4\cos t, z = 3t, 0 \le t \le \frac{\pi}{2}$. Then by Formula 9,

$$\int_C xy^3\,ds = \int_0^{\pi/2} (4\sin t)(4\cos t)^3 \sqrt{\left(\tfrac{dx}{dt}\right)^2 + \left(\tfrac{dy}{dt}\right)^2 + \left(\tfrac{dz}{dt}\right)^2}\,dt$$

$$= \int_0^{\pi/2} 4^4 \cos^3 t \sin t \sqrt{(4\cos t)^2 + (-4\sin t)^2 + (3)^2}\,dt$$

$$= \int_0^{\pi/2} 256 \cos^3 t \sin t \sqrt{16(\cos^2 t + \sin^2 t) + 9}\,dt$$

$$= 1280 \int_0^{\pi/2} \cos^3 t \sin t\,dt = -320 \cos^4 t \big]_0^{\pi/2} = 320$$

9. Parametric equations for C are $x = t, y = 2t, z = 3t, 0 \le t \le 1$. Then

$$\int_C xe^{yz}\,ds = \int_0^1 te^{(2t)(3t)} \sqrt{1^2 + 2^2 + 3^2}\,dt = \sqrt{14} \int_0^1 te^{6t^2}\,dt = \sqrt{14}\left[\tfrac{1}{12}e^{6t^2}\right]_0^1 = \tfrac{\sqrt{14}}{12}(e^6 - 1).$$

11. $\int_C x^2 y \sqrt{z}\,dz = \int_0^1 (t^3)^2 (t)\sqrt{t^2} \cdot 2t\,dt = \int_0^1 2t^9\,dt = \tfrac{1}{5}t^{10}\big]_0^1 = \tfrac{1}{5}$

13.

On C_1: $x = 1 + t \implies dx = dt, y = 3t \implies dy = 3\,dt, z = 1$

$\implies dz = 0\,dt, 0 \le t \le 1$.

On C_2: $x = 2 \implies dx = 0\,dt, y = 3 + 2t \implies$

$dy = 2\,dt, z = 1 + t \implies dz = dt, 0 \le t \le 1$.

Then $\int_C (x + yz)\,dx + 2x\,dy + xyz\,dz$

$$= \int_{C_1} (x + yz)\,dx + 2x\,dy + xyz\,dz + \int_{C_2} (x + yz)\,dx + 2x\,dy + xyz\,dz$$

$$= \int_0^1 (1 + t + (3t)(1))\,dt + 2(1 + t) \cdot 3\,dt + (1 + t)(3t)(1) \cdot 0\,dt$$

$$\quad + \int_0^1 (2 + (3 + 2t)(1 + t)) \cdot 0\,dt + 2(2) \cdot 2\,dt + (2)(3 + 2t)(1 + t)\,dt$$

$$= \int_0^1 (10t + 7)\,dt + \int_0^1 (4t^2 + 10t + 14)\,dt$$

$$= \left[5t^2 + 7t\right]_0^1 + \left[\tfrac{4}{3}t^3 + 5t^2 + 14t\right]_0^1 = 12 + \tfrac{61}{3} = \tfrac{97}{3}$$

15. (a) Along the line $x = -3$, the vectors of $\mathbf{F}$ have positive y-components, so since the path goes upward, the integrand $\mathbf{F} \cdot \mathbf{T}$ is always positive. Therefore $\int_{C_1} \mathbf{F} \cdot d\mathbf{r} = \int_{C_1} \mathbf{F} \cdot \mathbf{T}\,ds$ is positive.

(b) All of the (nonzero) field vectors along the circle with radius 3 are pointed in the clockwise direction, that is, opposite the direction to the path. So $\mathbf{F} \cdot \mathbf{T}$ is negative, and therefore $\int_{C_2} \mathbf{F} \cdot d\mathbf{r} = \int_{C_2} \mathbf{F} \cdot \mathbf{T}\,ds$ is negative.

17. $\mathbf{r}(t) = t^2\,\mathbf{i} - t^3\mathbf{j}$, so $\mathbf{F}(\mathbf{r}(t)) = (t^2)^2(-t^3)^3\mathbf{i} - (-t^3)\sqrt{t^2}\,\mathbf{j} = -t^{13}\,\mathbf{i} + t^4\,\mathbf{j}$ and $\mathbf{r}'(t) = 2t\,\mathbf{i} - 3t^2\,\mathbf{j}$.

Thus $\int_C \mathbf{F} \cdot d\mathbf{r} = \int_0^1 \mathbf{F}(\mathbf{r}(t)) \cdot \mathbf{r}'(t)\,dt = \int_0^1 (-2t^{14} - 3t^6)dt = \left[-\tfrac{2}{15}t^{15} - \tfrac{3}{7}t^7\right]_0^1 = -\tfrac{59}{105}$.

19. $\int_C \mathbf{F} \cdot d\mathbf{r} = \int_0^1 \langle \sin t^3, \cos(-t^2), t^4 \rangle \cdot \langle 3t^2, -2t, 1 \rangle\,dt$

$$= \int_0^1 (3t^2 \sin t^3 - 2t \cos t^2 + t^4)\,dt = \left[-\cos t^3 - \sin t^2 + \tfrac{1}{5}t^5\right]_0^1 = \tfrac{6}{5} - \cos 1 - \sin 1$$

21. We graph $\mathbf{F}(x,y) = (x - y)\,\mathbf{i} + xy\,\mathbf{j}$ and the curve C. We see that most of the vectors starting on C point in roughly the same direction as C, so for these portions of C the tangential component $\mathbf{F} \cdot \mathbf{T}$ is positive. Although some vectors in the third quadrant which start on C point in roughly the opposite direction, and hence give negative tangential components, it seems reasonable that the effect of these portions of C is outweighed by the positive tangential components. Thus, we would expect $\int_C \mathbf{F} \cdot d\mathbf{r} = \int_C \mathbf{F} \cdot \mathbf{T}\,ds$ to be positive.

To verify, we evaluate $\int_C \mathbf{F} \cdot d\mathbf{r}$. The curve C can be represented by $\mathbf{r}(t) = 2\cos t\,\mathbf{i} + 2\sin t\,\mathbf{j}$, $0 \le t \le \frac{3\pi}{2}$, so $\mathbf{F}(\mathbf{r}(t)) = (2\cos t - 2\sin t)\,\mathbf{i} + 4\cos t\sin t\,\mathbf{j}$ and $\mathbf{r}'(t) = -2\sin t\,\mathbf{i} + 2\cos t\,\mathbf{j}$. Then

$$\int_C \mathbf{F} \cdot d\mathbf{r} = \int_0^{3\pi/2} \mathbf{F}(\mathbf{r}(t)) \cdot \mathbf{r}'(t)\,dt$$

$$= \int_0^{3\pi/2} \left[-2\sin t (2\cos t - 2\sin t) + 2\cos t (4\cos t\sin t) \right] dt$$

$$= 4 \int_0^{3\pi/2} (\sin^2 t - \sin t\cos t + 2\sin t\cos^2 t)\,dt$$

$$= 3\pi + \tfrac{2}{3} \quad \text{[using a CAS]}$$

23. (a) $\int_C \mathbf{F} \cdot d\mathbf{r} = \int_0^1 \left\langle e^{t^2-1}, t^5 \right\rangle \cdot \left\langle 2t, 3t^2 \right\rangle dt = \int_0^1 \left(2te^{t^2-1} + 3t^7 \right) dt = \left[e^{t^2-1} + \tfrac{3}{8}t^8 \right]_0^1 = \tfrac{11}{8} - 1/e$

(b) $\mathbf{r}(0) = \mathbf{0}$, $\mathbf{F}(\mathbf{r}(0)) = \left\langle e^{-1}, 0 \right\rangle$;

$\mathbf{r}\!\left(\tfrac{1}{\sqrt{2}}\right) = \left\langle \tfrac{1}{2}, \tfrac{1}{2\sqrt{2}} \right\rangle$, $\mathbf{F}\!\left(\mathbf{r}\!\left(\tfrac{1}{\sqrt{2}}\right)\right) = \left\langle e^{-1/2}, \tfrac{1}{4\sqrt{2}} \right\rangle$;

$\mathbf{r}(1) = \langle 1, 1 \rangle$, $\mathbf{F}(\mathbf{r}(1)) = \langle 1, 1 \rangle$.

In order to generate the graph with Maple, we use the `PLOT` command (not to be confused with the `plot` command) to define each of the vectors. For example,

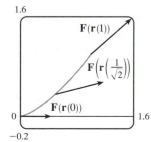

```
v1:=PLOT(CURVES([[0,0],[evalf(1/exp(1)),0]]));
```

generates the vector from the vector field at the point $(0,0)$ (but without an arrowhead) and gives it the name `v1`. To show everything on the same screen, we use the `display` command. In Mathematica, we use `ListPlot` (with the `PlotJoined -> True` option) to generate the vectors, and then `Show` to show everything on the same screen.

25. The part of the astroid that lies in the quadrant is parametrized by $x = \cos^3 t$, $y = \sin^3 t$, $0 \le t \le \frac{\pi}{2}$.

Now $\dfrac{dx}{dt} = 3\cos^2 t\,(-\sin t)$ and $\dfrac{dy}{dt} = 3\sin^2 t\cos t$, so

$$\sqrt{\left(\frac{dx}{dt}\right)^2 + \left(\frac{dy}{dt}\right)^2} = \sqrt{9\cos^4 t\sin^2 t + 9\sin^4 t\cos^2 t} = 3\cos t\sin t\sqrt{\cos^2 t + \sin^2 t} = 3\cos t\sin t.$$

Therefore $\int_C x^3 y^5\,ds = \int_0^{\pi/2} \cos^9 t\sin^{15} t\,(3\cos t\sin t)\,dt = \dfrac{945}{16{,}777{,}216}\pi.$

27. We use the parametrization $x = 2\cos t$, $y = 2\sin t$, $-\frac{\pi}{2} \le t \le \frac{\pi}{2}$. Then

$$ds = \sqrt{\left(\frac{dx}{dt}\right)^2 + \left(\frac{dy}{dt}\right)^2}\,dt = \sqrt{(-2\sin t)^2 + (2\cos t)^2}\,dt = 2\,dt, \text{ so } m = \int_C k\,ds = 2k \int_{-\pi/2}^{\pi/2} dt = 2k(\pi),$$

$\bar{x} = \dfrac{1}{2\pi k} \int_C xk\,ds = \dfrac{1}{2\pi} \int_{-\pi/2}^{\pi/2} (2\cos t)2\,dt = \dfrac{1}{2\pi} \left[4\sin t \right]_{-\pi/2}^{\pi/2} = \dfrac{4}{\pi}$, $\bar{y} = \dfrac{1}{2\pi k} \int_C yk\,ds = \dfrac{1}{2\pi} \int_{-\pi/2}^{\pi/2} (2\sin t)2\,dt = 0$. Hence $(\bar{x}, \bar{y}) = \left(\dfrac{4}{\pi}, 0 \right)$.

29. (a) $\overline{x} = \dfrac{1}{m} \displaystyle\int_C x\rho(x,y,z)\,ds$, $\overline{y} = \dfrac{1}{m} \displaystyle\int_C y\rho(x,y,z)\,ds$, $\overline{z} = \dfrac{1}{m} \displaystyle\int_C z\rho(x,y,z)\,ds$ where $m = \displaystyle\int_C \rho(x,y,z)\,ds$.

(b) $m = \displaystyle\int_C k\,ds = k\displaystyle\int_0^{2\pi}\sqrt{4\sin^2 t + 4\cos^2 t + 9}\,dt = k\sqrt{13}\displaystyle\int_0^{2\pi} dt = 2\pi k\sqrt{13}$,

$$\overline{x} = \dfrac{1}{2\pi k\sqrt{13}}\int_0^{2\pi} 2k\sqrt{13}\,\sin t\,dt = 0,\ \overline{y} = \dfrac{1}{2\pi k\sqrt{13}}\int_0^{2\pi} 2k\sqrt{13}\,\cos t\,dt = 0,$$

$$\overline{z} = \dfrac{1}{2\pi k\sqrt{13}}\int_0^{2\pi}\left(k\sqrt{13}\right)(3t)\,dt = \dfrac{3}{2\pi}\left(2\pi^2\right) = 3\pi.\ \text{Hence } (\overline{x},\overline{y},\overline{z}) = (0,0,3\pi).$$

31. From Example 3, $\rho(x,y) = k(1-y)$, $x = \cos t$, $y = \sin t$, and $ds = dt$, $0 \le t \le \pi$ $\Rightarrow$

$$I_x = \int_C y^2\rho(x,y)\,ds = \int_0^\pi \sin^2 t\,[k(1-\sin t)]\,dt = k\int_0^\pi (\sin^2 t - \sin^3 t)\,dt$$

$$= \tfrac{1}{2}k\int_0^\pi (1-\cos 2t)\,dt - k\int_0^\pi \left(1-\cos^2 t\right)\sin t\,dt \qquad \begin{bmatrix}\text{Let } u = \cos t,\ du = -\sin t\,dt \\ \text{in the second integral}\end{bmatrix}$$

$$= k\left[\tfrac{\pi}{2} + \int_1^{-1}(1-u^2)\,du\right] = k\left(\tfrac{\pi}{2} - \tfrac{4}{3}\right)$$

$$I_y = \int_C x^2\rho(x,y)\,ds = k\int_0^\pi \cos^2 t\,(1-\sin t)\,dt = \tfrac{k}{2}\int_0^\pi (1+\cos 2t)\,dt - k\int_0^\pi \cos^2 t\,\sin t\,dt$$

$$= k\left(\tfrac{\pi}{2} - \tfrac{2}{3}\right),\ \text{using the same substitution as above.}$$

33. $W = \displaystyle\int_C \mathbf{F}\cdot d\mathbf{r} = \displaystyle\int_0^{2\pi} \langle t-\sin t, 3-\cos t\rangle \cdot \langle 1-\cos t, \sin t\rangle\,dt$

$$= \int_0^{2\pi}(t - t\cos t - \sin t + \sin t\cos t + 3\sin t - \sin t\cos t)\,dt$$

$$= \int_0^{2\pi}(t - t\cos t + 2\sin t)\,dt = \left[\tfrac{1}{2}t^2 - (t\sin t + \cos t) - 2\cos t\right]_0^{2\pi} \qquad \begin{bmatrix}\text{by integrating by parts} \\ \text{in the second term}\end{bmatrix}$$

$$= 2\pi^2$$

35. $\mathbf{r}(t) = \langle 1+2t, 4t, 2t\rangle$, $0 \le t \le 1$,

$$W = \int_C \mathbf{F}\cdot d\mathbf{r} = \int_0^1 \langle 6t, 1+4t, 1+6t\rangle \cdot \langle 2,4,2\rangle\,dt = \int_0^1 (12t + 4(1+4t) + 2(1+6t))\,dt$$

$$= \int_0^1 (40t + 6)\,dt = \left[20t^2 + 6t\right]_0^1 = 26$$

37. Let $\mathbf{F} = 185\,\mathbf{k}$. To parametrize the staircase, let

$x = 20\cos t$, $y = 20\sin t$, $z = \dfrac{90}{6\pi}t = \dfrac{15}{\pi}t$, $0 \le t \le 6\pi$ $\Rightarrow$

$$W = \int_C \mathbf{F}\cdot d\mathbf{r} = \int_0^{6\pi} \langle 0,0,185\rangle \cdot \langle -20\sin t, 20\cos t, \tfrac{15}{\pi}\rangle\,dt = (185)\tfrac{15}{\pi}\int_0^{6\pi} dt = (185)(90) \approx 1.67\times 10^4 \text{ ft-lb}$$

39. (a) $\mathbf{r}(t) = \langle \cos t, \sin t\rangle$, $0 \le t \le 2\pi$, and let $\mathbf{F} = \langle a, b\rangle$. Then

$$W = \int_C \mathbf{F}\cdot d\mathbf{r} = \int_0^{2\pi} \langle a,b\rangle \cdot \langle -\sin t, \cos t\rangle\,dt = \int_0^{2\pi}(-a\sin t + b\cos t)\,dt = \left[a\cos t + b\sin t\right]_0^{2\pi}$$

$$= a + 0 - a + 0 = 0$$

(b) Yes. $\mathbf{F}(x,y) = k\,\mathbf{x} = \langle kx, ky\rangle$ and

$$W = \int_C \mathbf{F}\cdot d\mathbf{r} = \int_0^{2\pi} \langle k\cos t, k\sin t\rangle \cdot \langle -\sin t, \cos t\rangle\,dt = \int_0^{2\pi}(-k\sin t\cos t + k\sin t\cos t)\,dt = \int_0^{2\pi} 0\,dt = 0$$

41. The work done in moving the object is $\int_C \mathbf{F}\cdot d\mathbf{r} = \int_C \mathbf{F}\cdot \mathbf{T}\,ds$. We can approximate this integral by dividing C into 7 segments of equal length $\Delta s = 2$ and approximating $\mathbf{F}\cdot \mathbf{T}$, that is, the tangential component of force, at a point (x_i^*, y_i^*) on each segment. Since C is composed of straight line segments, $\mathbf{F}\cdot \mathbf{T}$ is the scalar projection of each force vector onto C. If we choose (x_i^*, y_i^*) to be the point on the segment closest to the origin, then the work done is

$$\int_C \mathbf{F}\cdot \mathbf{T}\,ds \approx \sum_{i=1}^{7} [\mathbf{F}(x_i^*, y_i^*)\cdot \mathbf{T}(x_i^*, y_i^*)]\,\Delta s = [2+2+2+2+1+1+1](2) = 22. \text{ Thus, we estimate the work done to}$$

be approximately 22 J.

13.3   The Fundamental Theorem for Line Integrals

1. C appears to be a smooth curve, and since ∇f is continuous, we know f is differentiable. Then Theorem 2 says that the value of $\int_C \nabla f \cdot d\mathbf{r}$ is simply the difference of the values of f at the terminal and initial points of C. From the graph, this is $50 - 10 = 40$.

3. $\partial(6x + 5y)/\partial y = 5 = \partial(5x + 4y)/\partial x$ and the domain of $\mathbf{F}$ is $\mathbb{R}^2$ which is open and simply-connected, so by Theorem 6 $\mathbf{F}$ is conservative. Thus, there exists a function f such that $\nabla f = \mathbf{F}$, that is, $f_x(x, y) = 6x + 5y$ and $f_y(x, y) = 5x + 4y$. But $f_x(x, y) = 6x + 5y$ implies $f(x, y) = 3x^2 + 5xy + g(y)$ and differentiating both sides of this equation with respect to y gives $f_y(x, y) = 5x + g'(y)$. Thus $5x + 4y = 5x + g'(y)$ so $g'(y) = 4y$ and $g(y) = 2y^2 + K$ where K is a constant. Hence $f(x, y) = 3x^2 + 5xy + 2y^2 + K$ is a potential function for $\mathbf{F}$.

5. $\partial(xe^y)/\partial y = xe^y$, $\partial(ye^x)/\partial x = ye^x$. Since these are not equal, $\mathbf{F}$ is not conservative.

7. $\partial(2x \cos y - y \cos x)/\partial y = -2x \sin y - \cos x = \partial(-x^2 \sin y - \sin x)/\partial x$ and the domain of $\mathbf{F}$ is $\mathbb{R}^2$. Hence $\mathbf{F}$ is conservative so there exists a function f such that $\nabla f = \mathbf{F}$. Then $f_x(x, y) = 2x \cos y - y \cos x$ implies $f(x, y) = x^2 \cos y - y \sin x + g(y)$ and $f_y(x, y) = -x^2 \sin y - \sin x + g'(y)$. But $f_y(x, y) = -x^2 \sin y - \sin x$ so $g'(y) = 0 \ \Rightarrow \ g(y) = K$. Then $f(x, y) = x^2 \cos y - y \sin x + K$ is a potential function for $\mathbf{F}$.

9. $\partial(ye^x + \sin y)/\partial y = e^x + \cos y = \partial(e^x + x \cos y)/\partial x$ and the domain of $\mathbf{F}$ is $\mathbb{R}^2$. Hence $\mathbf{F}$ is conservative so there exists a function f such that $\nabla f = \mathbf{F}$. Then $f_x(x, y) = ye^x + \sin y$ implies $f(x, y) = ye^x + x \sin y + g(y)$ and $f_y(x, y) = e^x + x \cos y + g'(y)$. But $f_y(x, y) = e^x + x \cos y$ so $g(y) = K$ and $f(x, y) = ye^x + x \sin y + K$ is a potential function for $\mathbf{F}$.

11. (a) $\mathbf{F}$ has continuous first-order partial derivatives and $\dfrac{\partial}{\partial y} \, 2xy = 2x = \dfrac{\partial}{\partial x} \, (x^2)$ on $\mathbb{R}^2$, which is open and simply-connected. Thus, $\mathbf{F}$ is conservative by Theorem 6. Then we know that the line integral of $\mathbf{F}$ is independent of path; in particular, the value of $\int_C \mathbf{F} \cdot d\mathbf{r}$ depends only on the endpoints of C. Since all three curves have the same initial and terminal points, $\int_C \mathbf{F} \cdot d\mathbf{r}$ will have the same value for each curve.

　　(b) We first find a potential function f, so that $\nabla f = \mathbf{F}$. We know $f_x(x, y) = 2xy$ and $f_y(x, y) = x^2$. Integrating $f_x(x, y)$ with respect to x, we have $f(x, y) = x^2 y + g(y)$. Differentiating both sides with respect to y gives $f_y(x, y) = x^2 + g'(y)$, so we must have $x^2 + g'(y) = x^2 \ \Rightarrow \ g'(y) = 0 \ \Rightarrow \ g(y) = K$, a constant. Thus $f(x, y) = x^2 y + K$. All three curves start at $(1, 2)$ and end at $(3, 2)$, so by Theorem 2, $\int_C \mathbf{F} \cdot d\mathbf{r} = f(3, 2) - f(1, 2) = 18 - 2 = 16$ for each curve.

13. (a) $f_x(x, y) = x^3 y^4$ implies $f(x, y) = \frac{1}{4}x^4 y^4 + g(y)$ and $f_y(x, y) = x^4 y^3 + g'(y)$. But $f_y(x, y) = x^4 y^3$ so $g'(y) = 0 \ \Rightarrow \ g(y) = K$, a constant. We can take $K = 0$, so $f(x, y) = \frac{1}{4}x^4 y^4$.

　　(b) The initial point of C is $\mathbf{r}(0) = (0, 1)$ and the terminal point is $\mathbf{r}(1) = (1, 2)$, so $\int_C \mathbf{F} \cdot d\mathbf{r} = f(1, 2) - f(0, 1) = 4 - 0 = 4$.

15. (a) $f_x(x, y, z) = yz$ implies $f(x, y, z) = xyz + g(y, z)$ and so $f_y(x, y, z) = xz + g_y(y, z)$. But $f_y(x, y, z) = xz$ so $g_y(y, z) = 0 \ \Rightarrow \ g(y, z) = h(z)$. Thus $f(x, y, z) = xyz + h(z)$ and $f_z(x, y, z) = xy + h'(z)$. But $f_z(x, y, z) = xy + 2z$, so $h'(z) = 2z \ \Rightarrow \ h(z) = z^2 + K$. Hence $f(x, y, z) = xyz + z^2$ (taking $K = 0$).

　　(b) $\int_C \mathbf{F} \cdot d\mathbf{r} = f(4, 6, 3) - f(1, 0, -2) = 81 - 4 = 77$.

17. (a) $f_x(x, y, z) = y^2 \cos z$ implies $f(x, y, z) = xy^2 \cos z + g(y, z)$ and so $f_y(x, y, z) = 2xy \cos z + g_y(y, z)$. But

$f_y(x, y, z) = 2xy \cos z$ so $g_y(y, z) = 0 \quad \Rightarrow \quad g(y, z) = h(z)$. Thus $f(x, y, z) = xy^2 \cos z + h(z)$ and

$f_z(x, y, z) = -xy^2 \sin z + h'(z)$. But $f_z(x, y, z) = -xy^2 \sin z$, so $h'(z) = 0 \quad \Rightarrow \quad h(z) = K$. Hence

$f(x, y, z) = xy^2 \cos z$ (taking $K = 0$).

(b) $\mathbf{r}(0) = \langle 0, 0, 0 \rangle$, $\mathbf{r}(\pi) = \langle \pi^2, 0, \pi \rangle$ so $\int_C \mathbf{F} \cdot d\mathbf{r} = f(\pi^2, 0, \pi) - f(0, 0, 0) = 0 - 0 = 0$.

19. Here $\mathbf{F}(x, y) = \tan y\,\mathbf{i} + x \sec^2 y\,\mathbf{j}$. Then $f(x, y) = x \tan y$ is a potential function for $\mathbf{F}$, that is, $\nabla f = \mathbf{F}$ so

$\mathbf{F}$ is conservative and thus its line integral is independent of path. Hence

$\int_C \tan y\, dx + x \sec^2 y\, dy = \int_C \mathbf{F} \cdot d\mathbf{r} = f\left(2, \frac{\pi}{4}\right) - f(1, 0) = 2 \tan \frac{\pi}{4} - \tan 0 = 2$.

21. $\mathbf{F}(x, y) = 2y^{3/2}\,\mathbf{i} + 3x\sqrt{y}\,\mathbf{j}$, $W = \int_C \mathbf{F} \cdot d\mathbf{r}$. Since $\partial(2y^{3/2})/\partial y = 3\sqrt{y} = \partial(3x\sqrt{y})/\partial x$, there exists a function f such

that $\nabla f = \mathbf{F}$. In fact, $f_x(x, y) = 2y^{3/2} \quad \Rightarrow \quad f(x, y) = 2xy^{3/2} + g(y) \quad \Rightarrow \quad f_y(x, y) = 3xy^{1/2} + g'(y)$. But

$f_y(x, y) = 3x\sqrt{y}$ so $g'(y) = 0$ or $g(y) = K$. We can take $K = 0 \quad \Rightarrow \quad f(x, y) = 2xy^{3/2}$. Thus

$W = \int_C \mathbf{F} \cdot d\mathbf{r} = f(2, 4) - f(1, 1) = 2(2)(8) - 2(1) = 30$.

23. We know that if the vector field (call it $\mathbf{F}$) is conservative, then around any closed path C, $\int_C \mathbf{F} \cdot d\mathbf{r} = 0$. But take C to be a

circle centered at the origin, oriented counterclockwise. All of the field vectors that start on C are roughly in the direction of

motion along C, so the integral around C will be positive. Therefore the field is not conservative.

25.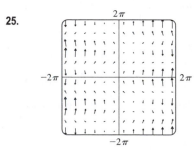

From the graph, it appears that $\mathbf{F}$ is conservative, since around all closed paths, the

number and size of the field vectors pointing in directions similar to that of the path

seem to be roughly the same as the number and size of the vectors pointing in the

opposite direction. To check, we calculate

$$\frac{\partial}{\partial y}(\sin y) = \cos y = \frac{\partial}{\partial x}(1 + x \cos y).$$ Thus $\mathbf{F}$ is conservative, by Theorem 6.

27. Since $\mathbf{F}$ is conservative, there exists a function f such that $\mathbf{F} = \nabla f$, that is, $P = f_x, Q = f_y$, and $R = f_z$. Since P, Q and R

have continuous first order partial derivatives, Clairaut's Theorem says that $\partial P/\partial y = f_{xy} = f_{yx} = \partial Q/\partial x$,

$\partial P/\partial z = f_{xz} = f_{zx} = \partial R/\partial x$, and $\partial Q/\partial z = f_{yz} = f_{zy} = \partial R/\partial y$.

29. $D = \{(x, y) \mid x > 0, y > 0\}$ = the first quadrant (excluding the axes).

(a) D is open because around every point in D we can put a disk that lies in D.

(b) D is connected because the straight line segment joining any two points in D lies in D.

(c) D is simply-connected because it's connected and has no holes.

31. $D = \{(x, y) \mid 1 < x^2 + y^2 < 4\}$ = the annular region between the circles with center $(0, 0)$ and radii 1 and 2.

(a) D is open.

(b) D is connected.

(c) D is not simply-connected. For example, $x^2 + y^2 = (1.5)^2$ is simple and closed and lies within D but encloses points that

are not in D. (Or we can say, D has a hole, so is not simply-connected.)

33. (a) $P = -\dfrac{y}{x^2 + y^2}$, $\dfrac{\partial P}{\partial y} = \dfrac{y^2 - x^2}{(x^2 + y^2)^2}$ and $Q = \dfrac{x}{x^2 + y^2}$, $\dfrac{\partial Q}{\partial x} = \dfrac{y^2 - x^2}{(x^2 + y^2)^2}$. Thus $\dfrac{\partial P}{\partial y} = \dfrac{\partial Q}{\partial x}$.

(b) C_1: $x = \cos t$, $y = \sin t$, $0 \le t \le \pi$, C_2: $x = \cos t$, $y = \sin t$, $t = 2\pi$ to $t = \pi$. Then

$$\int_{C_1} \mathbf{F} \cdot d\mathbf{r} = \int_0^\pi \frac{(-\sin t)(-\sin t) + (\cos t)(\cos t)}{\cos^2 t + \sin^2 t}\, dt = \int_0^\pi dt = \pi \quad \text{and} \quad \int_{C_2} \mathbf{F} \cdot d\mathbf{r} = \int_{2\pi}^\pi dt = -\pi$$

Since these aren't equal, the line integral of $\mathbf{F}$ isn't independent of path. (Or notice that $\int_{C_3} \mathbf{F} \cdot d\mathbf{r} = \int_0^{2\pi} dt = 2\pi$ where C_3 is the circle $x^2 + y^2 = 1$, and apply the contrapositive of Theorem 3.) This doesn't contradict Theorem 6, since the domain of $\mathbf{F}$, which is $\mathbb{R}^2$ except the origin, isn't simply-connected.

13.4 Green's Theorem

1. (a)

C_1: $x = t$ $\Rightarrow$ $dx = dt$, $y = 0$ $\Rightarrow$ $dy = 0\, dt$, $0 \le t \le 2$.

C_2: $x = 2$ $\Rightarrow$ $dx = 0\, dt$, $y = t$ $\Rightarrow$ $dy = dt$, $0 \le t \le 3$.

C_3: $x = 2 - t$ $\Rightarrow$ $dx = -dt$, $y = 3$ $\Rightarrow$ $dy = 0\, dt$, $0 \le t \le 2$.

C_4: $x = 0$ $\Rightarrow$ $dx = 0\, dt$, $y = 3 - t$ $\Rightarrow$ $dy = -dt$, $0 \le t \le 3$.

Thus $\oint_C xy^2\, dx + x^3\, dy = \oint_{C_1 + C_2 + C_3 + C_4} xy^2\, dx + x^3\, dy$

$$= \int_0^2 0\, dt + \int_0^3 8\, dt + \int_0^2 -9(2 - t)\, dt + \int_0^3 0\, dt$$

$$= 0 + 24 - 18 + 0 = 6$$

(b) $\oint_C xy^2\, dx + x^3\, dy = \iint_D \left[\frac{\partial}{\partial x}(x^3) - \frac{\partial}{\partial y}(xy^2) \right] dA = \int_0^2 \int_0^3 (3x^2 - 2xy)\, dy\, dx = \int_0^2 (9x^2 - 9x)\, dx = 24 - 18 = 6$

3. (a)

C_1: $x = t$ $\Rightarrow$ $dx = dt$, $y = 0$ $\Rightarrow$ $dy = 0\, dt$, $0 \le t \le 1$.

C_2: $x = 1$ $\Rightarrow$ $dx = 0\, dt$, $y = t$ $\Rightarrow$ $dy = dt$, $0 \le t \le 2$.

C_3: $x = 1 - t$ $\Rightarrow$ $dx = -dt$, $y = 2 - 2t$ $\Rightarrow$ $dy = -2\, dt$, $0 \le t \le 1$.

Thus $\oint_C xy\, dx + x^2 y^3\, dy = \oint_{C_1 + C_2 + C_3} xy\, dx + x^2 y^3\, dy$

$$= \int_0^1 0\, dt + \int_0^2 t^3\, dt + \int_0^1 \left[-(1 - t)(2 - 2t) - 2(1 - t)^2(2 - 2t)^3 \right] dt$$

$$= 0 + \left[\tfrac{1}{4} t^4 \right]_0^2 + \left[\tfrac{2}{3}(1 - t)^3 + \tfrac{8}{3}(1 - t)^6 \right]_0^1 = 4 - \tfrac{10}{3} = \tfrac{2}{3}$$

(b) $\oint_C xy\, dx + x^2 y^3\, dy = \iint_D \left[\frac{\partial}{\partial x}(x^2 y^3) - \frac{\partial}{\partial y}(xy) \right] dA = \int_0^1 \int_0^{2x} (2xy^3 - x)\, dy\, dx$

$$= \int_0^1 \left[\tfrac{1}{2} xy^4 - xy \right]_{y=0}^{y=2x} dx = \int_0^1 (8x^5 - 2x^2)\, dx = \tfrac{4}{3} - \tfrac{2}{3} = \tfrac{2}{3}$$

5. We can parametrize C as $x = \cos\theta$, $y = \sin\theta$, $0 \le \theta \le 2\pi$. Then the line integral is

$\oint_C P\, dx + Q\, dy = \int_0^{2\pi} \cos^4\theta \sin^5\theta\,(-\sin\theta)\, d\theta + \int_0^{2\pi} (-\cos^7\theta \sin^6\theta) \cos\theta\, d\theta = -\dfrac{29\pi}{1024}$, according to a CAS.

The double integral is

$$\iint_D \left(\frac{\partial Q}{\partial x} - \frac{\partial P}{\partial y} \right) dA = \int_{-1}^1 \int_{-\sqrt{1 - x^2}}^{\sqrt{1 - x^2}} (-7x^6 y^6 - 5x^4 y^4)\, dy\, dx = -\frac{29\pi}{1024},$$

verifying Green's Theorem in this case.

7. The region D enclosed by C is $[0, 1] \times [0, 1]$, so

$$\int_C e^y \, dx + 2xe^y \, dy = \iint_D \left[\frac{\partial}{\partial x} (2xe^y) - \frac{\partial}{\partial y} (e^y) \right] dA = \int_0^1 \int_0^1 (2e^y - e^y) \, dy \, dx$$

$$= \int_0^1 dx \int_0^1 e^y \, dy = (1)(e^1 - e^0) = e - 1$$

9. $\int_C \left(y + e^{\sqrt{x}} \right) dx + \left(2x + \cos y^2 \right) dy = \iint_D \left[\frac{\partial}{\partial x} (2x + \cos y^2) - \frac{\partial}{\partial y} \left(y + e^{\sqrt{x}} \right) \right] dA$

$$= \int_0^1 \int_{y^2}^{\sqrt{y}} (2 - 1) \, dx \, dy = \int_0^1 (y^{1/2} - y^2) \, dy = \tfrac{1}{3}$$

11. $\int_C y^3 \, dx - x^3 \, dy = \iint_D \left[\frac{\partial}{\partial x} (-x^3) - \frac{\partial}{\partial y} (y^3) \right] dA = \iint_D (-3x^2 - 3y^2) \, dA = \int_0^{2\pi} \int_0^2 (-3r^2) \, r \, dr \, d\theta$

$$= -3 \int_0^{2\pi} d\theta \int_0^2 r^3 \, dr = -3(2\pi)(4) = -24\pi$$

13. $\mathbf{F}(x, y) = \left\langle \sqrt{x} + y^3, x^2 + \sqrt{y} \right\rangle$ and the region D enclosed by C is given by $\{(x, y) \mid 0 \le x \le \pi, 0 \le y \le \sin x\}$.

C is traversed clockwise, so $-C$ gives the positive orientation.

$$\int_C \mathbf{F} \cdot d\mathbf{r} = -\int_{-C} \left(\sqrt{x} + y^3 \right) dx + \left(x^2 + \sqrt{y} \right) dy = -\iint_D \left[\frac{\partial}{\partial x} \left(x^2 + \sqrt{y} \right) - \frac{\partial}{\partial y} \left(\sqrt{x} + y^3 \right) \right] dA$$

$$= -\int_0^\pi \int_0^{\sin x} (2x - 3y^2) \, dy \, dx = -\int_0^\pi \left[2xy - y^3 \right]_{y=0}^{y=\sin x} dx$$

$$= -\int_0^\pi (2x \sin x - \sin^3 x) \, dx = -\int_0^\pi (2x \sin x - (1 - \cos^2 x) \sin x) \, dx$$

$$= -\left[2 \sin x - 2x \cos x + \cos x - \tfrac{1}{3} \cos^3 x \right]_0^\pi \qquad \text{[integrate by parts in the first term]}$$

$$= -\left(2\pi - 2 + \tfrac{2}{3} \right) = \tfrac{4}{3} - 2\pi$$

15. $\mathbf{F}(x, y) = \left\langle e^x + x^2 y, e^y - xy^2 \right\rangle$ and the region D enclosed by C is the disk $x^2 + y^2 \le 25$.

C is traversed clockwise, so $-C$ gives the positive orientation.

$$\int_C \mathbf{F} \cdot d\mathbf{r} = -\int_{-C} (e^x + x^2 y) \, dx + (e^y - xy^2) \, dy = -\iint_D \left[\frac{\partial}{\partial x} (e^y - xy^2) - \frac{\partial}{\partial y} (e^x + x^2 y) \right] dA$$

$$= -\iint_D (-y^2 - x^2) \, dA = \iint_D (x^2 + y^2) \, dA = \int_0^{2\pi} \int_0^5 (r^2) \, r \, dr \, d\theta = \int_0^{2\pi} d\theta \int_0^5 r^3 \, dr = 2\pi \left[\tfrac{1}{4} r^4 \right]_0^5 = \tfrac{625}{2}\pi$$

17. By Green's Theorem, $W = \int_C \mathbf{F} \cdot d\mathbf{r} = \int_C x(x + y) \, dx + xy^2 \, dy = \iint_D (y^2 - x) \, dy \, dx$ where C is the path described in the question and D is the triangle bounded by C. So

$$W = \int_0^1 \int_0^{1-x} (y^2 - x) \, dy \, dx = \int_0^1 \left[\tfrac{1}{3} y^3 - xy \right]_{y=0}^{y=1-x} dx = \int_0^1 \left(\tfrac{1}{3} (1 - x)^3 - x(1 - x) \right) dx$$

$$= \left[-\tfrac{1}{12} (1 - x)^4 - \tfrac{1}{2} x^2 + \tfrac{1}{3} x^3 \right]_0^1 = \left(-\tfrac{1}{2} + \tfrac{1}{3} \right) - \left(-\tfrac{1}{12} \right) = -\tfrac{1}{12}$$

19. Let C_1 be the arch of the cycloid from $(0, 0)$ to $(2\pi, 0)$, which corresponds to $0 \le t \le 2\pi$, and let C_2 be the segment from $(2\pi, 0)$ to $(0, 0)$, so C_2 is given by $x = 2\pi - t, y = 0, 0 \le t \le 2\pi$. Then $C = C_1 \cup C_2$ is traversed clockwise, so $-C$ is oriented positively. Thus $-C$ encloses the area under one arch of the cycloid and from (5) we have

$$A = -\oint_{-C} y \, dx = \int_{C_1} y \, dx + \int_{C_2} y \, dx = \int_0^{2\pi} (1 - \cos t)(1 - \cos t) \, dt + \int_0^{2\pi} 0 \, (-dt)$$

$$= \int_0^{2\pi} (1 - 2 \cos t + \cos^2 t) \, dt + 0 = \left[t - 2 \sin t + \tfrac{1}{2} t + \tfrac{1}{4} \sin 2t \right]_0^{2\pi} = 3\pi$$

21. (a) Using Equation 13.2.8, we write parametric equations of the line segment as $x = (1 - t)x_1 + tx_2, y = (1 - t)y_1 + ty_2$, $0 \le t \le 1$. Then $dx = (x_2 - x_1) \, dt$ and $dy = (y_2 - y_1) \, dt$, so

$$\int_C x \, dy - y \, dx = \int_0^1 [(1 - t)x_1 + tx_2](y_2 - y_1) \, dt + [(1 - t)y_1 + ty_2](x_2 - x_1) \, dt$$

$$= \int_0^1 (x_1(y_2 - y_1) - y_1(x_2 - x_1) + t[(y_2 - y_1)(x_2 - x_1) - (x_2 - x_1)(y_2 - y_1)]) \, dt$$

$$= \int_0^1 (x_1 y_2 - x_2 y_1) \, dt = x_1 y_2 - x_2 y_1$$

(b) We apply Green's Theorem to the path $C = C_1 \cup C_2 \cup \cdots \cup C_n$, where C_i is the line segment that joins (x_i, y_i) to (x_{i+1}, y_{i+1}) for $i = 1, 2, \ldots, n-1$, and C_n is the line segment that joins (x_n, y_n) to (x_1, y_1). From (5), $\frac{1}{2}\int_C x\,dy - y\,dx = \iint_D dA$, where D is the polygon bounded by C. Therefore

$$\text{area of polygon} = A(D) = \iint_D dA = \tfrac{1}{2}\int_C x\,dy - y\,dx$$

$$= \tfrac{1}{2}\left(\int_{C_1} x\,dy - y\,dx + \int_{C_2} x\,dy - y\,dx + \cdots + \int_{C_{n-1}} x\,dy - y\,dx + \int_{C_n} x\,dy - y\,dx\right)$$

To evaluate these integrals we use the formula from (a) to get

$$A(D) = \tfrac{1}{2}[(x_1 y_2 - x_2 y_1) + (x_2 y_3 - x_3 y_2) + \cdots + (x_{n-1} y_n - x_n y_{n-1}) + (x_n y_1 - x_1 y_n)].$$

(c) $A = \tfrac{1}{2}[(0 \cdot 1 - 2 \cdot 0) + (2 \cdot 3 - 1 \cdot 1) + (1 \cdot 2 - 0 \cdot 3) + (0 \cdot 1 - (-1) \cdot 2) + (-1 \cdot 0 - 0 \cdot 1)]$

$= \tfrac{1}{2}(0 + 5 + 2 + 2) = \tfrac{9}{2}$

23. Here $A = \tfrac{1}{2}(1)(1) = \tfrac{1}{2}$ and $C = C_1 + C_2 + C_3$, where C_1: $x = x$, $y = 0$, $0 \le x \le 1$;

C_2: $x = x$, $y = 1 - x$, $x = 1$ to $x = 0$; and C_3: $x = 0$, $y = 1$ to $y = 0$. Then

$$\bar{x} = \tfrac{1}{2A}\int_C x^2\,dy = \int_{C_1} x^2\,dy + \int_{C_2} x^2\,dy + \int_{C_3} x^2\,dy = 0 + \int_1^0 (x^2)(-dx) + 0 = \tfrac{1}{3}. \text{ Similarly,}$$

$$\bar{y} = -\tfrac{1}{2A}\int_C y^2\,dx = \int_{C_1} y^2\,dx + \int_{C_2} y^2\,dx + \int_{C_3} y^2\,dx = 0 + \int_1^0 (1-x)^2(-dx) + 0 = \tfrac{1}{3}.$$

Therefore $(\bar{x}, \bar{y}) = \left(\tfrac{1}{3}, \tfrac{1}{3}\right)$.

25. By Green's Theorem, $-\tfrac{1}{3}\rho \oint_C y^3\,dx = -\tfrac{1}{3}\rho \iint_D (-3y^2)\,dA = \iint_D y^2 \rho\,dA = I_x$ and

$$\tfrac{1}{3}\rho \oint_C x^3\,dy = \tfrac{1}{3}\rho \iint_D (3x^2)\,dA = \iint_D x^2 \rho\,dA = I_y.$$

27. Since C is a simple closed path which doesn't pass through or enclose the origin, there exists an open region that doesn't contain the origin but does contain D. Thus $P = -y/(x^2 + y^2)$ and $Q = x/(x^2 + y^2)$ have continuous partial derivatives on this open region containing D and we can apply Green's Theorem. But by Exercise 13.3.33(a), $\partial P/\partial y = \partial Q/\partial x$, so $\oint_C \mathbf{F} \cdot d\mathbf{r} = \iint_D 0\,dA = 0$.

29. Using the first part of (5), we have that $\iint_R dx\,dy = A(R) = \int_{\partial R} x\,dy$. But $x = g(u, v)$, and $dy = \dfrac{\partial h}{\partial u}\,du + \dfrac{\partial h}{\partial v}\,dv$, and we orient ∂S by taking the positive direction to be that which corresponds, under the mapping, to the positive direction along ∂R, so

$$\int_{\partial R} x\,dy = \int_{\partial S} g(u, v)\left(\frac{\partial h}{\partial u}\,du + \frac{\partial h}{\partial v}\,dv\right) = \int_{\partial S} g(u, v)\frac{\partial h}{\partial u}\,du + g(u, v)\frac{\partial h}{\partial v}\,dv$$

$$= \pm \iint_S \left[\frac{\partial}{\partial u}\left(g(u, v)\frac{\partial h}{\partial v}\right) - \frac{\partial}{\partial v}\left(g(u, v)\frac{\partial h}{\partial u}\right)\right] dA \quad \text{[using Green's Theorem in the } uv\text{-plane]}$$

$$= \pm \iint_S \left(\frac{\partial g}{\partial u}\frac{\partial h}{\partial v} + g(u, v)\frac{\partial^2 h}{\partial u\,\partial v} - \frac{\partial g}{\partial v}\frac{\partial h}{\partial u} - g(u, v)\frac{\partial^2 h}{\partial v\,\partial u}\right) dA \quad \text{[using the Chain Rule]}$$

$$= \pm \iint_S \left(\frac{\partial x}{\partial u}\frac{\partial y}{\partial v} - \frac{\partial x}{\partial v}\frac{\partial y}{\partial u}\right) dA \quad \text{[by the equality of mixed partials]} = \pm \iint_S \frac{\partial(x, y)}{\partial(u, v)}\,du\,dv$$

The sign is chosen to be positive if the orientation that we gave to ∂S corresponds to the usual positive orientation, and it is negative otherwise. In either case, since $A(R)$ is positive, the sign chosen must be the same as the sign of $\dfrac{\partial(x, y)}{\partial(u, v)}$. Therefore

$$A(R) = \iint_R dx\,dy = \iint_S \left|\frac{\partial(x, y)}{\partial(u, v)}\right| du\,dv.$$

13.5 Curl and Divergence

1. (a) curl $\mathbf{F} = \nabla \times \mathbf{F} = \begin{vmatrix} \mathbf{i} & \mathbf{j} & \mathbf{k} \\ \partial/\partial x & \partial/\partial y & \partial/\partial z \\ xyz & 0 & -x^2y \end{vmatrix} = (-x^2 - 0)\,\mathbf{i} - (-2xy - xy)\,\mathbf{j} + (0 - xz)\,\mathbf{k}$

$= -x^2\,\mathbf{i} + 3xy\,\mathbf{j} - xz\,\mathbf{k}$

(b) div $\mathbf{F} = \nabla \cdot \mathbf{F} = \dfrac{\partial}{\partial x}\,(xyz) + \dfrac{\partial}{\partial y}\,(0) + \dfrac{\partial}{\partial z}\,(-x^2y) = yz + 0 + 0 = yz$

3. (a) curl $\mathbf{F} = \nabla \times \mathbf{F} = \begin{vmatrix} \mathbf{i} & \mathbf{j} & \mathbf{k} \\ \partial/\partial x & \partial/\partial y & \partial/\partial z \\ 1 & x + yz & xy - \sqrt{z} \end{vmatrix} = (x - y)\,\mathbf{i} - (y - 0)\,\mathbf{j} + (1 - 0)\,\mathbf{k}$

$= (x - y)\,\mathbf{i} - y\,\mathbf{j} + \mathbf{k}$

(b) div $\mathbf{F} = \nabla \cdot \mathbf{F} = \dfrac{\partial}{\partial x}\,(1) + \dfrac{\partial}{\partial y}\,(x + yz) + \dfrac{\partial}{\partial z}\,(xy - \sqrt{z}) = z - \dfrac{1}{2\sqrt{z}}$

5. (a) curl $\mathbf{F} = \nabla \times \mathbf{F} = \begin{vmatrix} \mathbf{i} & \mathbf{j} & \mathbf{k} \\ \partial/\partial x & \partial/\partial y & \partial/\partial z \\ e^x \sin y & e^x \cos y & z \end{vmatrix} = (0 - 0)\,\mathbf{i} - (0 - 0)\,\mathbf{j} + (e^x \cos y - e^x \cos y)\,\mathbf{k} = \mathbf{0}$

(b) div $\mathbf{F} = \nabla \cdot \mathbf{F} = \dfrac{\partial}{\partial x}\,(e^x \sin y) + \dfrac{\partial}{\partial y}\,(e^x \cos y) + \dfrac{\partial}{\partial z}\,(z) = e^x \sin y - e^x \sin y + 1 = 1$

7. If the vector field is $\mathbf{F} = P\,\mathbf{i} + Q\,\mathbf{j} + R\,\mathbf{k}$, then we know $R = 0$. In addition, the x-component of each vector of $\mathbf{F}$ is 0, so

$P = 0$, hence $\dfrac{\partial P}{\partial x} = \dfrac{\partial P}{\partial y} = \dfrac{\partial P}{\partial z} = \dfrac{\partial R}{\partial x} = \dfrac{\partial R}{\partial y} = \dfrac{\partial R}{\partial z} = 0$. Q decreases as y increases, so $\dfrac{\partial Q}{\partial y} < 0$, but Q doesn't change

in the x- or z-directions, so $\dfrac{\partial Q}{\partial x} = \dfrac{\partial Q}{\partial z} = 0$.

(a) div $\mathbf{F} = \dfrac{\partial P}{\partial x} + \dfrac{\partial Q}{\partial y} + \dfrac{\partial R}{\partial z} = 0 + \dfrac{\partial Q}{\partial y} + 0 < 0$

(b) curl $\mathbf{F} = \left(\dfrac{\partial R}{\partial y} - \dfrac{\partial Q}{\partial z} \right)\mathbf{i} + \left(\dfrac{\partial P}{\partial z} - \dfrac{\partial R}{\partial x} \right)\mathbf{j} + \left(\dfrac{\partial Q}{\partial x} - \dfrac{\partial P}{\partial y} \right)\mathbf{k} = (0 - 0)\,\mathbf{i} + (0 - 0)\,\mathbf{j} + (0 - 0)\,\mathbf{k} = \mathbf{0}$

9. If the vector field is $\mathbf{F} = P\,\mathbf{i} + Q\,\mathbf{j} + R\,\mathbf{k}$, then we know $R = 0$. In addition, the y-component of each vector of $\mathbf{F}$ is 0, so

$Q = 0$, hence $\dfrac{\partial Q}{\partial x} = \dfrac{\partial Q}{\partial y} = \dfrac{\partial Q}{\partial z} = \dfrac{\partial R}{\partial x} = \dfrac{\partial R}{\partial y} = \dfrac{\partial R}{\partial z} = 0$. P increases as y increases, so $\dfrac{\partial P}{\partial y} > 0$, but P doesn't change in

the x- or z-directions, so $\dfrac{\partial P}{\partial x} = \dfrac{\partial P}{\partial z} = 0$.

(a) div $\mathbf{F} = \dfrac{\partial P}{\partial x} + \dfrac{\partial Q}{\partial y} + \dfrac{\partial R}{\partial z} = 0 + 0 + 0 = 0$

(b) curl $\mathbf{F} = \left(\dfrac{\partial R}{\partial y} - \dfrac{\partial Q}{\partial z} \right)\mathbf{i} + \left(\dfrac{\partial P}{\partial z} - \dfrac{\partial R}{\partial x} \right)\mathbf{j} + \left(\dfrac{\partial Q}{\partial x} - \dfrac{\partial P}{\partial y} \right)\mathbf{k}$

$= (0 - 0)\,\mathbf{i} + (0 - 0)\,\mathbf{j} + \left(0 - \dfrac{\partial P}{\partial y} \right)\mathbf{k} = -\dfrac{\partial P}{\partial y}\mathbf{k}$

Since $\dfrac{\partial P}{\partial y} > 0$, $-\dfrac{\partial P}{\partial y}\mathbf{k}$ is a vector pointing in the negative z-direction.

11. $\text{curl } \mathbf{F} = \nabla \times \mathbf{F} = \begin{vmatrix} \mathbf{i} & \mathbf{j} & \mathbf{k} \\ \partial/\partial x & \partial/\partial y & \partial/\partial z \\ yz & xz & xy \end{vmatrix} = (x-x)\mathbf{i} - (y-y)\mathbf{j} + (z-z)\mathbf{k} = \mathbf{0}$

and $\mathbf{F}$ is defined on all of $\mathbb{R}^3$ with component functions which have continuous partial derivatives, so by Theorem 4, $\mathbf{F}$ is

conservative. Thus, there exists a function f such that $\mathbf{F} = \nabla f$. Then $f_x(x, y, z) = yz$ implies $f(x, y, z) = xyz + g(y, z)$

and $f_y(x, y, z) = xz + g_y(y, z)$. But $f_y(x, y, z) = xz$, so $g(y, z) = h(z)$ and $f(x, y, z) = xyz + h(z)$. Thus

$f_z(x, y, z) = xy + h'(z)$ but $f_z(x, y, z) = xy$ so $h(z) = K$, a constant. Hence a potential function for $\mathbf{F}$ is

$f(x, y, z) = xyz + K$.

13. $\text{curl } \mathbf{F} = \nabla \times \mathbf{F} = \begin{vmatrix} \mathbf{i} & \mathbf{j} & \mathbf{k} \\ \partial/\partial x & \partial/\partial y & \partial/\partial z \\ 2xy & x^2 + 2yz & y^2 \end{vmatrix} = (2y-2y)\mathbf{i} - (0-0)\mathbf{j} + (2x-2x)\mathbf{k} = \mathbf{0}, \mathbf{F}$ is defined on all of $\mathbb{R}^3$,

and the partial derivatives of the component functions are continuous, so $\mathbf{F}$ is conservative. Thus there exists a function f such

that $\nabla f = \mathbf{F}$. Then $f_x(x, y, z) = 2xy$ implies $f(x, y, z) = x^2y + g(y, z)$ and $f_y(x, y, z) = x^2 + g_y(y, z)$. But

$f_y(x, y, z) = x^2 + 2yz$, so $g(y, z) = y^2z + h(z)$ and $f(x, y, z) = x^2y + y^2z + h(z)$. Thus $f_z(x, y, z) = y^2 + h'(z)$ but

$f_z(x, y, z) = y^2$ so $h(z) = K$ and $f(x, y, z) = x^2y + y^2z + K$.

15. $\text{curl } \mathbf{F} = \nabla \times \mathbf{F} = \begin{vmatrix} \mathbf{i} & \mathbf{j} & \mathbf{k} \\ \partial/\partial x & \partial/\partial y & \partial/\partial z \\ ye^{-x} & e^{-x} & 2z \end{vmatrix} = (0-0)\mathbf{i} - (0-0)\mathbf{j} + (-e^{-x} - e^{-x})\mathbf{k} = -2e^{-x}\mathbf{k} \neq \mathbf{0}$,

so $\mathbf{F}$ is not conservative.

17. No. Assume there is such a $\mathbf{G}$. Then $\text{div}(\text{curl } \mathbf{G}) = y^2 + z^2 + x^2 \neq 0$, which contradicts Theorem 11.

19. $\text{curl } \mathbf{F} = \begin{vmatrix} \mathbf{i} & \mathbf{j} & \mathbf{k} \\ \partial/\partial x & \partial/\partial y & \partial/\partial z \\ f(x) & g(y) & h(z) \end{vmatrix} = (0-0)\mathbf{i} + (0-0)\mathbf{j} + (0-0)\mathbf{k} = \mathbf{0}$.

Hence $\mathbf{F} = f(x)\mathbf{i} + g(y)\mathbf{j} + h(z)\mathbf{k}$ is irrotational.

For Exercises 21–27, let $\mathbf{F}(x, y, z) = P_1\mathbf{i} + Q_1\mathbf{j} + R_1\mathbf{k}$ and $\mathbf{G}(x, y, z) = P_2\mathbf{i} + Q_2\mathbf{j} + R_2\mathbf{k}$.

21. $\text{div}(\mathbf{F} + \mathbf{G}) = \dfrac{\partial(P_1 + P_2)}{\partial x} + \dfrac{\partial(Q_1 + Q_2)}{\partial y} + \dfrac{\partial(R_1 + R_2)}{\partial z}$

$= \left(\dfrac{\partial P_1}{\partial x} + \dfrac{\partial Q_1}{\partial y} + \dfrac{\partial R_1}{\partial z}\right) + \left(\dfrac{\partial P_2}{\partial x} + \dfrac{\partial Q_2}{\partial y} + \dfrac{\partial R_3}{\partial z}\right) = \text{div } \mathbf{F} + \text{div } \mathbf{G}$

23. $\text{div}(f\mathbf{F}) = \dfrac{\partial(fP_1)}{\partial x} + \dfrac{\partial(fQ_1)}{\partial y} + \dfrac{\partial(fR_1)}{\partial z}$

$= \left(f\dfrac{\partial P_1}{\partial x} + P_1\dfrac{\partial f}{\partial x}\right) + \left(f\dfrac{\partial Q_1}{\partial y} + Q_1\dfrac{\partial f}{\partial y}\right) + \left(f\dfrac{\partial R_1}{\partial z} + R_1\dfrac{\partial f}{\partial z}\right)$

$= f\left(\dfrac{\partial P_1}{\partial x} + \dfrac{\partial Q_1}{\partial y} + \dfrac{\partial R_1}{\partial z}\right) + \langle P_1, Q_1, R_1 \rangle \cdot \left\langle \dfrac{\partial f}{\partial x}, \dfrac{\partial f}{\partial y}, \dfrac{\partial f}{\partial z} \right\rangle = f\,\text{div } \mathbf{F} + \mathbf{F} \cdot \nabla f$

25. $\operatorname{div}(\mathbf{F} \times \mathbf{G}) = \nabla \cdot (\mathbf{F} \times \mathbf{G}) = \begin{vmatrix} \partial/\partial x & \partial/\partial y & \partial/\partial z \\ P_1 & Q_1 & R_1 \\ P_2 & Q_2 & R_2 \end{vmatrix} = \frac{\partial}{\partial x}\begin{vmatrix} Q_1 & R_1 \\ Q_2 & R_2 \end{vmatrix} - \frac{\partial}{\partial y}\begin{vmatrix} P_1 & R_1 \\ P_2 & R_2 \end{vmatrix} + \frac{\partial}{\partial z}\begin{vmatrix} P_1 & Q_1 \\ P_2 & Q_2 \end{vmatrix}$

$$= \left[Q_1 \frac{\partial R_2}{\partial x} + R_2 \frac{\partial Q_1}{\partial x} - Q_2 \frac{\partial R_1}{\partial x} - R_1 \frac{\partial Q_2}{\partial x} \right]$$

$$- \left[P_1 \frac{\partial R_2}{\partial y} + R_2 \frac{\partial P_1}{\partial y} - P_2 \frac{\partial R_1}{\partial y} - R_1 \frac{\partial P_2}{\partial y} \right]$$

$$+ \left[P_1 \frac{\partial Q_2}{\partial z} + Q_2 \frac{\partial P_1}{\partial z} - P_2 \frac{\partial Q_1}{\partial z} - Q_1 \frac{\partial P_2}{\partial z} \right]$$

$$= \left[P_2 \left(\frac{\partial R_1}{\partial y} - \frac{\partial Q_1}{\partial z} \right) + Q_2 \left(\frac{\partial P_1}{\partial z} - \frac{\partial R_1}{\partial x} \right) + R_2 \left(\frac{\partial Q_1}{\partial x} - \frac{\partial P_1}{\partial y} \right) \right]$$

$$- \left[P_1 \left(\frac{\partial R_2}{\partial y} - \frac{\partial Q_2}{\partial z} \right) + Q_1 \left(\frac{\partial P_2}{\partial z} - \frac{\partial R_2}{\partial x} \right) + R_1 \left(\frac{\partial Q_2}{\partial x} - \frac{\partial P_2}{\partial y} \right) \right]$$

$$= \mathbf{G} \cdot \operatorname{curl} \mathbf{F} - \mathbf{F} \cdot \operatorname{curl} \mathbf{G}$$

27. $\operatorname{curl}(\operatorname{curl} \mathbf{F}) = \nabla \times (\nabla \times \mathbf{F}) = \begin{vmatrix} \mathbf{i} & \mathbf{j} & \mathbf{k} \\ \partial/\partial x & \partial/\partial y & \partial/\partial z \\ \partial R_1/\partial y - \partial Q_1/\partial z & \partial P_1/\partial z - \partial R_1/\partial x & \partial Q_1/\partial x - \partial P_1/\partial y \end{vmatrix}$

$$= \left(\frac{\partial^2 Q_1}{\partial y \partial x} - \frac{\partial^2 P_1}{\partial y^2} - \frac{\partial^2 P_1}{\partial z^2} + \frac{\partial^2 R_1}{\partial z \partial x} \right) \mathbf{i} + \left(\frac{\partial^2 R_1}{\partial z \partial y} - \frac{\partial^2 Q_1}{\partial z^2} - \frac{\partial^2 Q_1}{\partial x^2} + \frac{\partial^2 P_1}{\partial x \partial y} \right) \mathbf{j}$$

$$+ \left(\frac{\partial^2 P_1}{\partial x \partial z} - \frac{\partial^2 R_1}{\partial x^2} - \frac{\partial^2 R_1}{\partial y^2} + \frac{\partial^2 Q_1}{\partial y \partial z} \right) \mathbf{k}$$

Now let's consider $\operatorname{grad}(\operatorname{div} \mathbf{F}) - \nabla^2 \mathbf{F}$ and compare with the above.

(Note that $\nabla^2 \mathbf{F}$ is defined on page 944.)

$$\operatorname{grad}(\operatorname{div} \mathbf{F}) - \nabla^2 \mathbf{F} = \left[\left(\frac{\partial^2 P_1}{\partial x^2} + \frac{\partial^2 Q_1}{\partial x \partial y} + \frac{\partial^2 R_1}{\partial x \partial z} \right) \mathbf{i} + \left(\frac{\partial^2 P_1}{\partial y \partial x} + \frac{\partial^2 Q_1}{\partial y^2} + \frac{\partial^2 R_1}{\partial y \partial z} \right) \mathbf{j} \right.$$

$$\left. + \left(\frac{\partial^2 P_1}{\partial z \partial x} + \frac{\partial^2 Q_1}{\partial z \partial y} + \frac{\partial^2 R_1}{\partial z^2} \right) \mathbf{k} \right]$$

$$- \left[\left(\frac{\partial^2 P_1}{\partial x^2} + \frac{\partial^2 P_1}{\partial y^2} + \frac{\partial^2 P_1}{\partial z^2} \right) \mathbf{i} + \left(\frac{\partial^2 Q_1}{\partial x^2} + \frac{\partial^2 Q_1}{\partial y^2} + \frac{\partial^2 Q_1}{\partial z^2} \right) \mathbf{j} \right.$$

$$\left. + \left(\frac{\partial^2 R_1}{\partial x^2} + \frac{\partial^2 R_1}{\partial y^2} + \frac{\partial^2 R_1}{\partial z^2} \right) \mathbf{k} \right]$$

$$= \left(\frac{\partial^2 Q_1}{\partial x \partial y} + \frac{\partial^2 R_1}{\partial x \partial z} - \frac{\partial^2 P_1}{\partial y^2} - \frac{\partial^2 P_1}{\partial z^2} \right) \mathbf{i} + \left(\frac{\partial^2 P_1}{\partial y \partial x} + \frac{\partial^2 R_1}{\partial y \partial z} - \frac{\partial^2 Q_1}{\partial x^2} - \frac{\partial^2 Q_1}{\partial z^2} \right) \mathbf{j}$$

$$+ \left(\frac{\partial^2 P_1}{\partial z \partial x} + \frac{\partial^2 Q_1}{\partial z \partial y} - \frac{\partial^2 R_1}{\partial x^2} - \frac{\partial^2 R_2}{\partial y^2} \right) \mathbf{k}$$

Then applying Clairaut's Theorem to reverse the order of differentiation in the second partial derivatives as needed and comparing, we have $\operatorname{curl} \operatorname{curl} \mathbf{F} = \operatorname{grad} \operatorname{div} \mathbf{F} - \nabla^2 \mathbf{F}$ as desired.

29. (a) $\nabla r = \nabla \sqrt{x^2 + y^2 + z^2} = \dfrac{x}{\sqrt{x^2 + y^2 + z^2}}\mathbf{i} + \dfrac{y}{\sqrt{x^2 + y^2 + z^2}}\mathbf{j} + \dfrac{z}{\sqrt{x^2 + y^2 + z^2}}\mathbf{k} = \dfrac{x\mathbf{i} + y\mathbf{j} + z\mathbf{k}}{\sqrt{x^2 + y^2 + z^2}} = \dfrac{\mathbf{r}}{r}$

(b) $\nabla \times \mathbf{r} = \begin{vmatrix} \mathbf{i} & \mathbf{j} & \mathbf{k} \\ \dfrac{\partial}{\partial x} & \dfrac{\partial}{\partial y} & \dfrac{\partial}{\partial z} \\ x & y & z \end{vmatrix} = \left[\dfrac{\partial}{\partial y}(z) - \dfrac{\partial}{\partial z}(y)\right]\mathbf{i} + \left[\dfrac{\partial}{\partial z}(x) - \dfrac{\partial}{\partial x}(z)\right]\mathbf{j} + \left[\dfrac{\partial}{\partial x}(y) - \dfrac{\partial}{\partial y}(x)\right]\mathbf{k} = \mathbf{0}$

(c) $\nabla\left(\dfrac{1}{r}\right) = \nabla\left(\dfrac{1}{\sqrt{x^2 + y^2 + z^2}}\right)$

$= \dfrac{-\dfrac{1}{2\sqrt{x^2 + y^2 + z^2}}(2x)}{x^2 + y^2 + z^2}\mathbf{i} - \dfrac{\dfrac{1}{2\sqrt{x^2 + y^2 + z^2}}(2y)}{x^2 + y^2 + z^2}\mathbf{j} - \dfrac{\dfrac{1}{2\sqrt{x^2 + y^2 + z^2}}(2z)}{x^2 + y^2 + z^2}\mathbf{k}$

$= -\dfrac{x\mathbf{i} + y\mathbf{j} + z\mathbf{k}}{(x^2 + y^2 + z^2)^{3/2}} = -\dfrac{\mathbf{r}}{r^3}$

(d) $\nabla \ln r = \nabla \ln(x^2 + y^2 + z^2)^{1/2} = \tfrac{1}{2}\nabla \ln(x^2 + y^2 + z^2)$

$= \dfrac{x}{x^2 + y^2 + z^2}\mathbf{i} + \dfrac{y}{x^2 + y^2 + z^2}\mathbf{j} + \dfrac{z}{x^2 + y^2 + z^2}\mathbf{k} = \dfrac{x\mathbf{i} + y\mathbf{j} + z\mathbf{k}}{x^2 + y^2 + z^2} = \dfrac{\mathbf{r}}{r^2}$

31. By (13), $\oint_C f(\nabla g) \cdot \mathbf{n}\,ds = \iint_D \operatorname{div}(f\nabla g)\,dA = \iint_D [f\operatorname{div}(\nabla g) + \nabla g \cdot \nabla f]\,dA$ by Exercise 23. But $\operatorname{div}(\nabla g) = \nabla^2 g$.

Hence $\iint_D f\nabla^2 g\,dA = \oint_C f(\nabla g) \cdot \mathbf{n}\,ds - \iint_D \nabla g \cdot \nabla f\,dA$.

33. Let $f(x, y) = 1$. Then $\nabla f = \mathbf{0}$ and Green's first identity (see Exercise 31) says

$\iint_D \nabla^2 g\,dA = \oint_C (\nabla g) \cdot \mathbf{n}\,ds - \iint_D \mathbf{0} \cdot \nabla g\,dA \quad \Rightarrow \quad \iint_D \nabla^2 g\,dA = \oint_C \nabla g \cdot \mathbf{n}\,ds$. But g is harmonic on D, so

$\nabla^2 g = 0 \quad \Rightarrow \quad \oint_C \nabla g \cdot \mathbf{n}\,ds = 0$ and $\oint_C D_{\mathbf{n}} g\,ds = \oint_C (\nabla g \cdot \mathbf{n})\,ds = 0$.

35. (a) We know that $\omega = v/d$, and from the diagram $\sin\theta = d/r \quad \Rightarrow \quad v = d\omega = (\sin\theta)r\omega = |\mathbf{w} \times \mathbf{r}|$. But $\mathbf{v}$ is perpendicular to both $\mathbf{w}$ and $\mathbf{r}$, so that $\mathbf{v} = \mathbf{w} \times \mathbf{r}$.

(b) From (a), $\mathbf{v} = \mathbf{w} \times \mathbf{r} = \begin{vmatrix} \mathbf{i} & \mathbf{j} & \mathbf{k} \\ 0 & 0 & \omega \\ x & y & z \end{vmatrix} = (0 \cdot z - \omega y)\mathbf{i} + (\omega x - 0 \cdot z)\mathbf{j} + (0 \cdot y - x \cdot 0)\mathbf{k} = -\omega y\,\mathbf{i} + \omega x\,\mathbf{j}$

(c) $\operatorname{curl}\mathbf{v} = \nabla \times \mathbf{v} = \begin{vmatrix} \mathbf{i} & \mathbf{j} & \mathbf{k} \\ \partial/\partial x & \partial/\partial y & \partial/\partial z \\ -\omega y & \omega x & 0 \end{vmatrix}$

$= \left[\dfrac{\partial}{\partial y}(0) - \dfrac{\partial}{\partial z}(\omega x)\right]\mathbf{i} + \left[\dfrac{\partial}{\partial z}(-\omega y) - \dfrac{\partial}{\partial x}(0)\right]\mathbf{j} + \left[\dfrac{\partial}{\partial x}(\omega x) - \dfrac{\partial}{\partial y}(-\omega y)\right]\mathbf{k}$

$= [\omega - (-\omega)]\mathbf{k} = 2\omega\,\mathbf{k} = 2\mathbf{w}$

37. For any continuous function f on $\mathbb{R}^3$, define a vector field $\mathbf{G}(x, y, z) = \langle g(x, y, z), 0, 0\rangle$ where $g(x, y, z) = \int_0^x f(t, y, z)\,dt$.

Then $\operatorname{div}\mathbf{G} = \dfrac{\partial}{\partial x}(g(x, y, z)) + \dfrac{\partial}{\partial y}(0) + \dfrac{\partial}{\partial z}(0) = \dfrac{\partial}{\partial x}\int_0^x f(t, y, z)\,dt = f(x, y, z)$ by the Fundamental Theorem of

Calculus. Thus every continuous function f on $\mathbb{R}^3$ is the divergence of some vector field.

13.6 Surface Integrals

1. Each face of the cube has surface area $2^2 = 4$, and the points P_{ij}^* are the points where the cube intersects the coordinate axes. Here, $f(x, y, z) = \sqrt{x^2 + 2y^2 + 3z^2}$, so by Definition 1,

$$\iint_S f(x, y, z)\, dS \approx [f(1, 0, 0)](4) + [f(-1, 0, 0)](4) + [f(0, 1, 0)](4) + [f(0, -1, 0)](4)$$
$$+ [f(0, 0, 1)](4) + [f(0, 0, -1)](4)$$
$$= 4\left(1 + 1 + 2\sqrt{2} + 2\sqrt{3}\right) = 8\left(1 + \sqrt{2} + \sqrt{3}\right) \approx 33.170$$

3. We can use the xz- and yz-planes to divide H into four patches of equal size, each with surface area equal to $\frac{1}{8}$ the surface area of a sphere with radius $\sqrt{50}$, so $\Delta S = \frac{1}{8}(4)\pi\left(\sqrt{50}\right)^2 = 25\pi$. Then $(\pm 3, \pm 4, 5)$ are sample points in the four patches, and using a Riemann sum as in Definition 1, we have

$$\iint_H f(x, y, z)\, dS \approx f(3, 4, 5)\, \Delta S + f(3, -4, 5)\, \Delta S + f(-3, 4, 5)\, \Delta S + f(-3, -4, 5)\, \Delta S$$
$$= (7 + 8 + 9 + 12)(25\pi) = 900\pi \approx 2827$$

5. $\mathbf{r}(u, v) = u^2\,\mathbf{i} + u\sin v\,\mathbf{j} + u\cos v\,\mathbf{k}$, $0 \le u \le 1$, $0 \le v \le \pi/2$ and

$\mathbf{r}_u \times \mathbf{r}_v = (2u\,\mathbf{i} + \sin v\,\mathbf{j} + \cos v\,\mathbf{k}) \times (u\cos v\,\mathbf{j} - u\sin v\,\mathbf{k}) = -u\,\mathbf{i} + 2u^2\sin v\,\mathbf{j} + 2u^2\cos v\,\mathbf{k}$ and

$|\mathbf{r}_u \times \mathbf{r}_v| = \sqrt{u^2 + 4u^4\sin^2 v + 4u^4\cos^2 v} = \sqrt{u^2 + 4u^4(\sin^2 v + \cos^2 v)} = u\sqrt{1 + 4u^2}$ (since $u \ge 0$). Then

$$\iint_S yz\, dS = \int_0^{\pi/2}\int_0^1 (u\sin v)(u\cos v) \cdot u\sqrt{1 + 4u^2}\, du\, dv = \int_0^1 u^3\sqrt{1 + 4u^2}\, du \int_0^{\pi/2} \sin v\cos v\, dv$$

$$\left[\text{let } t = 1 + 4u^2 \quad \Rightarrow \quad u^2 = \tfrac{1}{4}(t - 1) \text{ and } \tfrac{1}{8}\, dt = u\, du\right]$$

$$= \int_1^5 \tfrac{1}{8} \cdot \tfrac{1}{4}(t - 1)\sqrt{t}\, dt \int_0^{\pi/2} \sin v\cos v\, dv = \tfrac{1}{32}\int_1^5 \left(t^{3/2} - \sqrt{t}\right) dt \int_0^{\pi/2} \sin v\cos v\, dv$$

$$= \tfrac{1}{32}\left[\tfrac{2}{5}t^{5/2} - \tfrac{2}{3}t^{3/2}\right]_1^5 \left[\tfrac{1}{2}\sin^2 v\right]_0^{\pi/2} = \tfrac{1}{32}\left(\tfrac{2}{5}(5)^{5/2} - \tfrac{2}{3}(5)^{3/2} - \tfrac{2}{5} + \tfrac{2}{3}\right) \cdot \tfrac{1}{2}(1 - 0)$$

$$= \tfrac{5}{48}\sqrt{5} + \tfrac{1}{240}$$

7. $z = 1 + 2x + 3y$ so $\dfrac{\partial z}{\partial x} = 2$ and $\dfrac{\partial z}{\partial y} = 3$. Then by Formula 4,

$$\iint_S x^2 yz\, dS = \iint_D x^2 yz\sqrt{\left(\frac{\partial z}{\partial x}\right)^2 + \left(\frac{\partial z}{\partial y}\right)^2 + 1}\, dA = \int_0^3\int_0^2 x^2 y(1 + 2x + 3y)\sqrt{4 + 9 + 1}\, dy\, dx$$

$$= \sqrt{14}\int_0^3\int_0^2 (x^2 y + 2x^3 y + 3x^2 y^2)\, dy\, dx = \sqrt{14}\int_0^3 \left[\tfrac{1}{2}x^2 y^2 + x^3 y^2 + x^2 y^3\right]_{y=0}^{y=2}\, dx$$

$$= \sqrt{14}\int_0^3 (10x^2 + 4x^3)\, dx = \sqrt{14}\left[\tfrac{10}{3}x^3 + x^4\right]_0^3 = 171\sqrt{14}$$

9. S is the part of the plane $z = 1 - x - y$ over the region $D = \{(x, y) \mid 0 \le x \le 1, 0 \le y \le 1 - x\}$. Thus
$$\iint_S yz\, dS = \iint_D y(1 - x - y)\sqrt{(-1)^2 + (-1)^2 + 1}\, dA = \sqrt{3}\int_0^1\int_0^{1-x} (y - xy - y^2)\, dy\, dx$$

$$= \sqrt{3}\int_0^1 \left[\tfrac{1}{2}y^2 - \tfrac{1}{2}xy^2 - \tfrac{1}{3}y^3\right]_{y=0}^{y=1-x}\, dx = \sqrt{3}\int_0^1 \tfrac{1}{6}(1 - x)^3\, dx = -\tfrac{\sqrt{3}}{24}(1 - x)^4\Big|_0^1 = \tfrac{\sqrt{3}}{24}$$

11. S is the portion of the cone $z^2 = x^2 + y^2$ for $1 \le z \le 3$, or equivalently, S is the part of the surface $z = \sqrt{x^2 + y^2}$ over the region $D = \{(x, y) \mid 1 \le x^2 + y^2 \le 9\}$. Thus

$$\iint_S x^2 z^2\, dS = \iint_D x^2(x^2 + y^2)\sqrt{\left(\frac{x}{\sqrt{x^2 + y^2}}\right)^2 + \left(\frac{y}{\sqrt{x^2 + y^2}}\right)^2 + 1}\, dA$$

$$= \iint_D x^2(x^2 + y^2)\sqrt{\frac{x^2 + y^2}{x^2 + y^2} + 1}\, dA = \iint_D \sqrt{2}\, x^2(x^2 + y^2)\, dA = \sqrt{2}\int_0^{2\pi}\int_1^3 (r\cos\theta)^2(r^2)\, r\, dr\, d\theta$$

$$= \sqrt{2}\int_0^{2\pi} \cos^2\theta\, d\theta \int_1^3 r^5\, dr = \sqrt{2}\left[\tfrac{1}{2}\theta + \tfrac{1}{4}\sin 2\theta\right]_0^{2\pi}\left[\tfrac{1}{6}r^6\right]_1^3 = \sqrt{2}\,(\pi) \cdot \tfrac{1}{6}(3^6 - 1) = \frac{364\sqrt{2}}{3}\pi$$

13. Using x and z as parameters, we have $\mathbf{r}(x, z) = x\,\mathbf{i} + (x^2 + z^2)\,\mathbf{j} + z\,\mathbf{k}$, $x^2 + z^2 \leq 4$. Then

$\mathbf{r}_x \times \mathbf{r}_z = (\mathbf{i} + 2x\,\mathbf{j}) \times (2z\,\mathbf{j} + \mathbf{k}) = 2x\,\mathbf{i} - \mathbf{j} + 2z\,\mathbf{k}$ and $|\mathbf{r}_x \times \mathbf{r}_z| = \sqrt{4x^2 + 1 + 4z^2} = \sqrt{1 + 4(x^2 + z^2)}$. Thus

$$\iint_S y\,dS = \iint_{x^2 + z^2 \leq 4} (x^2 + z^2)\sqrt{1 + 4(x^2 + z^2)}\,dA = \int_0^{2\pi}\int_0^2 r^2\sqrt{1 + 4r^2}\,r\,dr\,d\theta$$

$$= \int_0^{2\pi} d\theta \int_0^2 r^2\sqrt{1 + 4r^2}\,r\,dr = 2\pi\int_0^2 r^2\sqrt{1 + 4r^2}\,r\,dr$$

$$\text{[let } u = 1 + 4r^2 \;\Rightarrow\; r^2 = \tfrac{1}{4}(u - 1) \text{ and } \tfrac{1}{8}\,du = r\,dr]$$

$$= 2\pi\int_1^{17} \tfrac{1}{4}(u - 1)\sqrt{u} \cdot \tfrac{1}{8}\,du = \tfrac{1}{16}\pi\int_1^{17}(u^{3/2} - u^{1/2})\,du$$

$$= \tfrac{1}{16}\pi\left[\tfrac{2}{5}u^{5/2} - \tfrac{2}{3}u^{3/2}\right]_1^{17} = \tfrac{1}{16}\pi\left[\tfrac{2}{5}(17)^{5/2} - \tfrac{2}{3}(17)^{3/2} - \tfrac{2}{5} + \tfrac{2}{3}\right] = \frac{\pi}{60}(391\sqrt{17} + 1)$$

15. Using spherical coordinates and Example 12.6.1 we have $\mathbf{r}(\phi, \theta) = 2\sin\phi\cos\theta\,\mathbf{i} + 2\sin\phi\sin\theta\,\mathbf{j} + 2\cos\phi\,\mathbf{k}$ and

$|\mathbf{r}_\phi \times \mathbf{r}_\theta| = 4\sin\phi$. Then $\iint_S(x^2 z + y^2 z)\,dS = \int_0^{2\pi}\int_0^{\pi/2}(4\sin^2\phi)(2\cos\phi)(4\sin\phi)\,d\phi\,d\theta = 16\pi\sin^4\phi\Big]_0^{\pi/2} = 16\pi$.

17. Using cylindrical coordinates, we have $\mathbf{r}(\theta, z) = 3\cos\theta\,\mathbf{i} + 3\sin\theta\,\mathbf{j} + z\,\mathbf{k}$, $0 \leq \theta \leq 2\pi$, $0 \leq z \leq 2$,

and $|\mathbf{r}_\theta \times \mathbf{r}_z| = 3$.

$$\iint_S(x^2 y + z^2)\,dS = \int_0^{2\pi}\int_0^2(27\cos^2\theta\sin\theta + z^2)\,3\,dz\,d\theta = \int_0^{2\pi}(162\cos^2\theta\sin\theta + 8)\,d\theta = 16\pi$$

19. $\mathbf{F}(x, y, z) = xy\,\mathbf{i} + yz\,\mathbf{j} + zx\,\mathbf{k}$, $z = g(x, y) = 4 - x^2 - y^2$, and D is the square $[0, 1] \times [0, 1]$, so by Equation 10

$$\iint_S \mathbf{F} \cdot d\mathbf{S} = \iint_D[-xy(-2x) - yz(-2y) + zx]\,dA = \int_0^1\int_0^1[2x^2 y + 2y^2(4 - x^2 - y^2) + x(4 - x^2 - y^2)]\,dy\,dx$$

$$= \int_0^1\left(\tfrac{1}{3}x^2 + \tfrac{11}{3}x - x^3 + \tfrac{34}{15}\right)dx = \frac{713}{180}$$

21. $\mathbf{F}(x, y, z) = xze^y\,\mathbf{i} - xze^y\,\mathbf{j} + z\,\mathbf{k}$, $z = g(x, y) = 1 - x - y$, and $D = \{(x, y) \mid 0 \leq x \leq 1, 0 \leq y \leq 1 - x\}$. Since S has downward orientation, we have

$$\iint_S \mathbf{F} \cdot d\mathbf{S} = -\iint_D[-xze^y(-1) - (-xze^y)(-1) + z]\,dA = -\int_0^1\int_0^{1-x}(1 - x - y)\,dy\,dx$$

$$= -\int_0^1\left(\tfrac{1}{2}x^2 - x + \tfrac{1}{2}\right)dx = -\tfrac{1}{6}$$

23. $\mathbf{F}(x, y, z) = x\,\mathbf{i} - z\,\mathbf{j} + y\,\mathbf{k}$, $z = g(x, y) = \sqrt{4 - x^2 - y^2}$ and D is the quarter disk

$\{(x, y) \mid 0 \leq x \leq 2, 0 \leq y \leq \sqrt{4 - x^2}\}$. S has downward orientation, so by Formula 10,

$$\iint_S \mathbf{F} \cdot d\mathbf{S} = -\iint_D\left[-x \cdot \tfrac{1}{2}(4 - x^2 - y^2)^{-1/2}(-2x) - (-z) \cdot \tfrac{1}{2}(4 - x^2 - y^2)^{-1/2}(-2y) + y\right]dA$$

$$= -\iint_D\left(\frac{x^2}{\sqrt{4 - x^2 - y^2}} - \sqrt{4 - x^2 - y^2} \cdot \frac{y}{\sqrt{4 - x^2 - y^2}} + y\right)dA$$

$$= -\iint_D x^2(4 - (x^2 + y^2))^{-1/2}\,dA = -\int_0^{\pi/2}\int_0^2(r\cos\theta)^2(4 - r^2)^{-1/2}\,r\,dr\,d\theta$$

$$= -\int_0^{\pi/2}\cos^2\theta\,d\theta\int_0^2 r^3(4 - r^2)^{-1/2}\,dr \quad \text{[let } u = 4 - r^2 \;\Rightarrow\; r^2 = 4 - u \text{ and } -\tfrac{1}{2}\,du = r\,dr]$$

$$= -\int_0^{\pi/2}\left(\tfrac{1}{2} + \tfrac{1}{2}\cos 2\theta\right)d\theta\int_4^0 -\tfrac{1}{2}(4 - u)(u)^{-1/2}\,du$$

$$= -\left[\tfrac{1}{2}\theta + \tfrac{1}{4}\sin 2\theta\right]_0^{\pi/2}\left(-\tfrac{1}{2}\right)\left[8\sqrt{u} - \tfrac{2}{3}u^{3/2}\right]_4^0 = -\tfrac{\pi}{4}\left(-\tfrac{1}{2}\right)\left(-16 + \tfrac{16}{3}\right) = -\tfrac{4}{3}\pi$$

25. Let S_1 be the paraboloid $y = x^2 + z^2$, $0 \leq y \leq 1$ and S_2 the disk $x^2 + z^2 \leq 1$, $y = 1$. Since S is a closed surface, we use the outward orientation. On S_1: $\mathbf{F}(\mathbf{r}(x, z)) = (x^2 + z^2)\,\mathbf{j} - z\,\mathbf{k}$ and $\mathbf{r}_x \times \mathbf{r}_z = 2x\,\mathbf{i} - \mathbf{j} + 2z\,\mathbf{k}$ (since the $\mathbf{j}$-component must be negative on S_1). Then

$$\iint_{S_1} \mathbf{F} \cdot d\mathbf{S} = \iint_{x^2 + z^2 \leq 1}[-(x^2 + z^2) - 2z^2]\,dA = -\int_0^{2\pi}\int_0^1(r^2 + 2r^2\cos^2\theta)\,r\,dr\,d\theta$$

$$= -\int_0^{2\pi}\tfrac{1}{4}(1 + 2\cos^2\theta)\,d\theta = -\left(\tfrac{\pi}{2} + \tfrac{\pi}{2}\right) = -\pi$$

On S_2: $\mathbf{F}(\mathbf{r}(x, z)) = \mathbf{j} - z\,\mathbf{k}$ and $\mathbf{r}_z \times \mathbf{r}_x = \mathbf{j}$. Then $\iint_{S_2} \mathbf{F} \cdot d\mathbf{S} = \iint_{x^2 + z^2 \leq 1}(1)\,dA = \pi$. Hence $\iint_S \mathbf{F} \cdot d\mathbf{S} = -\pi + \pi = 0$.

27. Here S consists of four surfaces: S_1, the top surface (a portion of the circular cylinder $y^2 + z^2 = 1$); S_2, the bottom surface (a portion of the xy-plane); S_3, the front half-disk in the plane $x = 2$, and S_4, the back half-disk in the plane $x = 0$.

On S_1: The surface is $z = \sqrt{1 - y^2}$ for $0 \le x \le 2$, $-1 \le y \le 1$ with upward orientation, so

$$\iint_{S_1} \mathbf{F} \cdot d\mathbf{S} = \int_0^2 \int_{-1}^1 \left[-x^2 (0) - y^2 \left(-\frac{y}{\sqrt{1-y^2}} \right) + z^2 \right] dy \, dx = \int_0^2 \int_{-1}^1 \left(\frac{y^3}{\sqrt{1-y^2}} + 1 - y^2 \right) dy \, dx$$

$$= \int_0^2 \left[-\sqrt{1-y^2} + \tfrac{1}{3}(1-y^2)^{3/2} + y - \tfrac{1}{3}y^3 \right]_{y=-1}^{y=1} dx = \int_0^2 \tfrac{4}{3} \, dx = \tfrac{8}{3}$$

On S_2: The surface is $z = 0$ with downward orientation, so

$$\iint_{S_2} \mathbf{F} \cdot d\mathbf{S} = \int_0^2 \int_{-1}^1 (-z^2) \, dy \, dx = \int_0^2 \int_{-1}^1 (0) \, dy \, dx = 0$$

On S_3: The surface is $x = 2$ for $-1 \le y \le 1$, $0 \le z \le \sqrt{1-y^2}$, oriented in the positive x-direction. Regarding y and z as parameters, we have $\mathbf{r}_y \times \mathbf{r}_z = \mathbf{i}$ and

$$\iint_{S_3} \mathbf{F} \cdot d\mathbf{S} = \int_{-1}^1 \int_0^{\sqrt{1-y^2}} x^2 \, dz \, dy = \int_{-1}^1 \int_0^{\sqrt{1-y^2}} 4 \, dz \, dy = 4A(S_3) = 2\pi$$

On S_4: The surface is $x = 0$ for $-1 \le y \le 1$, $0 \le z \le \sqrt{1-y^2}$, oriented in the negative x-direction. Regarding y and z as parameters, we use $-(\mathbf{r}_y \times \mathbf{r}_z) = -\mathbf{i}$ and

$$\iint_{S_4} \mathbf{F} \cdot d\mathbf{S} = \int_{-1}^1 \int_0^{\sqrt{1-y^2}} x^2 \, dz \, dy = \int_{-1}^1 \int_0^{\sqrt{1-y^2}} (0) \, dz \, dy = 0$$

Thus $\iint_S \mathbf{F} \cdot d\mathbf{S} = \tfrac{8}{3} + 0 + 2\pi + 0 = 2\pi + \tfrac{8}{3}$.

29. We use Formula 4 with $z = 3 - 2x^2 - y^2 \;\Rightarrow\; \partial z/\partial x = -4x$, $\partial z/\partial y = -2y$. The boundaries of the region $3 - 2x^2 - y^2 \ge 0$ are $-\sqrt{\tfrac{3}{2}} \le x \le \sqrt{\tfrac{3}{2}}$ and $-\sqrt{3-2x^2} \le y \le \sqrt{3-2x^2}$, so we use a CAS (with precision reduced to seven or fewer digits; otherwise the calculation takes a very long time) to calculate

$$\iint_S x^2 y^2 z^2 \, dS = \int_{-\sqrt{3/2}}^{\sqrt{3/2}} \int_{-\sqrt{3-2x^2}}^{\sqrt{3-2x^2}} x^2 y^2 (3 - 2x^2 - y^2)^2 \sqrt{16x^2 + 4y^2 + 1} \, dy \, dx \approx 3.4895$$

31. If S is given by $y = h(x, z)$, then S is also the level surface $f(x, y, z) = y - h(x, z) = 0$.

$$\mathbf{n} = \frac{\nabla f(x, y, z)}{|\nabla f(x, y, z)|} = \frac{-h_x \, \mathbf{i} + \mathbf{j} - h_z \, \mathbf{k}}{\sqrt{h_x^2 + 1 + h_z^2}}, \text{ and } -\mathbf{n} \text{ is the unit normal that points to the left. Now we proceed as in the}$$

derivation of (10), using Formula 4 to evaluate

$$\iint_S \mathbf{F} \cdot d\mathbf{S} = \iint_S \mathbf{F} \cdot \mathbf{n} \, dS = \iint_D (P\mathbf{i} + Q\mathbf{j} + R\mathbf{k}) \, \frac{-\dfrac{\partial h}{\partial x} \mathbf{i} - \mathbf{j} + \dfrac{\partial h}{\partial z} \mathbf{k}}{\sqrt{\left(\dfrac{\partial h}{\partial x}\right)^2 + 1 + \left(\dfrac{\partial h}{\partial z}\right)^2}} \sqrt{\left(\dfrac{\partial h}{\partial x}\right)^2 + 1 + \left(\dfrac{\partial h}{\partial z}\right)^2} \, dA$$

where D is the projection of $f(x, y, z)$ onto the xz-plane. Therefore

$$\iint_S \mathbf{F} \cdot d\mathbf{S} = \iint_D \left(P \frac{\partial h}{\partial x} - Q + R \frac{\partial h}{\partial z} \right) dA$$

33. $m = \iint_S K \, dS = K \cdot 4\pi\left(\tfrac{1}{2}a^2\right) = 2\pi a^2 K$; by symmetry $M_{xz} = M_{yz} = 0$, and

$M_{xy} = \iint_S zK \, dS = K \int_0^{2\pi} \int_0^{\pi/2} (a \cos \phi)(a^2 \sin \phi) \, d\phi \, d\theta = 2\pi K a^3 \left[-\tfrac{1}{4} \cos 2\phi\right]_0^{\pi/2} = \pi K a^3$.

Hence $(\overline{x}, \overline{y}, \overline{z}) = \left(0, 0, \tfrac{1}{2}a\right)$.

35. (a) $I_z = \iint_S (x^2 + y^2)\rho(x, y, z) \, dS$

(b) $I_z = \iint_S (x^2 + y^2)\left(10 - \sqrt{x^2 + y^2}\right) dS = \iint_{1 \le x^2 + y^2 \le 16} (x^2 + y^2)\left(10 - \sqrt{x^2 + y^2}\right) \sqrt{2} \, dA$

$$= \int_0^{2\pi} \int_1^4 \sqrt{2}\,(10r^3 - r^4) \, dr \, d\theta = 2\sqrt{2}\,\pi\left(\tfrac{4329}{10}\right) = \tfrac{4329}{5}\sqrt{2}\,\pi$$

37. The rate of flow through the cylinder is the flux $\iint_S \rho \mathbf{v} \cdot \mathbf{n}\, dS = \iint_S \rho \mathbf{v} \cdot d\mathbf{S}$. We use the parametric representation $\mathbf{r}(u, v) = 2\cos u\, \mathbf{i} + 2\sin u\, \mathbf{j} + v\, \mathbf{k}$ for S, where $0 \le u \le 2\pi$, $0 \le v \le 1$, so $\mathbf{r}_u = -2\sin u\, \mathbf{i} + 2\cos u\, \mathbf{j}$, $\mathbf{r}_v = \mathbf{k}$, and the outward orientation is given by $\mathbf{r}_u \times \mathbf{r}_v = 2\cos u\, \mathbf{i} + 2\sin u\, \mathbf{j}$. Then

$$\iint_S \rho \mathbf{v} \cdot d\mathbf{S} = \rho \int_0^{2\pi} \int_0^1 \left(v\, \mathbf{i} + 4\sin^2 u\, \mathbf{j} + 4\cos^2 u\, \mathbf{k}\right) \cdot (2\cos u\, \mathbf{i} + 2\sin u\, \mathbf{j})\, dv\, du$$

$$= \rho \int_0^{2\pi} \int_0^1 \left(2v\cos u + 8\sin^3 u\right) dv\, du = \rho \int_0^{2\pi} \left(\cos u + 8\sin^3 u\right) du$$

$$= \rho\left[\sin u + 8\left(-\tfrac{1}{3}\right)\left(2 + \sin^2 u\right)\cos u\right]_0^{2\pi} = 0 \text{ kg/s}$$

39. S consists of the hemisphere S_1 given by $z = \sqrt{a^2 - x^2 - y^2}$ and the disk S_2 given by $0 \le x^2 + y^2 \le a^2$, $z = 0$. On S_1: $\mathbf{E} = a\sin\phi\cos\theta\, \mathbf{i} + a\sin\phi\sin\theta\, \mathbf{j} + 2a\cos\phi\, \mathbf{k}$, $\mathbf{T}_\phi \times \mathbf{T}_\theta = a^2\sin^2\phi\cos\theta\, \mathbf{i} + a^2\sin^2\phi\sin\theta\, \mathbf{j} + a^2\sin\phi\cos\phi\, \mathbf{k}$. Thus

$$\iint_{S_1} \mathbf{E} \cdot d\mathbf{S} = \int_0^{2\pi} \int_0^{\pi/2}(a^3\sin^3\phi + 2a^3\sin\phi\cos^2\phi)\, d\phi\, d\theta$$

$$= \int_0^{2\pi} \int_0^{\pi/2}(a^3\sin\phi + a^3\sin\phi\cos^2\phi)\, d\phi\, d\theta = (2\pi)a^3\left(1 + \tfrac{1}{3}\right) = \tfrac{8}{3}\pi a^3$$

On S_2: $\mathbf{E} = x\, \mathbf{i} + y\, \mathbf{j}$, and $\mathbf{r}_y \times \mathbf{r}_x = -\mathbf{k}$ so $\iint_{S_2} \mathbf{E} \cdot d\mathbf{S} = 0$. Hence the total charge is $q = \varepsilon_0 \iint_S \mathbf{E} \cdot d\mathbf{S} = \tfrac{8}{3}\pi a^3 \varepsilon_0$.

41. $K\nabla u = 6.5(4y\, \mathbf{j} + 4z\, \mathbf{k})$. S is given by $\mathbf{r}(x, \theta) = x\, \mathbf{i} + \sqrt{6}\,\cos\theta\, \mathbf{j} + \sqrt{6}\,\sin\theta\, \mathbf{k}$ and since we want the inward heat flow, we use $\mathbf{r}_x \times \mathbf{r}_\theta = -\sqrt{6}\,\cos\theta\, \mathbf{j} - \sqrt{6}\,\sin\theta\, \mathbf{k}$. Then the rate of heat flow inward is given by

$$\iint_S (-K\nabla u) \cdot d\mathbf{S} = \int_0^{2\pi} \int_0^4 -(6.5)(-24)\, dx\, d\theta = (2\pi)(156)(4) = 1248\pi.$$

43. Let S be a sphere of radius a centered at the origin. Then $|\mathbf{r}| = a$ and $\mathbf{F}(\mathbf{r}) = c\mathbf{r}/|\mathbf{r}|^3 = (c/a^3)(x\, \mathbf{i} + y\, \mathbf{j} + z\, \mathbf{k})$. A parametric representation for S is $\mathbf{r}(\phi, \theta) = a\sin\phi\cos\theta\, \mathbf{i} + a\sin\phi\sin\theta\, \mathbf{j} + a\cos\phi\, \mathbf{k}$, $0 \le \phi \le \pi$, $0 \le \theta \le 2\pi$. Then $\mathbf{r}_\phi = a\cos\phi\cos\theta\, \mathbf{i} + a\cos\phi\sin\theta\, \mathbf{j} - a\sin\phi\, \mathbf{k}$, $\mathbf{r}_\theta = -a\sin\phi\sin\theta\, \mathbf{i} + a\sin\phi\cos\theta\, \mathbf{j}$, and the outward orientation is given by $\mathbf{r}_\phi \times \mathbf{r}_\theta = a^2\sin^2\phi\cos\theta\, \mathbf{i} + a^2\sin^2\phi\sin\theta\, \mathbf{j} + a^2\sin\phi\cos\phi\, \mathbf{k}$. The flux of $\mathbf{F}$ across S is

$$\iint_S \mathbf{F} \cdot d\mathbf{S} = \int_0^\pi \int_0^{2\pi} \frac{c}{a^3}\left(a\sin\phi\cos\theta\, \mathbf{i} + a\sin\phi\sin\theta\, \mathbf{j} + a\cos\phi\, \mathbf{k}\right)$$

$$\cdot \left(a^2\sin^2\phi\cos\theta\, \mathbf{i} + a^2\sin^2\phi\sin\theta\, \mathbf{j} + a^2\sin\phi\cos\phi\, \mathbf{k}\right) d\theta\, d\phi$$

$$= \frac{c}{a^3} \int_0^\pi \int_0^{2\pi} a^3\left(\sin^3\phi + \sin\phi\cos^2\phi\right) d\theta\, d\phi = c \int_0^\pi \int_0^{2\pi} \sin\phi\, d\theta\, d\phi = 4\pi c$$

Thus the flux does not depend on the radius a.

13.7 Stokes' Theorem

1. Both H and P are oriented piecewise-smooth surfaces that are bounded by the simple, closed, smooth curve $x^2 + y^2 = 4$, $z = 0$ (which we can take to be oriented positively for both surfaces). Then H and P satisfy the hypotheses of Stokes' Theorem, so by (3) we know $\iint_H \operatorname{curl} \mathbf{F} \cdot d\mathbf{S} = \int_C \mathbf{F} \cdot d\mathbf{r} = \iint_P \operatorname{curl} \mathbf{F} \cdot d\mathbf{S}$ (where C is the boundary curve).

3. The boundary curve C is the circle $x^2 + y^2 = 4$, $z = 0$ oriented in the counterclockwise direction. The vector equation is $\mathbf{r}(t) = 2\cos t\, \mathbf{i} + 2\sin t\, \mathbf{j}$, $0 \le t \le 2\pi$, so $\mathbf{r}'(t) = -2\sin t\, \mathbf{i} + 2\cos t\, \mathbf{j}$ and $\mathbf{F}(\mathbf{r}(t)) = (2\cos t)^2 e^{(2\sin t)(0)}\, \mathbf{i} + (2\sin t)^2 e^{(2\cos t)(0)}\, \mathbf{j} + (0)^2 e^{(2\cos t)(2\sin t)}\, \mathbf{k} = 4\cos^2 t\, \mathbf{i} + 4\sin^2 t\, \mathbf{j}$. Then, by Stokes' Theorem,

$$\iint_S \operatorname{curl} \mathbf{F} \cdot d\mathbf{S} = \int_C \mathbf{F} \cdot d\mathbf{r} = \int_0^{2\pi} \mathbf{F}(\mathbf{r}(t)) \cdot \mathbf{r}'(t)\, dt = \int_0^{2\pi}(-8\cos^2 t\sin t + 8\sin^2 t\cos t)\, dt$$

$$= 8\left[\tfrac{1}{3}\cos^3 t + \tfrac{1}{3}\sin^3 t\right]_0^{2\pi} = 0$$

5. C is the square in the plane $z = -1$. By (3), $\iint_{S_1} \text{curl } \mathbf{F} \cdot d\mathbf{S} = \oint_C \mathbf{F} \cdot d\mathbf{r} = \iint_{S_2} \text{curl } \mathbf{F} \cdot d\mathbf{S}$ where S_1 is the original cube

without the bottom and S_2 is the bottom face of the cube. $\text{curl } \mathbf{F} = x^2 z \, \mathbf{i} + (xy - 2xyz) \, \mathbf{j} + (y - xz) \, \mathbf{k}$. For S_2, we choose

$\mathbf{n} = \mathbf{k}$ so that C has the same orientation for both surfaces. Then $\text{curl } \mathbf{F} \cdot \mathbf{n} = y - xz = x + y$ on S_2, where $z = -1$. Thus

$\iint_{S_2} \text{curl } \mathbf{F} \cdot d\mathbf{S} = \int_{-1}^{1} \int_{-1}^{1} (x + y) \, dx \, dy = 0$ so $\iint_{S_1} \text{curl } \mathbf{F} \cdot d\mathbf{S} = 0$.

7. $\text{curl } \mathbf{F} = -2z \, \mathbf{i} - 2x \, \mathbf{j} - 2y \, \mathbf{k}$ and we take the surface S to be the planar region enclosed by C, so S is the portion of the plane

$x + y + z = 1$ over $D = \{(x, y) \mid 0 \le x \le 1, 0 \le y \le 1 - x\}$. Since C is oriented counterclockwise, we orient S upward.

Using Equation 13.6.10, we have $z = g(x, y) = 1 - x - y$, $P = -2z$, $Q = -2x$, $R = -2y$, and

$$\int_C \mathbf{F} \cdot d\mathbf{r} = \iint_S \text{curl } \mathbf{F} \cdot d\mathbf{S} = \iint_D [-(-2z)(-1) - (-2x)(-1) + (-2y)] \, dA$$
$$= \int_0^1 \int_0^{1-x} (-2) \, dy \, dx = -2 \int_0^1 (1 - x) \, dx = -1$$

9. $\text{curl } \mathbf{F} = (xe^{xy} - 2x) \, \mathbf{i} - (ye^{xy} - y) \, \mathbf{j} + (2z - z) \, \mathbf{k}$ and we take S to be the disk $x^2 + y^2 \le 16$, $z = 5$. Since C is oriented

counterclockwise (from above), we orient S upward. Then $\mathbf{n} = \mathbf{k}$ and $\text{curl } \mathbf{F} \cdot \mathbf{n} = 2z - z$ on S, where $z = 5$. Thus

$$\oint \mathbf{F} \cdot d\mathbf{r} = \iint_S \text{curl } \mathbf{F} \cdot \mathbf{n} \, dS = \iint_S (2z - z) \, dS = \iint_S (10 - 5) \, dS = 5(\text{area of } S) = 5(\pi \cdot 4^2) = 80\pi$$

11. (a) The curve of intersection is an ellipse in the plane $x + y + z = 1$ with unit normal $\mathbf{n} = \frac{1}{\sqrt{3}}(\mathbf{i} + \mathbf{j} + \mathbf{k})$,

$\text{curl } \mathbf{F} = x^2 \, \mathbf{j} + y^2 \, \mathbf{k}$, and $\text{curl } \mathbf{F} \cdot \mathbf{n} = \frac{1}{\sqrt{3}}(x^2 + y^2)$. Then

$$\oint_C \mathbf{F} \cdot d\mathbf{r} = \iint_S \frac{1}{\sqrt{3}}(x^2 + y^2) \, dS = \iint_{x^2 + y^2 \le 9} (x^2 + y^2) \, dx \, dy = \int_0^{2\pi} \int_0^3 r^3 \, dr \, d\theta = 2\pi \left(\frac{81}{4}\right) = \frac{81\pi}{2}$$

(b)

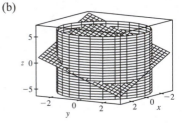

(c) One possible parametrization is $x = 3 \cos t$, $y = 3 \sin t$,

$z = 1 - 3 \cos t - 3 \sin t$, $0 \le t \le 2\pi$.

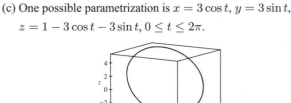

13. The boundary curve C is the circle $x^2 + y^2 = 1$, $z = 1$ oriented in the counterclockwise direction as viewed from above.

We can parametrize C by $\mathbf{r}(t) = \cos t \, \mathbf{i} + \sin t \, \mathbf{j} + \mathbf{k}$, $0 \le t \le 2\pi$, and then $\mathbf{r}'(t) = -\sin t \, \mathbf{i} + \cos t \, \mathbf{j}$. Thus

$\mathbf{F}(\mathbf{r}(t)) = \sin^2 t \, \mathbf{i} + \cos t \, \mathbf{j} + \mathbf{k}$, $\mathbf{F}(\mathbf{r}(t)) \cdot \mathbf{r}'(t) = \cos^2 t - \sin^3 t$, and

$$\oint_C \mathbf{F} \cdot d\mathbf{r} = \int_0^{2\pi} (\cos^2 t - \sin^3 t) \, dt = \int_0^{2\pi} \tfrac{1}{2}(1 + \cos 2t) \, dt - \int_0^{2\pi} (1 - \cos^2 t) \sin t \, dt$$

$$= \tfrac{1}{2} \left[t + \tfrac{1}{2} \sin 2t \right]_0^{2\pi} - \left[-\cos t + \tfrac{1}{3} \cos^3 t \right]_0^{2\pi} = \pi$$

Now $\text{curl } \mathbf{F} = (1 - 2y) \, \mathbf{k}$, and the projection D of S on the xy-plane is the disk $x^2 + y^2 \le 1$, so by

Equation 13.6.10 with $z = g(x, y) = x^2 + y^2$ we have

$$\iint_S \text{curl } \mathbf{F} \cdot d\mathbf{S} = \iint_D (1 - 2y) \, dA = \int_0^{2\pi} \int_0^1 (1 - 2r \sin \theta) \, r \, dr \, d\theta = \int_0^{2\pi} \left(\tfrac{1}{2} - \tfrac{2}{3} \sin \theta \right) d\theta = \pi$$

15. The boundary curve C is the circle $x^2 + z^2 = 1$, $y = 0$ oriented in the counterclockwise direction as viewed from the positive

y-axis. Then C can be described by $\mathbf{r}(t) = \cos t\,\mathbf{i} - \sin t\,\mathbf{k}$, $0 \le t \le 2\pi$, and $\mathbf{r}'(t) = -\sin t\,\mathbf{i} - \cos t\,\mathbf{k}$. Thus

$\mathbf{F}(\mathbf{r}(t)) = -\sin t\,\mathbf{j} + \cos t\,\mathbf{k}$, $\mathbf{F}(\mathbf{r}(t)) \cdot \mathbf{r}'(t) = -\cos^2 t$, and $\oint_C \mathbf{F} \cdot d\mathbf{r} = \int_0^{2\pi} -\cos^2 t\,dt = -\frac{1}{2}t - \frac{1}{4}\sin 2t\big]_0^{2\pi} = -\pi$.

Now curl $\mathbf{F} = -\mathbf{i} - \mathbf{j} - \mathbf{k}$, and S can be parametrized (see Example 12.6.1) by

$\mathbf{r}(\phi, \theta) = \sin\phi\cos\theta\,\mathbf{i} + \sin\phi\sin\theta\,\mathbf{j} + \cos\phi\,\mathbf{k}$, $0 \le \theta \le \pi$, $0 \le \phi \le \pi$. Then

$\mathbf{r}_\phi \times \mathbf{r}_\theta = \sin^2\phi\cos\theta\,\mathbf{i} + \sin^2\phi\sin\theta\,\mathbf{j} + \sin\phi\cos\phi\,\mathbf{k}$ and

$$\iint_S \operatorname{curl}\mathbf{F} \cdot d\mathbf{S} = \iint_{x^2+z^2 \le 1} \operatorname{curl}\mathbf{F} \cdot (\mathbf{r}_\phi \times \mathbf{r}_\theta)\,dA$$

$$= \int_0^\pi \int_0^\pi (-\sin^2\phi\cos\theta - \sin^2\phi\sin\theta - \sin\phi\cos\phi)\,d\theta\,d\phi$$

$$= \int_0^\pi (-2\sin^2\phi - \pi\sin\phi\cos\phi)\,d\phi = \left[\tfrac{1}{2}\sin 2\phi - \phi - \tfrac{\pi}{2}\sin^2\phi\right]_0^\pi = -\pi$$

17. It is easier to use Stokes' Theorem than to compute the work directly. Let S be the planar region enclosed by the path of the

particle, so S is the portion of the plane $z = \frac{1}{2}y$ for $0 \le x \le 1$, $0 \le y \le 2$, with upward orientation.

curl $\mathbf{F} = 8y\,\mathbf{i} + 2z\,\mathbf{j} + 2y\,\mathbf{k}$ and

$$\oint_C \mathbf{F} \cdot d\mathbf{r} = \iint_S \operatorname{curl}\mathbf{F} \cdot d\mathbf{S} = \iint_D \left[-8y(0) - 2z\left(\tfrac{1}{2}\right) + 2y\right] dA = \int_0^1 \int_0^2 \left(2y - \tfrac{1}{2}y\right) dy\,dx$$

$$= \int_0^1 \int_0^2 \tfrac{3}{2}y\,dy\,dx = \int_0^1 \left[\tfrac{3}{4}y^2\right]_{y=0}^{y=2} dx = \int_0^1 3\,dx = 3$$

19. Assume S is centered at the origin with radius a and let H_1 and H_2 be the upper and lower hemispheres, respectively, of S.

Then $\iint_S \operatorname{curl}\mathbf{F} \cdot d\mathbf{S} = \iint_{H_1} \operatorname{curl}\mathbf{F} \cdot d\mathbf{S} + \iint_{H_2} \operatorname{curl}\mathbf{F} \cdot d\mathbf{S} = \oint_{C_1} \mathbf{F} \cdot d\mathbf{r} + \oint_{C_2} \mathbf{F} \cdot d\mathbf{r}$ by Stokes' Theorem. But C_1 is the

circle $x^2 + y^2 = a^2$ oriented in the counterclockwise direction while C_2 is the same circle oriented in the clockwise direction.

Hence $\oint_{C_2} \mathbf{F} \cdot d\mathbf{r} = -\oint_{C_1} \mathbf{F} \cdot d\mathbf{r}$ so $\iint_S \operatorname{curl}\mathbf{F} \cdot d\mathbf{S} = 0$ as desired.

13.8 The Divergence Theorem

1. div $\mathbf{F} = 3 + x + 2x = 3 + 3x$, so

$\iiint_E \operatorname{div}\mathbf{F}\,dV = \int_0^1 \int_0^1 \int_0^1 (3x + 3)\,dx\,dy\,dz = \frac{9}{2}$ (notice the triple integral is

three times the volume of the cube plus three times $\bar{x}$).

To compute $\iint_S \mathbf{F} \cdot d\mathbf{S}$, on S_1: $\mathbf{n} = \mathbf{i}$, $\mathbf{F} = 3\mathbf{i} + y\mathbf{j} + 2z\mathbf{k}$, and

$\iint_{S_1} \mathbf{F} \cdot d\mathbf{S} = \iint_{S_1} 3\,dS = 3$;

S_2: $\mathbf{F} = 3x\,\mathbf{i} + x\,\mathbf{j} + 2xz\,\mathbf{k}$, $\mathbf{n} = \mathbf{j}$ and $\iint_{S_2} \mathbf{F} \cdot d\mathbf{S} = \iint_{S_2} x\,dS = \frac{1}{2}$;

S_3: $\mathbf{F} = 3x\,\mathbf{i} + xy\,\mathbf{j} + 2x\,\mathbf{k}$, $\mathbf{n} = \mathbf{k}$ and $\iint_{S_3} \mathbf{F} \cdot d\mathbf{S} = \iint_{S_3} 2x\,dS = 1$;

S_4: $\mathbf{F} = 0$, $\iint_{S_4} \mathbf{F} \cdot d\mathbf{S} = 0$; S_5: $\mathbf{F} = 3x\,\mathbf{i} + 2x\,\mathbf{k}$, $\mathbf{n} = -\mathbf{j}$ and $\iint_{S_5} \mathbf{F} \cdot d\mathbf{S} = \iint_{S_5} 0\,dS = 0$;

S_6: $\mathbf{F} = 3x\,\mathbf{i} + xy\,\mathbf{j}$, $\mathbf{n} = -\mathbf{k}$ and $\iint_{S_6} \mathbf{F} \cdot d\mathbf{S} = \iint_{S_6} 0\,dS = 0$. Thus $\iint_S \mathbf{F} \cdot d\mathbf{S} = \frac{9}{2}$.

3. div $\mathbf{F} = x + y + z$, so

$$\iiint_E \text{div } \mathbf{F} \, dV = \int_0^{2\pi} \int_0^1 \int_0^1 (r\cos\theta + r\sin\theta + z)\, r \, dz \, dr \, d\theta = \int_0^{2\pi} \int_0^1 \left(r^2\cos\theta + r^2\sin\theta + \tfrac{1}{2}r\right) dr \, d\theta$$

$$= \int_0^{2\pi} \left(\tfrac{1}{3}\cos\theta + \tfrac{1}{3}\sin\theta + \tfrac{1}{4}\right) d\theta = \tfrac{1}{4}(2\pi) = \tfrac{\pi}{2}$$

Let S_1 be the top of the cylinder, S_2 the bottom, and S_3 the vertical edge. On S_1, $z = 1$, $\mathbf{n} = \mathbf{k}$, and $\mathbf{F} = xy\,\mathbf{i} + y\,\mathbf{j} + x\,\mathbf{k}$, so

$$\iint_{S_1} \mathbf{F} \cdot d\mathbf{S} = \iint_{S_1} \mathbf{F} \cdot \mathbf{n}\, dS = \iint_{S_1} x \, dS = \int_0^{2\pi} \int_0^1 (r\cos\theta)\, r \, dr \, d\theta = \left[\sin\theta\right]_0^{2\pi} \left[\tfrac{1}{3}r^3\right]_0^1 = 0.$$ On S_2, $z = 0$, $\mathbf{n} = -\mathbf{k}$, and

$\mathbf{F} = xy\,\mathbf{i}$ so $\iint_{S_2} \mathbf{F} \cdot d\mathbf{S} = \iint_{S_2} 0 \, dS = 0$. S_3 is given by $\mathbf{r}(\theta, z) = \cos\theta\,\mathbf{i} + \sin\theta\,\mathbf{j} + z\,\mathbf{k}$, $0 \le \theta \le 2\pi$, $0 \le z \le 1$. Then

$\mathbf{r}_\theta \times \mathbf{r}_z = \cos\theta\,\mathbf{i} + \sin\theta\,\mathbf{j}$ and

$$\iint_{S_3} \mathbf{F} \cdot d\mathbf{S} = \iint_D \mathbf{F} \cdot (\mathbf{r}_\theta \times \mathbf{r}_z)\, dA = \int_0^{2\pi} \int_0^1 (\cos^2\theta\sin\theta + z\sin^2\theta)\, dz \, d\theta$$

$$= \int_0^{2\pi} \left(\cos^2\theta\sin\theta + \tfrac{1}{2}\sin^2\theta\right) d\theta = \left[-\tfrac{1}{3}\cos^3\theta + \tfrac{1}{4}\left(\theta - \tfrac{1}{2}\sin 2\theta\right)\right]_0^{2\pi} = \tfrac{\pi}{2}$$

Thus $\iint_S \mathbf{F} \cdot d\mathbf{S} = 0 + 0 + \tfrac{\pi}{2} = \tfrac{\pi}{2}$.

5. div $\mathbf{F} = \frac{\partial}{\partial x}(e^x\sin y) + \frac{\partial}{\partial y}(e^x\cos y) + \frac{\partial}{\partial z}(yz^2) = e^x\sin y - e^x\sin y + 2yz = 2yz$, so by the Divergence Theorem,

$$\iint_S \mathbf{F} \cdot d\mathbf{S} = \iiint_E \text{div } \mathbf{F} \, dV = \int_0^1 \int_0^1 \int_0^2 2yz \, dz \, dy \, dx = 2\int_0^1 dx \int_0^1 y \, dy \int_0^1 z \, dz$$

$$= 2\left[x\right]_0^1 \left[\tfrac{1}{2}y^2\right]_0^1 \left[\tfrac{1}{2}z^2\right]_0^2 = 2$$

7. div $\mathbf{F} = 3y^2 + 0 + 3z^2$, so using cylindrical coordinates with $y = r\cos\theta$, $z = r\sin\theta$, $x = x$ we have

$$\iint_S \mathbf{F} \cdot d\mathbf{S} = \iiint_E (3y^2 + 3z^2)\, dV = \int_0^{2\pi} \int_0^1 \int_{-1}^2 (3r^2\cos^2\theta + 3r^2\sin^2\theta)\, r \, dx \, dr \, d\theta$$

$$= 3\int_0^{2\pi} d\theta \int_0^1 r^3 \, dr \int_{-1}^2 dx = 3(2\pi)\left(\tfrac{1}{4}\right)(3) = \tfrac{9\pi}{2}$$

9. div $\mathbf{F} = y\sin z + 0 - y\sin z = 0$, so by the Divergence Theorem, $\iint_S \mathbf{F} \cdot d\mathbf{S} = \iiint_E 0 \, dV = 0$.

11. div $\mathbf{F} = y^2 + 0 + x^2 = x^2 + y^2$ so

$$\iint_S \mathbf{F} \cdot d\mathbf{S} = \iiint_E (x^2 + y^2)\, dV = \int_0^{2\pi} \int_0^2 \int_{r^2}^4 r^2 \cdot r \, dz \, dr \, d\theta = \int_0^{2\pi} \int_0^2 r^3(4 - r^2)\, dr \, d\theta$$

$$= \int_0^{2\pi} d\theta \int_0^2 (4r^3 - r^5)\, dr = 2\pi\left[r^4 - \tfrac{1}{6}r^6\right]_0^2 = \tfrac{32}{3}\pi$$

13. div $\mathbf{F} = 12x^2z + 12y^2z + 12z^3$ so

$$\iint_S \mathbf{F} \cdot d\mathbf{S} = \iiint_E 12z(x^2 + y^2 + z^2)\, dV = \int_0^{2\pi} \int_0^\pi \int_0^R 12(\rho\cos\phi)(\rho^2)\rho^2\sin\phi \, d\rho \, d\phi \, d\theta$$

$$= 12\int_0^{2\pi} d\theta \int_0^\pi \sin\phi\cos\phi \, d\phi \int_0^R \rho^5 \, d\rho = 12(2\pi)\left[\tfrac{1}{2}\sin^2\phi\right]_0^\pi \left[\tfrac{1}{6}\rho^6\right]_0^R = 0$$

15. $\iint_S \mathbf{F} \cdot d\mathbf{S} = \iiint_E \sqrt{3 - x^2} \, dV = \int_{-1}^1 \int_{-1}^1 \int_0^{2 - x^4 - y^4} \sqrt{3 - x^2} \, dz \, dy \, dx = \tfrac{341}{60}\sqrt{2} + \tfrac{81}{20}\sin^{-1}\left(\tfrac{\sqrt{3}}{3}\right)$

17. For S_1 we have $\mathbf{n} = -\mathbf{k}$, so $\mathbf{F} \cdot \mathbf{n} = \mathbf{F} \cdot (-\mathbf{k}) = -x^2z - y^2 = -y^2$ (since $z = 0$ on S_1). So if D is the unit disk, we get

$\iint_{S_1} \mathbf{F} \cdot d\mathbf{S} = \iint_{S_1} \mathbf{F} \cdot \mathbf{n}\, dS = \iint_D (-y^2)\, dA = -\int_0^{2\pi} \int_0^1 r^2(\sin^2\theta)\, r \, dr \, d\theta = -\tfrac{1}{4}\pi$. Now since S_2 is closed, we can use

the Divergence Theorem. Since div $\mathbf{F} = \frac{\partial}{\partial x}(z^2x) + \frac{\partial}{\partial y}\left(\tfrac{1}{3}y^3 + \tan z\right) + \frac{\partial}{\partial z}(x^2z + y^2) = z^2 + y^2 + x^2$, we use spherical

coordinates to get $\iint_{S_2} \mathbf{F} \cdot d\mathbf{S} = \iiint_E \text{div } \mathbf{F} \, dV = \int_0^{2\pi} \int_0^{\pi/2} \int_0^1 \rho^2 \cdot \rho^2\sin\phi \, d\rho \, d\phi \, d\theta = \tfrac{2}{5}\pi$. Finally

$\iint_S \mathbf{F} \cdot d\mathbf{S} = \iint_{S_2} \mathbf{F} \cdot d\mathbf{S} - \iint_{S_1} \mathbf{F} \cdot d\mathbf{S} = \tfrac{2}{5}\pi - \left(-\tfrac{1}{4}\pi\right) = \tfrac{13}{20}\pi$.

19. The vectors that end near P_1 are longer than the vectors that start near P_1, so the net flow is inward near P_1 and div $\mathbf{F}(P_1)$ is negative. The vectors that end near P_2 are shorter than the vectors that start near P_2, so the net flow is outward near P_2 and div $\mathbf{F}(P_2)$ is positive.

21.

From the graph it appears that for points above the x-axis, vectors starting near a particular point are longer than vectors ending there, so divergence is positive. The opposite is true at points below the x-axis, where divergence is negative.

$$\mathbf{F}(x, y) = \langle xy, x + y^2 \rangle \quad \Rightarrow$$

div $\mathbf{F} = \frac{\partial}{\partial x}(xy) + \frac{\partial}{\partial y}(x + y^2) = y + 2y = 3y$. Thus div $\mathbf{F} > 0$ for $y > 0$, and div $\mathbf{F} < 0$ for $y < 0$.

23. Since $\dfrac{\mathbf{x}}{|\mathbf{x}|^3} = \dfrac{x\,\mathbf{i} + y\,\mathbf{j} + z\,\mathbf{k}}{(x^2 + y^2 + z^2)^{3/2}}$ and $\dfrac{\partial}{\partial x}\left(\dfrac{x}{(x^2 + y^2 + z^2)^{3/2}}\right) = \dfrac{(x^2 + y^2 + z^2) - 3x^2}{(x^2 + y^2 + z^2)^{5/2}}$ with similar expressions for

$\dfrac{\partial}{\partial y}\left(\dfrac{y}{(x^2 + y^2 + z^2)^{3/2}}\right)$ and $\dfrac{\partial}{\partial z}\left(\dfrac{z}{(x^2 + y^2 + z^2)^{3/2}}\right)$, we have

div $\left(\dfrac{\mathbf{x}}{|\mathbf{x}|^3}\right) = \dfrac{3(x^2 + y^2 + z^2) - 3(x^2 + y^2 + z^2)}{(x^2 + y^2 + z^2)^{5/2}} = 0$, except at $(0, 0, 0)$ where it is undefined.

25. $\iint_S \mathbf{a} \cdot \mathbf{n}\, dS = \iiint_E \operatorname{div} \mathbf{a}\, dV = 0$ since div $\mathbf{a} = 0$.

27. $\iint_S \operatorname{curl} \mathbf{F} \cdot d\mathbf{S} = \iiint_E \operatorname{div}(\operatorname{curl} \mathbf{F})\, dV = 0$ by Theorem 13.5.11.

29. $\iint_S (f\nabla g) \cdot \mathbf{n}\, dS = \iiint_E \operatorname{div}(f\nabla g)\, dV = \iiint_E (f\nabla^2 g + \nabla g \cdot \nabla f)\, dV$ by Exercise 13.5.23.

31. If $\mathbf{c} = c_1\,\mathbf{i} + c_2\,\mathbf{j} + c_3\,\mathbf{k}$ is an arbitrary constant vector, we define $\mathbf{F} = f\mathbf{c} = fc_1\,\mathbf{i} + fc_2\,\mathbf{j} + fc_3\,\mathbf{k}$. Then

div $\mathbf{F} = \operatorname{div} f\mathbf{c} = \dfrac{\partial f}{\partial x}c_1 + \dfrac{\partial f}{\partial y}c_2 + \dfrac{\partial f}{\partial z}c_3 = \nabla f \cdot \mathbf{c}$ and the Divergence Theorem says $\iint_S \mathbf{F} \cdot d\mathbf{S} = \iiint_E \operatorname{div} \mathbf{F}\, dV \quad \Rightarrow$

$\iint_S \mathbf{F} \cdot \mathbf{n}\, dS = \iiint_E \nabla f \cdot \mathbf{c}\, dV$. In particular, if $\mathbf{c} = \mathbf{i}$ then $\iint_S f\mathbf{i} \cdot \mathbf{n}\, dS = \iiint_E \nabla f \cdot \mathbf{i}\, dV \quad \Rightarrow$

$\iint_S fn_1\, dS = \iiint_E \dfrac{\partial f}{\partial x}\, dV$ (where $\mathbf{n} = n_1\,\mathbf{i} + n_2\,\mathbf{j} + n_3\,\mathbf{k}$). Similarly, if $\mathbf{c} = \mathbf{j}$ we have $\iint_S fn_2\, dS = \iiint_E \dfrac{\partial f}{\partial y}\, dV$,

and $\mathbf{c} = \mathbf{k}$ gives $\iint_S fn_3\, dS = \iiint_E \dfrac{\partial f}{\partial z}\, dV$. Then

$$\iint_S f\mathbf{n}\, dS = \left(\iint_S fn_1\, dS\right)\mathbf{i} + \left(\iint_S fn_2\, dS\right)\mathbf{j} + \left(\iint_S fn_3\, dS\right)\mathbf{k}$$
$$= \left(\iiint_E \frac{\partial f}{\partial x}\, dV\right)\mathbf{i} + \left(\iiint_E \frac{\partial f}{\partial y}\, dV\right)\mathbf{j} + \left(\iiint_E \frac{\partial f}{\partial z}\, dV\right)\mathbf{k}$$
$$= \iiint_E \left(\frac{\partial f}{\partial x}\mathbf{i} + \frac{\partial f}{\partial y}\mathbf{j} + \frac{\partial f}{\partial z}\mathbf{k}\right) dV = \iiint_E \nabla f\, dV$$

as desired.

13 Review

1. See Definitions 1 and 2 in Section 13.1. A vector field can represent, for example, the wind velocity at any location in space, the speed and direction of the ocean current at any location, or the force vectors of Earth's gravitational field at a location in space.

2. (a) A conservative vector field $\mathbf{F}$ is a vector field which is the gradient of some scalar function f.

   (b) The function f in part (a) is called a potential function for $\mathbf{F}$, that is, $\mathbf{F} = \nabla f$.

3. (a) See Definition 13.2.2.

   (b) We normally evaluate the line integral using Formula 13.2.3.

   (c) The mass is $m = \int_C \rho(x, y) \, ds$, and the center of mass is $(\overline{x}, \overline{y})$ where $\overline{x} = \frac{1}{m} \int_C x\rho(x, y) \, ds, \overline{y} = \frac{1}{m} \int_C y\rho(x, y) \, ds$.

   (d) See (5) and (6) in Section 13.2 for plane curves; we have similar definitions when C is a space curve [see the equation preceding (10) in Section 13.2].

   (e) For plane curves, see Equations 13.2.7. We have similar results for space curves [see the equation preceding (10) in Section 13.2].

4. (a) See Definition 13.2.13.

   (b) If $\mathbf{F}$ is a force field, $\int_C \mathbf{F} \cdot d\mathbf{r}$ represents the work done by $\mathbf{F}$ in moving a particle along the curve C.

   (c) $\int_C \mathbf{F} \cdot d\mathbf{r} = \int_C P \, dx + Q \, dy + R \, dz$

5. See Theorem 13.3.2.

6. (a) $\int_C \mathbf{F} \cdot d\mathbf{r}$ is independent of path if the line integral has the same value for any two curves that have the same initial and terminal points.

   (b) See Theorem 13.3.4.

7. See the statement of Green's Theorem on page 933.

8. See Equations 13.4.5.

9. (a) $\operatorname{curl} \mathbf{F} = \left(\dfrac{\partial R}{\partial y} - \dfrac{\partial Q}{\partial z} \right) \mathbf{i} + \left(\dfrac{\partial P}{\partial z} - \dfrac{\partial R}{\partial x} \right) \mathbf{j} + \left(\dfrac{\partial Q}{\partial x} - \dfrac{\partial P}{\partial y} \right) \mathbf{k} = \nabla \times \mathbf{F}$

   (b) $\operatorname{div} \mathbf{F} = \dfrac{\partial P}{\partial x} + \dfrac{\partial Q}{\partial y} + \dfrac{\partial R}{\partial z} = \nabla \cdot \mathbf{F}$

   (c) For curl $\mathbf{F}$, see the discussion accompanying Figure 1 on page 943 as well as Figure 6 and the accompanying discussion on page 963. For div $\mathbf{F}$, see the discussion following Example 5 on page 944 as well as the discussion preceding (8) on page 970.

10. See Theorem 13.3.6; see Theorem 13.5.4.

11. (a) See (1) in Section 13.6.

    (b) We normally evaluate the surface integral using Formula 13.6.2.

    (c) See Formula 13.6.4.

    (d) The mass is $m = \iint_S \rho(x, y, z) \, dS$ and the center of mass is $(\overline{x}, \overline{y}, \overline{z})$ where $\overline{x} = \frac{1}{m} \iint_S x\rho(x, y, z) \, dS$, $\overline{y} = \frac{1}{m} \iint_S y\rho(x, y, z) \, dS, \overline{z} = \frac{1}{m} \iint_S z\rho(x, y, z) \, dS$.

12. (a) See Figures 6 and 7 and the accompanying discussion in Section 13.6. A Möbius strip is a nonorientable surface; see Figures 4 and 5 and the accompanying discussion on page 952.

(b) See Definition 13.6.8.

(c) See Formula 13.6.9.

(d) See Formula 13.6.10.

13. See the statement of Stokes' Theorem on page 959.

14. See the statement of the Divergence Theorem on page 966.

15. In each theorem, we have an integral of a "derivative" over a region on the left side, while the right side involves the values of the original function only on the boundary of the region.

TRUE-FALSE QUIZ

1. False; div $\mathbf{F}$ is a scalar field.

3. True, by Theorem 13.5.3 and the fact that div $\mathbf{0} = 0$.

5. False. See Exercise 13.3.33. (But the assertion is true if D is simply-connected; see Theorem 13.3.6.)

7. True. Apply the Divergence Theorem and use the fact that div $\mathbf{F} = 0$.

EXERCISES

1. (a) Vectors starting on C point in roughly the direction opposite to C, so the tangential component $\mathbf{F} \cdot \mathbf{T}$ is negative. Thus $\int_C \mathbf{F} \cdot d\mathbf{r} = \int_C \mathbf{F} \cdot \mathbf{T} \, ds$ is negative.

(b) The vectors that end near P are shorter than the vectors that start near P, so the net flow is outward near P and div $\mathbf{F}(P)$ is positive.

3. $\int_C yz \cos x \, ds = \int_0^\pi (3\cos t)(3\sin t)\cos t \sqrt{(1)^2 + (-3\sin t)^2 + (3\cos t)^2} \, dt = \int_0^\pi (9\cos^2 t \sin t)\sqrt{10} \, dt$

$= 9\sqrt{10}\left(-\frac{1}{3}\cos^3 t\right)\Big]_0^\pi = -3\sqrt{10}(-2) = 6\sqrt{10}$

5. $\int_C y^3 \, dx + x^2 \, dy = \int_{-1}^1 \left[y^3(-2y) + (1-y^2)^2 \right] dy = \int_{-1}^1 (-y^4 - 2y^2 + 1) \, dy$

$= \left[-\frac{1}{5}y^5 - \frac{2}{3}y^3 + y \right]_{-1}^1 = -\frac{1}{5} - \frac{2}{3} + 1 - \frac{1}{5} - \frac{2}{3} + 1 = \frac{4}{15}$

7. $C: x = 1+2t \;\Rightarrow\; dx = 2\,dt, y = 4t \;\Rightarrow\; dy = 4\,dt, z = -1+3t \;\Rightarrow\; dz = 3\,dt, 0 \le t \le 1.$

$\int_C xy \, dx + y^2 \, dy + yz \, dz = \int_0^1 \left[(1+2t)(4t)(2) + (4t)^2(4) + (4t)(-1+3t)(3) \right] dt$

$= \int_0^1 (116t^2 - 4t) \, dt = \left[\frac{116}{3}t^3 - 2t^2 \right]_0^1 = \frac{116}{3} - 2 = \frac{110}{3}$

9. $\mathbf{F}(\mathbf{r}(t)) = e^{-t}\mathbf{i} + t^2(-t)\mathbf{j} + (t^2 + t^3)\mathbf{k}, \mathbf{r}'(t) = 2t\mathbf{i} + 3t^2\mathbf{j} - \mathbf{k}$ and $\int_C \mathbf{F} \cdot d\mathbf{r} = \int_0^1 (2te^{-t} - 3t^5 - (t^2 + t^3)) \, dt = \left[-2te^{-t} - 2e^{-t} - \frac{1}{2}t^6 - \frac{1}{3}t^3 - \frac{1}{4}t^4 \right]_0^1 = \frac{11}{12} - \frac{4}{e}.$

11. $\frac{\partial}{\partial y}\left[(1+xy)e^{xy}\right] = 2xe^{xy} + x^2ye^{xy} = \frac{\partial}{\partial x}\left[e^y + x^2e^{xy}\right]$ and the domain of $\mathbf{F}$ is $\mathbb{R}^2$, so $\mathbf{F}$ is conservative. Thus there

exists a function f such that $\mathbf{F} = \nabla f$. Then $f_y(x,y) = e^y + x^2e^{xy}$ implies $f(x,y) = e^y + xe^{xy} + g(x)$ and then

$f_x(x,y) = xye^{xy} + e^{xy} + g'(x) = (1+xy)e^{xy} + g'(x)$. But $f_x(x,y) = (1+xy)e^{xy}$, so $g'(x) = 0 \implies g(x) = K$.

Thus $f(x,y) = e^y + xe^{xy} + K$ is a potential function for $\mathbf{F}$.

13. Since $\frac{\partial}{\partial y}\left(4x^3y^2 - 2xy^3\right) = 8x^3y - 6xy^2 = \frac{\partial}{\partial x}\left(2x^4y - 3x^2y^2 + 4y^3\right)$ and the domain of $\mathbf{F}$ is $\mathbb{R}^2$, $\mathbf{F}$ is conservative.

Furthermore $f(x,y) = x^4y^2 - x^2y^3 + y^4$ is a potential function for $\mathbf{F}$. $t = 0$ corresponds to the point $(0,1)$ and $t = 1$

corresponds to $(1,1)$, so $\int_C \mathbf{F} \cdot d\mathbf{r} = f(1,1) - f(0,1) = 1 - 1 = 0$.

15.

C_1: $\mathbf{r}(t) = t\mathbf{i} + t^2\mathbf{j}, -1 \le t \le 1$;

C_2: $\mathbf{r}(t) = -t\mathbf{i} + \mathbf{j}, -1 \le t \le 1$.

Then

$$\int_C xy^2\,dx - x^2y\,dy = \int_{-1}^{1}(t^5 - 2t^5)\,dt + \int_{-1}^{1}t\,dt$$

$$= \left[-\tfrac{1}{6}t^6\right]_{-1}^{1} + \left[\tfrac{1}{2}t^2\right]_{-1}^{1} = 0$$

Using Green's Theorem, we have

$$\int_C xy^2\,dx - x^2y\,dy = \iint_D \left[\frac{\partial}{\partial x}(-x^2y) - \frac{\partial}{\partial y}(xy^2)\right]dA = \iint_D (-2xy - 2xy)\,dA = \int_{-1}^{1}\int_{x^2}^{1}-4xy\,dy\,dx$$

$$= \int_{-1}^{1}\left[-2xy^2\right]_{y=x^2}^{y=1}\,dx = \int_{-1}^{1}(2x^5 - 2x)\,dx = \left[\tfrac{1}{3}x^6 - x^2\right]_{-1}^{1} = 0$$

17. $\int_C x^2y\,dx - xy^2\,dy = \iint\limits_{x^2+y^2 \le 4}\left[\frac{\partial}{\partial x}(-xy^2) - \frac{\partial}{\partial y}(x^2y)\right]dA = \iint\limits_{x^2+y^2 \le 4}(-y^2 - x^2)\,dA = -\int_0^{2\pi}\int_0^2 r^3\,dr\,d\theta = -8\pi$

19. If we assume there is such a vector field $\mathbf{G}$, then $\text{div}(\text{curl}\,\mathbf{G}) = 2 + 3z - 2xz$. But $\text{div}(\text{curl}\,\mathbf{F}) = 0$ for all vector fields $\mathbf{F}$.

Thus such a $\mathbf{G}$ cannot exist.

21. For any piecewise-smooth simple closed plane curve C bounding a region D, we can apply Green's Theorem to

$\mathbf{F}(x,y) = f(x)\mathbf{i} + g(y)\mathbf{j}$ to get $\int_C f(x)\,dx + g(y)\,dy = \iint_D\left[\frac{\partial}{\partial x}g(y) - \frac{\partial}{\partial y}f(x)\right]dA = \iint_D 0\,dA = 0$.

23. $\nabla^2 f = 0$ means that $\frac{\partial^2 f}{\partial x^2} + \frac{\partial^2 f}{\partial y^2} = 0$. Now if $\mathbf{F} = f_y\mathbf{i} - f_x\mathbf{j}$ and C is any closed path in D, then applying Green's

Theorem, we get

$$\int_C \mathbf{F}\cdot d\mathbf{r} = \int_C f_y\,dx - f_x\,dy = \iint_D\left[\frac{\partial}{\partial x}(-f_x) - \frac{\partial}{\partial y}(f_y)\right]dA = -\iint_D(f_{xx} + f_{yy})\,dA = -\iint_D 0\,dA = 0$$

Therefore the line integral is independent of path, by Theorem 13.3.3.

25. $z = f(x,y) = x^2 + y^2$ with $0 \le x^2 + y^2 \le 4$ so $\mathbf{r}_x \times \mathbf{r}_y = -2x\mathbf{i} - 2y\mathbf{j} + \mathbf{k}$ (using upward orientation). Then

$$\iint_S z\,dS = \iint\limits_{x^2+y^2 \le 4}(x^2 + y^2)\sqrt{4x^2 + 4y^2 + 1}\,dA = \int_0^{2\pi}\int_0^2 r^3\sqrt{1 + 4r^2}\,dr\,d\theta = \tfrac{1}{60}\pi\left(391\sqrt{17} + 1\right)$$

(Substitute $u = 1 + 4r^2$ and use tables.)

27. Since the sphere bounds a simple solid region, the Divergence Theorem applies and

$\iint_S \mathbf{F} \cdot d\mathbf{S} = \iiint_E (z - 2)\, dV = \iiint_E z\, dV - 2\iiint_E dV = m\overline{z} - 2\left(\frac{4}{3}\pi 2^3\right) = -\frac{64}{3}\pi.$

Alternate solution: $\mathbf{F}(\mathbf{r}(\phi, \theta)) = 4\sin\phi\cos\theta\cos\phi\,\mathbf{i} - 4\sin\phi\sin\theta\,\mathbf{j} + 6\sin\phi\cos\theta\,\mathbf{k},$

$\mathbf{r}_\phi \times \mathbf{r}_\theta = 4\sin^2\phi\cos\theta\,\mathbf{i} + 4\sin^2\phi\sin\theta\,\mathbf{j} + 4\sin\phi\cos\phi\,\mathbf{k},$ and

$\mathbf{F} \cdot (\mathbf{r}_\phi \times \mathbf{r}_\theta) = 16\sin^3\phi\cos^2\theta\cos\phi - 16\sin^3\phi\sin^2\theta + 24\sin^2\phi\cos\phi\cos\theta.$ Then

$$\iint_S F \cdot dS = \int_0^{2\pi}\int_0^\pi (16\sin^3\phi\cos\phi\cos^2\theta - 16\sin^3\phi\sin^2\theta + 24\sin^2\phi\cos\phi\cos\theta)\, d\phi\, d\theta$$

$$= \int_0^{2\pi} \frac{4}{3}(-16\sin^2\theta)\, d\theta = -\frac{64}{3}\pi$$

29. Since $\operatorname{curl}\mathbf{F} = \mathbf{0}$, $\iint_S (\operatorname{curl}\mathbf{F}) \cdot d\mathbf{S} = 0$. We parametrize C: $\mathbf{r}(t) = \cos t\,\mathbf{i} + \sin t\,\mathbf{j}$, $0 \le t \le 2\pi$ and

$\oint_C \mathbf{F} \cdot d\mathbf{r} = \int_0^{2\pi}(-\cos^2 t\sin t + \sin^2 t\cos t)\, dt = \frac{1}{3}\cos^3 t + \frac{1}{3}\sin^3 t\Big]_0^{2\pi} = 0.$

31. The surface is given by $x + y + z = 1$ or $z = 1 - x - y$, $0 \le x \le 1$, $0 \le y \le 1 - x$ and $\mathbf{r}_x \times \mathbf{r}_y = \mathbf{i} + \mathbf{j} + \mathbf{k}$. Then

$\oint_C \mathbf{F} \cdot d\mathbf{r} = \iint_S \operatorname{curl}\mathbf{F} \cdot d\mathbf{S} = \iint_D(-y\,\mathbf{i} - z\,\mathbf{j} - x\,\mathbf{k}) \cdot (\mathbf{i} + \mathbf{j} + \mathbf{k})\, dA = \iint_D(-1)\, dA = -(\text{area of } D) = -\frac{1}{2}$

33. $\iiint_E \operatorname{div}\mathbf{F}\, dV = \iiint\limits_{x^2 + y^2 + z^2 \le 1} 3\, dV = 3(\text{volume of sphere}) = 4\pi.$ Then

$\mathbf{F}(\mathbf{r}(\phi, \theta)) \cdot (\mathbf{r}_\phi \times \mathbf{r}_\theta) = \sin^3\phi\cos^2\theta + \sin^3\phi\sin^2\theta + \sin\phi\cos^2\phi = \sin\phi$ and

$\iint_S \mathbf{F} \cdot d\mathbf{S} = \int_0^{2\pi}\int_0^\pi \sin\phi\, d\phi\, d\theta = (2\pi)(2) = 4\pi.$

35. Because $\operatorname{curl}\mathbf{F} = \mathbf{0}$, $\mathbf{F}$ is conservative, and if $f(x, y, z) = x^3yz - 3xy + z^2$, then $\nabla f = \mathbf{F}$. Hence

$\int_C \mathbf{F} \cdot d\mathbf{r} = \int_C \nabla f \cdot d\mathbf{r} = f(0, 3, 0) - f(0, 0, 2) = 0 - 4 = -4.$

37. By the Divergence Theorem, $\iint_S \mathbf{F} \cdot \mathbf{n}\, dS = \iiint_E \operatorname{div}\mathbf{F}\, dV = 3(\text{volume of } E) = 3(8 - 1) = 21.$

□ FOCUS ON PROBLEM SOLVING

1. Let S_1 be the portion of $\Omega(S)$ between $S(a)$ and S, and let ∂S_1 be its boundary. Also let S_L be the lateral surface of S_1 [that

is, the surface of S_1 except S and $S(a)$]. Applying the Divergence Theorem we have $\iint_{\partial S_1} \dfrac{\mathbf{r} \cdot \mathbf{n}}{r^3}\, dS = \iiint_{S_1} \nabla \cdot \dfrac{\mathbf{r}}{r^3}\, dV$.

But

$$\nabla \cdot \frac{\mathbf{r}}{r^3} = \left\langle \frac{\partial}{\partial x}, \frac{\partial}{\partial y}, \frac{\partial}{\partial z} \right\rangle \cdot \left\langle \frac{x}{(x^2+y^2+z^2)^{3/2}}, \frac{y}{(x^2+y^2+z^2)^{3/2}}, \frac{z}{(x^2+y^2+z^2)^{3/2}} \right\rangle$$

$$= \frac{(x^2+y^2+z^2-3x^2)+(x^2+y^2+z^2-3y^2)+(x^2+y^2+z^2-3z^2)}{(x^2+y^2+z^2)^{5/2}} = 0$$

$\Rightarrow \iint_{\partial S_1} \dfrac{\mathbf{r} \cdot \mathbf{n}}{r^3}\, dS = \iiint_{S_1} 0 \, dV = 0$. On the other hand, notice that for the surfaces of ∂S_1 other than $S(a)$ and S,

$\mathbf{r} \cdot \mathbf{n} = 0 \ \Rightarrow$

$$0 = \iint_{\partial S_1} \frac{\mathbf{r} \cdot \mathbf{n}}{r^3}\, dS = \iint_{S} \frac{\mathbf{r} \cdot \mathbf{n}}{r^3}\, dS + \iint_{S(a)} \frac{\mathbf{r} \cdot \mathbf{n}}{r^3}\, dS + \iint_{S_L} \frac{\mathbf{r} \cdot \mathbf{n}}{r^3}\, dS$$

$$= \iint_{S} \frac{\mathbf{r} \cdot \mathbf{n}}{r^3}\, dS + \iint_{S(a)} \frac{\mathbf{r} \cdot \mathbf{n}}{r^3}\, dS$$

$\Rightarrow \iint_{S} \dfrac{\mathbf{r} \cdot \mathbf{n}}{r^3}\, dS = -\iint_{S(a)} \dfrac{\mathbf{r} \cdot \mathbf{n}}{r^3}\, dS$. Notice that on $S(a)$, $r = a \ \Rightarrow \ \mathbf{n} = -\dfrac{\mathbf{r}}{r} = -\dfrac{\mathbf{r}}{a}$ and $\mathbf{r} \cdot \mathbf{r} = r^2 = a^2$, so that

$$-\iint_{S(a)} \frac{\mathbf{r} \cdot \mathbf{n}}{r^3}\, dS = \iint_{S(a)} \frac{\mathbf{r} \cdot \mathbf{r}}{a^4}\, dS = \iint_{S(a)} \frac{a^2}{a^4}\, dS = \frac{1}{a^2} \iint_{S(a)} dS = \frac{\text{area of } S(a)}{a^2} = |\Omega(S)|. \text{ Therefore}$$

$|\Omega(S)| = \iint_{S} \dfrac{\mathbf{r} \cdot \mathbf{n}}{r^3}\, dS.$

3. Let $\mathbf{F} = \mathbf{a} \times \mathbf{r} = \langle a_1, a_2, a_3 \rangle \times \langle x, y, z \rangle = \langle a_2 z - a_3 y, a_3 x - a_1 z, a_1 y - a_2 x \rangle$. Then $\operatorname{curl} \mathbf{F} = \langle 2a_1, 2a_2, 2a_3 \rangle = 2\mathbf{a}$, and

$$\iint_{S} 2\mathbf{a} \cdot d\mathbf{S} = \iint_{S} \operatorname{curl} \mathbf{F} \cdot d\mathbf{S} = \int_{C} \mathbf{F} \cdot d\mathbf{r} = \int_{C} (\mathbf{a} \times \mathbf{r}) \cdot d\mathbf{r}$$

by Stokes' Theorem.

5. The given line integral $\frac{1}{2} \int_C (bz - cy)\, dx + (cx - az)\, dy + (ay - bx)\, dz$ can be expressed as $\int_C \mathbf{F} \cdot d\mathbf{r}$ if we define the vector

field $\mathbf{F}$ by $\mathbf{F}(x, y, z) = P\mathbf{i} + Q\mathbf{j} + R\mathbf{k} = \frac{1}{2}(bz - cy)\mathbf{i} + \frac{1}{2}(cx - az)\mathbf{j} + \frac{1}{2}(ay - bx)\mathbf{k}$. Then define S to be the planar

interior of C, so S is an oriented, smooth surface. Stokes' Theorem says $\int_C \mathbf{F} \cdot d\mathbf{r} = \iint_S \operatorname{curl} \mathbf{F} \cdot d\mathbf{S} = \iint_S \operatorname{curl} \mathbf{F} \cdot \mathbf{n}\, dS$.

Now

$$\operatorname{curl} \mathbf{F} = \left(\frac{\partial R}{\partial y} - \frac{\partial Q}{\partial z} \right) \mathbf{i} + \left(\frac{\partial P}{\partial z} - \frac{\partial R}{\partial x} \right) \mathbf{j} + \left(\frac{\partial Q}{\partial x} - \frac{\partial P}{\partial y} \right) \mathbf{k}$$

$$= \left(\tfrac{1}{2}a + \tfrac{1}{2}a \right) \mathbf{i} + \left(\tfrac{1}{2}b + \tfrac{1}{2}b \right) \mathbf{j} + \left(\tfrac{1}{2}c + \tfrac{1}{2}c \right) \mathbf{k} = a\mathbf{i} + b\mathbf{j} + c\mathbf{k} = \mathbf{n}$$

so $\operatorname{curl} \mathbf{F} \cdot \mathbf{n} = \mathbf{n} \cdot \mathbf{n} = |\mathbf{n}|^2 = 1$, hence $\iint_S \operatorname{curl} \mathbf{F} \cdot \mathbf{n}\, dS = \iint_S dS$ which is simply the surface area of S. Thus,

$\int_C \mathbf{F} \cdot d\mathbf{r} = \frac{1}{2} \int_C (bz - cy)\, dx + (cx - az)\, dy + (ay - bx)\, dz$ is the plane area enclosed by C.

☐ APPENDIXES

D PRECISE DEFINITIONS OF LIMITS

1. (*Note:* This is Exercise 21 in the full version of the text.)

Let $\varepsilon > 0$. We want to find $\delta > 0$ such that

$$\left| \frac{xy}{\sqrt{x^2 + y^2}} - 0 \right| < \varepsilon \qquad \text{whenever} \qquad 0 < \sqrt{x^2 + y^2} < \delta$$

that is,

$$\frac{|xy|}{\sqrt{x^2 + y^2}} < \varepsilon \qquad \text{whenever} \qquad 0 < \sqrt{x^2 + y^2} < \delta$$

But $|x| = \sqrt{x^2} \le \sqrt{x^2 + y^2}$ and $|y| = \sqrt{y^2} \le \sqrt{x^2 + y^2}$, so

$$\frac{|xy|}{\sqrt{x^2 + y^2}} \le \frac{\left(\sqrt{x^2 + y^2} \right)^2}{\sqrt{x^2 + y^2}} = \sqrt{x^2 + y^2}$$

Thus, if we choose $\delta = \varepsilon$ and let $0 < \sqrt{x^2 + y^2} < \delta$, then

$$\left| \frac{xy}{\sqrt{x^2 + y^2}} - 0 \right| \le \sqrt{x^2 + y^2} < \delta = \varepsilon$$

Hence, by Definition 1 (Definition 5 in the full version of the text),

$$\lim_{(x,y) \to (0,0)} \frac{xy}{\sqrt{x^2 + y^2}} = 0$$

H POLAR COORDINATES

H.1 Curves in Polar Coordinates

1. (a) By adding 2π to $\frac{\pi}{2}$, we obtain the point $\left(1, \frac{5\pi}{2}\right)$. The direction opposite $\frac{\pi}{2}$ is $\frac{3\pi}{2}$, so $\left(-1, \frac{3\pi}{2}\right)$ is a point that satisfies the $r < 0$ requirement.

(b) $\left(-2, \frac{\pi}{4}\right)$

(c) $(3, 2)$

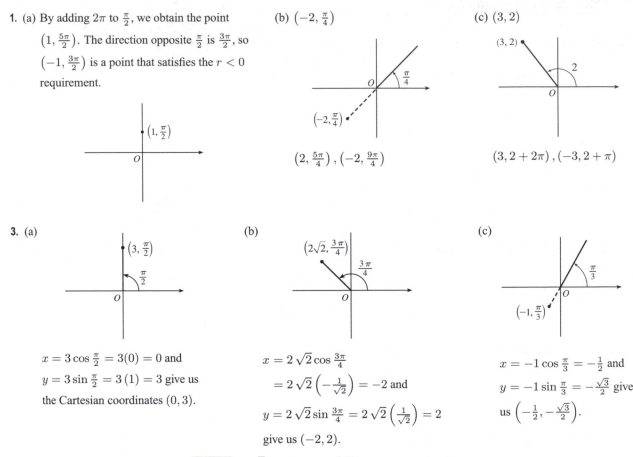

$\left(2, \frac{5\pi}{4}\right), \left(-2, \frac{9\pi}{4}\right)$

$(3, 2 + 2\pi), (-3, 2 + \pi)$

3. (a)

(b)

(c)

$x = 3\cos\frac{\pi}{2} = 3(0) = 0$ and $y = 3\sin\frac{\pi}{2} = 3(1) = 3$ give us the Cartesian coordinates $(0, 3)$.

$x = 2\sqrt{2}\cos\frac{3\pi}{4}$
$= 2\sqrt{2}\left(-\frac{1}{\sqrt{2}}\right) = -2$ and
$y = 2\sqrt{2}\sin\frac{3\pi}{4} = 2\sqrt{2}\left(\frac{1}{\sqrt{2}}\right) = 2$
give us $(-2, 2)$.

$x = -1\cos\frac{\pi}{3} = -\frac{1}{2}$ and
$y = -1\sin\frac{\pi}{3} = -\frac{\sqrt{3}}{2}$ give
us $\left(-\frac{1}{2}, -\frac{\sqrt{3}}{2}\right)$.

5. (a) $x = 1$ and $y = 1 \Rightarrow r = \sqrt{1^2 + 1^2} = \sqrt{2}$ and $\theta = \tan^{-1}\left(\frac{1}{1}\right) = \frac{\pi}{4}$. Since $(1, 1)$ is in the first quadrant, the polar coordinates are (i) $\left(\sqrt{2}, \frac{\pi}{4}\right)$ and (ii) $\left(-\sqrt{2}, \frac{5\pi}{4}\right)$.

(b) $x = 2\sqrt{3}$ and $y = -2 \Rightarrow r = \sqrt{\left(2\sqrt{3}\right)^2 + (-2)^2} = \sqrt{12 + 4} = \sqrt{16} = 4$ and $\theta = \tan^{-1}\left(-\frac{2}{2\sqrt{3}}\right) = \tan^{-1}\left(-\frac{1}{\sqrt{3}}\right) = -\frac{\pi}{6}$. Since $\left(2\sqrt{3}, -2\right)$ is in the fourth quadrant and $0 \le \theta \le 2\pi$, the polar coordinates are (i) $\left(4, \frac{11\pi}{6}\right)$ and (ii) $\left(-4, \frac{5\pi}{6}\right)$.

7. The curves $r = 1$ and $r = 2$ represent circles with center O and radii 1 and 2. The region in the plane satisfying $1 \le r \le 2$ consists of both circles and the shaded region between them in the figure.

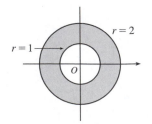

9. The region satisfying $0 \le r < 4$ and

$-\pi/2 \le \theta < \pi/6$ does not include the circle $r = 4$

nor the line $\theta = \frac{\pi}{6}$.

11. $2 < r < 3$, $\frac{5\pi}{3} \le \theta \le \frac{7\pi}{3}$

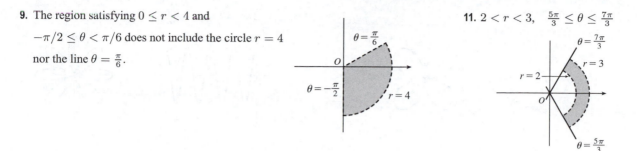

13. $r = 3\sin\theta \Rightarrow r^2 = 3r\sin\theta \Leftrightarrow x^2 + y^2 = 3y \Leftrightarrow x^2 + \left(y - \frac{3}{2}\right)^2 = \left(\frac{3}{2}\right)^2$, a circle of radius $\frac{3}{2}$ centered at $\left(0, \frac{3}{2}\right)$.
The first two equations are actually equivalent since $r^2 = 3r\sin\theta \Rightarrow r(r - 3\sin\theta) = 0 \Rightarrow r = 0$ or $r = 3\sin\theta$. But
$r = 3\sin\theta$ gives the point $r = 0$ (the pole) when $\theta = 0$. Thus, the single equation $r = 3\sin\theta$ is equivalent to the compound
condition $(r = 0 \text{ or } r = 3\sin\theta)$.

15. $r = \csc\theta \Leftrightarrow r = \dfrac{1}{\sin\theta} \Leftrightarrow r\sin\theta = 1 \Leftrightarrow y = 1$, a horizontal line 1 unit above the x-axis.

17. $x = -y^2 \Leftrightarrow r\cos\theta = -r^2\sin^2\theta \Leftrightarrow \cos\theta = -r\sin^2\theta \Leftrightarrow r = -\dfrac{\cos\theta}{\sin^2\theta} = -\cot\theta\csc\theta$.

19. $x^2 + y^2 = 2cx \Leftrightarrow r^2 = 2cr\cos\theta \Leftrightarrow r^2 - 2cr\cos\theta = 0 \Leftrightarrow r(r - 2c\cos\theta) = 0 \Leftrightarrow r = 0 \text{ or } r = 2c\cos\theta$.
$r = 0$ is included in $r = 2c\cos\theta$ when $\theta = \frac{\pi}{2} + n\pi$, so the curve is represented by the single equation $r = 2c\cos\theta$.

21. (a) The description leads immediately to the polar equation $\theta = \frac{\pi}{6}$, and the Cartesian equation $y = \tan\left(\frac{\pi}{6}\right) x = \frac{1}{\sqrt{3}} x$ is
slightly more difficult to derive.

(b) The easier description here is the Cartesian equation $x = 3$.

23. $\theta = -\pi/6$

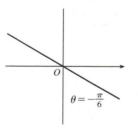

25. $r = \sin\theta \Leftrightarrow r^2 = r\sin\theta \Leftrightarrow x^2 + y^2 = y \Leftrightarrow$
$x^2 + \left(y - \frac{1}{2}\right)^2 = \left(\frac{1}{2}\right)^2$. The reasoning here is the
same as in Exercise 13. This is a circle of radius $\frac{1}{2}$
centered at $\left(0, \frac{1}{2}\right)$.

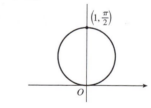

27. $r = 2(1 - \sin\theta)$. This curve is a cardioid.

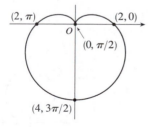

29. $r = \theta$, $\theta \ge 0$

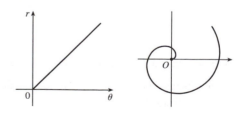

31. $r = \sin 2\theta$

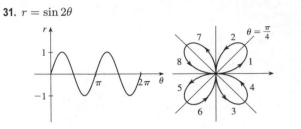

33. $r = 2\cos 4\theta$

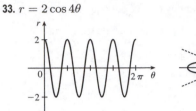
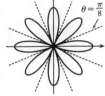

35. $r^2 = 4\cos 2\theta$

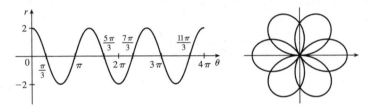

37. $r = 2\cos\left(\frac{3}{2}\theta\right)$

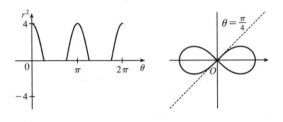

39. $r = 1 + 2\cos 2\theta$

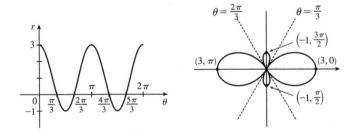

41. For $\theta = 0$, π, and 2π, r has its minimum value of about 0.5. For $\theta = \frac{\pi}{2}$ and $\frac{3\pi}{2}$, r attains its maximum value of 2. We see that the graph has a similar shape for $0 \le \theta \le \pi$ and $\pi \le \theta \le 2\pi$.

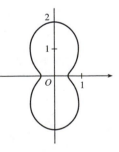

43. $x = (r)\cos\theta = (4 + 2\sec\theta)\cos\theta = 4\cos\theta + 2$. Now, $r \to \infty \implies$

$(4 + 2\sec\theta) \to \infty \implies \theta \to \left(\frac{\pi}{2}\right)^-$ or $\theta \to \left(\frac{3\pi}{2}\right)^+$ (since we need only consider

$0 \le \theta < 2\pi$), so $\lim\limits_{r \to \infty} x = \lim\limits_{\theta \to \pi/2^-} (4\cos\theta + 2) = 2$. Also, $r \to -\infty \implies$

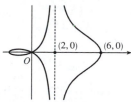

$(4 + 2\sec\theta) \to -\infty \implies \theta \to \left(\frac{\pi}{2}\right)^+$ or $\theta \to \left(\frac{3\pi}{2}\right)^-$, so

$\lim\limits_{r \to -\infty} x = \lim\limits_{\theta \to \pi/2^+} (4\cos\theta + 2) = 2$. Therefore, $\lim\limits_{r \to \pm\infty} x = 2 \implies x = 2$ is a vertical asymptote.

45. (a) We see that the curve crosses itself at the origin, where $r = 0$ (in fact the inner loop corresponds to negative r-values,) so

we solve the equation of the limaçon for $r = 0 \iff c\sin\theta = -1 \iff \sin\theta = -1/c$. Now if $|c| < 1$, then this

equation has no solution and hence there is no inner loop. But if $c < -1$, then on the interval $(0, 2\pi)$ the equation has the

two solutions $\theta = \sin^{-1}(-1/c)$ and $\theta = \pi - \sin^{-1}(-1/c)$, and if $c > 1$, the solutions are $\theta = \pi + \sin^{-1}(1/c)$ and

$\theta = 2\pi - \sin^{-1}(1/c)$. In each case, $r < 0$ for θ between the two solutions, indicating a loop.

(b) For $0 < c < 1$, the dimple (if it exists) is characterized by the fact that y has a local maximum at $\theta = \frac{3\pi}{2}$. So we determine

for what c-values $\dfrac{d^2y}{d\theta^2}$ is negative at $\theta = \frac{3\pi}{2}$, since by the Second Derivative Test this indicates a maximum:

$y = r\sin\theta = \sin\theta + c\sin^2\theta \implies \dfrac{dy}{d\theta} = \cos\theta + 2c\sin\theta\cos\theta = \cos\theta + c\sin 2\theta \implies \dfrac{d^2y}{d\theta^2} = -\sin\theta + 2c\cos 2\theta$.

At $\theta = \frac{3\pi}{2}$, this is equal to $-(-1) + 2c(-1) = 1 - 2c$, which is negative only for $c > \frac{1}{2}$. A similar argument shows that

for $-1 < c < 0$, y only has a local minimum at $\theta = \frac{\pi}{2}$ (indicating a dimple) for $c < -\frac{1}{2}$.

47. $r = 2\sin\theta \implies x = r\cos\theta = 2\sin\theta\cos\theta = \sin 2\theta$, $y = r\sin\theta = 2\sin^2\theta \implies$

$$\frac{dy}{dx} = \frac{dy/d\theta}{dx/d\theta} = \frac{2 \cdot 2\sin\theta\cos\theta}{\cos 2\theta \cdot 2} = \frac{\sin 2\theta}{\cos 2\theta} = \tan 2\theta$$

When $\theta = \dfrac{\pi}{6}$, $\dfrac{dy}{dx} = \tan\left(2 \cdot \dfrac{\pi}{6}\right) = \tan\dfrac{\pi}{3} = \sqrt{3}$.

49. $r = 1/\theta \implies x = r\cos\theta = (\cos\theta)/\theta$, $y = r\sin\theta = (\sin\theta)/\theta \implies$

$$\frac{dy}{dx} = \frac{dy/d\theta}{dx/d\theta} = \frac{\sin\theta(-1/\theta^2) + (1/\theta)\cos\theta}{\cos\theta(-1/\theta^2) - (1/\theta)\sin\theta} \cdot \frac{\theta^2}{\theta^2} = \frac{-\sin\theta + \theta\cos\theta}{-\cos\theta - \theta\sin\theta}$$

When $\theta = \pi$, $\dfrac{dy}{dx} = \dfrac{-0 + \pi(-1)}{-(-1) - \pi(0)} = \dfrac{-\pi}{1} = -\pi$.

51. $r = 3\cos\theta \implies x = r\cos\theta = 3\cos\theta\cos\theta$, $y = r\sin\theta = 3\cos\theta\sin\theta \implies$

$dy/d\theta = -3\sin^2\theta + 3\cos^2\theta = 3\cos 2\theta = 0 \implies 2\theta = \frac{\pi}{2}$ or $\frac{3\pi}{2} \iff \theta = \frac{\pi}{4}$ or $\frac{3\pi}{4}$. So the tangent is horizontal at

$\left(\frac{3}{\sqrt{2}}, \frac{\pi}{4}\right)$ and $\left(-\frac{3}{\sqrt{2}}, \frac{3\pi}{4}\right)$ $\left[\text{same as } \left(\frac{3}{\sqrt{2}}, -\frac{\pi}{4}\right)\right]$. $dx/d\theta = -6\sin\theta\cos\theta = -3\sin 2\theta = 0 \implies 2\theta = 0$ or $\pi \iff$

$\theta = 0$ or $\frac{\pi}{2}$. So the tangent is vertical at $(3, 0)$ and $\left(0, \frac{\pi}{2}\right)$.

53. $r = 1 + \cos\theta \Rightarrow x = r\cos\theta = \cos\theta(1 + \cos\theta), y = r\sin\theta = \sin\theta(1 + \cos\theta) \Rightarrow$

$dy/d\theta = (1 + \cos\theta)\cos\theta - \sin^2\theta = 2\cos^2\theta + \cos\theta - 1 = (2\cos\theta - 1)(\cos\theta + 1) = 0 \Rightarrow$

$\cos\theta = \frac{1}{2}$ or $-1 \Rightarrow \theta = \frac{\pi}{3}, \pi,$ or $\frac{5\pi}{3} \Rightarrow$ horizontal tangent at $\left(\frac{3}{2}, \frac{\pi}{3}\right), (0, \pi),$ and $\left(\frac{3}{2}, \frac{5\pi}{3}\right)$.

$dx/d\theta = -(1 + \cos\theta)\sin\theta - \cos\theta\sin\theta = -\sin\theta(1 + 2\cos\theta) = 0 \Rightarrow \sin\theta = 0$ or $\cos\theta = -\frac{1}{2} \Rightarrow$

$\theta = 0, \pi, \frac{2\pi}{3},$ or $\frac{4\pi}{3} \Rightarrow$ vertical tangent at $(2, 0), \left(\frac{1}{2}, \frac{2\pi}{3}\right),$ and $\left(\frac{1}{2}, \frac{4\pi}{3}\right)$.

Note that the tangent is horizontal, not vertical when $\theta = \pi$, since $\lim_{\theta \to \pi} \dfrac{dy/d\theta}{dx/d\theta} = 0$.

55. $r = a\sin\theta + b\cos\theta \Rightarrow r^2 = ar\sin\theta + br\cos\theta \Rightarrow x^2 + y^2 = ay + bx \Rightarrow$

$x^2 - bx + \left(\frac{1}{2}b\right)^2 + y^2 - ay + \left(\frac{1}{2}a\right)^2 = \left(\frac{1}{2}b\right)^2 + \left(\frac{1}{2}a\right)^2 \Rightarrow \left(x - \frac{1}{2}b\right)^2 + \left(y - \frac{1}{2}a\right)^2 = \frac{1}{4}\left(a^2 + b^2\right)$, and this is a circle

with center $\left(\frac{1}{2}b, \frac{1}{2}a\right)$ and radius $\frac{1}{2}\sqrt{a^2 + b^2}$.

Note for Exercises 57–59: Maple is able to plot polar curves using the `polarplot` command, or using the `coords=polar` option in a regular `plot` command. In Mathematica, use `PolarPlot`. In Derive, change to `Polar` under `Options State`. If your graphing device cannot plot polar equations, you must convert to parametric equations. For example, in Exercise 57, $x = r\cos\theta = \left[e^{\sin\theta} - 2\cos(4\theta)\right]\cos\theta$,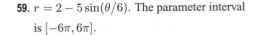

$y = r\sin\theta = \left[e^{\sin\theta} - 2\cos(4\theta)\right]\sin\theta$.

57. $r = e^{\sin\theta} - 2\cos(4\theta)$. The parameter interval is $[0, 2\pi]$.

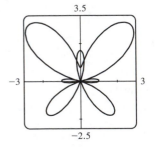

59. $r = 2 - 5\sin(\theta/6)$. The parameter interval is $[-6\pi, 6\pi]$.

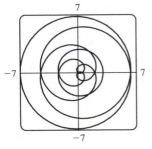

61.

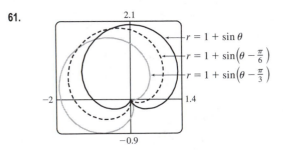

It appears that the graph of $r = 1 + \sin\left(\theta - \frac{\pi}{6}\right)$ is the same shape as the graph of $r = 1 + \sin\theta$, but rotated counterclockwise about the origin by $\frac{\pi}{6}$. Similarly, the graph of $r = 1 + \sin\left(\theta - \frac{\pi}{3}\right)$ is rotated by $\frac{\pi}{3}$. In general, the graph of $r = f(\theta - \alpha)$ is the same shape as that of $r = f(\theta)$, but rotated counterclockwise through α about the origin. That is, for any point (r_0, θ_0) on the curve $r = f(\theta)$, the point $(r_0, \theta_0 + \alpha)$ is on the curve $r = f(\theta - \alpha)$, since $r_0 = f(\theta_0) = f((\theta_0 + \alpha) - \alpha)$.

63. (a) $r = \sin n\theta$. From the graphs, it seems that when n is even, the number of loops in the curve (called a rose) is $2n$, and when

n is odd, the number of loops is simply n.

This is because in the case of n odd, every point on the graph is traversed twice, due to the fact that

$$r(\theta + \pi) = \sin\left[n(\theta + \pi)\right] = \sin n\theta \cos n\pi + \cos n\theta \sin n\pi = \begin{cases} \sin n\theta & \text{if } n \text{ is even} \\ -\sin n\theta & \text{if } n \text{ is odd} \end{cases}$$

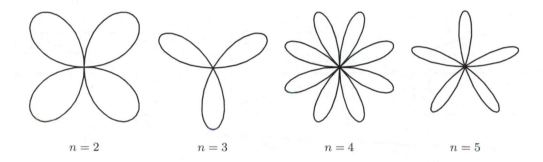

$n = 2$ $n = 3$ $n = 4$ $n = 5$

(b) The graph of $r = |\sin n\theta|$ has $2n$ loops whether n is odd or even, since $r(\theta + \pi) = r(\theta)$.

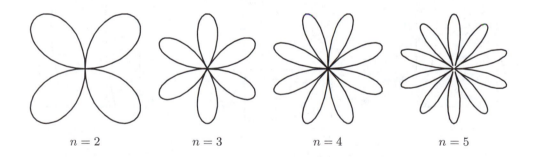

$n = 2$ $n = 3$ $n = 4$ $n = 5$

65. $r = \dfrac{1 - a\cos\theta}{1 + a\cos\theta}$. We start with $a = 0$, since in this case the curve is simply the circle $r = 1$.

As a increases, the graph moves to the left, and its right side becomes flattened. As a increases through about 0.4, the right side seems to grow a dimple, which upon closer investigation (with narrower θ-ranges) seems to appear at $a \approx 0.42$ (the actual value is $\sqrt{2} - 1$). As $a \to 1$, this dimple becomes more pronounced, and the curve begins to stretch out horizontally, until at $a = 1$ the denominator vanishes at $\theta = \pi$, and the dimple becomes an actual cusp. For $a > 1$ we must choose our parameter interval carefully, since $r \to \infty$ as $1 + a\cos\theta \to 0 \iff \theta \to \pm\cos^{-1}(-1/a)$. As a increases from 1, the curve splits into two parts. The left part has a loop, which grows larger as a increases, and the right part grows broader vertically, and its left tip develops a dimple when $a \approx 2.42$ (actually, $\sqrt{2} + 1$). As a increases, the dimple grows more and more pronounced.

If $a < 0$, we get the same graph as we do for the corresponding positive a-value, but with a rotation through π about the pole, as happened when c was replaced with $-c$ in Exercise 64.

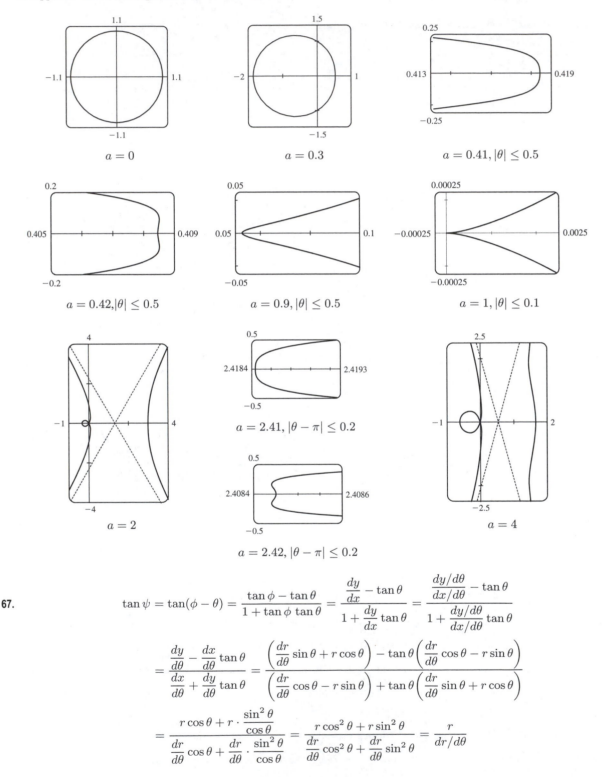

$a = 0$

$a = 0.3$

$a = 0.41, |\theta| \leq 0.5$

$a = 0.42, |\theta| \leq 0.5$

$a = 0.9, |\theta| \leq 0.5$

$a = 1, |\theta| \leq 0.1$

$a = 2$

$a = 2.41, |\theta - \pi| \leq 0.2$

$a = 2.42, |\theta - \pi| \leq 0.2$

$a = 4$

67.

$$\tan \psi = \tan(\phi - \theta) = \frac{\tan \phi - \tan \theta}{1 + \tan \phi \tan \theta} = \frac{\dfrac{dy}{dx} - \tan \theta}{1 + \dfrac{dy}{dx} \tan \theta} = \frac{\dfrac{dy/d\theta}{dx/d\theta} - \tan \theta}{1 + \dfrac{dy/d\theta}{dx/d\theta} \tan \theta}$$

$$= \frac{\dfrac{dy}{d\theta} - \dfrac{dx}{d\theta} \tan \theta}{\dfrac{dx}{d\theta} + \dfrac{dy}{d\theta} \tan \theta} = \frac{\left(\dfrac{dr}{d\theta} \sin \theta + r \cos \theta\right) - \tan \theta \left(\dfrac{dr}{d\theta} \cos \theta - r \sin \theta\right)}{\left(\dfrac{dr}{d\theta} \cos \theta - r \sin \theta\right) + \tan \theta \left(\dfrac{dr}{d\theta} \sin \theta + r \cos \theta\right)}$$

$$= \frac{r \cos \theta + r \cdot \dfrac{\sin^2 \theta}{\cos \theta}}{\dfrac{dr}{d\theta} \cos \theta + \dfrac{dr}{d\theta} \cdot \dfrac{\sin^2 \theta}{\cos \theta}} = \frac{r \cos^2 \theta + r \sin^2 \theta}{\dfrac{dr}{d\theta} \cos^2 \theta + \dfrac{dr}{d\theta} \sin^2 \theta} = \frac{r}{dr/d\theta}$$

H.2   Areas and Lengths in Polar Coordinates

1. $r = \sqrt{\theta}, 0 \leq \theta \leq \frac{\pi}{4}$. $A = \int_0^{\pi/4} \frac{1}{2} r^2 \, d\theta = \int_0^{\pi/4} \frac{1}{2} \left(\sqrt{\theta}\right)^2 d\theta = \int_0^{\pi/4} \frac{1}{2} \theta \, d\theta = \left[\frac{1}{4} \theta^2\right]_0^{\pi/4} = \frac{1}{64} \pi^2$

3. $r = \sin\theta, \frac{\pi}{3} \leq \theta \leq \frac{2\pi}{3}$.

$$A = \int_{\pi/3}^{2\pi/3} \frac{1}{2} \sin^2\theta \, d\theta = \frac{1}{4} \int_{\pi/3}^{2\pi/3} (1 - \cos 2\theta) \, d\theta = \frac{1}{4} \left[\theta - \frac{1}{2}\sin 2\theta\right]_{\pi/3}^{2\pi/3}$$

$$= \frac{1}{4}\left[\frac{2\pi}{3} - \frac{1}{2}\sin\frac{4\pi}{3} - \frac{\pi}{3} + \frac{1}{2}\sin\frac{2\pi}{3}\right] = \frac{1}{4}\left[\frac{2\pi}{3} - \frac{1}{2}\left(-\frac{\sqrt{3}}{2}\right) - \frac{\pi}{3} + \frac{1}{2}\left(\frac{\sqrt{3}}{2}\right)\right] = \frac{1}{4}\left(\frac{\pi}{3} + \frac{\sqrt{3}}{2}\right) = \frac{\pi}{12} + \frac{\sqrt{3}}{8}$$

5. $r = \theta, 0 \leq \theta \leq \pi$. $A = \int_0^{\pi} \frac{1}{2} \theta^2 \, d\theta = \left[\frac{1}{6} \theta^3\right]_0^{\pi} = \frac{1}{6} \pi^3$

7. $r = 4 + 3\sin\theta, -\frac{\pi}{2} \leq \theta \leq \frac{\pi}{2}$.

$$A = \int_{-\pi/2}^{\pi/2} \frac{1}{2}(4 + 3\sin\theta)^2 d\theta = \frac{1}{2}\int_{-\pi/2}^{\pi/2} \left(16 + 24\sin\theta + 9\sin^2\theta\right) d\theta$$

$$= \frac{1}{2}\int_{-\pi/2}^{\pi/2} \left(16 + 9\sin^2\theta\right) d\theta \quad \text{[by Theorem 5.5.6(b)]}$$

$$= \frac{1}{2} \cdot 2 \int_0^{\pi/2} \left[16 + 9 \cdot \frac{1}{2}(1 - \cos 2\theta)\right] d\theta \quad \text{[by Theorem 5.5.6(a)]}$$

$$= \int_0^{\pi/2} \left(\frac{41}{2} - \frac{9}{2}\cos 2\theta\right) d\theta = \left[\frac{41}{2}\theta - \frac{9}{4}\sin 2\theta\right]_0^{\pi/2} = \left(\frac{41\pi}{4} - 0\right) - (0 - 0) = \frac{41\pi}{4}$$

9. The curve goes through the pole when $\theta = \pi/4$, so we'll find the area for
$0 \leq \theta \leq \pi/4$ and multiply it by 4.

$A = 4 \int_0^{\pi/4} \frac{1}{2} r^2 \, d\theta = 2 \int_0^{\pi/4} (4\cos 2\theta) \, d\theta$

$= 8 \int_0^{\pi/4} \cos 2\theta \, d\theta = 4\left[\sin 2\theta\right]_0^{\pi/4} = 4$

11. One-sixth of the area lies above the polar axis and is bounded by the curve

$r = 2\cos 3\theta$ for $\theta = 0$ to $\theta = \pi/6$.

$A = 6 \int_0^{\pi/6} \frac{1}{2}(2\cos 3\theta)^2 \, d\theta = 12 \int_0^{\pi/6} \cos^2 3\theta \, d\theta$

$= \frac{12}{2} \int_0^{\pi/6} (1 + \cos 6\theta) d\theta$

$= 6\left[\theta + \frac{1}{6}\sin 6\theta\right]_0^{\pi/6} = 6\left(\frac{\pi}{6}\right) = \pi$

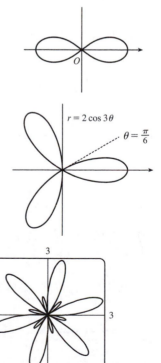

$r = 2\cos 3\theta$

$\theta = \frac{\pi}{6}$

13. $A = \int_0^{2\pi} \frac{1}{2}(1 + 2\sin 6\theta)^2 \, d\theta = \frac{1}{2}\int_0^{2\pi} (1 + 4\sin 6\theta + 4\sin^2 6\theta) d\theta$

$= \frac{1}{2}\int_0^{2\pi} \left[1 + 4\sin 6\theta + 4 \cdot \frac{1}{2}(1 - \cos 12\theta)\right] d\theta$

$= \frac{1}{2}\int_0^{2\pi} (3 + 4\sin 6\theta - 2\cos 12\theta) d\theta$

$= \frac{1}{2}\left[3\theta - \frac{2}{3}\cos 6\theta - \frac{1}{6}\sin 12\theta\right]_0^{2\pi}$

$= \frac{1}{2}\left[(6\pi - \frac{2}{3} - 0) - (0 - \frac{2}{3} - 0)\right] = 3\pi.$

15. The shaded loop is traced out from $\theta = 0$ to $\theta = \pi/2$.

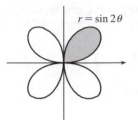

$r = \sin 2\theta$

$$A = \int_0^{\pi/2} \tfrac{1}{2} r^2 \, d\theta = \tfrac{1}{2} \int_0^{\pi/2} \sin^2 2\theta \, d\theta$$

$$= \tfrac{1}{2} \int_0^{\pi/2} \tfrac{1}{2}(1 - \cos 4\theta) \, d\theta = \tfrac{1}{4} \left[\theta - \tfrac{1}{4} \sin 4\theta\right]_0^{\pi/2} = \tfrac{1}{4}\left(\tfrac{\pi}{2}\right) = \tfrac{\pi}{8}$$

17.

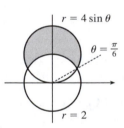

$r = 1 + 2\sin\theta$ (rect.)

$\left(3, \frac{\pi}{2}\right)$ $r = 1 + 2\sin\theta$

$\left(-1, \frac{3\pi}{2}\right)$

$\theta = \frac{7\pi}{6}$ $\theta = \frac{11\pi}{6}$

This is a limaçon, with inner loop traced out between $\theta = \frac{7\pi}{6}$ and $\frac{11\pi}{6}$ [found by solving $r = 0$].

$$A = 2\int_{7\pi/6}^{3\pi/2} \tfrac{1}{2}(1 + 2\sin\theta)^2 \, d\theta = \int_{7\pi/6}^{3\pi/2} \left(1 + 4\sin\theta + 4\sin^2\theta\right) d\theta = \int_{7\pi/6}^{3\pi/2} \left[1 + 4\sin\theta + 4 \cdot \tfrac{1}{2}(1 - \cos 2\theta)\right] d\theta$$

$$= [\theta - 4\cos\theta + 2\theta - \sin 2\theta]_{7\pi/6}^{3\pi/2} = \left(\tfrac{9\pi}{2}\right) - \left(\tfrac{7\pi}{2} + 2\sqrt{3} - \tfrac{\sqrt{3}}{2}\right) = \pi - \tfrac{3\sqrt{3}}{2}$$

19. $4\sin\theta = 2 \Leftrightarrow \sin\theta = \tfrac{1}{2} \Leftrightarrow \theta = \tfrac{\pi}{6}$ or $\tfrac{5\pi}{6}$ (for $0 \le \theta \le 2\pi$). We'll subtract the unshaded area from the shaded area for $\pi/6 \le \theta \le \pi/2$ and double that value.

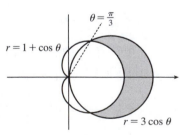

$r = 4\sin\theta$

$\theta = \frac{\pi}{6}$

$r = 2$

$$A = 2\int_{\pi/6}^{\pi/2} \tfrac{1}{2}(4\sin\theta)^2 \, d\theta - 2\int_{\pi/6}^{\pi/2} \tfrac{1}{2}(2)^2 \, d\theta = 2\int_{\pi/6}^{\pi/2} \tfrac{1}{2}\left[(4\sin\theta)^2 - 2^2\right] d\theta$$

$$= \int_{\pi/6}^{\pi/2} \left(16\sin^2\theta - 4\right) d\theta = \int_{\pi/6}^{\pi/2} [8(1 - \cos 2\theta) - 4] \, d\theta$$

$$= \int_{\pi/6}^{\pi/2} (4 - 8\cos 2\theta) \, d\theta = [4\theta - 4\sin 2\theta]_{\pi/6}^{\pi/2} = (2\pi - 0) - \left(\tfrac{2\pi}{3} - 4 \cdot \tfrac{\sqrt{3}}{2}\right) = \tfrac{4}{3}\pi + 2\sqrt{3}$$

21. $3\cos\theta = 1 + \cos\theta \Leftrightarrow \cos\theta = \tfrac{1}{2} \Rightarrow \theta = \tfrac{\pi}{3}$ or $-\tfrac{\pi}{3}$.

$\theta = \frac{\pi}{3}$

$r = 1 + \cos\theta$

$r = 3\cos\theta$

$$A = 2\int_0^{\pi/3} \tfrac{1}{2}\left[(3\cos\theta)^2 - (1 + \cos\theta)^2\right] d\theta$$

$$= \int_0^{\pi/3} \left(8\cos^2\theta - 2\cos\theta - 1\right) d\theta = \int_0^{\pi/3} [4(1 + \cos 2\theta) - 2\cos\theta - 1] \, d\theta$$

$$= \int_0^{\pi/3} (3 + 4\cos 2\theta - 2\cos\theta) \, d\theta = [3\theta + 2\sin 2\theta - 2\sin\theta]_0^{\pi/3}$$

$$= \pi + \sqrt{3} - \sqrt{3} = \pi$$

23. $A = 2\int_0^{\pi/4} \frac{1}{2}\sin^2\theta\, d\theta = \int_0^{\pi/4} \frac{1}{2}(1 - \cos 2\theta)\, d\theta$

$= \frac{1}{2}\left[\theta - \frac{1}{2}\sin 2\theta\right]_0^{\pi/4} = \frac{1}{2}\left[\left(\frac{\pi}{4} - \frac{1}{2}\cdot 1\right) - (0 - 0)\right]$

$= \frac{1}{8}\pi - \frac{1}{4}$

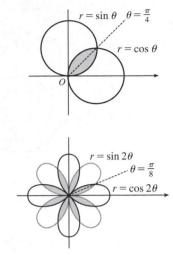

25. $\sin 2\theta = \cos 2\theta \quad\Rightarrow\quad \dfrac{\sin 2\theta}{\cos 2\theta} = 1 \quad\Rightarrow\quad \tan 2\theta = 1 \quad\Rightarrow\quad 2\theta = \frac{\pi}{4} \quad\Rightarrow$

$\theta = \frac{\pi}{8} \quad\Rightarrow$

$A = 8\cdot 2\int_0^{\pi/8} \frac{1}{2}\sin^2 2\theta\, d\theta = 8\int_0^{\pi/8} \frac{1}{2}(1 - \cos 4\theta)\, d\theta$

$= 4\left[\theta - \frac{1}{4}\sin 4\theta\right]_0^{\pi/8} = 4\left(\frac{\pi}{8} - \frac{1}{4}\cdot 1\right) = \frac{1}{2}\pi - 1$

27. The darker shaded region (from $\theta = 0$ to $\theta = 2\pi/3$) represents $\frac{1}{2}$ of the desired area plus $\frac{1}{2}$ of the area of the inner loop. From this area, we'll subtract $\frac{1}{2}$ of the area of the inner loop (the lighter shaded region from $\theta = 2\pi/3$ to $\theta = \pi$), and then double that difference to obtain the desired area.

$A = 2\left[\int_0^{2\pi/3} \frac{1}{2}\left(\frac{1}{2} + \cos\theta\right)^2 d\theta - \int_{2\pi/3}^{\pi} \frac{1}{2}\left(\frac{1}{2} + \cos\theta\right)^2 d\theta\right]$

$= \int_0^{2\pi/3}\left(\frac{1}{4} + \cos\theta + \cos^2\theta\right) d\theta - \int_{2\pi/3}^{\pi}\left(\frac{1}{4} + \cos\theta + \cos^2\theta\right) d\theta$

$= \int_0^{2\pi/3}\left[\frac{1}{4} + \cos\theta + \frac{1}{2}(1 + \cos 2\theta)\right] d\theta$

$\qquad\qquad - \int_{2\pi/3}^{\pi}\left[\frac{1}{4} + \cos\theta + \frac{1}{2}(1 + \cos 2\theta)\right] d\theta$

$= \left[\dfrac{\theta}{4} + \sin\theta + \dfrac{\theta}{2} + \dfrac{\sin 2\theta}{4}\right]_0^{2\pi/3} - \left[\dfrac{\theta}{4} + \sin\theta + \dfrac{\theta}{2} + \dfrac{\sin 2\theta}{4}\right]_{2\pi/3}^{\pi}$

$= \left(\frac{\pi}{6} + \frac{\sqrt{3}}{2} + \frac{\pi}{3} - \frac{\sqrt{3}}{8}\right) - \left(\frac{\pi}{4} + \frac{\pi}{2}\right) + \left(\frac{\pi}{6} + \frac{\sqrt{3}}{2} + \frac{\pi}{3} - \frac{\sqrt{3}}{8}\right)$

$= \frac{\pi}{4} + \frac{3}{4}\sqrt{3} = \frac{1}{4}\left(\pi + 3\sqrt{3}\right)$

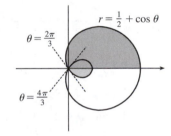

29. The curves intersect at the pole since $\left(0, \frac{\pi}{2}\right)$ satisfies $r = \cos\theta$ and $(0, 0)$ satisfies $r = 1 - \cos\theta$. Now $\cos\theta = 1 - \cos\theta \quad\Rightarrow\quad 2\cos\theta = 1 \quad\Rightarrow$ $\cos\theta = \frac{1}{2} \quad\Rightarrow\quad \theta = \frac{\pi}{3}$ or $\frac{5\pi}{3} \quad\Rightarrow$ the other intersection points are $\left(\frac{1}{2}, \frac{\pi}{3}\right)$ and $\left(\frac{1}{2}, \frac{5\pi}{3}\right)$.

31. The pole is a point of intersection. $\sin\theta = \sin 2\theta = 2\sin\theta\cos\theta \quad\Leftrightarrow$ $\sin\theta\,(1 - 2\cos\theta) = 0 \quad\Leftrightarrow\quad \sin\theta = 0$ or $\cos\theta = \frac{1}{2} \quad\Rightarrow$ $\theta = 0, \pi, \frac{\pi}{3}$, or $-\frac{\pi}{3} \quad\Rightarrow$ the other intersection points are $\left(\frac{\sqrt{3}}{2}, \frac{\pi}{3}\right)$ and $\left(\frac{\sqrt{3}}{2}, \frac{2\pi}{3}\right)$ (by symmetry).

33.

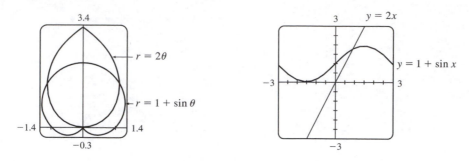

From the first graph, we see that the pole is one point of intersection. By zooming in or using the cursor, we find the θ-values of the intersection points to be $\alpha \approx 0.88786 \approx 0.89$ and $\pi - \alpha \approx 2.25$. (The first of these values may be more easily estimated by plotting $y = 1 + \sin x$ and $y = 2x$ in rectangular coordinates; see the second graph.)

By symmetry, the total area contained is twice the area contained in the first quadrant, that is,

$$A = 2 \int_0^\alpha \tfrac{1}{2}(2\theta)^2 \, d\theta + 2 \int_\alpha^{\pi/2} \tfrac{1}{2}(1 + \sin\theta)^2 \, d\theta = \int_0^\alpha 4\theta^2 \, d\theta + \int_\alpha^{\pi/2} \left[1 + 2\sin\theta + \tfrac{1}{2}(1 - \cos 2\theta)\right] d\theta$$

$$= \left[\tfrac{4}{3}\theta^3\right]_0^\alpha + \left[\theta - 2\cos\theta + \left(\tfrac{1}{2}\theta - \tfrac{1}{4}\sin 2\theta\right)\right]_\alpha^{\pi/2}$$

$$= \tfrac{4}{3}\alpha^3 + \left[\left(\tfrac{\pi}{2} + \tfrac{\pi}{4}\right) - \left(\alpha - 2\cos\alpha + \tfrac{1}{2}\alpha - \tfrac{1}{4}\sin 2\alpha\right)\right] \approx 3.4645$$

35. $L = \int_a^b \sqrt{r^2 + (dr/d\theta)^2} \, d\theta = \int_0^{\pi/3} \sqrt{(3\sin\theta)^2 + (3\cos\theta)^2} \, d\theta = \int_0^{\pi/3} \sqrt{9(\sin^2\theta + \cos^2\theta)} \, d\theta$

$$= 3 \int_0^{\pi/3} d\theta = 3\left[\theta\right]_0^{\pi/3} = 3\left(\tfrac{\pi}{3}\right) = \pi.$$

As a check, note that the circumference of a circle with radius $\tfrac{3}{2}$ is $2\pi\left(\tfrac{3}{2}\right) = 3\pi$, and since $\theta = 0$ to $\pi = \tfrac{\pi}{3}$ traces out $\tfrac{1}{3}$ of the circle (from $\theta = 0$ to $\theta = \pi$), $\tfrac{1}{3}(3\pi) = \pi$.

37. $L = \int_a^b \sqrt{r^2 + (dr/d\theta)^2} \, d\theta = \int_0^{2\pi} \sqrt{(\theta^2)^2 + (2\theta)^2} \, d\theta = \int_0^{2\pi} \sqrt{\theta^4 + 4\theta^2} \, d\theta$

$$= \int_0^{2\pi} \sqrt{\theta^2(\theta^2 + 4)} \, d\theta = \int_0^{2\pi} \theta\sqrt{\theta^2 + 4} \, d\theta$$

Now let $u = \theta^2 + 4$, so that $du = 2\theta \, d\theta$ $\left[\theta \, d\theta = \tfrac{1}{2} \, du\right]$ and

$$\int_0^{2\pi} \theta\sqrt{\theta^2 + 4} \, d\theta = \int_4^{4\pi^2 + 4} \tfrac{1}{2}\sqrt{u} \, du = \tfrac{1}{2} \cdot \tfrac{2}{3}\left[u^{3/2}\right]_4^{4(\pi^2+1)} = \tfrac{1}{3}\left[4^{3/2}(\pi^2 + 1)^{3/2} - 4^{3/2}\right] = \tfrac{8}{3}\left[(\pi^2 + 1)^{3/2} - 1\right]$$

39. The curve $r = 3\sin 2\theta$ is completely traced with $0 \le \theta \le 2\pi$. $r^2 + \left(\tfrac{dr}{d\theta}\right)^2 = (3\sin 2\theta)^2 + (6\cos 2\theta)^2 \Rightarrow$

$L = \int_0^{2\pi} \sqrt{9\sin^2 2\theta + 36\cos^2 2\theta} \, d\theta \approx 29.0653$

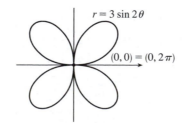

$r = 3\sin 2\theta$

$(0, 0) = (0, 2\pi)$

I   COMPLEX NUMBERS

1. $(5 - 6i) + (3 + 2i) = (5 + 3) + (-6 + 2)i = 8 + (-4)i = 8 - 4i$

3. $(2 + 5i)(4 - i) = 2(4) + 2(-i) + (5i)(4) + (5i)(-i) = 8 - 2i + 20i - 5i^2$
$$= 8 + 18i - 5(-1) = 8 + 18i + 5 = 13 + 18i$$

5. $\overline{12 + 7i} = 12 - 7i$

7. $\dfrac{1 + 4i}{3 + 2i} = \dfrac{1 + 4i}{3 + 2i} \cdot \dfrac{3 - 2i}{3 - 2i} = \dfrac{3 - 2i + 12i - 8(-1)}{3^2 + 2^2} = \dfrac{11 + 10i}{13} = \dfrac{11}{13} + \dfrac{10}{13}i$

9. $\dfrac{1}{1 + i} = \dfrac{1}{1 + i} \cdot \dfrac{1 - i}{1 - i} = \dfrac{1 - i}{1 - (-1)} = \dfrac{1 - i}{2} = \dfrac{1}{2} - \dfrac{1}{2}i$

11. $i^3 = i^2 \cdot i = (-1)i = -i$

13. $\sqrt{-25} = \sqrt{25}\,i = 5i$

15. $\overline{12 - 5i} = 12 + 15i$ and $|12 - 15i| = \sqrt{12^2 + (-5)^2} = \sqrt{144 + 25} = \sqrt{169} = 13$

17. $\overline{-4i} = \overline{0 - 4i} = 0 + 4i = 4i$ and $|-4i| = \sqrt{0^2 + (-4)^2} = \sqrt{16} = 4$

19. $4x^2 + 9 = 0 \iff 4x^2 = -9 \iff x^2 = -\frac{9}{4} \iff x = \pm\sqrt{-\frac{9}{4}} = \pm\sqrt{\frac{9}{4}}\,i = \pm\frac{3}{2}i.$

21. By the quadratic formula, $x^2 + 2x + 5 = 0 \iff x = \dfrac{-2 \pm \sqrt{2^2 - 4(1)(5)}}{2(1)} = \dfrac{-2 \pm \sqrt{-16}}{2} = \dfrac{-2 \pm 4i}{2} = -1 \pm 2i.$

23. By the quadratic formula, $z^2 + z + 2 = 0 \iff z = \dfrac{-1 \pm \sqrt{1^2 - 4(1)(2)}}{2(1)} = \dfrac{-1 \pm \sqrt{-7}}{2} = -\dfrac{1}{2} \pm \dfrac{\sqrt{7}}{2}i.$

25. For $z = -3 + 3i$, $r = \sqrt{(-3)^2 + 3^2} = 3\sqrt{2}$ and $\tan\theta = \frac{3}{-3} = -1 \Rightarrow \theta = \frac{3\pi}{4}$ (since z lies in the second quadrant).

Therefore, $-3 + 3i = 3\sqrt{2}\left(\cos\frac{3\pi}{4} + i\sin\frac{3\pi}{4}\right).$

27. For $z = 3 + 4i$, $r = \sqrt{3^2 + 4^2} = 5$ and $\tan\theta = \frac{4}{3} \Rightarrow \theta = \tan^{-1}\left(\frac{4}{3}\right)$ (since z lies in the first quadrant). Therefore,

$3 + 4i = 5\left[\cos\left(\tan^{-1}\frac{4}{3}\right) + i\sin\left(\tan^{-1}\frac{4}{3}\right)\right].$

29. For $z = \sqrt{3} + i$, $r = \sqrt{\left(\sqrt{3}\right)^2 + 1^2} = 2$ and $\tan\theta = \frac{1}{\sqrt{3}} \Rightarrow \theta = \frac{\pi}{6} \Rightarrow z = 2\left(\cos\frac{\pi}{6} + i\sin\frac{\pi}{6}\right).$

For $w = 1 + \sqrt{3}\,i$, $r = 2$ and $\tan\theta = \sqrt{3} \Rightarrow \theta = \frac{\pi}{3} \Rightarrow w = 2\left(\cos\frac{\pi}{3} + i\sin\frac{\pi}{3}\right).$

Therefore, $zw = 2 \cdot 2\left[\cos\left(\frac{\pi}{6} + \frac{\pi}{3}\right) + i\sin\left(\frac{\pi}{6} + \frac{\pi}{3}\right)\right] = 4\left(\cos\frac{\pi}{2} + i\sin\frac{\pi}{2}\right),$

$z/w = \frac{2}{2}\left[\cos\left(\frac{\pi}{6} - \frac{\pi}{3}\right) + i\sin\left(\frac{\pi}{6} - \frac{\pi}{3}\right)\right] = \cos\left(-\frac{\pi}{6}\right) + i\sin\left(-\frac{\pi}{6}\right)$, and $1 = 1 + 0i = 1(\cos 0 + i\sin 0) \Rightarrow$

$1/z = \frac{1}{2}\left[\cos\left(0 - \frac{\pi}{6}\right) + i\sin\left(0 - \frac{\pi}{6}\right)\right] = \frac{1}{2}\left[\cos\left(-\frac{\pi}{6}\right) + i\sin\left(-\frac{\pi}{6}\right)\right].$ For $1/z$, we could also use the formula that precedes

Example 5 to obtain $1/z = \frac{1}{2}\left(\cos\frac{\pi}{6} - i\sin\frac{\pi}{6}\right).$

31. For $z = 2\sqrt{3} - 2i$, $r = \sqrt{\left(2\sqrt{3}\right)^2 + (-2)^2} = 4$ and $\tan\theta = \frac{-2}{2\sqrt{3}} = -\frac{1}{\sqrt{3}} \Rightarrow \theta = -\frac{\pi}{6} \Rightarrow$

$z = 4\left[\cos\left(-\frac{\pi}{6}\right) + i\sin\left(-\frac{\pi}{6}\right)\right]$. For $w = -1 + i$, $r = \sqrt{2}$, $\tan\theta = \frac{1}{-1} = -1 \Rightarrow \theta = \frac{3\pi}{4} \Rightarrow$

$w = \sqrt{2}\left(\cos\frac{3\pi}{4} + i\sin\frac{3\pi}{4}\right)$. Therefore, $zw = 4\sqrt{2}\left[\cos\left(-\frac{\pi}{6} + \frac{3\pi}{4}\right) + i\sin\left(-\frac{\pi}{6} + \frac{3\pi}{4}\right)\right] = 4\sqrt{2}\left(\cos\frac{7\pi}{12} + i\sin\frac{7\pi}{12}\right)$,

$z/w = \frac{4}{\sqrt{2}}\left[\cos\left(-\frac{\pi}{6} - \frac{3\pi}{4}\right) + i\sin\left(-\frac{\pi}{6} - \frac{3\pi}{4}\right)\right] = \frac{4}{\sqrt{2}}\left[\cos\left(-\frac{11\pi}{12}\right) + i\sin\left(-\frac{11\pi}{12}\right)\right]$

$\quad = 2\sqrt{2}\left(\cos\frac{13\pi}{12} + i\sin\frac{13\pi}{12}\right)$, and

$1/z = \frac{1}{4}\left[\cos\left(-\frac{\pi}{6}\right) - i\sin\left(-\frac{\pi}{6}\right)\right] = \frac{1}{4}\left(\cos\frac{\pi}{6} + i\sin\frac{\pi}{6}\right)$.

33. For $z = 1 + i$, $r = \sqrt{2}$ and $\tan\theta = \frac{1}{1} = 1 \Rightarrow \theta = \frac{\pi}{4} \Rightarrow z = \sqrt{2}\left(\cos\frac{\pi}{4} + i\sin\frac{\pi}{4}\right)$. So by De Moivre's Theorem,

$$(1 + i)^{20} = \left[\sqrt{2}\left(\cos\frac{\pi}{4} + i\sin\frac{\pi}{4}\right)\right]^{20} = \left(2^{1/2}\right)^{20}\left(\cos\frac{20 \cdot \pi}{4} + i\sin\frac{20 \cdot \pi}{4}\right)$$

$$= 2^{10}(\cos 5\pi + i\sin 5\pi) = 2^{10}[-1 + i(0)] = -2^{10} = -1024$$

35. For $z = 2\sqrt{3} + 2i$, $r = \sqrt{\left(2\sqrt{3}\right)^2 + 2^2} = \sqrt{16} = 4$ and $\tan\theta = \frac{2}{2\sqrt{3}} = \frac{1}{\sqrt{3}} \Rightarrow \theta = \frac{\pi}{6} \Rightarrow z = 4\left(\cos\frac{\pi}{6} + i\sin\frac{\pi}{6}\right)$.

So by De Moivre's Theorem,

$$\left(2\sqrt{3} + 2i\right)^5 = \left[4\left(\cos\frac{\pi}{6} + i\sin\frac{\pi}{6}\right)\right]^5 = 4^5\left(\cos\frac{5\pi}{6} + i\sin\frac{5\pi}{6}\right) = 1024\left[-\frac{\sqrt{3}}{2} + \frac{1}{2}i\right] = -512\sqrt{3} + 512i$$

37. $1 = 1 + 0i = 1(\cos 0 + i\sin 0)$. Using Equation 3 with $r = 1$, $n = 8$, and $\theta = 0$, we have

$w_k = 1^{1/8}\left[\cos\left(\frac{0 + 2k\pi}{8}\right) + i\sin\left(\frac{0 + 2k\pi}{8}\right)\right] = \cos\frac{k\pi}{4} + i\sin\frac{k\pi}{4}$, where $k = 0, 1, 2, \ldots, 7$.

$w_0 = 1(\cos 0 + i\sin 0) = 1$, $w_1 = 1\left(\cos\frac{\pi}{4} + i\sin\frac{\pi}{4}\right) = \frac{1}{\sqrt{2}} + \frac{1}{\sqrt{2}}i$,

$w_2 = 1\left(\cos\frac{\pi}{2} + i\sin\frac{\pi}{2}\right) = i$, $w_3 = 1\left(\cos\frac{3\pi}{4} + i\sin\frac{3\pi}{4}\right) = -\frac{1}{\sqrt{2}} + \frac{1}{\sqrt{2}}i$,

$w_4 = 1(\cos\pi + i\sin\pi) = -1$, $w_5 = 1\left(\cos\frac{5\pi}{4} + i\sin\frac{5\pi}{4}\right) = -\frac{1}{\sqrt{2}} - \frac{1}{\sqrt{2}}i$,

$w_6 = 1\left(\cos\frac{3\pi}{2} + i\sin\frac{3\pi}{2}\right) = -i$, $w_7 = 1\left(\cos\frac{7\pi}{4} + i\sin\frac{7\pi}{4}\right) = \frac{1}{\sqrt{2}} - \frac{1}{\sqrt{2}}i$

39. $i = 0 + i = 1\left(\cos\frac{\pi}{2} + i\sin\frac{\pi}{2}\right)$. Using Equation 3 with $r = 1$, $n = 3$, and $\theta = \frac{\pi}{2}$, we have

$w_k = 1^{1/3}\left[\cos\left(\frac{\frac{\pi}{2} + 2k\pi}{3}\right) + i\sin\left(\frac{\frac{\pi}{2} + 2k\pi}{3}\right)\right]$, where $k = 0, 1, 2$.

$w_0 = \left(\cos\frac{\pi}{6} + i\sin\frac{\pi}{6}\right) = \frac{\sqrt{3}}{2} + \frac{1}{2}i$

$w_1 = \left(\cos\frac{5\pi}{6} + i\sin\frac{5\pi}{6}\right) = -\frac{\sqrt{3}}{2} + \frac{1}{2}i$

$w_2 = \left(\cos\frac{9\pi}{6} + i\sin\frac{9\pi}{6}\right) = -i$

41. Using Euler's formula (6) with $y = \frac{\pi}{2}$, we have $e^{i\pi/2} = \cos\frac{\pi}{2} + i\sin\frac{\pi}{2} = 0 + 1i = i$.

43. Using Euler's formula (6) with $y = \dfrac{\pi}{3}$, we have $e^{i\pi/3} = \cos\dfrac{\pi}{3} + i\sin\dfrac{\pi}{3} = \dfrac{1}{2} + \dfrac{\sqrt{3}}{2}i$.

45. Using Equation 7 with $x = 2$ and $y = \pi$, we have $e^{2+i\pi} = e^2 e^{i\pi} = e^2(\cos\pi + i\sin\pi) = e^2(-1 + 0) = -e^2$.

47. Take $r = 1$ and $n = 3$ in De Moivre's Theorem to get

$$[1(\cos\theta + i\sin\theta)]^3 = 1^3(\cos 3\theta + i\sin 3\theta)$$

$$(\cos\theta + i\sin\theta)^3 = \cos 3\theta + i\sin 3\theta$$

$$\cos^3\theta + 3(\cos^2\theta)(i\sin\theta) + 3(\cos\theta)(i\sin\theta)^2 + (i\sin\theta)^3 = \cos 3\theta + i\sin 3\theta$$

$$\cos^3\theta + (3\cos^2\theta\,\sin\theta)i - 3\cos\theta\,\sin^2\theta - (\sin^3\theta)i = \cos 3\theta + i\sin 3\theta$$

$$(\cos^3\theta - 3\sin^2\theta\,\cos\theta) + (3\sin\theta\,\cos^2\theta - \sin^3\theta)i = \cos 3\theta + i\sin 3\theta$$

Equating real and imaginary parts gives

$$\cos 3\theta = \cos^3\theta - 3\sin^2\theta\,\cos\theta \quad \text{and} \quad \sin 3\theta = 3\sin\theta\,\cos^2\theta - \sin^3\theta$$

49. $F(x) = e^{rx} = e^{(a+bi)x} = e^{ax+bxi} = e^{ax}(\cos bx + i\sin bx) = e^{ax}\cos bx + i(e^{ax}\sin bx) \quad \Rightarrow$

$$
\begin{aligned}
F'(x) &= (e^{ax}\cos bx)' + i(e^{ax}\sin bx)' \\
&= (ae^{ax}\cos bx - be^{ax}\sin bx) + i(ae^{ax}\sin bx + be^{ax}\cos bx) \\
&= a\left[e^{ax}(\cos bx + i\sin bx)\right] + b\left[e^{ax}(-\sin bx + i\cos bx)\right] \\
&= ae^{rx} + b\left[e^{ax}(i^2\sin bx + i\cos bx)\right] \\
&= ae^{rx} + bi\left[e^{ax}(\cos bx + i\sin bx)\right] = ae^{rx} + bie^{rx} = (a + bi)e^{rx} = re^{rx}
\end{aligned}
$$